Reviewers' comments about *Vegas 4 Editing Workshop*:

Douglas Spotted Eagle has been a key beta tester, idea man, and evangelist for Vegas since the earliest pre-beta concepting stages of version 1. His continuous feedback over the years has been immensely valuable to us as Vegas has evolved. We're all forever grateful for his comments, suggestions, and those many long and sometimes loud phone calls during ship week too. Douglas knows all the Vegas techniques and tricks better than any user on the planet—everybody will learn something from this book!

—Dave Hill (aka Dr. Dropout), Engineering Manager, Vegas

Vegas 4 Editing Workshop is not only a complete handbook for new and advanced users of Vegas 4.0, it shows you how to set up your facility so you can efficiently polish your video and audio work from start to finish. Author, Grammy-winning musician, and pro editor Douglas Spotted Eagle, the foremost guru for this great audio and video editing software, turns complex technology into easy-to-understand concepts with his accessible and graceful writing style. *Vegas 4 Editing Workshop* gives you everything you need to get the most out of Vegas 4.0, with special attention paid to groundbreaking new features like its sophisticated color correction tools and powerful DVD Architect. If you're looking for the best, easiest way to learn how to use Vegas 4.0 for professional-level video and audio production, look no further—this is the definitive guide for Vegas 4.0 users.

—Charlie White, Executive Producer, Digital Media Net

Finally the definitive [A]udio to [V]ideo guide to conquering this powerful NLE that doesn't leave out the WXYZ either. This is an exhaustive, richly detailed resource that no Vegas user should be without. Douglas shares his vast knowledge of every Vegas video and audio feature through easy-to-understand explanations, step-by-step tutorials, ample illustrations, a CD-ROM packed with useful goodies (including dozens of sample .VEG files), and real-world workflow scenarios. The hardest thing is balancing the book in one hand, mouse in the other, and working through these ideas yourself. Once you master these techniques, your audio and video projects will look gorgeous, sound bigger and better, and WOW! your audiences. Highly recommended!

—Jeffrey P. Fisher, Music/Audio/Video producer and writer

Vegas 4 Editing Workshop

Douglas Spotted Eagle

Published by CMP Books
an imprint of CMP Media LLC
Main office: 600 Harrison Street, San Francisco, CA 94107 USA
Tel: 415-947-6615; fax: 415-947-6015
Editorial office: 4601 West 6th Street, Suite B, Lawrence, KS 66049 USA
www.cmpbooks.com
email: books@cmp.com

Acquisitions editor: Dorothy Cox
Technical editor: David Hague
Layout design, composition, and project management: Happenstance Type-O-Rama
Cover design: Damien Castaneda

Distributed to the book trade in the U.S. by:
Publishers Group West
1700 Fourth Street
Berkeley, CA 94710
1-800-788-3123

Distributed in Canada by:
Jaguar Book Group
100 Armstrong Avenue
Georgetown, Ontario M6K 3E7 Canada
905-877-4483

For individual orders and for information on special discounts for quantity orders, please contact:
CMP Books Distribution Center, 6600 Silacci Way, Gilroy, CA 95020
Tel: 1-800-500-6875 or 408-848-3854; fax: 408-848-5784
email: cmp@rushorder.com; Web: www.cmpbooks.com

Figures 1.24, 1.34 and 1.35 adapted from *Audio Postproduction for Digital Video* © 2002 Jay Rose, CMP Books

Printed in the United States of America
03 04 05 06 07 5 4 3 2 1

ISBN: 1-57820-219-1

To my son Joshua for giving me the drive to complete this book.
I wish only that you could have seen it finished.

To Amanda, for showing me courage in the face of intense adversity,
and to Linda, for her constant support.

Table of Contents

Want to download the source code or get periodic updates? Send a blank email to `vegas4@news.cmpbooks.com`. **If you have a suggestion or correction, please send your comments to** `info@sundancemediagroup.com`.

Foreword

In an age where there are so many heated debates and headlines focusing on the issues of illegal downloading and piracy, filmmakers and musicians have been quietly embracing and taking advantage of the digital revolution. Today's young independent filmmakers are honing their craft and creating visually exhilarating movies with inexpensive high-quality digital video cameras and affordable audio/video editing software, such as Vegas 4.0, on just about any consumer-level desktop or laptop.

These same tools are also available to any musician who wants to expand their visibility and level the playing field against artists who are signed to major record labels by creating professional-quality promotional videos and making them accessible as streaming videos on the Internet.

This book is a must-have tool for those who want to take full advantage of what Sonic Foundry's Vegas 4.0, the most powerful and intuitive version of any digital audio/video editing software, has to offer. I've been using Vegas since its first version, and it's enabled me to create promotional videos of our band, Quiet Riot, by compiling live Jumbotron footage from a series of live shows and then digitizing them and editing it all in a few easy steps on my laptop while on the road. I've even been able to create my own instructional bass DVD with just my digital video camera and Vegas.

Even though Vegas and the rest of Sonic Foundry's line of software have an extremely easy learning curve, I wouldn't have been able to create such professional-quality footage in such a short amount of time without the help of my Digital Guru, Douglas Spotted Eagle. Any time I have a question on how to improve my project, I phone Douglas, and he's always kind enough to take time from his busy schedule to answer my questions and give me insight into Vegas that goes beyond what the manual has to offer. I've learned more about Vegas in a five-minute phone conversation with Douglas than I have in a week's worth of reading the manual.

If you've had the opportunity to catch Douglas in action on one of the V.A.S.S.T. Tour seminars, then you've already experienced his ability to clearly explain step-by-step all the great features Vegas 4.0 has to offer and to inspire you to take your creative vision to the extreme.

This book is the next best thing to having Douglas' phone number on your speed dial.

—Rudy Sarzo
Bassist
Quiet Riot, Whitesnake, and Ozzy Osbourne

Editor's Note

When I was asked if I would like to technically edit this book, I was pleasantly surprised, which became high excitement as the idea sank in. How many people get the opportunity to work professionally with someone of the stature and reputation as Douglas Spotted Eagle?

I had worked with Douglas before as a host on the Creative Cow and Digital Media Net Vegas forums. We had also swapped many emails discussing everything under the sun from didgeridoos to prawns versus shrimps to the meaning of such Australianisms as arvo and Chrissy and to what a gorilla was. But we had never met—in fact at the time of writing, we still haven't and have only even spoken once on the phone, albeit for about 30 seconds when Douglas was walking down some obscure street in New York while I was firmly ensconced at home here in the northern beaches of Sydney, 12,000 miles away and half a day in the future!

Over this period though, I have come to know Douglas probably as well as I could know anyone under the same circumstances and have gained a huge respect for him and value him as a very close friend.

On the way through, one of the major traits I have learned about Douglas is the fact that he loves to impart to others the knowledge he has accumulated. You cannot stop him! In fact, so eager is he to get this information out to an waiting world that I have never come across someone with such a prodigious writing output!

I know you will enjoy this book and, more importantly, learn greatly from it and have fun in the process. Douglas is a truly great educator, and I am proud to have been able to work with him, even if in such a small way, and in the process learn more than I could ever have dreamed of!

Thanks Spot!
David Hague
Sydney, Australia

Acknowledgments

Although my name is the one that appears on the cover of this book, this book wouldn't exist without the assistance, advice, and help of many people, most of whom have no idea that they were a part. I bear responsibility for any errors contained within, but maybe by thanking them I can distribute the blame (or maybe not).

First and foremost, Linda deserves much credit for making sure that there are always cold beverages and clean clothes, because during the writing of this book, most of which took place in hotel rooms and airplanes, I forgot the normal responsibilities of life.

Special appreciation for my partner and friend at Sundance Media Group and V.A.S.S.T., Mannie Frances, for answering nearly every call that came my way and keeping the phones at bay, making sure I kept on schedule for the chapters, even though some of the chapters were outdated with each different build of Vegas 4.0 (over 30 in all).

Of course, Dave Hill, Richard Kim, Dennis Adams, Caleb Pourchot, Brian Orr, Dave Chaimson, Bob, Steve, and Curt, among so many more at Sonic Foundry and Sony, too many to name, are tremendously appreciated for being patient with my constant requests for updates, answers, and readings for technical accuracy. I've also appreciated your friendships much over the years.

David Hague, the man who is half a world away from Utah, technically edited and vetted the entire book even in the midst of working on his own writings and suggested topics, rewrites, and repairs. Thank you David! Australia ROCKS!

Lonnie Bates of Thomson-Grass Valley has taught me a lot about the analog world and high definition as we traded information about DV and SD. He's forgotten more about calibration, scopes, and displays than most will ever know.

A lot of my editing tricks are inspired by requests, suggestions, questions, and arguments from folks in the various forums I've been a part of. There are literally hundreds, if not thousands, of people who have taught me a lot or forced me to find answers to difficult questions. Thanks to you this book was inspired. Without the folks at DMN, Sonic Foundry, and the WWUG forums, this book never would have been. Lou, Frank, Michael, and DV.com, thank you for providing me a venue to meet new people and to see the questions that are hopefully answered in this book.

Of course, I can't forget to thank Windham Hill Records who refused to finance my first music video, causing me to start editing on my own many years ago. From those days of analog, I've been forced to learn to do things that I couldn't afford to have others do. That refusal was perhaps the best thing that ever happened to me. And the video went on to be a top video later that year.

Sterling Johnson, Brian Morris, Ric Burns, Ken Burns, Jeff Spitz, Andy Dillon, James Cameron, Steven Seagal, Jerry Lonn, and Tom Bee, you've all taught me to be better behind the camera and smarter at the editing console. Thank you.

Bruce Braunstein, Paul, Louise, Bill, and Amanda, thank you for loaning your images to this book as well. Earl, thanks for keeping the computers running.

Rudy Sarzo, you are the most awesome and humble rock star I've ever known. I wanna be just like you when I grow up. Thanks for the constant calls with words of encouragement.

To all the manufacturers that have supported Sundance Media Group, V.A.S.S.T., and the endeavors, thank you. There are too many to mention; most are mentioned in this book and/or found on the included CD. I do have to mention Tim Wilson/Boris, Michael Feerer/Pixelan, Steve Pitzel/Intel, Tim and Mike at Canon, Phil Shaw at Matrox, Chris Hurd, Tim Kolb, Mike Gunter, and the VASST team. Phil, Bob, Julie, and the gang at Artbeats, your support is immense! (Nearly every image in this book comes from an Artbeats library.) Thank you. Appreciation to the guys at Canopus and ADS for providing me with immense knowledge and assistance.

Finally, many thanks to Dorothy, Paul, and the team at CMP for providing this opportunity to share my knowledge. May this be the first of several.

Introduction

Interestingly enough, this book started out as a small guide to what became Vegas 4.0. The moniker of Vegas Video was dropped in the middle of the project, and the same happened with the Vegas Audio name, thereby creating a product known simply as Vegas.

As an audio engineer, I have used Vegas as my tool of choice for a long time, thus inspiring my audio-based tutorials for Vegas. In becoming a videographer/editor to further my musical vision many years ago, Vegas Video 2.0 soon became my application of choice when I had the opportunity to see it in an alpha version in Madison, Wisconsin, back in 1999. It became the perfect meld of my two favorite passions in one application.

Therefore, I wrote this book really intending to please two different groups of users: one being the audio user, and the other being the video user. Audio users rarely think of the video user, and that's fine. Video users, however, absolutely must become familiar with the tools of the audio user; hence, the fairly intense audio for video section in this book. Moreover, this book has many tips, tricks, and techniques that I've learned over the years as an audio engineer, compressionist, and videographer/editor, some of which I felt were critical to share. Others merely appear because I felt they were of interest.

The intent of this book is to teach and inform, yet also to provide reminders to professionals who have "done it all" and "heard it all" in the course of their years of experience. While I consider myself an experienced audio and video editor, nearly every engineer I've met has dropped some small golden nugget that I've been able to learn from. It is my hope that this book does the same for you.

The CD in this book contains video, VEG files, and test media I felt important for users to more completely grasp the concepts presented in this book. VEG files can also be found in-depth at `http://www.sundancemediagroup.com/help/kb`. These files are for your educational use. If you use them as a significant part of what you create with your video project, please give credit where it's due, if it's due. That's part of the honor system of Vegas users. Scripts can also be found in this book as well and on the Sundance Media Group website.

If this book seems too technical, I suggest that you go back and re-read a passage. I've done my best to keep descriptions and explanations as simple as possible in this book, for the purpose of getting information across, not to demonstrate my technical skills or lack thereof. Multimedia is an articulate beast; anything combining the eye and ear are bound to sorely display any errors, whether you or others see them. Hopefully this book will teach you to avoid some of the pitfalls that you otherwise might step into. One of my favorite sayings is "I've screwed up more times than most people attempt to try." And I feel this is true. I've probably made every mistake that can be made in audio and video, multiple times. This book is a result of the knowledge that I've gleaned in the process of discovery.

Some of the information is necessarily technical and might be over the heads of some readers. If you sit with the book and read through a chapter once or twice, coupled with a little work in front of a computer, however, the information will make sense at some point. This book is absolutely not a white paper on any subject. It's a teaching tool for the uninitiated and a reminder/reference for the professional. A basic understanding of your camera, a computer, and general education are all that is needed to complete the tutorials found in this book. If I've missed something or you feel I didn't reach enough depth, I can be reached at dse@sundancemedia-group.com. My personal website, outlining some of my work, can be viewed at http://www.spottedeagle.com. Keeping my music career while expanding my video career has definitely been difficult, but pleasurable. One has benefited the other significantly. I hope this book does the same for you.

Spot

Computer Setup

Setting Up Your Computer in Vegas 4.0

Vegas 4.0 is an application that runs on the Microsoft Windows platform and does not perform well on Macintosh PC emulators due to the intensive processor load and translation requirements, so only PC computers are referred to in this book.

Computers have come a long way since the advent of the desktop editor and no longer require proprietary capture cards and expensive outboard equipment to create broadcast-quality video. Digital video editing, however, is very processor and resource intensive, so setting up the computer properly is a critical exercise in getting ready to edit.

Although many big-brand makers build off-the-shelf computers, few of them are configured out of the box for professional-level digital video editing. This is not to suggest that these machines cannot be used for editing, but rather that they need to be optimized for digital video editing. Usually, these machines come equipped with all kinds of software that either is loaded into the startup menu or, at a minimum, uses system resources in the background, stealing precious memory and resource space from Vegas' operational needs.

1.1 CMX 300 editing system, circa 1972.

1.2 Editing media in the early days of entertainment.

Hardware

Vegas 4.0 can operate on a Pentium II 400 with only 128MB of RAM. These are the minimum operating requirements and not the recommended system specifications. Minimum requirements more or less says, "prepare for a painful experience, but it will work" or "you can use a pencil for a flyswatter," when it comes to digital video (DV). Faster, bigger, and more is nearly always better when working with video at any level. While Sonic Foundry has not released anything stating that computer systems should be build faster, bigger, or better than any preset numbers, it's clear that faster processors, faster or more RAM, SCSI hard drives, and any number of system components can provide for a better video editing experience. Furthermore, when completing a project and rendering to print to tape or DVD, faster processors absolutely make waiting for the finished product more bearable.

Table 1.1 System Requirements

400 MHz processor
CD-ROM drive
7200 RPM hard disk drive
24-bit color display recommended
40MB hard disk space for program installation
Windows-compatible sound card
Supported CD-Recordable drive *(for CD burning only)*
OHCI-compliant IEEE-1394/DV capture card (for DV capture and print-to-tape tools only)
128MB RAM
Microsoft Windows 98SE, ME, 2000, or XP

Processors

Any of the Pentium III 800 MHz processors perform adequately, but stepping up to a 3.0 GHz or faster processor is better by far. *Dual processing* means that the motherboard contains two processors that share the load. This phrase doesn't mean that the processors share the load equally nor does it mean that the system is twice as fast. It does mean that the system is more able to handle numerous instructions simultaneously, including background rendering, browsing the web, writing a document, or listening to music. Vegas is multithreaded, and version 4.0 is more dual-processor–ready than ever before. Vegas does not run twice as fast, however, on a dual-processor system than it runs on a single-processor system. Rather, Vegas splits the load by sending video processing to one processor and audio processing to the other processor. Benefits of a dual processor are most noticeable when editing video with lots of audio tracks and several video tracks running at the same time. For the average editing situation consisting of three-to-four tracks of video and three-to-four tracks of audio, dual processors are consequently of little benefit.

Vegas 4.0 is optimized for Intel's Hyperthreaded (HT) Pentium 4 processors and operates at peak efficiency on these new CPUs.

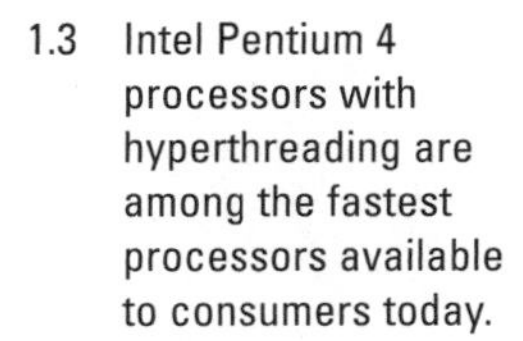

1.3 Intel Pentium 4 processors with hyperthreading are among the fastest processors available to consumers today.

Adding more efficiency are the newer frontside bus speeds of 533 MHz and soon to be faster. The frontside bus (FSB) is the weakest link in the overall scheme of processing media. The faster the FSB is, the faster the process is. Many systems now support extremely high speeds in the FSB. It is predicted that by the end of 2004, FSB speeds will exceed 1 GHz.

RAM

Random access memory (RAM) is a critical component for DV and Vegas, and while minimum requirements are only 128MB of RAM, it's not going to be a pleasant editing session. Users find that 512MB is the actual minimum. Having a gigabyte of RAM is even better and not only allows for faster, smoother editing, but makes the best use of Vegas' ability to render to RAM. RAM is directly related to the computer's video and graphics abilities, so don't skimp here.

RAM comes in different forms: Synchronous Dynamic RAM (SDRAM), Double Data Rate RAM (DDR RAM), and Rambus Dynamic RAM (RDRAM). All three work with Vegas. Various motherboard configurations determine the type of RAM required for that motherboard. DDR333 is a newer form of RAM that is very fast, rivaling RDRAM in speed, and is clocking excellent times with various non-linear editors (NLEs).

Display Adapters

The *display adapter* is also known as the *video card* in most computer systems. Vegas requires a VGA card; a much better experience should be had with a high-quality video card. Low-end cards often yield slow and poor color displays, making color-sensitive edits very difficult to work with. Any of the newer video display cards are sufficient for video editing at a high level. Accelerated graphics port (AGP) cards are usually the best choice, as AGP allows for virtually unlimited video memory. AGP cards also reduce strain on the processor, again providing for a faster experience.

Display cards are also available in dual-head modes, which allow multiple computer monitors to be connected to the video display card. Vegas can be spread across multiple monitors, allowing the various docking points of the application to be laid out in a custom setting. Generally, two 15-inch monitors are preferable for video editing rather than one 21-inch monitor. Some video editing setups with Vegas consist of three and four monitors!

1.4 Dual monitors make editing much more efficient and pleasant. These monitors are connected to the Matrox 550 Series card, a standard in the industry. Shown here is an "A" room at Sundance Media Group.

CD-ROM and DVD Drives and Burners

Although CD burners and DVD burners are constantly growing faster in their ability to burn data, video, and audio, high speeds also encourage error rates. You should be more concerned about playback speed than burn speed. Make sure your DVD burner complies with general standards or you might just find yourself with a high-priced boat anchor. The format war for DVD-R, DVD-RAM, DVD-RW, DVD+RW, and DVD+R will rage on for a while. It should not be a concern for most editors, however, as by the time the format wars settle out, either most DVD burners will accept and burn all formats, or the current machines will be obsolete and it will be time for a new burner.

Sound Cards

Vegas works with nearly any sound card, but to work with 5.1 surround sound, you need to have a card capable of 5.1 output. This requirement means that six speakers must be connected to the sound card if surround mixing and listening are desired. However, a standard two-channel card will suffice. Sound cards dictate a large part of the final output quality and are discussed in depth in the configuring hardware and software sections later in this chapter.

Hard Drives

One of the most difficult areas to make decisions regarding DV is the requirement for hard drive space. First, it should be said that video must live on a different drive than the operating system and the applications. In other words, for even minimum DV editing in any application, at least two hard drives are strongly recommended. The main, or boot, drive can be nearly any size, as it contains only the operating system, applications, and software related to other aspects of the computer's operation outside the DV editing space. DV consumes space at an average rate of 3.6 MB per second, or roughly 200 MB per minute, so it doesn't take long to fill up a hard drive. A ten-minute file uses over 2GB all by itself! This doesn't include any extra media that the finished ten-minute file was built from, or space for photos, audio, graphic, or other files related to the finished product. In a professional setting, one hour of tape yields ten minutes of usable material on average.

Table 1.2 Throughput/Storage of Audio at Various Sampling Rates

Number of Audio Tracks/Sample Rate	Throughput at 16 bit
One track @44.1 KHz	5 MB per minute
Two tracks (stereo) @44.1 KHz	10 MB per minute
Eight tracks @44.1 KHz	53 MB per minute
One track @48 KHz	6 MB per minute
Two tracks @48 KHz	12 MB per minute
Number of Audio Tracks/Sample Rate	**Throughput at 24 bit**
One track (mono) @44.1 KHz	8 MB per minute
Two track (stereo) @44.1 KHz	16 MB per minute
Eight tracks (mono) @44.1 KHz	64 MB per minute

Not only does a drive need to be large, it must be fast. A 7,200 rpm drive is the preference, although a 5,400 rpm drive provides minimal ability to edit. By the same token, two 9GB drives are preferable to one 18GB drive, due to the way the media lives on the drive, as well as the seek times. Integrated development environment (IDE) drives and buses are getting faster and faster every day. ATA-133 drives promise incredible speed in the near future and will boost video editing to a new level if they meet the much-anticipated numbers. *Small computer system interface* (SCSI) drives are still the best choice, albeit more expensive. SCSI drives are available in speeds of 10,000 and 15,000 rpm, plus they have faster seek times than do IDE drives. When using high-definition media, SCSI absolutely improves the editing experience, particularly when using the newer dual-channel 320 SCSI controllers.

Systems using *redundant array of inexpensive disks* (RAID) systems are the fastest of all and can be built around either SCSI or IDE controllers. RAID systems allow for the processor to break information down into blocks, which are then written to different parts of different drives, allowing for increased bandwidth, less heat generation, faster access times, and lower cost. While many types of RAID setups and protocols exist, RAID 0 is the most common form of RAID for video editing, as it offers the best price/performance ratio. However, it offers no protection, meaning when one drive goes down, all data is lost.

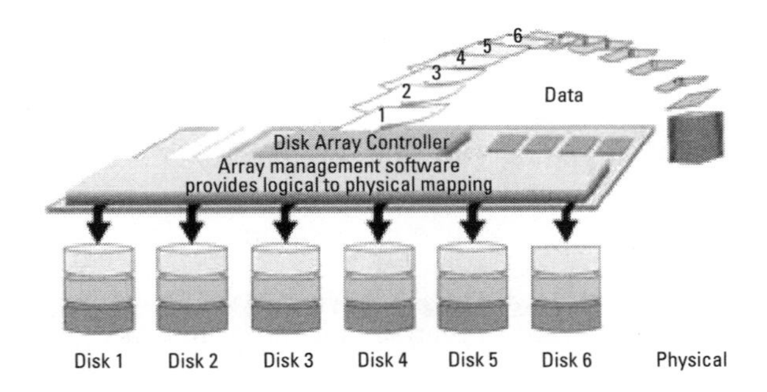

1.5 RAID Level 0: "Disk striping" high I/O performance.

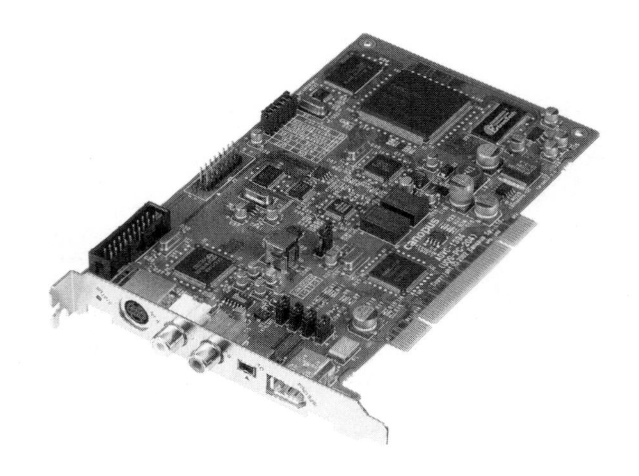

1.6 Canopus ADVC 1394.

Table 1.3 Throughput/Storage of Video at Various Sampling Rates

Video Tracks	DV 25 (5:1 compression)
One track	200 MB per minute
Two tracks	400 MB per minute
Eight tracks	8 GB per minute
Video Tracks	**Uncompressed**
One track	900 MB per minute
Two tracks	18 GB per minute
Eight tracks	7.2 GB per minute

A minimum of three disks are needed for a RAID, plus a controller. Some motherboards have RAID controllers built in; others are available as PCI bus controllers. Adaptec has long been known for their quality RAID controllers and have just recently released both software-based controllers and a new dual-channel 320 MBps (megabytes per second) controller.

IDE drives can be configured into a RAID as well, using hardware or software controllers, allowing for multiple IDE drives to be built into a system for high volumes of storage while keeping cost down.

OHCI Cards (Capture Cards)

The *open host controller interface* (OHCI) card, or 1394 card, is what connects the DV camera or video deck and transfers data into the computer. To be truly accurate, this card should not be called a capture card because it is merely a data transfer method from the camera to the hard drive, no different from a SCSI controller, parallel port, universal serial bus (USB) port, or serial port on the computer. A true capture card is a proprietary card that actually compresses the video as the video is transferred from the camera to the hard drive. On playback, the true capture card has a reader built into a processor on the card that reads and decodes video as it streams off the hard drive. These cards reduce processor load and often contain pre-built or pre-optimized effects, titling tools, and other enhancements that speed up the rendering, editing, or viewing process. These cards can be connected to nothing but a camera.

An OHCI card does none of this. It merely acts as a conduit from the camera to the hard drive. OHCI cards also can control scanners, networks, hard drives, video cameras, and other devices. No compression occurs, and no additional information is written to the file by these generic cards. These cards, also known as FireWire, i.Link, or 1394, can rapidly become confusing to the uninitiated. Current standards for 1394(a) are data transfers of up to 400 Mbps (megabits per second), yet upcoming standards provide for speeds of up to 3,200 Mbps! DVCam moves data at 25 Mbps, while uncompressed video moves at 170 Mbps.

Tip

If purchasing an OHCI card bundle that contains a driver disk, do not load the drivers that come with the disk. Doing so might load proprietary codecs or drivers that can corrupt or subvert Vegas drivers. OHCI requires no custom drivers to be loaded on a Microsoft Windows 98 SE, Windows 2000, Windows ME, or Windows XP system.

Once considered the budget alternative, OHCI is gaining ground for digital video editors everywhere. Vegas was the first of the OHCI-optimized editors, but now every editing system in the under-$5,000 price range has jumped on the bandwagon. Cards from ADS Technologies, Canopus ADVC1394, SIIG, Unibrain, and Orange have been successfully tested to be compliant with all OHCI requirements and work well with Vegas.

Additional Hardware

Having a front-panel connection bay is a valuable and helpful piece of hardware that is relatively inexpensive. The hardware moves the sound card, 1394, microphone, headphone, and musical instrument digital interface (MIDI) connections to the front of the computer and sits in an empty hard drive bay. The hardware connects to any sound card configuration, as it's essentially an extension of the connections already found on the sound card that pass through the internal part of the computer.

Another hardware device that is valuable for running Vegas efficiently is the Shuttle Pro from Contour Design. This piece of hardware allows users to program shortcuts on the device and minimizes hand movement from the keyboard to the mouse. Many users are able to completely avoid using keyboard and mouse shortcuts by using the Shuttle Pro. A jog shuttle wheel provides for smooth shuttling of the cursor on the timeline, moving either frame-by-frame or at high speed just like the jog/shuttle on a high-end video machine or remote. The device has 13 user-definable keys that allow buttons to be programmed for any shortcut/keystroke found in Vegas and other Sonic Foundry applications, such as Forge and Acid. Contour Design also manufactures the Contour Express, a smaller version of the Contour Shuttle.

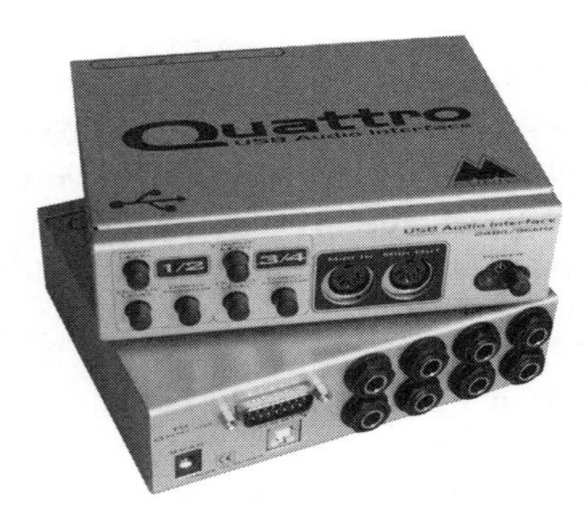

1.7 Quattro USB Audio Interface.

1.8 Shuttle Pro from Contour Design.

Configuring Hardware

Hardware configuration is critical to running Vegas or any other NLE system smoothly. Hardware placement and configuration often determine how the BIOS and operating system allocate resources. For instance, it's generally best to keep the first peripheral component interconnect (PCI) slot empty, as it shares an interrupt request (IRQ) with the AGP slot. PCI slots are usually white, while AGP slots are typically brown or gray.

1.9 Contour Express, a smaller version of the Contour Shuttle.

In setting up the hardware, make sure that the sound card, 1394 card, and the video card don't share an IRQ. Window's advanced configuration and power interface (ACPI) often addresses these to the same IRQ; consult your motherboard owner's manual and BIOS settings to change the IRQ settings on the motherboard. Shared IRQs can lead to application failure, due to restricted resources being paged or requested at the same time, and in some lower-end motherboards and BIOS systems, only patient testing results in correct IRQ allocation. Be wary of VIA KT 133 and 233 chipsets on low-end motherboards, as these chipsets are well documented as being troublesome in multimedia authoring systems.

When installing and setting up hardware, heat transfer is an important consideration. Heat is the enemy of the processor and hard drive setup. The cooler the machine is, the more stable and faster the system is. Systems should have a minimum of two fans—one for intake of outside air and one for exhaust of inside air. Many systems integrators use multiple fans, including fans directly mounted to hard drives, blowers that act as air intakes and blow directed air onto the processor fan, and heat sink. In addition, systems with several intake and exhaust fans are common. If noise is a consideration, refrigeration kits and water-cooling kits are also available, but make sure that an experienced installer installs these. Failure of one of these systems can easily lead to irrecoverable damage to the motherboard, processor, hard drives, and anything else that happens to be internal to the computer case.

Be sure hard drives are all in *direct memory access* (DMA) mode when operating Vegas. DMA can be enabled by going to the MY COMPUTER | PROPERTIES | DEVICE MANAGER | IDE CONTROLLERS | ADVANCED SETTINGS dialog and viewing the transfer mode. Each hard drive on the controller has a Transfer Mode setting, which should be set to DMA if available for optimum speed. Change all hard drives to DMA mode, and you'll see the benefit of the faster data transfer rate. Click OK for all changes and reboot to make the changes take effect.

1.10 Keep the system cool with plenty of airflow.

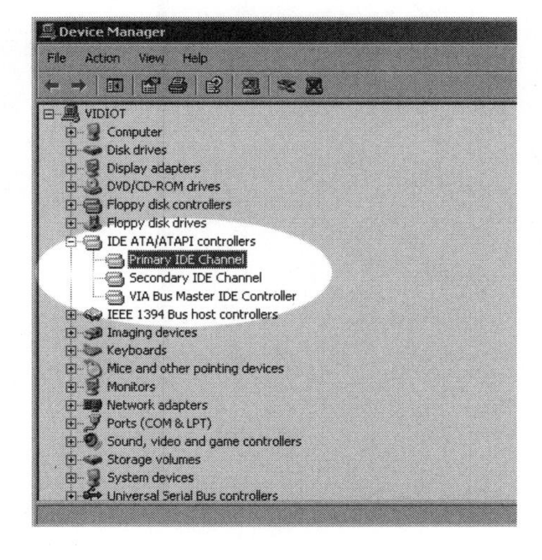

1.11 Be certain that the hard drive controllers are set to DMA mode, regardless of the number of drives. This setting is found in the **CONTROL PANEL | SYSTEM | HARDWARE** dialog.

With some systems, graphics processors can be problematic if they are slower, and setting hardware acceleration to less than full usually provides best results, with little or no effect on the quality of the preview window in Vegas. Newer cards should be configured with the hardware acceleration at 100 percent. To make changes to the graphics card settings, right-click in the desktop space, choose Properties in the dialog, choose the Settings tab, click the Advanced button, and select the Troubleshoot tab. Experiment with the settings of the graphics acceleration. Older cards might need to have the acceleration shut off completely.

The installation of a 1394 card can cause Windows 2000 and XP to install networking bridges for these cards. This issue can cause failures during use of 1394 hard drives and external previews from Vegas and should be disabled. To disable a 1394 network bridge, go to the START | CONTROL PANEL | NETWORK CONNECTIONS dialog, and right-click the 1394 connection, and select Disable. This process won't affect video editing in Vegas; it merely prevents the networking applications from paging for the 1394 network card and affecting system resources.

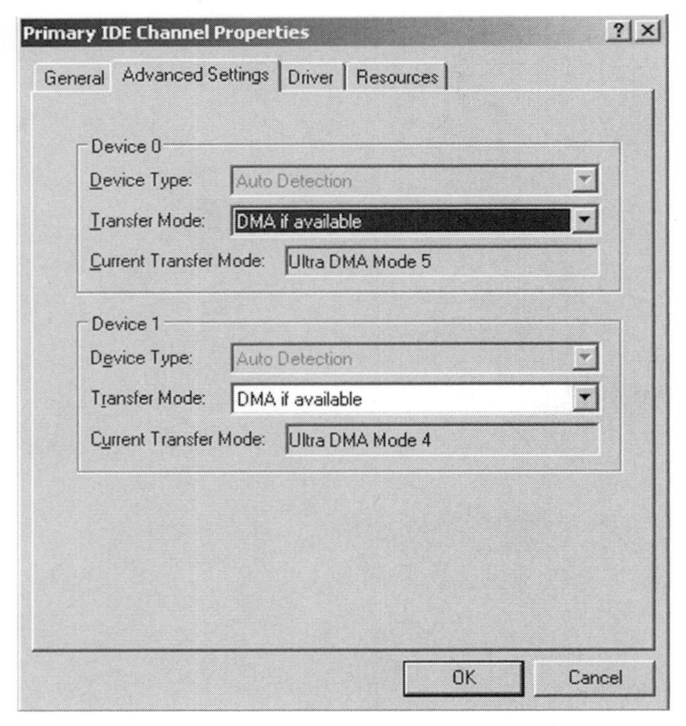

1.12 Setting the hard drive controller modes.

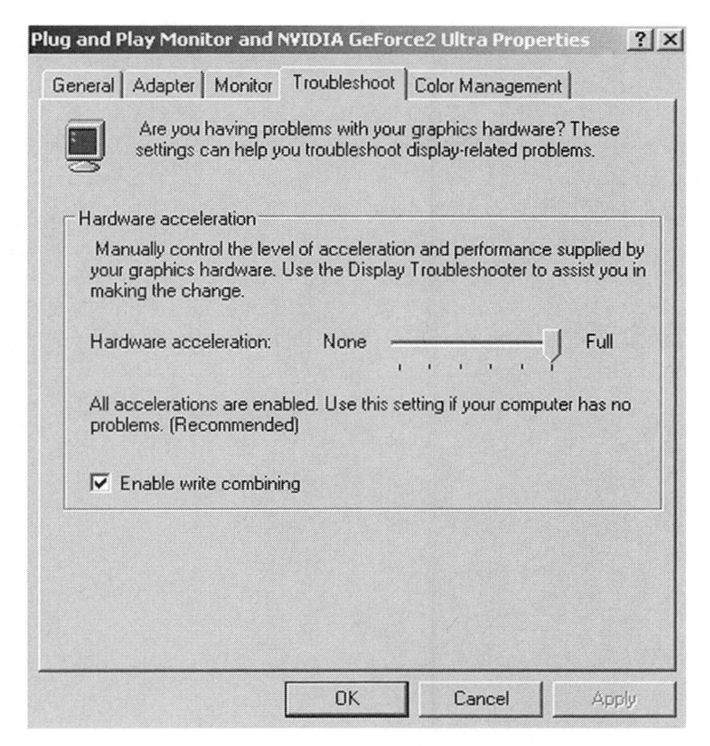

1.13 Adjusting the graphics processor can sometimes repair system glitches, particularly with lower quality or blended modes video cards.

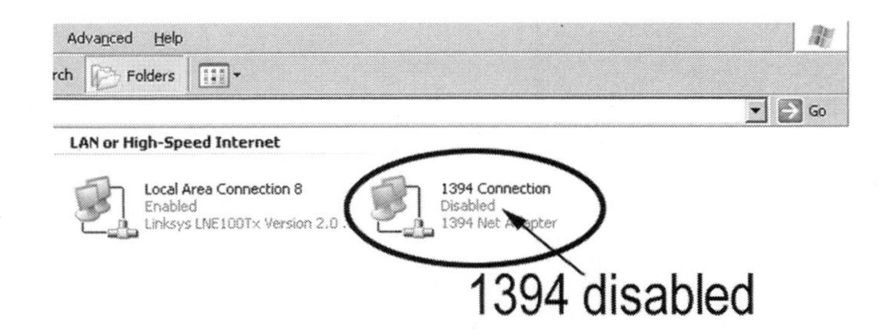

1.14 Disable the 1394 card in the network connections dialog.

Configuring Software

Video editing is very processor and resource intensive, and any software that doesn't need to be running shouldn't be. Some applications load into the Start menu and aren't visible without looking for them, so you might not even know they are running. Applications such as Microsoft Office, anti-virus apps, color profiling tools, and others often can't be seen in the system tray without specially looking for them. There are several ways to shut down applications that are running in the background, and they all vary with the Windows 98 SE, ME, 2000, and XP operating systems that Vegas runs within. With Windows XP for instance, right-clicking an empty space in the task bar brings up a dialog box. Select Properties and then Task Bar and Start Menu Properties. From here, it's a simple task to determine what will and won't be part of the Start menu and system.

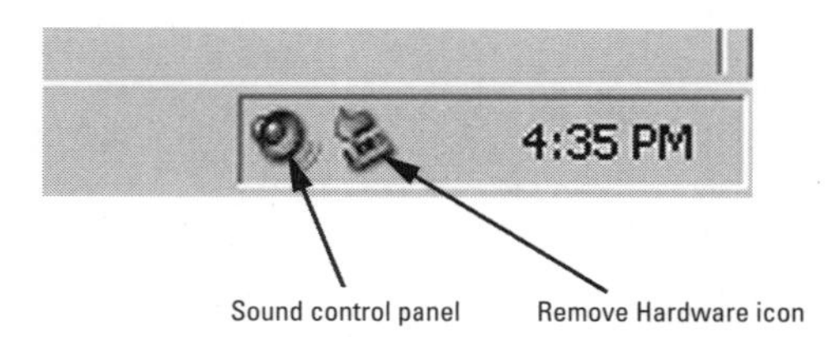

1.15 Typical applications running in the background.

1.16 Echo Layla.

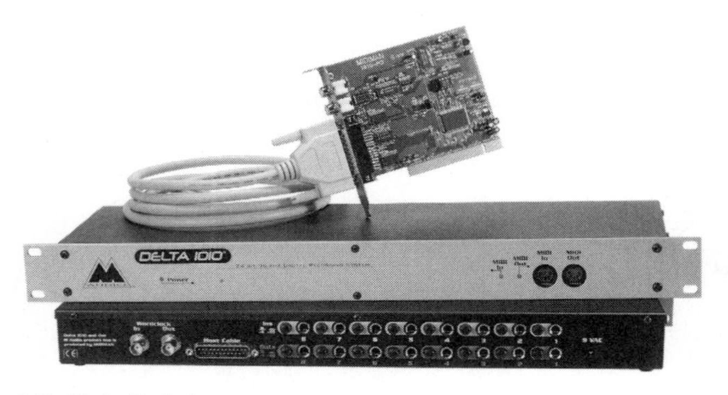

1.17 M-Audio Delta.

Tip

For best system results, defrag drives after the completion of every project or before beginning a new one. This process keeps files together as much as possible, keeps Vegas and other applications running at top speed, and assists in identifying drives that are beginning to fail.

One of the quickest, safest, and most surefire ways to shut down background applications is with the EndItAll product, which is available for all operating systems. This application closes all Windows networking apps that don't need to be running; it also closes any additional applications that don't need to be running in order for the computer to function correctly. EndItAll was written by Neil J. Rubenking, technical editor for *PC Magazine*, and can be found on the `www.pcmag.com` site.

Anti-virus software running in the background of any NLE system can create problems, due to its constant access to system resources. These applications should always be disabled when editing, regardless of the editing system. The same should be said for any maintenance software, such as scheduled defragmentation, backup, or error-checking software. Starting a defrag process in the middle of editing or data transfer/capture from the camera can result in crashed applications, damaged files, and frustration.

Networking software also presents problems on some systems, as the Admin Services tools that run in the background of Windows XP and Windows 2000 use a large chunk of system resources. Again, various operating systems treat disabling Admin Services differently, so consult with a networking specialist or do research on the web for hundreds of sites demonstrating how to shut this feature off.

When using external monitor views from Vegas, system resources are exceptionally tasked. All resources must be optimized for best experiences. While faster processors and more RAM are important considerations, background events, resource allocation, and bus speed are the more important considerations.

Setup for Multitrack Audio

Multitrack audio in its simplest form is merely multiple individual tracks recorded and played in synchronous form, so that the resulting audio sounds like one event, composition, or moment in action. When a movie soundtrack is recorded, the sound design, orchestration, dialog, walla, and effects are all recorded at different times in different places. These eventually come together on a multitrack recorder and are stitched together to create one single sound event that blends. Technically, a stereo recording is multitrack because it contains a right signal and a left signal, played in sync, which result in a sound that appears to belong together.

Recording equipment has come a long way since the early days of Les Paul and Mary Ford. Even the Beatles only had a four-track recorder in the later years. Vegas has unlimited tracks, restricted only by hard drive space and processor speed. Hiss and noise, once the most important considerations in the analog audio world, are all but dead in the digital world because hiss and noise can only be generated by external devices. In the audio realm, the primary concern is timing, and technology has all but eliminated that concern as well.

Vegas was originally written as a multitrack recording application before it became a video editing application. The audio legacy lives on and has grown much with Vegas 4.0. In fact, many producers use Vegas as their multitrack audio tool, never using or even seeing the video toolsets found in Vegas.

In the multitrack environment, it's critical to monitor sounds already recorded during the recording of new sounds. In order to do this, a *full-duplex sound card* is required. A full-duplex card allows for audio to be bi-directional—passing out of the sound card so the recording artist can hear what's coming off the hard drive as previously recorded tracks, while at the same time recording the artist's performance in time with the existing audio tracks. Be aware that most laptop computers only have half-duplex sound cards, allowing sound to pass in or out but not both at the same time. If a laptop is to be used, a separate USB, FireWire, or CardBus-based sound card must be used to monitor audio while recording. M-Audio, Echo Audio, Mark of the Unicorn, and many other companies make professional grade hardware that allows for up to 24 simultaneous inputs and outputs from Vegas.

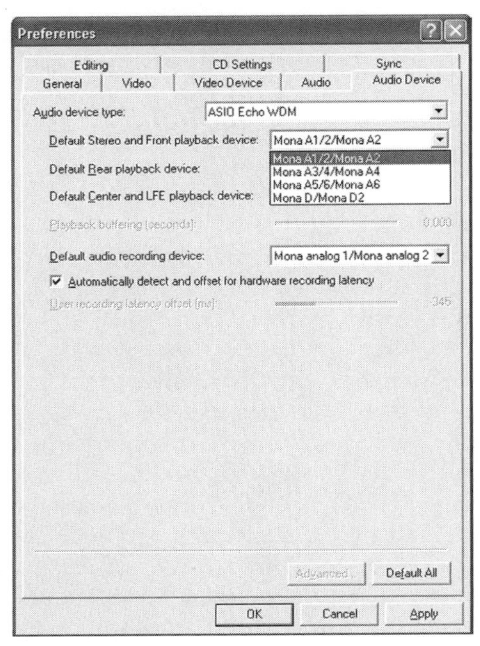

1.18 Setting up the multichannel audio card is fast and nearly plug-and-play in Vegas 4.0 with the new audio tools interface. With Vegas' new ASIO driver support, input monitoring is low-latency even with compressors and reverb applied to an input channel.

These same devices are available for the desktop computer as well and are highly recommended if surround sound or multichannel sound mixing is your goal. While low-end audio cards may indeed be used for monitoring and mixing multiple channels, the noise floor of these inexpensive cards renders them all but useless.

Setting up a multichannel system, either for multitrack recording or for surround mixing, involves the same equipment. The only difference is that a multichannel card must have at least six outputs for 5.1 surround mixing. A two-channel sound card from a high-end manufacturer, however, allows for monitoring audio with audio stream I/O (ASIO) drivers at the same time as audio is being recorded but doesn't allow for surround mixing.

Recording audio onto the computer's hard drive requires at least one microphone. The microphone(s) connect to the inputs of the sound card, and Vegas sees the audio coming into the sound card. If a standard two-channel sound card is used, nothing more needs to be set. In the event that a multichannel sound card is used, Vegas must be configured to see the sound card correctly.

After installing the sound card in the computer, open Vegas 4.0. Select TOOLS | PREFERENCES | AUDIO DEVICE and choose the Microsoft Classic Drivers or the ASIO drivers, depending on the brand of sound card installed. In the default choices, set up the individual devices for each input/output. Up to 26 input/outputs may be specified in Vegas. (26 I/O channels correspond to 26 letters in the alphabet.)

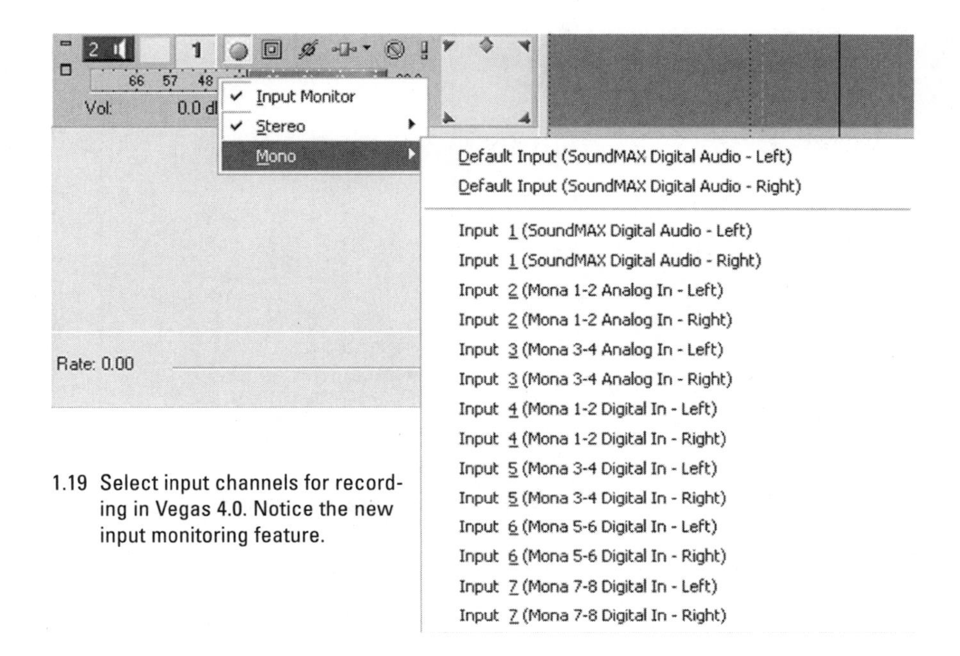

1.19 Select input channels for recording in Vegas 4.0. Notice the new input monitoring feature.

A folder for recorded audio must be specified, unless Vegas' default folder of `C:\Program Files\Sonic Foundry\Vegas 4.0` is the desired destination for recorded audio files. To change this default folder to another hard drive or location, go to the **FILE | PROJECT PROPERTIES** dialog and select the Audio tab. Click the Browse button and navigate to the folder or hard drive where recorded audio should be routed to. Alternatively, press **ALT+ENTER** to call the Project Properties dialog. Finally, Project Properties can be accessed via the Project Audio Properties button found in the Audio Master pane.

If Vegas 4.0 is required to synchronize to an external device, such as a tape machine, video tape recorder (VTR), or keyboard, then a timecode master device that has MIDI timecode (MTC) is required. These boxes are available from JL Cooper, Tascam, Lynx, and many other suppliers. Most multichannel audio cards or multichannel MIDI distribution boxes now have the ability to output MTC. MTC is used to keep Vegas in sync, or clocked, to the external device. When Vegas 4.0 is following an external device, Vegas 4.0 is the slave to the master. Vegas 4.0 also has the ability to generate timecode to cause other external devices to follow Vegas. When Vegas 4.0 is the tool controlling external devices, Vegas is the master, and the external devices are slaves. Vegas 4.0 generates timecode based on either the computer's clock or on the clock generated by the sound card. This choice is user-selectable. (See the "Recording Audio in Vegas 4.0" section in Chapter 7 for more information.)

Vegas is also capable of recording and playback of audio at different sample rates. Rates of up to 24 bit/192 KHz can be selected by going to the **PROJECT PROPERTIES | AUDIO** dialog and selecting the bit rate and sample rate desired. Please be aware that not all sound cards can support all bit and sample rates, particularly the higher rates. Check the documentation that came with the sound card for more information.

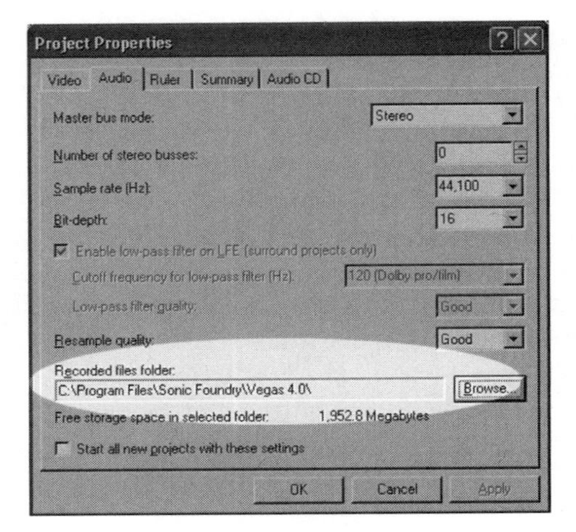

1.20 Specify where audio is to be recorded. Best results are obtained when using a drive that is different from the boot/OS drive.

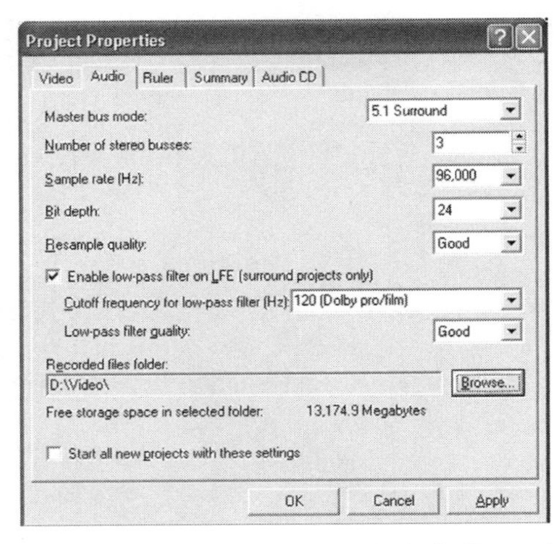

1.21 Set the desired sample rate in Audio Properties.

Most of the newer sound cards support the 24/96 rate, and some support sample rates up to 24 bit/192 KHz, known as *high definition audio*. CD-quality audio requires that final output files be 44.1 KHz and 16 bit for burning to a CD; however, audio can be recorded at any sample rate/bit depth. Be aware that audio at 24/96 consumes hard drive space at double the rate of a 16 bit/44.1 KHz recording. 24 bit/192 KHz uses even greater disk space. 24/96 audio files also require faster hard drives with greater cache, as well as fast processors and adequate RAM to handle the file size and throughput/data rates.

For voice-overs, general recording, and dialog, 16 bit/44.1 KHz should generally be the preference set. When working with highly detailed sounds, such as Foley work, musical instruments (particularly acoustic instruments), or sound design, the higher clarity of 24/48 or 24/96 audio might be a better choice, particularly when working with sounds that have higher frequency information. More air becomes apparent, and the sound is smoother. When the mixing is done in these frequencies for final output to CD, the information is dithered to a lower sampling rate compatible with Redbook CDs. A small degree of information is lost in the dithering process in which bits are thrown away; however, if CD is the final destination for the media, there currently is no choice. Some digital audio tape (DAT) machines accept the 24-bit files, allowing a duplicator or mastering house to keep files in their pristine state. At the time of this writing, however, 48 KHz is the highest sample rate available to a standard DAT machine.

Vegas 4.0 may also be connected to a tape machine via analog or digital connections, using MTC to lock the application and machine together in time. Sample accurate transfers may be accomplished if a sound card has TDIF (Tascam Digital Interface), Lightpipe, or other digital interfacing. Media may be transferred from the tape machine to Vegas' timeline, edited as desired, and placed back on the tape for later mixing, archiving, or sharing with other studios. This workflow is common in today's technologically advanced world.

A CD burner needs to be selected if CDs are to be created using Vegas 4.0. CD burner selection is accessed via the OPTIONS | PREFERENCES | CD SETTINGS dialog. Select the CD you wish to use as the master burning tool.

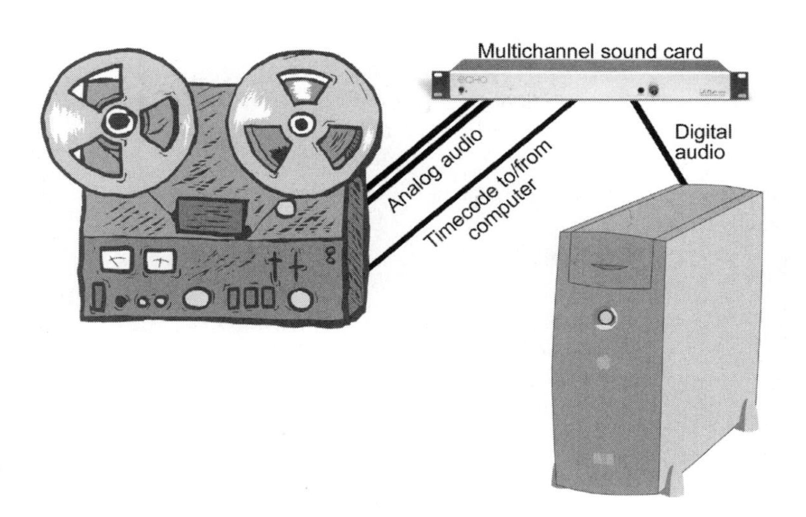

1.22 Connecting a multitrack to Vegas 4.0 is easy with a multichannel soundcard.

If CD burning is not required or if no CD burner is present on the system, this option may be ignored. When choosing a CDR for burning Redbook CDs, speed should not be a consideration as burned CDs destined for replication should be burned at single or double speeds to ensure minimum errors. While all CD burns have errors, fewer errors on the disk make for better replication and fewer problems. If you'd like to check your CDs for errors, a utility is available at `http://www.elpros.si/CDCheck/news.php`. CD Check is a checking utility that tests for errors and display the errors on any CD. This utility is also useful for testing various brands and types of media.

Audio Flow in Vegas 4.0

Any instrument, voice, device, or other means of producing sound in a room or other environment produces vibration. Vibrations carry a pure tone plus a wide variety of harmonics, the number and order of which are determined by the instrument, voice, or device generating the sound. This vibration generates air movement. These same vibrations and pressures on the air are what the human ear senses and translates into information that the brain can comprehend. These vibrations and subsequent air movement can also be captured by a microphone. Microphones convert air movement and vibrations (sound energy) into electrical energy by means of a diaphragm that moves when vibrations strike it. Varying qualities of microphones have different sizes and thicknesses of diaphragms and are made from a variety of materials.

The diaphragm converts acoustic energy into voltage. This voltage is sent down the microphone cable and into a preamplifier or amplifier. Most mixers have preamps for each channel built in.

The preamplifier makes the voltage greater, so that it reaches standard output levels for recording or input to a public address system.

In the computer or hard disk recording environment, the sound card can have a preamplifier built in. In the case of a professional-quality breakout box, some boxes have preamplifiers built in, but others do not. If your breakout box doesn't have a preamplification circuit built into it, a preamplifier is required. Devices such as the Echo Audio Layla or M-Audio Delta 1010 boxes do not have preamplifiers built into them.

Analog audio that is input to the sound card or breakout box is converted to bits, which the computer can process. The conversion process is the most important part of any recording project, as poor quality converters, called digital audio converters (DACs), result in poor quality audio that can neither be repaired nor significantly improved.

Audio passes from the DAC to the processor and hard drive, where the media is stored, until it's recalled for editing. The same sound card or breakout box is used for monitoring audio during recording. DACs vary from two-channel devices to eight-channel devices, allowing for up to eight inputs per breakout box at a time. Multiple cards may be combined for creating up to 24 inputs and outputs. Generally eight channels are sufficient for most recording situations.

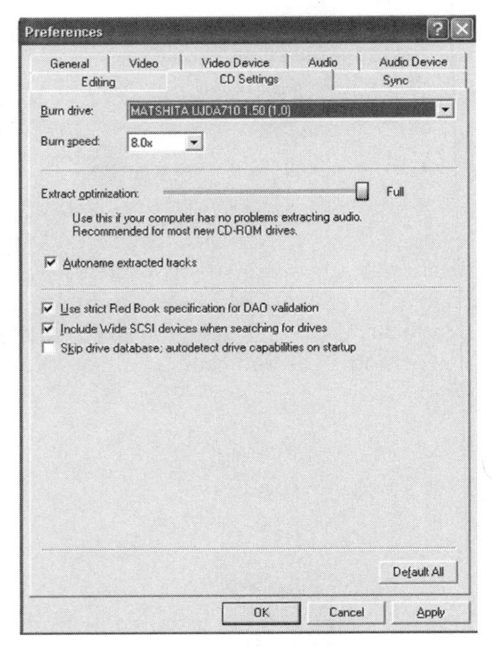

1.23 Vegas sees the DVD and CD burners on the system. You can select which drive (if multiple drives are available) is desired for burning audio.

Audio is monitored from the computer via the sound card or breakout box. An amplifier or amplified speakers are required for monitoring. Headphones allow for monitoring as well but should never be used to create a mix, as headphones do not present an accurate reference to what's really being presented in the recording. Mixes created through headphones tend to be bass-light due to the headphone interaction with the skull and bone structure of the human head. Worse still is the body's lack of sense of power as generated by moving large amounts of air in a room. The ear canal is capable of perceiving higher frequencies more quickly than lower frequencies due to the proximity to the speaker and conductor in the earphone housing. All sounds are channeled straight into the ear, rather than having the benefit of reflecting around the mix room or listening environment. This issue further creates a false sense of what's being heard. Some diffusion-oriented headphones are available at great cost and somewhat artificially simulate diffusion in a room. However, these headphones are modeled after the shape of the human head, and because no two heads are alike, various individuals perceive sound differently. Use monitor speakers in nearly all instances for accurate representation of what the recording sounds like.

Setting up the listening environment is equally important. A midsize room is best, such as a 12 × 12 room similar to a bedroom. A room with a high ceiling is preferable to reduce the compression of bass frequencies. At lower listening levels, however, standard-height ceilings do not present any obstacle. If monitoring at high volume is necessary, either a large room or sufficient room treatment are required.

Rooms that are too large or too small both present difficulties in the average recording and monitoring environment. A room that is too large can portray reverberations and time delay of unique frequencies, causing the ear to hear false colorations in the sound. A room that is too small can cause bounce and, even at nominal listening levels, inaccurately reproduce sound as well. Both large and small rooms benefit greatly from relatively inexpensive absorption and/or reflective treatment.

Treating walls and ceilings is easy, fairly inexpensive, and very effective in reducing reflections in the typical room. Using carpet and egg-crate foam is not very effective in preventing any but extremely high frequencies from bouncing around a room. Particularly in commercial or public environments, egg-crate foam designed for use under mattresses or for packing is even illegal, as this kind of foam carries a high fire danger in addition to being less effective.

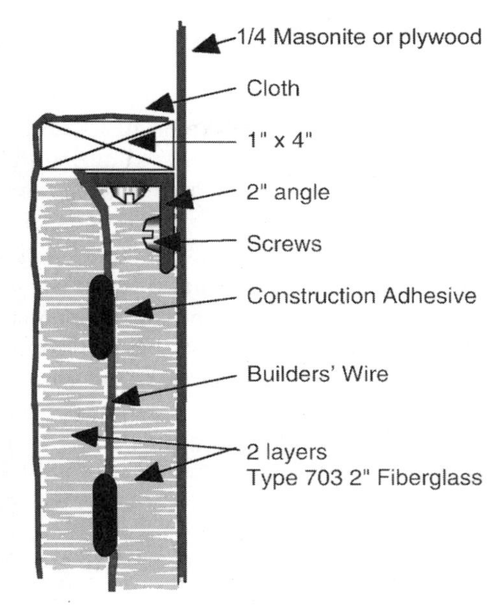

1.24 GOBO construction.

Ceilings can be brought under control by using drop-in panels, available from most music or professional audio retailers. Heavy density foam or fiberglass panel treatment also can suffice. Gobos can be manufactured from frames of 2 × 2, filled with fiberglass insulation, and covered with any loose-weave fabric. Gobos effectively kill reflections, predominantly behind the listening space or sweet spot, and provide for a tighter, more accurate representation of what's taking place in the mix.

1.25 Auralex Pyramids and Metro foam used in Sundance Media Group's "A" Room to control audio reflection. Note Metro control on the room's ceiling.

Large speakers that are too close don't accurately reproduce sound at the listening point or sweet spot because the sound waves have not had enough distance to develop properly. Speakers that have amplifiers built in are a great choice for many reasons. Most importantly, the amplifier is matched to the enclosure and components, reducing concerns about matching a system. Built-in amplifiers reduce the amount of cable found laying on the floor and reduce concerns for noise induction to the system.

Monitors that are shielded for video use are critical. Otherwise the monitor might not operate to spec, and video monitors can suffer in quality from the nearby magnets of the speaker monitor. Most near-field monitors are video ready. For program material that will be musically intense or sound-design heavy, for a recording studio, or for the 5.1 surround sound system, a subwoofer is necessary for small monitors.

When shopping for monitors, bring along a CD that you know well and have listened to in a number of environments. This addition can help choose monitors better. Also make sure that if you don't want a subwoofer, one is not connected during the monitor-listening session. Choose monitors that are pleasing to the ear. Staying with name brands, such as JBL, MAudio, Genelec, KRK, Event, and Mackie, ensures that only the personal opinions matter, as they are made to a predominantly similar quality. Some monitors have well-balanced response; others are top-end or bottom-end heavy, while still others can be oriented to the mid-range. There is no right or wrong in this part of choosing studio equipment; it's all about personal preferences. All of the bigger brand-name monitors will suffice in a properly treated room. None of them will sound their best in a non-treated or poorly treated room, regardless of the primary goal of the editor.

1.26 Mackie 626 THX-certified monitor.

Monitor speakers should be placed at a height that brings the high-frequency drivers, or *tweeters,* to the height of the ear when sitting at the mixing desk, editing desk, or listening environment. Otherwise, the speakers are what are known as off-axis, and this point is critical in the listening experience. Off-axis sound is colored differently than on-axis sound. Therefore, when sitting in the sweet spot, the tweeters should be pointing towards the ear, at ear height. By natural view, it will look to your eyes as if the tweeter is staring you right in the eye. Having the monitors on-axis also ensures correct depth of field in the listening space and is critical to evaluating placement of instruments or voices in the mix. Off-axis systems are nearly impossible to gauge the relationship of right and left audio correctly.

The distance between the monitors should be the same distance as it is to the listening position. In other words, an equidistant triangle should be formed between the listener and each monitor speaker. In the event of a surround-sound system, such as a 5.1 surround system to author for DVD, similar rules apply. All speakers should be isolated as best as possible from the environment. Center speaker, front left, and front right should all be equal in distance to each other. This means that rear speakers can be closer than normally expected or planned for. Rear speakers may be set back from the standard distance to the sweet spot, while care must be taken to ensure that the rear monitor speakers are on axis. The center channel should be at the same height as the front-right and front-left speakers but set back in space so that it is the same distance to the ear as for the remaining four speakers. This process is difficult to do in a small home studio or office environment because distances of at least ten feet measuring front to back are required to create a spatially accurate 5.1 mix. In the ideal situation, the center speaker is equidistant from the front-right and front-left monitors. The home living room is rarely ideal, however, and most home entertainment systems have the same depth of field for the center speaker as the front right and front left. When mixing for the average listening environment, it is generally better to mix for the environment in which the audio will be heard, rather than mixing in a mathematically correct room.

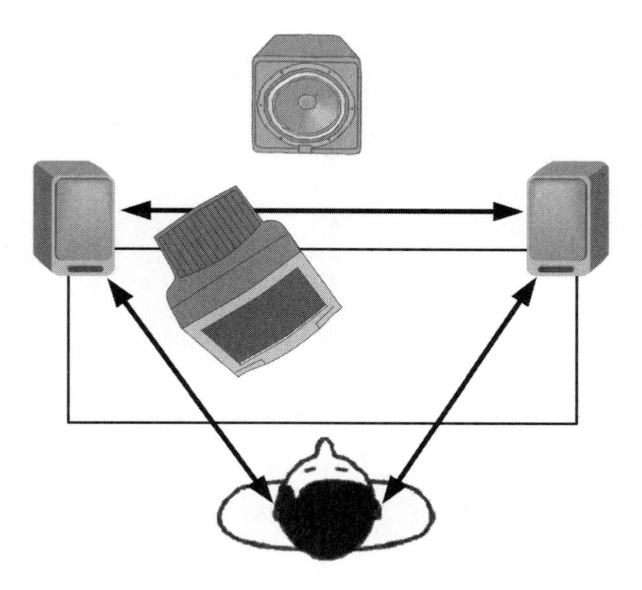

1.27 Correct monitor placement for near-field monitors. Maintain equal distance and have the monitors relative to ear-level for the best monitoring experiences.

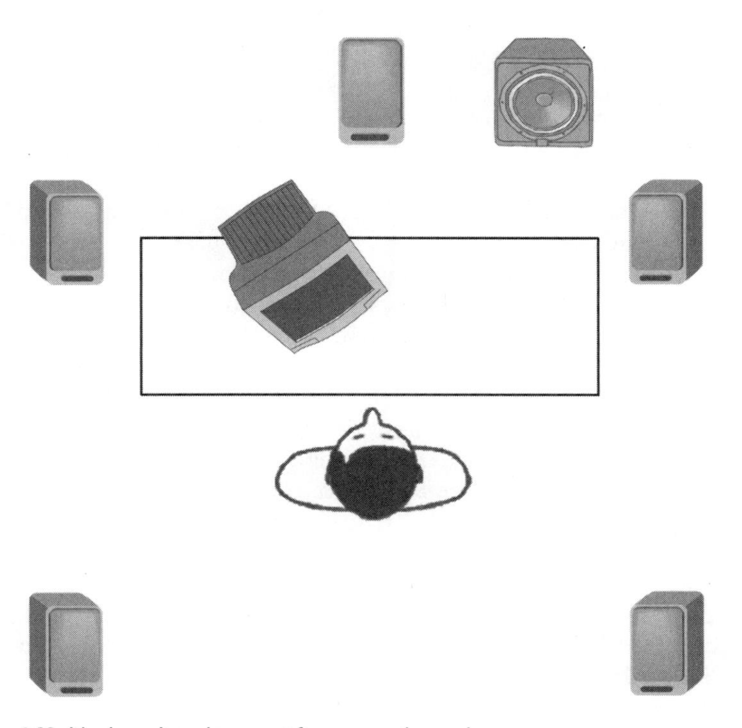

1.28 Ideal monitor placement for surround sound.

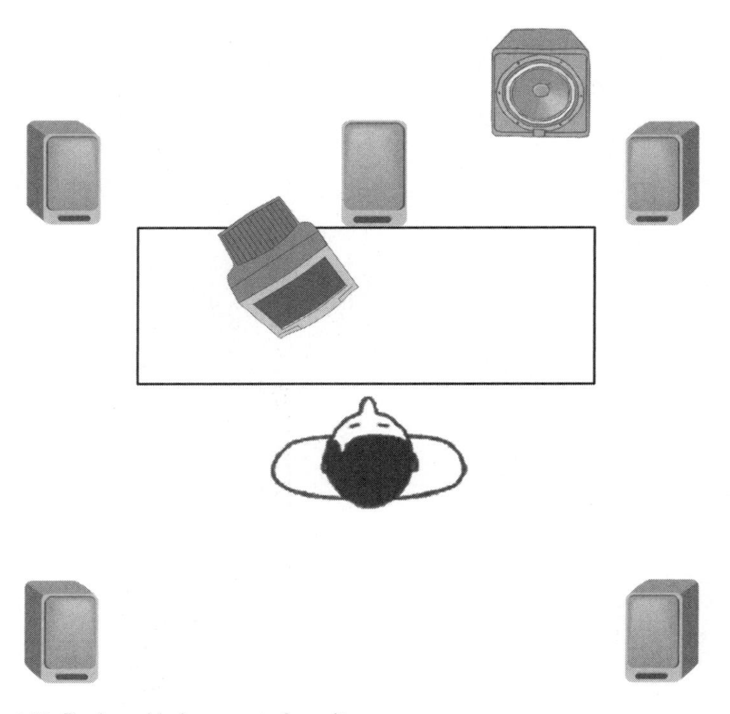

1.29 Real-world placement of monitors.

Isolating the monitors from other surfaces is beneficial in reducing unwanted vibrations, preventing the monitor from acoustically coupling with the surface on which it is mounted, and isolating the monitor enclosure itself. This process can be done in a number of ways. Speaker stands are available from a variety of sources and are common choices. Often these stands are filled with sand to make them acoustically dead or non-resonant.

Auralex manufactures a product called Mo-Pad that is designed to dampen the acoustic coupling that takes place in all studios, as well as to reduce vibrations. This process can also be accomplished by using dense foam, cut to hold the speaker in place, which provides at least an inch of isolation from the bottom of the monitor to the surface of the desktop.

Make sure all shelves, microphones, or anything else in the monitoring environment are battened down or at least foamed to prevent vibration. All things vibrate; it's just a matter of what frequency they vibrate at, so minimizing these issues up front makes for much better mixes in the end. While all of these details might seem like standard fare in a recording studio and overkill in a video-editing suite, always remember that sound is the most important aspect of any production environment.

Connect the monitors to the sound card using either XLR or TRS cables. These cables are balanced, low impedance, and may be run for great distances. If high-impedance cables are used, such as a typical 1/8-inch connector found on all low-end sound cards, short cable runs should be used, or hum or buzz can be induced. Balanced cables, properly connected, prevent hum or buzz from entering the cable path. Use balanced cable everywhere possible. Some sound cards

1.30 Auralex MoPads.

do not accept a balanced input. In the event that your sound card does not allow for connection of a balanced line, obtain a direct box and a short (6″ × 12″) patch cable with a male stereo mini plug on one end and a male ¼-inch plug on the other. The microphone connects to the direct box via balanced cable, run as far as it needs to be run. The direct box sits next to the computer, with the ¼-inch plug connected to the direct box and the male mini plug connected to the sound card input. This setup eliminates hum and buzz normally induced by running long lengths of high-impedance or unbalanced cables. At no time should speaker cables be used for microphone lines, and just the opposite applies. Line cable should never be used as speaker cable.

Run cables at a 90-degree angle over power cables. Running any cable parallel to a power cable is an invitation for trouble. Ensure that audio lines are far away from wall warts or AC adapters that are common in the studio. Also, keep cables tied in neatly, so that similar cables are running with each other. Power cables run in one tie system, speaker, microphone, line, patch, and other similar low-voltage cables run in another. If the studio layout has not yet been decided or equipment not yet installed, lay down the AC/electrical lines first. This procedure helps keep things more organized in the end and ensures that cables are not laid over AC paths. AC hum presents itself as a 60 Hz buzz that is constant, so you just need to isolate the source of the hum and eliminate it, which can be difficult. Start by ensuring that all equipment is properly grounded and connected with balanced cables. *Never remove the grounding plug or tab from a power cable.* While this process occasionally removes hum by removing the ground from the system, it is potentially *lethal* and masks the real problem.

Fluorescent lights are another source of hum in a recording environment; fluorescent lights are also problematic for the video-editing environment as it's difficult to color correct in a room lit with fluorescent lighting. Use incandescent lighting whenever possible. If accurate color viewing is critical for video editing, consider low-wattage halogen lighting.

Another common source of hum in a recording environment are dimmer controls connected to lighting. These dimmers contain silicon-controlled resistors (SCRs) that cause induced interference that translates to buzz in the system. The buzz rises or falls in volume as the dimmer is turned up or down. Dimmers seem to come only in two types: really good dimmers with noise filtering that are electrically quiet and are fairly expensive and really bad dimmers that are cheap and have no noise filtering (or anything else) built into them. Get good dimmers if dimmed lights are necessary.

If at all possible, keep the power to the walls of the room separate from the power to the lighting and other office equipment. This procedure helps ensure clean power and sound. For the recording environment, tools, such as Furman or Equitek power conditioners, are available and can power almost any NLE studio equipment while assuring spike-free, clean power output. The use of one of these power distribution systems doesn't guarantee pure, hum-free audio, but it goes a long way towards getting there.

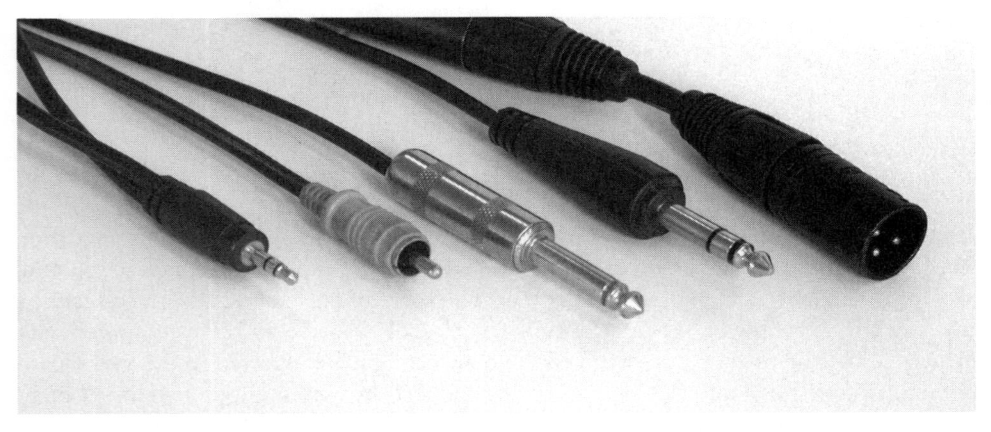

1.31 Mini, RCA, unbalanced quarter inch, balanced quarter inch (TRS), and XLR connectors.

Setting Up for Professional Video Editing

Cameras

With DV being so prevalent in today's world, just about anything can be classified as broadcast quality. For example, the cameras used in the DV space are predominantly higher quality than the Handicams of just a few years ago, which were used in news broadcasts and for some television shows. Any of the higher-end cameras that cross the professional and consumer price points and features (hence the term *pro-sumer*) may be used for professionals and video hobbyists. Canon, JVC, Sony, and Panasonic all make sub-$10,000 cameras and even the sub-$5,000 cameras that are used in broadcast, corporate, training, and home video today. The first footage of the attack on September 11, 2001, was captured on a DV camera and rivals the quality of some of the finest broadcast cameras.

1.32 Canon XL1S.

Armed with a DV cam, a reasonably fast computer, Vegas 4.0, a decent television monitor, some editing skills, and an understanding of the strengths and limitations of the DV format, anyone can create broadcast-quality media. Key benefits of Vegas 4.0 for the professional or hobby video editor are that Vegas contains in-depth professional audio editing tools that have never been found in any video editing application, as well as containing incredible video editing tools, but also contains in-depth professional audio editing tools that have never been found in any video editing application. These features alone set Vegas 4.0 above any video NLE system in its class. The key points for interfacing with professional equipment are significant and may be costly if top-level equipment is desired.

OHCI Card

An OHCI-compliant card is needed to transfer video from the DV camera to the hard drive. This requirement does not mean that owners of Canopus, Matrox, Pinnacle, Targa, Viewcast, or other proprietary capture cards cannot work with Vegas. They can! Because Vegas is format agnostic and resolution independent, media captured with other capture applications and hardware can be imported and edited in Vegas 4.0 and transferred to tape via the original capture tool. One primary advantage of Vegas is that it accepts nearly any kind of media on the timeline, whether it's AVI, QuickTime, MPEG, or other media format. Vegas accepts nearly everything from uncompressed SDI media to small compressed Windows MEdia Video (WMV) on the timeline. However, only OHCI DV devices interface directly with Vegas' capture utility. (Vegas does have support for wavelength division multiplexing (WDM) capture devices as well.)

DV Decks and Tape Machines

Using a DV deck as opposed to using a camera for capture is a good thing to do because DV cameras are not designed for the shuttling back and forth of tape. DV tape machines, on the other hand, are designed to do just that. DV decks usually are capable of offering visual timecode in some kind of display and offer analog inputs for converting analog tape media to digital, superior transports, faster rewind and fast forward, multiple FireWire outputs and inputs. Some handle different types of DV tape, as well.

Tape machines also can function as converters for viewing media on an external monitor. Tape decks for DV start in the $700 range, and prices are always dropping. eBay is an excellent source for used decks, and there are many used-equipment clearing houses in the back of most DV magazines.

Panasonic, Sony, and JVC all make DV decks in a variety of configurations. JVC even has a deck that plays and records both mini-DV and SVHS on the same machine.

Monitors

Having an external monitor is critical if the video being edited is ever going to be seen on television, projection, DVD, or anywhere else that is not a computer monitor. Gamma differences between a computer screen and a television screen are significant; colors in a video will not look like the same on both computer and television screens. Color correction is all but impossible to perform accurately without an external monitor.

Mid-level monitors, starting at $400 and up, can include phase, color, and brightness controls, and perhaps control of the color guns. Inputs are usually either BNC or composite on the less-expensive or lower-quality monitors. At least 300 lines are necessary for quality preview of video.

High-quality monitors have BNC and S-Video/Y/C inputs and a choice of component/RGB inputs, as well as composite inputs. *Component inputs* allow for separate cabling for each color—red, green, and blue (RGB). *Composite input* is just as the name implies—a composite of the three colors—and is of lesser quality. The yellow-colored RCA plugs on the back of consumer VCRs are composite inputs. Some of the better consumer VCRs, receivers, and televisions now have component inputs as well. Most high-end monitors also provide an underscan capability that allows for the full signal to be seen without masking from the edges of the monitor housing.

Y/C inputs, which are also called *S-Video inputs*, are a combination of two signals—luminance (Y) and chrominance (color). The quality of Y/C input falls between composite and component. Component is better and consequently more expensive. Composite is cheaper and, as a result, is lower quality. Most DV cams and decks have both composite and Y/C output/inputs on them. Use Y/C whenever possible. Converters that have RGB input/output on them are very expensive and are generally not used by any but the top-end professionals. The Laird Blue Flame Professional is one of these units.

When analog media must be captured or brought into the digital realm, the professional and hobbyist have a few choices. ADS Technology, Canopus, Laird, and Sony all make analog-to-DV conversion boxes that are very high quality and are acceptable at the professional levels. Keep in mind, the codec that the device uses to convert media is critical. Therefore, any device that uses a substandard DV codec is not truly of professional caliber, as media appears soft and contains less color depth. Devices, such as the Hollywood Dazzle card and Micron's conversion box, should be avoided. The Canopus device uses Canopus' own codec to compress the analog media, and the quality is quite high. Laird Telemedia's device converts using proprietary codec information to compress the analog media and is very, very good.

Analog video brought in via converter box is converted from the original resolution to the National Television Standards Committee (NTSC) DV standard of 720 × 480 or, for phase alternate line (PAL), converted to 720 × 576. Approximately ten percent of the signal seen in the computer monitor is not seen on the television, regardless of whether its original source is DV or analog. Be aware, therefore, that banded edges or top or bottom screen chatter might be seen in the computer monitor but won't be seen in the television monitor. All television monitors at consumer or professional grades mask portions of the television signal. Underscan-capable monitors show masked media areas not intended to be seen in broadcast.

Microphones and Voice-Over Boxes

Audio for the video is critically important. Audiences will accept poor quality video and editing but will not tolerate bad sound. Our ears are substantially more sensitive than our eyes, and therefore we perceive much of what we see based on what the ears tell the brain that it should be seeing. Audio is frequently used to mask editing errors, transitions, and changes in color. Therefore, having good audio equipment is a must.

1.33 Various microphones.

Start with a good microphone. Audio Technica's shotgun series, while typically used in the field for audio during the video process, is also excellent in the studio for voice-over (V/O) work. Audix's SCR 25 microphones are also excellent for V/O work. ElectroVoice's PL/RE20 is a common choice, as is the Shure SM 58. SM 58s have been around forever, you can nearly hammer nails with them, and they are cheap at $99 new. Mics need not be new to achieve great sound. A recent search of eBay turned up a huge number of very high-quality microphones at incredibly low prices. These are all balanced microphones, which is important in the video studio because of all the potential noise interference that video and audio equipment can generate. Microphones that have cords attached to the body are usually quite low in quality. Beware of microphones made of plastic rather than heavy brass. Plastic microphones tend to resonate and have unacceptable frequency responses.

The microphone is only half the task in achieving great V/O or dialog recording; the room must be tuned as well. Absent the sound treatment mentioned previously, every video editor doing V/O work should have a voice-over booth, box, or at least enclosure. While a bedroom closet can suffice for this, a portable, foldable V/O box can be built for a small cost and tremendous value.

Using foam core from the local hardware or craft store and Metrofoam from Auralex or similar manufacturers, mount the acoustic control panels to the foam core with spray cement. Facing the foam inward, build a four-sided box consisting of two sides, a back, and a top, holding the sides together with plastic foam core hinges or with duct tape.

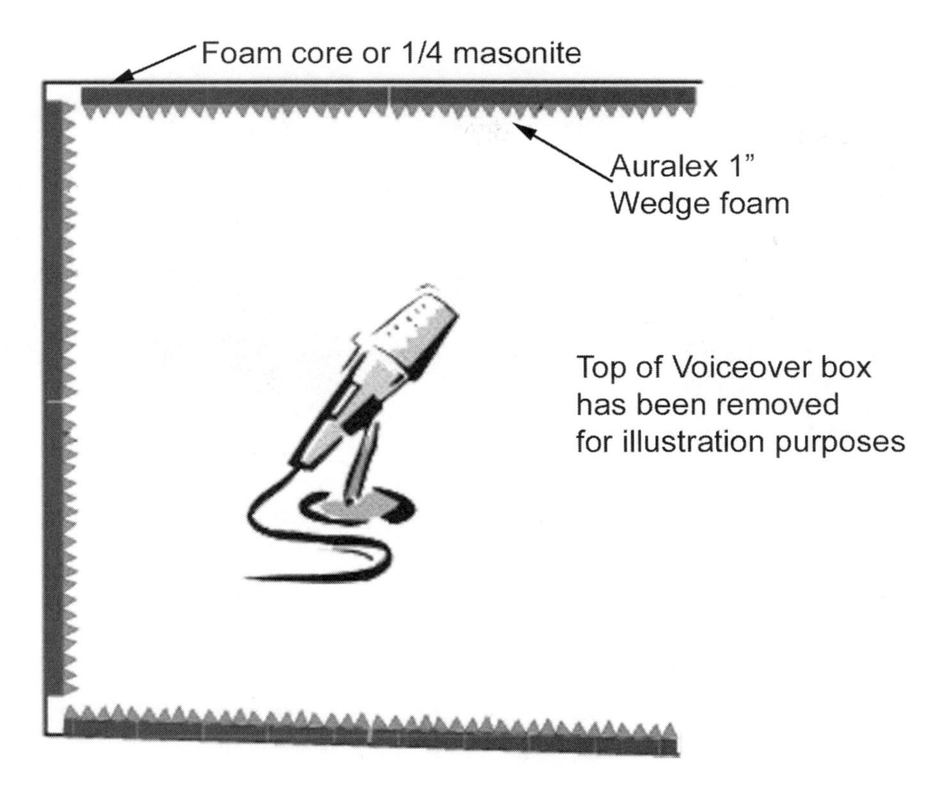

1.34 V/O box made from Auralex foam.

Place the microphone inside the box so that the front of the mic is just slightly behind the edges of the box. Run the cable under the edge of the box.

This box eliminates reflections in the room and gives the best direct mic sound overall. This technique is often used to mic instruments and elements in nature. Our ears are remarkable instruments, capable of filtering out sounds that are around us. The folds in our ears also help tune our ears to what's in front of our heads, rather than behind. As an example, most of us don't notice the sound of air moving around a room, even though it's actually very, very loud. A microphone is a dumb instrument; it cannot discern the difference between a human speaking and the sound of air rushing around a room. It cannot distinguish the sound of noise caused by a running refrigerator from the sound of a pair of hands clapping. Our ears can. So even though a room might seem good to record in, after the recording is made any number of background noises can be heard. These sometimes can be electronically removed, but it's much better to have the sound go to tape or hard drive correctly in the first place, isn't it? Using an isolation device helps enormously.

MiniDisk Players and DAT Machines

The same holds true for recording in the field. Isolating the camera from the mic in any way guarantees better audio. In fact, the mic built onto any camera is effectively useless in all but the most raucous situations. The on-camera mic is great for soccer games, concerts, and any event where no one is actually speaking or imparting an important message. Because the camera mic can't discern the difference between the direct sound and the reflected sounds of a room or other environmental space, a separate mic should always be used when a message from a lecture or other information source is being recorded. Whether using a wired mic that runs directly to the camera, a wireless mic that does the same, or using a MiniDisk (MD) player or DAT machine with a mixer, direct sound is always best. Weddings with a pastor, bride and groom, plus any other persons talking, can be audio-captured using wireless mics, a boom operator, or by slipping an MD player into the pastor's pocket with a lavaliere mic on his/her lapel. This setup usually suffices to pick up the bride and groom, plus the pastor, during the ceremony. Audio can then be transferred to the computer for later editing.

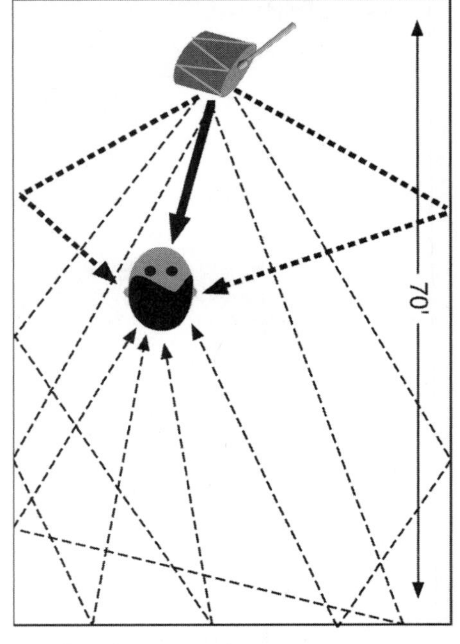

1.35 Room reflections.

Table 1.4 DV Editing Suite Equipment List

Computer:

- Two hard drives
- 512MB RAM minimum

OHCI card

Microphone

External monitor

DV camera, deck, or converter

Voiceover box

MD player/DAT machine

Under no circumstances should a cassette machine be used for capturing audio that is to be synchronized to video later. Cassette players use analog motors and nearly always have wow and flutter that speed up or slow down audio over time. This cannot be synchronized to video unless the audio and video are cut into very short (ten second or so) segments and hand-synchronized and even then this is difficult to achieve. Digital devices, such as MD players or DAT machines, use a clock source so the time of the recording doesn't shift and can easily be locked to video at a later time. These devices can also usually capture audio at a variety of sampling rates. If so, capture audio at a sampling rate of 48 KHz if it's to be synchronized back to video. DV has two standard audio sampling rates, and 48 KHz is the standard at which most DV cams work. Vegas is capable of taking any sample rate and upsampling on the fly to the 48 KHz standard, but, if you are working with other NLE systems, audio must be resampled to 48 KHz before editing if the audio is not captured at that sampling standard.

Vegas has automated mixing built in, so a mixing device is not necessary unless a preamplification device is needed for microphone input. (See the microphones section in Chapter 7.)

Because Vegas has waveform monitoring (WFM) and video monitoring displays built in, no external scopes or other measurement equipment is needed in the editing suite. Vegas can also generate uncompressed files for transfer between various editing applications and hardware, eliminating the need for other external equipment.

Networking and Vegas 4.0

Vegas works well with networked systems, and several client machines on the network can somewhat share media over the network. The faster giga-networks allow video to be pulled from a video server and to be edited in Vegas. Drives that are mapped on a network are displayed in the Vegas Explorer.

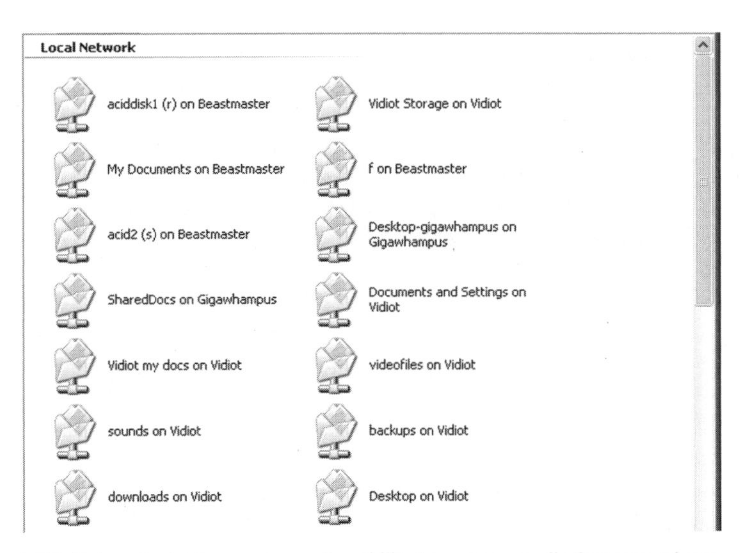

1.36 Networked drives open up the ability to access media from anywhere on the system. With the right script, Vegas can render on any machine on the network.

Bear in mind that DV requires 3.6 MBps to function, and drawing video over a standard 10/100 local area network (LAN) can cause stuttering or gapping in audio or video. Depending on the setup of the operating system, a network can potentially interfere with other system resources and cause instability during editing. If ACPI has been disabled and if IRQs are properly allocated in the system BIOS, this problem should not present itself.

The primary advantage of Vegas' ability to work on a networked system is that several users may have access to the same media, allowing for teamwork in a normal working environment. One machine can be recording an audio session in one room, while at the same time another machine in a different room is editing audio for noise, processing, cuts, fades, or mastering on the same project. Conceivably, a client could be recording music in an "A" room, and an engineer mixing the songs as they hit the hard drives in a "B" room, while still another is mastering and doing the CD burn preparations in a "C" room. By the time the session is complete, rough or finished mixes could be nearly ready to take from the studio on a CD! The same features apply with video, as one piece of media is being edited for broadcast, another room can be working on the same files for a webcast, while yet another system is editing for DVD or CD-ROM output. All this happening at the same time!

A secondary advantage to having Vegas 4.0 on a network is the potential use of scripts to create a *render farm* in which idle machines are used by the editing machine to render in the background and to edit on the primary machine in the foreground. For example, make a selection or region on the timeline and run the render farm script. Vegas then instructs one of the idle or less busy machines to render the region and, once rendered, to place the rendered region or file in place on the editing timeline.

Vegas' ability to share media on a network presents unprecedented strength in a working environment. Only the most expensive editing systems have had this ability in the past. Now workspace and time can be streamlined for maximum efficiency.

In a home studio, it's common to find more than one computer. In this event, a one-person show can easily manipulate several files at once. While one machine is rendering one section of multi-media, a second machine can easily be doing an encode to MPEG or can be used for mixing a song, while another machine is burning a CD, rendering video, or any other task that Vegas might be called on to accomplish.

One aspect of Vegas that is little known however is the ability to have multiple instances of Vegas running at once, allowing one machine to accomplish several tasks simultaneously. A single machine can be rendering, prerendering, burning a CD, logging, editing, and capturing all at the same time! (It's not advisable to capture during simultaneous render processes.) This capability reduces the need for Vegas to be running across a network in a single-operator environment in even the most demanding of times, as one machine with good hardware can do it all concurrently. However, using network resources with Vegas 4.0 means that the network becomes an extension of the editing machine during time-critical renders, as well as letting users share media across a network.

Basics of DV

Understanding the DV Format

In order to edit and output digital video (DV), it's important for users to comprehend the basics of DV. In the past, the standards for editors have been NTSC-based (or elsewhere, such as Europe and Australia, based on PAL) as DV is now, but the standards are different because of a variety of new standardizations for the DV workspace. While NTSC still controls the broadcast space in the United States, due to the formatting of DV, a number of unique aspects exist that editors need to be aware of. Those new to the DV editing space can benefit greatly from learning some of the basics of video, regardless of format.

DV compresses media at a ratio of 5:1 at the camera. This compression results in some color and light loss to the information, but, for the most part, it's not visible to the human eye.

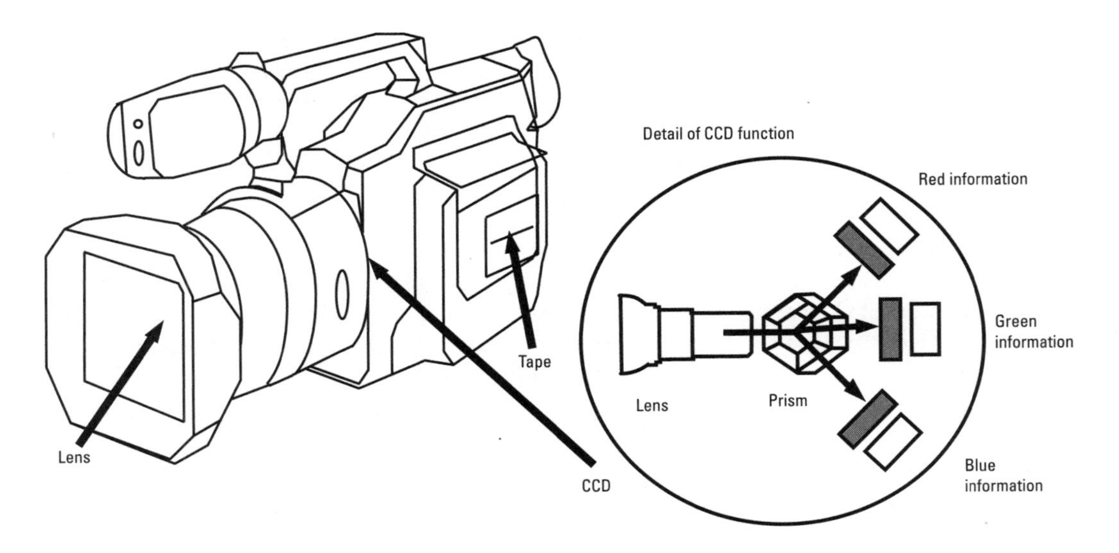

2.1 How a DV camera works.

The camera lens sees the image and transfers the image to the charge-coupled device (CCD). The CCD then converts the analog image into an expression of mathematical equations represented as bits, or ones and zeros. The ones and zeros are interpreted as to their color value and arrangement as pixels, and the data is stored on digital tape. Better cameras have three CCDs on them—one for each color channel (red, green, blue). Three-chip cameras generally have less sensitivity to light but offer more accurate color interpretation. Lower-quality cameras have one CCD that processes all three colors. Single-chip cameras generally are used for family home videos, as the cameras lose sharpness in less-than-good light and are very sensitive in dim light.

Table 2.1 DV Standards to Know

NTSC-DV	720 × 480	29.97 frames per second
PAL-DV	720 × 576	25.00 frames per second

The media is compressed at the camera so that the media can fit on to a digital tape and sustain a data rate that allows the tape to store and transfer the media. DV25, which is the most common DV format, sustains an average data rate of 25 Mbps. Other DV variants, such as DV50 and DV100, carry higher data rates.

When the media is stored on the DV tape, it is no longer media that resembles what went into the camera. It's merely data, converted to a preset series of numbers and pixel locations allocated to grids. The data is transferred via FireWire from the camera to the hard drive, where the data lives during the capture, editing, and print to tape, DVD, or Internet stages. Therefore, it's somewhat inaccurate to consider an OHCI card as a capture card, as it's merely a conduit for data and is no different than a serial port, parallel port, SCSI connector, IDE bus, or other data transfer protocol.

Resolution and Frame Rate

The NTSC-DV standard has a resolution of 720 × 480, which means that the DV image on a computer screen is 720 pixels in width and 480 pixels in height. Europe's PAL-DV standard has a resolution of 720 × 576. A typical analog capture card captures media at 640 × 480, which is a square pixel representation of the analog signal and differs from the non-square pixels used in DV images.

NTSC-DV has a frame rate of 29.97 frames per second (fps), whereas PAL-DV has a frame rate of 25.00 fps. By comparison, acetate film has a frame rate of 24.00 fps, which is why many DV producers use PAL cameras for simulating film, even in the United States. In recent times, several manufacturers have released or announced camera models that have the ability to shoot at 24 fps and in a progressive scan format instead of an interlaced format. This capability provides videographers and editors with the opportunity to shoot images that are more film-like.

A faster frame rate typically results in smoother motion and a sharper picture. Film shot at 24 fps may have a softer appearance than does NTSC video. NTSC arrives at its frame timing based on the 60 Hz electrical frequency that electricity in the United States operates within. With two fields being drawn at 1/60th of a second for each field, electricity becomes the synchronizing signal. A portion of the signal is required to carry timing data, hence the dropped frames that take the signal from a pure 30 fps to 29.97[1]. This issue is also why DV frame rate is referred to as *29.97 drop-frame*.

Tip

Because video is captured at the camera at a frame rate of 29.97 fps or 25.00 fps and because computer monitors typically refresh at rates of 72 to 85 Hz, a black band or bar appears to be rolling up the preview screen in the camera. The banding can be dealt with in most cases by setting the refresh rate on the monitor to a rate of 60 Hz. Some cameras allow shutter speeds to be adjusted as well in clearscan modes, which are typically found in the shutter settings of the camera. Either method generally eliminates or significantly reduces the banding/blanking seen on a computer monitor. If clean screen shots are required, shoot LDC monitors, which have no refresh rate. The camera can be zoomed tight on the screen. If video is being shot for educational or training purposes, shooting an LCD screen is preferable.

Aspect Ratio

Applications of Aspect in the DV Realm

The aspect ratio of a standard television is 4:3. For every four image elements in width, the picture is three image elements in height, if we're to reduce this formula to the most basic level. Wide-screen television has a large variety of aspect ratios, but the most commonly used and understood aspect in the wide-screen world is the 16:9 aspect ratio, measuring for every nine picture elements in height, 16 elements in width, or nearly double the width compared to the height of the image. Pixels are generally presented at an aspect of 1:2121 in this format. In the 4:3 aspect ratio, pixels are presented at .9091 (.909 for simplicity's sake).

Nearly all DV cameras shoot in the 4:3 aspect ratio. Very few shoot a true 16:9 aspect but rather provide a stretched aspect format called *anamorphic wide-screen*. This aspect is not a true cinematic aspect, but functions for most of the pro-sumer needs. The DV spec does not actually have a provision for 16:9, so 16:9 images are created by squeezing pixels in the DV format. There is no additional pixel count.

Several high-end digital cameras from Thomson, Sony, and Panasonic shoot a true 16:9 aspect ratio, but they are very expensive. These cameras use a combination of electronic programming and optics to generate a true 16:9 image.

2.2 A 4:3 aspect ratio is 4 units wide by 3 units high.

2.3 A 16:9 aspect ratio is 16 units wide by 9 units high.

The 16:9 aspect ratio is still quite misunderstood by many shooters/editors. Many flavors of the 16:9 aspect exist, depending on who is editing and presenting the film.[2] It is rapidly becoming the standard of television and film for myriad reasons and allows viewers to see all of a picture as it was shot for film. It's very close in aspect to the popular cinematic aspect of 1:85. If your work is eventually destined for broadcast, consider shooting in 16:9 to take advantage of future broadcast format. This aspect will be supported more strongly in the future, but it is still an evolving format.

2.4 Notice the slight squeezing of the picture in a simulated 16:9 aspect ratio applied by a camera not shooting with anamorphic lens (simulated to match 2.3/2.4).

Some cameras can shoot native 16:9. To ascertain if your camera can, select the 16:9 switch on the camera. If the picture is squished in the viewfinder or if it grows wider in view from what it was in the 4:3 mode but has letterboxing above and below the image, chances are that the camera can shoot 16:9 correctly. If switching the aspect at the camera results in the picture being the same width but letterboxing has been added to the picture, all that is occurring is that the top and bottom of the image is being cut off and is not true 16:9. When this process happens, the camera is only generating or using 75 percent of the scan lines available. In other words, rather than using the full 480 lines available, the camera is using only 350 lines, resulting in a softer picture.

Cameras such as the Canon XL-1s, GL-1, and Sony VX 1000/2000 are all simulated 16:9 format cameras. However, tests have shown that in most instances, shooting at a simulated 16:9 provides a higher resolution than shooting in 4:3 and then cropping 25 percent of the image away. This issue is thought to be true because the image is compressed at the normal 720 × 480 size, so the encoding chip sees a normal image even though it was shot at 720 × 360. With PAL cameras, this feature is even more advantageous, as PAL cameras use the full-chip resolution, which usually results in a much sharper picture.

Another means of achieving 16:9, and by far the best way without a native 16:9 camera, is to rent or buy an anamorphic lens from Century Optics or Optex. Anamorphic lenses price themselves from $500 and up at the time of this writing.

One of the most important DV characteristics to understand is the aspect ratio of the pixel. In the computer graphics space, pixels are typically 1.0, or square. In the DV world, pixel aspect ratio is usually non-square, often seen at .909 or rectangular. You might view various pixel aspect ratios in the Vegas 4.0 Project Properties drop-down menu. The pixel aspect ratio is important to understand particularly when combining media from different sources, such as photographs, titles from third-party applications, graphics, Microsoft PowerPoint presentations,

2.5 DV pixels are non-square, while computer graphics pixels are square.

and pre-compressed media, such as WMV or QuickTime files. If graphics for a DV presentation are created in a graphics editor, such as Adobe Photoshop or Corel Paint programs, and then imported to Vegas, the graphics should be adapted to the resolution of DV in the graphic editing/creation stage. Vegas can handle any resolution and aspect, but it's best if the graphic is created correctly in the first place. (See Chapter 4 for more information on importing stills.)

Interlace vs. Progressive Scan

When an image shows up on a television screen, the image is made of a series of individual pictures, called *frames*. These frames are drawn at a vertical resolution of 480 lines, at a rate of 29.97 frames per second, when using NTSC as an example. When the frame is drawn on the television, it's drawn a half a frame at a time. It takes two fields per frame to draw a complete picture.

Each field is scanned separately, one after the other, and there are odd fields and even fields. Odd fields are sometimes called the *top field*, and even fields are sometimes called the *bottom field*. Generally, DV nearly always draws the bottom field first.

Interlacing stems from the early days of television when not enough bandwidth was available to draw the entire image at one time. Lines moving horizontally down the screen faded out as the phosphor reached the other side. Two updates per frame occur every $1/60^{th}$ of a second, which allows a balanced colored image to cover the entire screen. This explanation is a fairly simplistic one for a complex process.

2.6 A progressive scanned image provides a smooth still image.

2.7 Notice the interlaced lines running through the image where the image is highly contrasted; these lines are particularly visible in the shirt stripes and above the cup in the image.

With a *progressive scan*, the entire frame is drawn at one time on the screen, updating once for every frame and drawing twice the information on a screen to compensate for requiring half the number of frames. Rather than seeing 30 images per second as seen with interlaced media, 60 images per second are seen. The result is that the picture appears sharper and is reputed to be less stressful for the human eye to interpret.

Color Sampling in DV

When a DV camera captures an image through the lens, not all of the pixels that the image generates are captured and placed on tape. Mini-DV or DV-25 registers at a sampling rate of 4:1:1. With a 4:1:1 sample, four luminance (Y) pixels are recorded for every four pixels of a signal. However, only one pixel from blue (Cb) and only one pixel from red (Cr) are sampled. This point means that a slight degradation in the color of the recording occurs as compared to the actual image being recorded. The biggest loss comes to color saturation, but a well-lit shot captured with a high-quality camera is still an impressive picture. In fact, the color sampling of a DV cam is not substantially different from the color-sampling scheme of a Beta SP camera, if the Beta were to be characterized in a digital expression. (Beta SP is an analog format.)

A 4:2:2 color sample uses the same formula in which four pixels of information are recorded—two pixels of Cb and two pixels of Cr. The additional pixels allow for greater chroma fidelity and more accurate recording of the image.

The sampling formula can be expressed as follows:

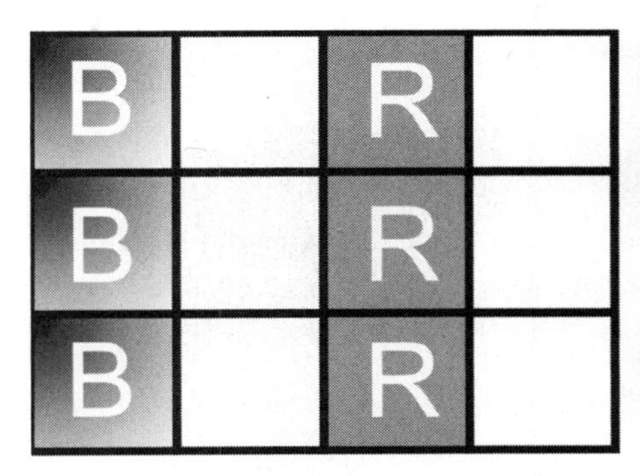

2.8 A 4:1:1 color sampling, with one pixel of blue and one pixel of red for each set of four luminance pixels.

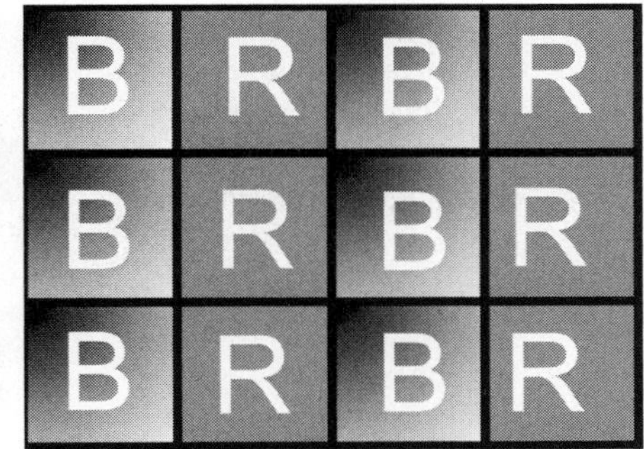

2.9 A 4:2:2 color sampling, with two pixels of red and two pixels of blue for each set of four luminance pixels.

Human eyes are a great deal more sensitive to brightness/luminance than to changes in chroma. To take advantage of this lesser sensitivity to changes in chroma, sampling schemes have been developed based around greater sampling of lumen and lesser sampling of chroma.

Various camera formats use the 4:1:1 sampling scheme. While the numbers don't add up to much (less than 50 percent of existing image color becomes part of the digital information in a 4:1:1 sample), it is currently the more prevalent sampling mode, partially out of cost but partially out of the enormous breadth of hardware that uses this sampling scheme. The human eye, however, perceives less of a color loss than is actually there.

Table 2.2 Another Means of Expressing Color Sampling in DV

4:1:1 Signal Sampling	4:2:2 Signal Sampling
YC Y Y Y	YC Y YC Y
YC Y Y Y	YC Y YC Y
YC Y Y Y	YC Y YC Y
YC Y Y Y	YC Y YC Y

YC = Luminance/Color

Y = Luminance

PAL-DV cameras sample color at 4:2:0, which some people believe maintains better color conversion to MPEG than does NTSC-DV. Less color information and smaller bandwidth of NTSC, however, make for a faster and easier encode to MPEG.

As Table 2.3 demonstrates, many cameras in broadcast use the 4:1:1 sampling scheme. One of the few places that the loss in colors might be noticed is when chroma key is necessary for the edited media. The more colors present, the easier it is to key those colors. Fewer colors give the computer's algorithms less information with which to work, and consequently a potential exists for jagged edges, soft colors, and poor separation between the masked image and background image. Some hardware configurations can upsample the 4:1:1 signal to a 4:2:2 or 4:4:4 ratio. Much discussion occurs about the merits in the upsampling/downsampling of the video in these systems—some in favor and some not. Anytime media is downsampled from its native format, some degradation of information occurs. Vegas 4.0 has significant abilities in pixel sub-sampling, and high-quality keys are easily achieved with a small learning curve and practice.

Table 2.3 Sampling Schemes of Cameras in Broadcast

Sampling Scheme	Cameras
4:1:1	Mini-DV, DVC Pro, DVCam
4:2:2	DVC Pro-50, Digibeta, BetaSX, D1, Digital-S

Connections at the Camera

DV cameras have an output that is a four-pin FireWire connector. This connector is what connects the camera to the computer's OHCI card. The connector is standard regardless of whether the camera is PAL or NTSC. Keep this connection covered when shooting, unless you are using a DV drive kit to capture video during shooting. While it's not fragile, it's easy to get dirt inside the connector, fouling the contacts and potentially preventing the camera from communicating with the computer.

New formats are coming on the market, and it won't be long with USB 2.0's faster interface up to 480 Mbps that some cameras will have the USB 2.0 Type B connection right alongside a FireWire connection. Cameras are also available that record MPEG video directly to a disk that fits in the camera. The disk drops into the computer's CD-ROM, and media can be transferred directly to Vegas. While most program stream MPEG files can be dropped on the timeline in Vegas, Vegas always recompresses MPEG files, which can result in loss of image quality. MPEG media can be choppy and rough in playback until rendered to an AVI file format. Vegas is optimized to work with AVI files, even though nearly any media format may be placed on the timeline in Vegas.

2.10 FireWire connection on Canon GL-1.

Video captured in Vegas is captured as an AVI file, or, more accurately, the data transferred from the video camera is given an AVI header in the file and packaged as an AVI. There are two types of AVI files: Type 1 and Type 2. Both can be used and accessed in Vegas 4.0.

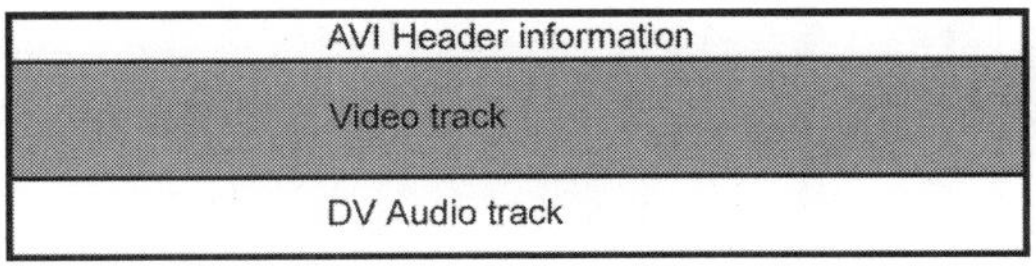

2.11 Type 1 AVI stream.

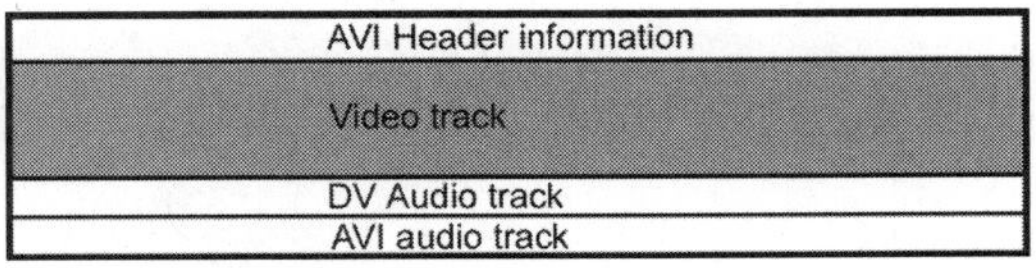

2.12 Type 2 AVI stream.

Type 1 files are the typical AVI format found in nearly every NLE. Audio and video are interleaved or packaged together and remain locked together at all times. The camera effectively captures Type 1 files, as audio is interleaved or packaged with the video in each frame.

Type 2 files are separate streams but are still locked together by timecode, allowing users to split audio and video into separate files. This process allows the CPU to handle each one more efficiently. Only a few applications can read Type 2 AVI files containing DV and may not be capable of opening AVIs generated in Vegas unless the user is aware of the benefits and limitations of Type 2 AVIs. Type 2 files are more advanced, are optimized for input/output (I/O) purposes, and only open in a limited number of updated products, such as After Effects or Commotion. AVI files can use a formatting language called Open Data Manipulation Language (Open DML). Type 2 files are more efficient, because in the editing stage, the audio is not stripped out of the AVI file, edited, and then replaced on a frame-by-frame basis as occurs with Type 1 files. The same applies to video; video is not stripped from the audio during editing and then replaced because it's already a separate stream of data. Many applications capture as Type 1 files and then split out the information to a Type 2 file. This process is less I/O intensive, although it does take greater processing time to split the interleaved information into a Type 2 file.

The Type 2-DV stream carries slightly more information than does the Type 1 stream, causing the Type 2-DV to be approximately five percent larger, but again, is much easier for an NLE system to cope with, because the system is reading audio and video separately.

More information on the DV data/AVI stream format is located at Microsoft's website, `www.microsoft.com/hwdev/tech/stream/vidcap/dvavi.asp`. Due to greater layer/information storage, it may be that in the near future capture utilities will have the option to store media in a non-recompressed WMV file format, which will allow for faster and more in-depth cataloging of video files. Vegas always captures and renders Type 2 DML AVI files by default.

Endnotes

[1] See Zettl, Herbert. *Television Production Handbook* (Florence, KY: Wadsworth Publishing, 2003). The actual timing is 30,000/1,001.

[2] The complex topic of pixel aspect ratio is well detailed in Charles Poynton's *Digital Video and HDTV* (San Francisco, CA: Morgan Kaufmann, 2003).

Getting Started with Vegas

Installing Vegas

Installing Vegas 4.0 is no different from loading any application, but certain settings, steps, and habits can make operating Vegas more pleasant and efficient.

Before loading Vegas, make sure that any anti-virus software installed has been temporarily disabled. If a great deal of editing is to be done on this machine, it's worth considering building a separate user profile that keeps the anti-virus off at all times during the capture, editing, rendering, and printing-to-tape processes. Anti-virus applications are notorious for stealing system resources at will and for popping up when least wanted.

Vegas 4.0 can be loaded either from a boxed-product CD or from a downloaded executable file. If Vegas is being loaded from a CD that came in a box, place the CD into the CD drive, close it, and follow the instructions as they appear on the computer screen. Vegas auto-starts unless this feature has been disabled in the Control Panel.

Complete the instructions given by Vegas' installation dialog. Vegas can be installed to any drive; it's generally best to install to the `C:\` (or boot) drive, unless the application will be used on a network boot. (Special licensing is required to run Vegas on multiple machines at one time.)

3.1 Vegas asks where it should be loaded/installed. In most cases, choose the boot drive.

3.2 Be sure to register Vegas 4.0 so that Sonic Foundry can notify you of updates and upgrades to the product and so that you gain access to the Get Media website.

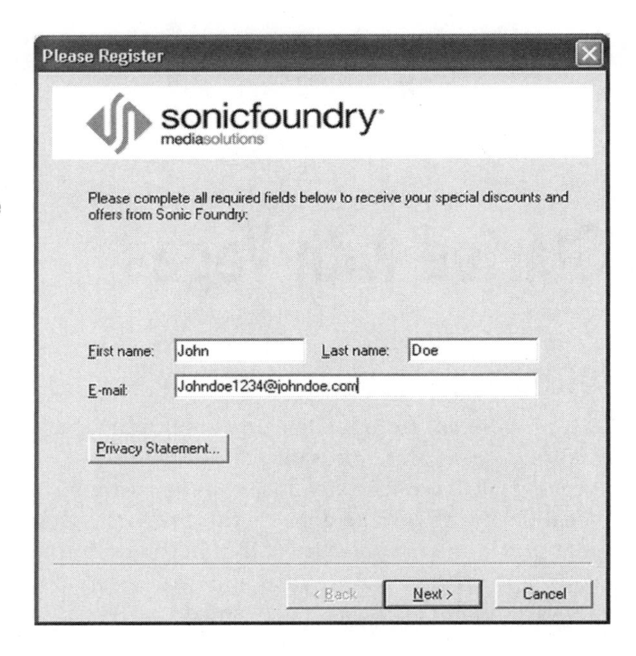

Register Vegas when prompted to do so; by registering, you can access media for free that is on Sonic Foundry's Get Media website, which is updated regularly for Vegas users. This media is generally royalty-free and worth having, particularly at the cost of registering the Sonic Foundry software.

Registration can be accomplished via the Internet, telephone, or another computer. Registration also gives users 60 days of free technical support and special discounts.

3.3 Vegas provides a successful registration screen after registration is completed.

The Basic Tools in Vegas 4.0

When Vegas launches for the first time, it's a blank screen with no media in it and makes no assumptions on how the user will be editing with Vegas.

The three primary levels at which audio, video, stills, and graphics are edited and processed are as follows:

- **Event**—can be viewed as individual pieces of media on a track. Many events can be on a single track.
- **Track**—is a single line on a project timeline, on which all events are placed. A project can have multiple tracks.
- **Project**—is a culmination of all tracks that contain events.

These terms shall be used throughout this book and the Vegas owners' manual.

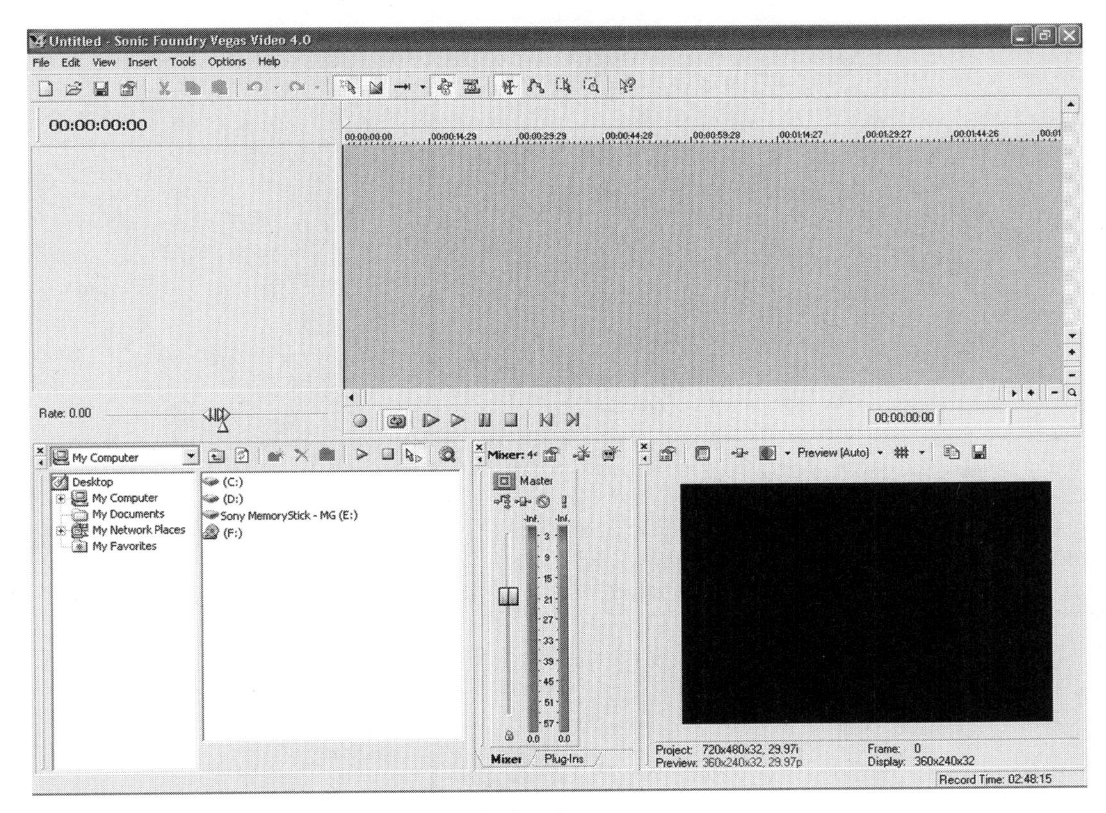

3.4 Opening the Main Window in Vegas 4.0 for the first time.

Vegas is essentially broken up into three main workspaces:

- Track header/control
- Timeline/track view
- Docked windows

Track Control Pane

The Track Control pane is where track names are listed and track-focused effects are added. Users can also define special behaviors of the individual track, such as routing to external audio cards for sound control/mixing, and compositing behaviors for video. The Track Control pane is also where audio tracks and video tracks are inserted. Tracks can be inserted one of three different ways:

- Select INSERT | AUDIO TRACK or INSERT | VIDEO TRACK from the menu bar at the top of the screen
- Use a shortcut key: CTRL+Q (audio tracks) or CTRL+SHIFT+Q (video tracks)
- Drag media from the Explorer to the blank timeline

The Track Control pane can be expanded or contracted, depending on the desired workspace and information space. This feature is particularly valuable on single-monitor systems where screen space is at a premium. Hovering the cursor over the line that divides the Track Control pane and the timeline changes the cursor to a resize icon. Clicking the left mouse button when this icon appears allows the Track Control pane to be expanded or contracted.

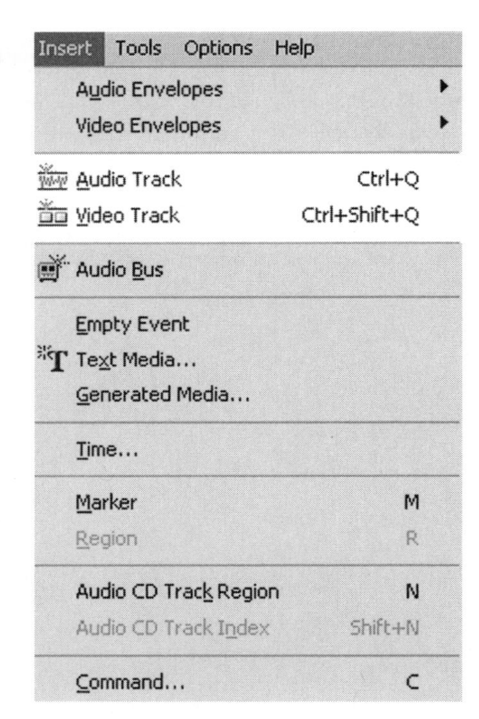

3.5 Tracks can be inserted from the menu, as shown here, by dragging, or by using keyboard shortcuts.

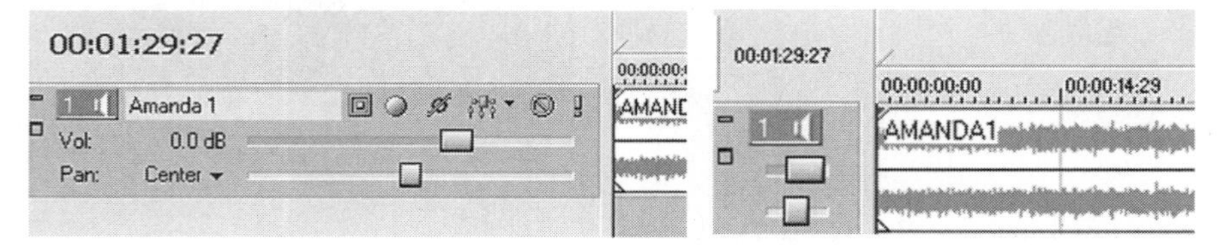

3.6 The Track Control pane can be expanded or contracted, depending on the amount of space desired for viewing either the track pane or timeline.

The Track Control pane can be completely removed by using the shortcut keys SHIFT+F11. This step removes the control pane completely, so be aware that it's easy to forget that it's been hidden. Using the UP ARROW and DOWN ARROW keys makes it fast and easy to move above or below a selected track. The currently selected track will be a darker shade than the tracks above or below. Multiple tracks can be selected by holding the SHIFT key and clicking each track chosen for selection. When multiple tracks are selected, any changes made to the controls on one track affects all tracks. This feature is useful when multiple audio tracks need to be raised or lowered in volume or for panning or bus routing. When multiple video tracks require changes to opacity or changes made in compositing modes, selecting multiple tracks is a fast way to affect all tracks simultaneously. Changes made to affect parameters will not affect multiple audio or video tracks, nor will track motion assigned to one track carry over to additional tracks when multiple tracks are selected.

Each track can be given a name for reference purposes. The track name becomes the default prefix name of the media stored on the timeline when the audio is recorded.

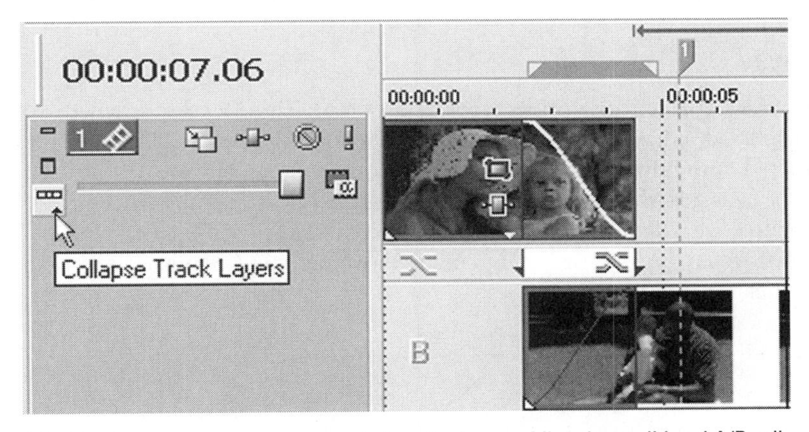

3.7 Track panes can also be expanded vertically, providing the traditional A/B roll view found in most other NLE systems. This feature uses twice the amount of screen space, which is the primary reason Vegas defaults to a collapsed track view. Expanding this view can help you become more familiar with editing in Vegas if you are coming from another NLE system.

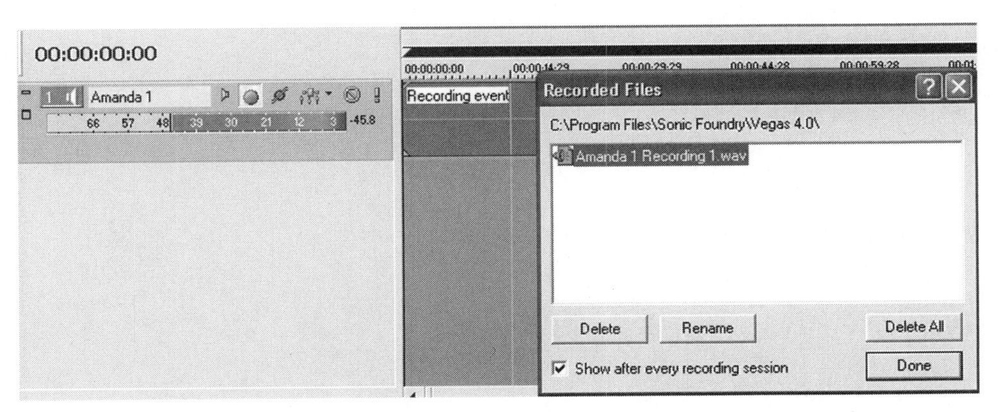

3.8 If a track is named before recording audio, the name on that track becomes part of the file naming process.

Adding a name value to a track after recording audio does not affect the filename properties of the audio. Adding a name value to a video track does not affect the name properties of the video on the named track. Adding a name value to audio ripped from a CD or imported does not change the name properties of the audio file.

Tip

It's often said in the digital world that "if it ain't labeled, it don't exist." This is somewhat true in that cataloging media correctly at all levels is critical to locating and maintaining files. In today's digital world, files accumulate very quickly. Be sure to get in the habit of good track, file, and project management.

The Track Control pane is also where volume, panning, and bus controls take place in an audio track, as well as track-level audio processing. Each track has an FX button on it, where either default processing or custom plug-ins can be inserted. (See Chapter 6 for more details on plug-ins.)

Timeline/Track View

The timeline/track view is where the majority of the work takes place. This location is where all audio and all video are displayed, edited, and viewed during the editing process. The timeline can be expanded or contracted both horizontally and vertically to suit the preferences of the user. This area displays all media found in the project and can be zoomed in for sample accurate or frame accurate edits or out for overall views of the entire project regardless of length.

In addition, the controls for starting and stopping cursor movement and playback in Vegas are found in the timeline/track view.

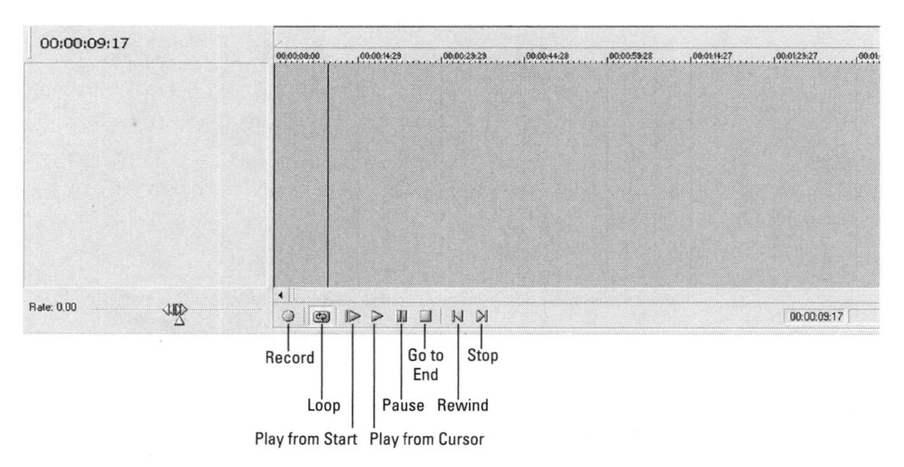

3.9 The Transport bar controls shuttling the cursor on the timeline.

These controls can also be controlled by these shortcuts:

- Record—CTRL+R
- Loop—Q
- Play from Start—SHIFT+SPACEBAR
- Play—SPACEBAR
- Pause—ENTER
- Rewind—W
- Go to End—CTRL+END

These controls can also be programmed in third-party hardware, such as the Contour ShuttlePro.

Next to the Transport bar control is a small yellow triangle that is used as a shuttle/scrub control. While playing media from the timeline, this control can be slid to the right to scrub through audio/video during playback. The indicator will stay at the faster or slower speed until the yellow triangle is double-clicked. Double-clicking will restore the speed to the normal playback rate.

Scrub can also be achieved by pressing the J, K, or L keys during playback. J reverses audio and incrementally increases reverse speed if the key is continually held. K pauses playback, and L scrubs forward while incrementally increasing speed as the key is held. Speeds of up to 20x are possible within Vegas. To control the incremental speed adjustment and how fast it implements, choose OPTIONS | PREFERENCES | EDITING and select the JKL/Shuttle Speed option to suit personal preference.

The Contour ShuttlePro and the Space Station AV can both be used for incremental shuttle control. Setting the shuttle speed to High provides the greatest use of the shuttle wheels on these hardware tools.

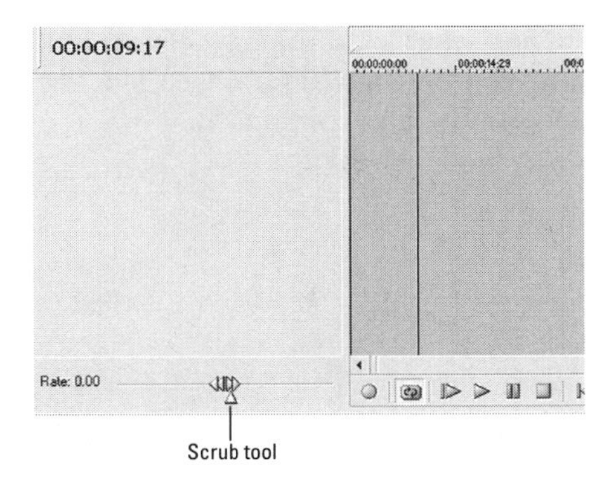

3.10 The Scrub tool can be locked at forward or reverse speeds. Double-click to restore playback to its normal rate.

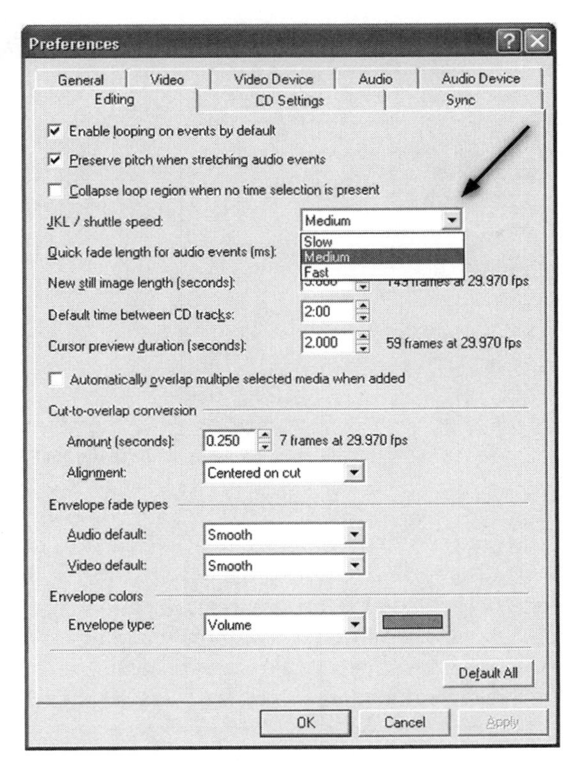

3.11 Setting the JKL/shuttle speed.

Above the timeline is a series of tools on a toolbar that can be customized for individual use. These tools can be removed from the top of the toolbar by selecting OPTIONS | CUSTOMIZE TOOLBAR to add/remove them from the view. The toolbar space can also be double-clicked to call up the Customize Toolbar dialog.

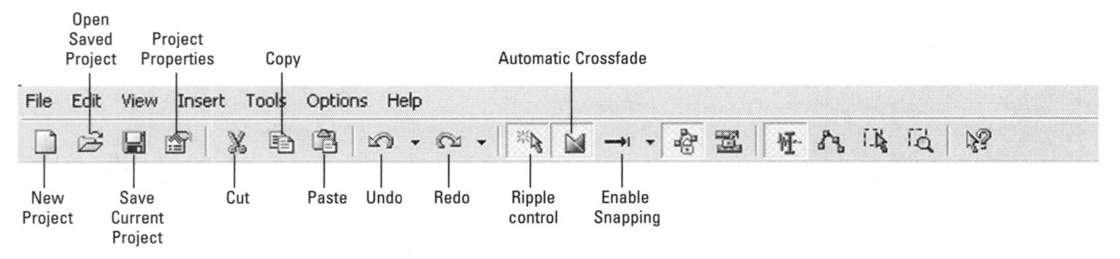

3.12 The main toolbar in Vegas. Buttons can be added to this toolbar depending on editing needs.

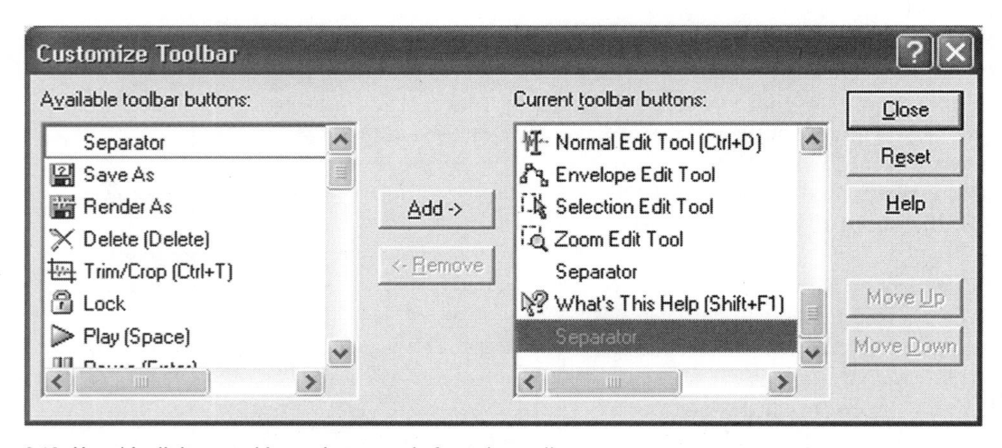

3.13 Use this dialog to add or subtract tools from the toolbar.

In the Add/Remove Tools view, select tools for display and click the Add button. Choose the tools to remove from the toolbar and click the Remove button. Audio editors might not wish to see tools related to video, just as video editors might or might not wish to see tools relating to audio. These tools are shortcuts to common editing tasks. All of these tools found on the toolbar can be accessed using keyboard shortcuts as well.

The dockable workspaces are found at the bottom of the screen in a default setting. Because these areas are dockable, they can be arranged to suit any specific need or desired appearance for efficient workflow. This feature greatly benefits users who have multiple monitors, as both the timeline can be stretched across multiple monitors and the docking windows can be stripped off and placed on any monitor. Vegas supports multiple monitors.

Vegas Windows

The three primary windows used in Vegas are the Explorer window, the Mixer window, and the Preview window. Clicking the View menu at the top of the toolbar opens several dockable windows. A checked box next to the window name means that the window is open.

Dockable Windows

The exception to this rule is that the Timing window always defaults to the top-left of the Track View window. The Timing window can be docked or floated freely at any point, based on personal preference. As with all docking windows, the Time Display window can be expanded or contracted. Unlike other docking windows, however, the Time Display window can also be changed in background or number color by right-clicking the display and selecting Custom. This feature is significant in a dark studio where the time display needs to be large and not distracting.

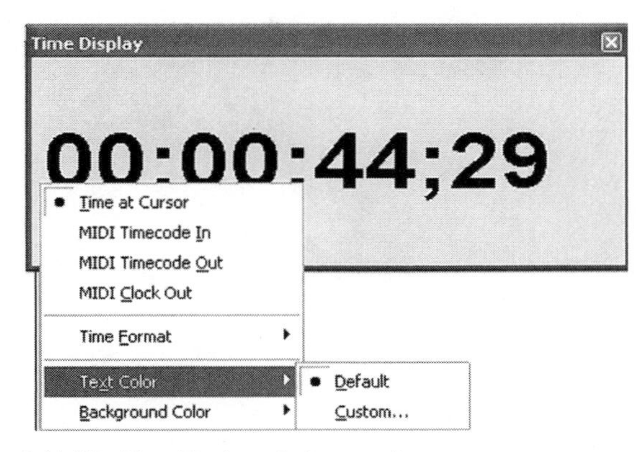

3.14 The Time Display window can be undocked and moved around the screen but cannot be removed.

For video editors, some of these windows can be irrelevant. For audio-only studio use, many of the dockable windows might be superfluous as well, so it can be more visually efficient to hide the video-related windows. This feature is valuable to both, as it allows for custom views to be created for each individual.

The following is a list of the dockable windows, their shortcuts, and a brief description:

- The Explorer window (ALT+1) provides a view of all media on the computer system and network.

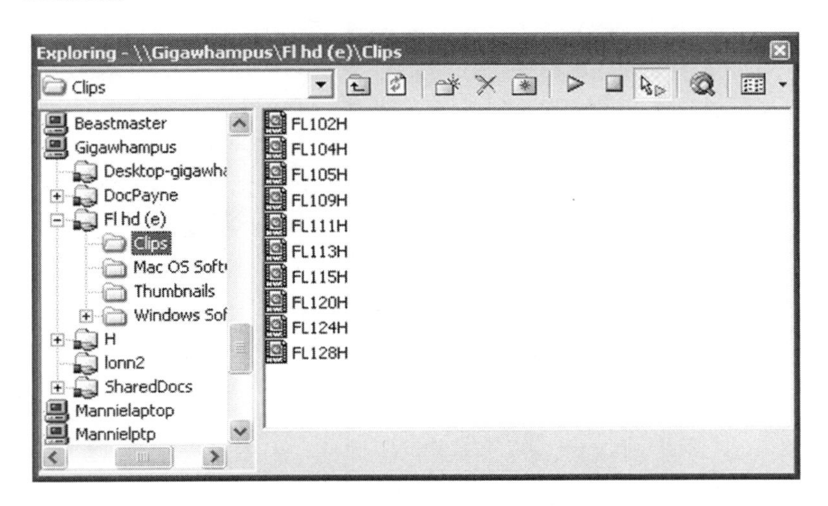

- The Trimmer window (ALT+2) allows media to be trimmed prior to placement on the timeline.

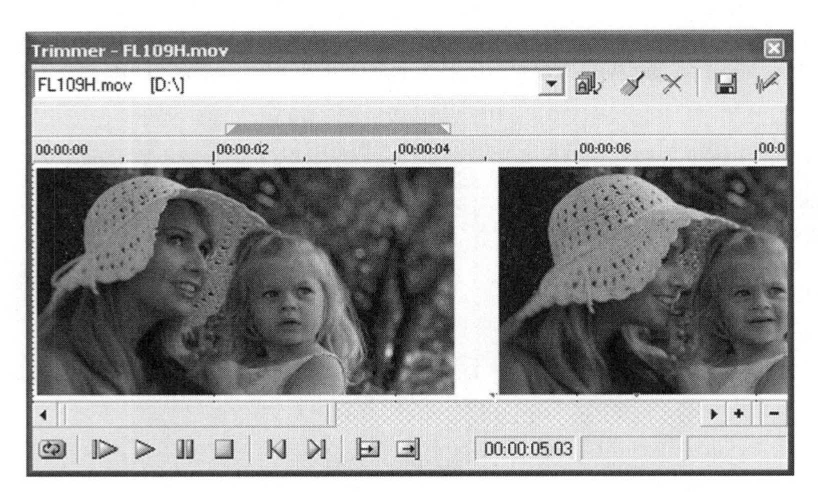

- The Mixer window (ALT+3) mixes audio, FX buses, and Bus send/returns.

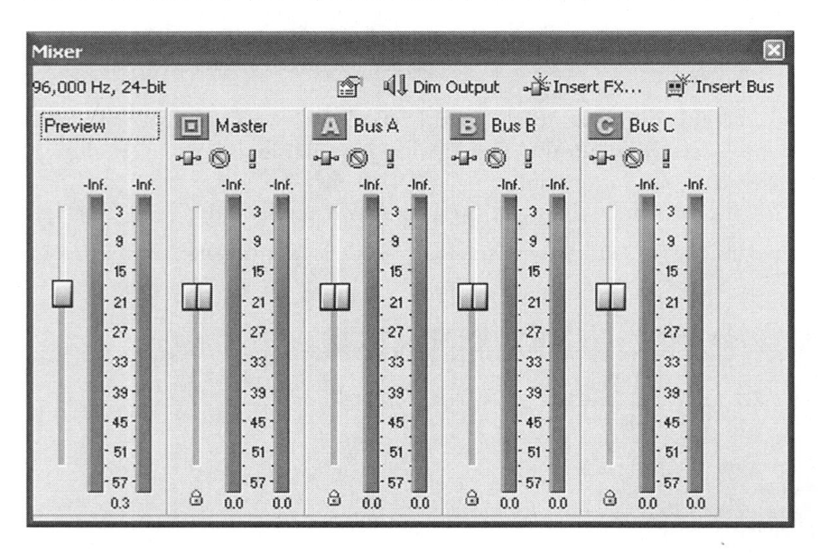

- The Edit Details window (ALT+4) displays all editing details pertaining to the current time-line view. Resort, import, and view information in this window.

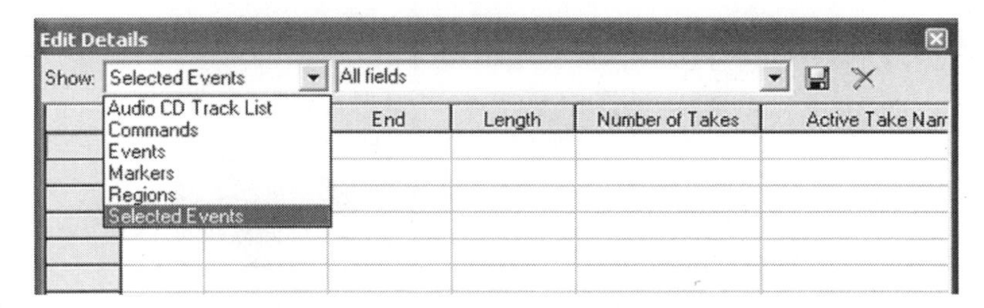

- The Media Pool window (ALT+5) shows all media used in the current project, length of use, number of use, and allows for filters to be applied to large numbers of events.

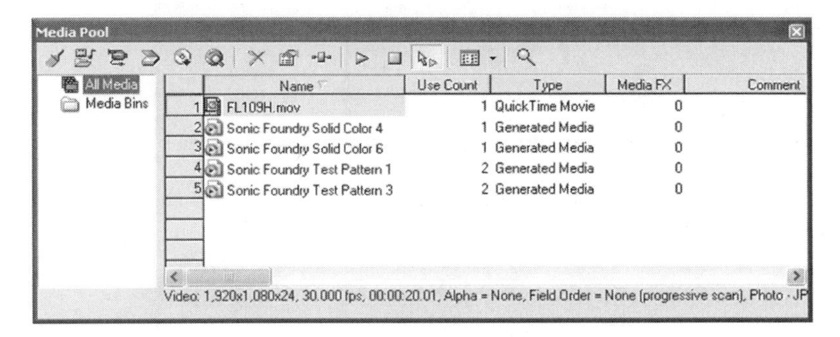

- The Video Preview window (ALT+6) provides preview of video on the timeline.

- The Transitions window (ALT+7) displays all transitions on the system, with motion thumbnail views of how the transition will function.

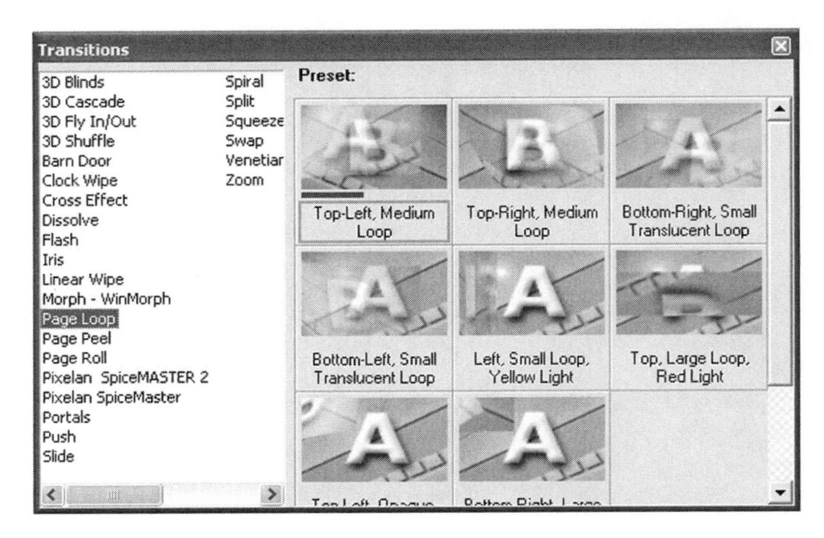

- The Video FX window (ALT+8) displays all filters on the system, with motion thumbnail views of how the filter will function.

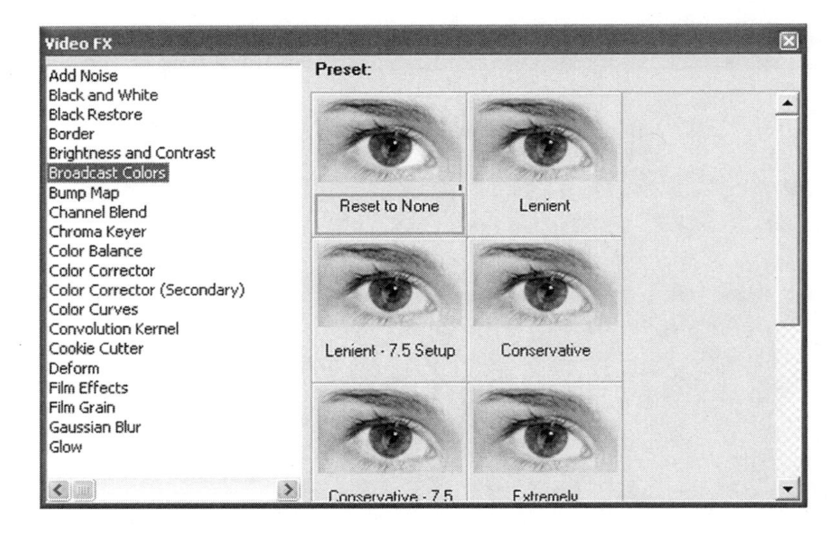

- The Media Generators window (ALT+9) generates media as simple as a title, or as complex as gradient or noise generation. Test patterns are also found here.

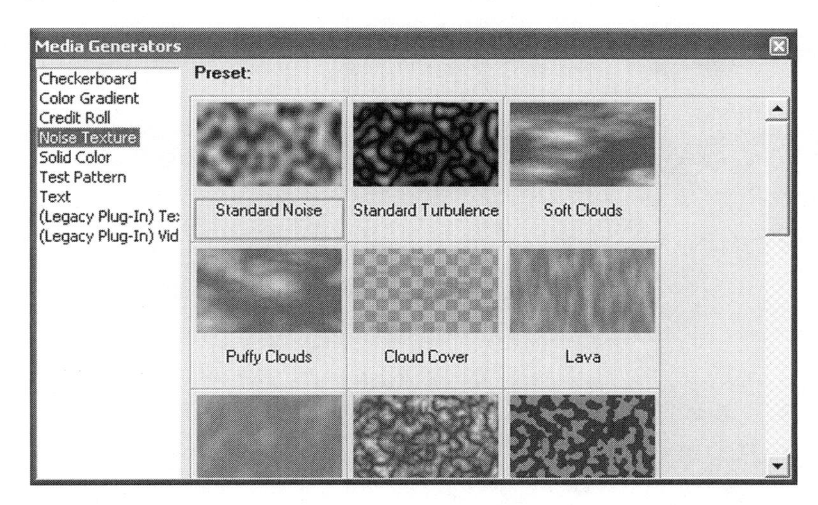

- The All Plug-ins window (CTRL+ALT+1) displays selected video or audio plug-ins that exist on the system, viewable by menu choice.

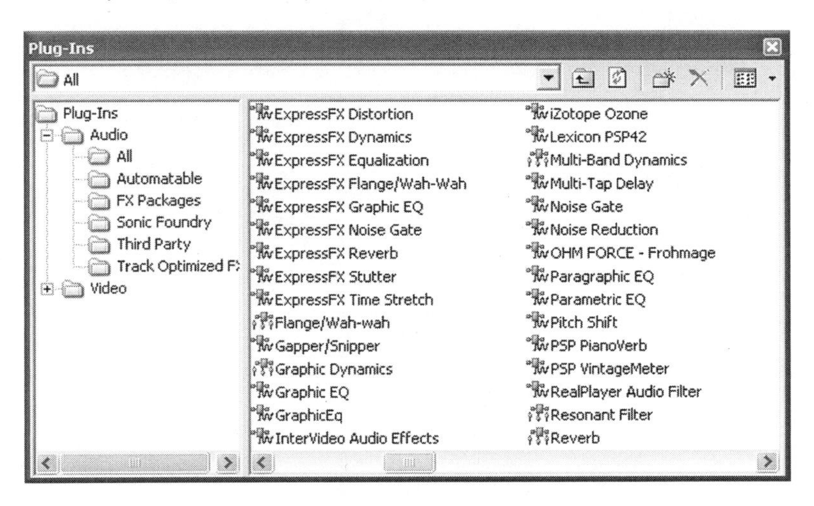

- Video displays and the Vectorscope may be opened from the Video Scopes window (CTRL+ALT+2), providing color/luminance monitoring and matching tools.

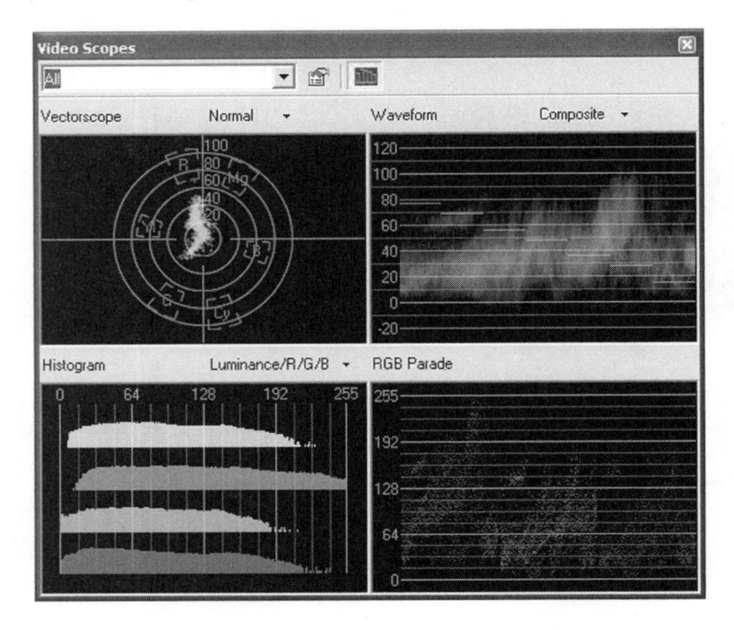

- The Surround Panner window (CTRL+ALT+3) is used to create surround sound keyframes and monitor the aural position of audio in a 5.1 mix.

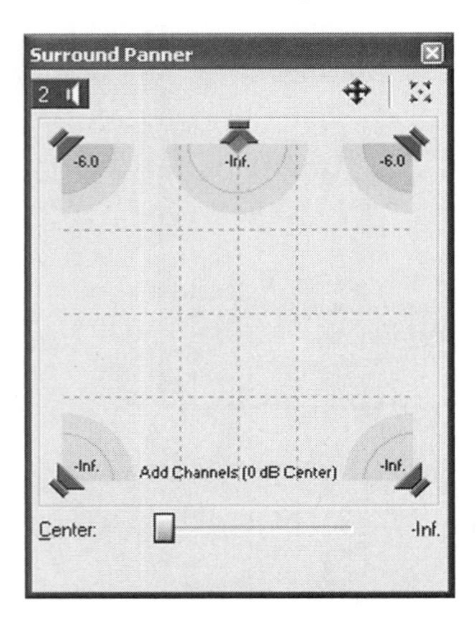

To move a dockable window, grab the vertical bar on the left side of the window you want to move and click with the mouse. Hold the mouse button down and drag the window to the desired space. Release the mouse button.

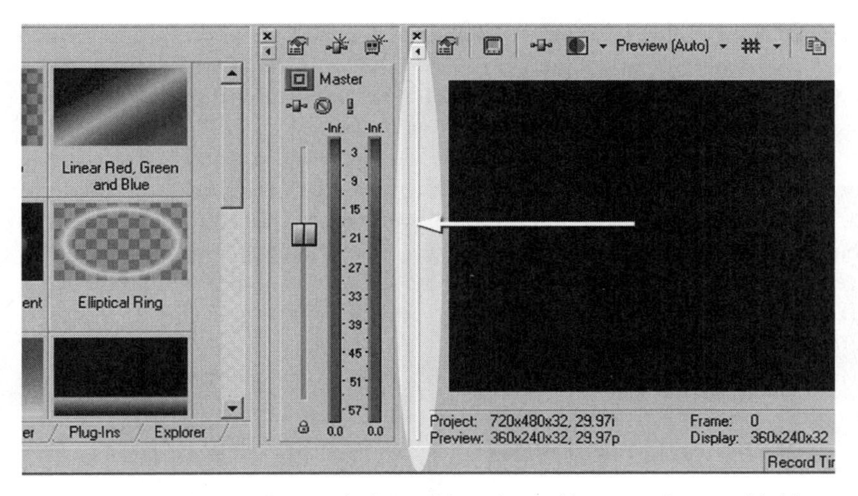

3.15 Use the docking handle to undock a tool from the docking pane. Press and hold CTRL down while moving the window back to the docking area to prevent the window from redocking.

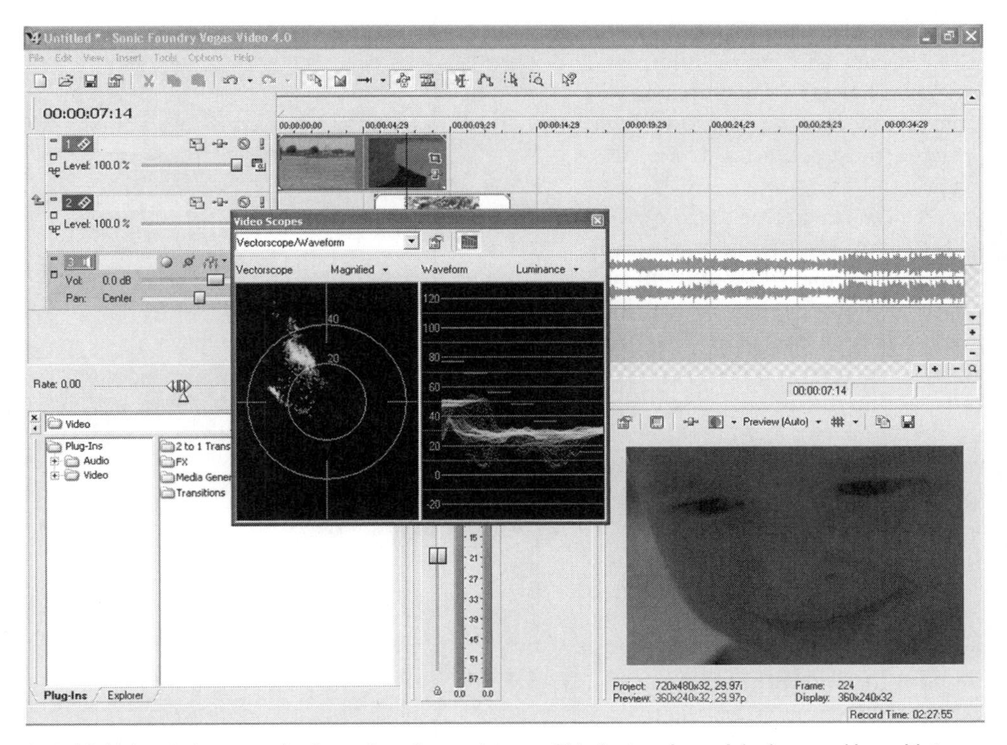

3.16 Multiple windows can be floated on the workspace. This feature is useful when working with two monitors.

Windows can be stripped from the lower section of Vegas and dropped anywhere onto the timeline, another monitor, or above the main work area. If a window is dropped too close to the lower section, however, the window will attempt to dock itself. Holding the CTRL key prevents this problem and allows for precise placement of floating/dockable windows.

As many windows as desired can be docked. If space is limited, some windows can dock on top of others. Windows can float or stay left in place and then selected with the tabs at the bottom of the docking area. Dockable windows can also be expanded or reduced in size.

To place a dockable window in the lower section of Vegas, drag the dockable window to the lower part of the screen and release. The window will automatically dock and fit within the window docking area.

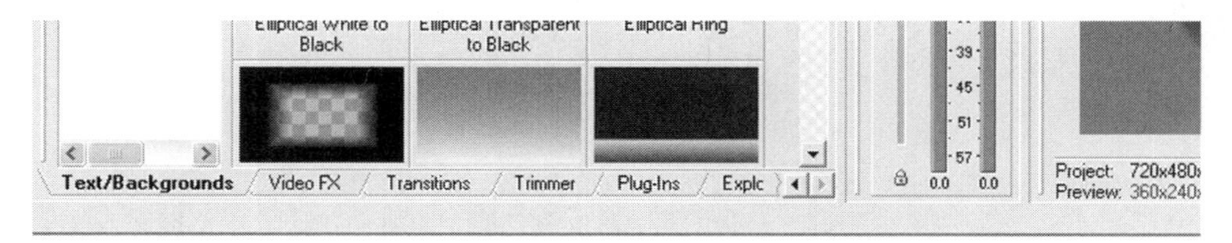

3.17 When multiple windows are docked in the docking area, each window has a tab for fast access.

The entire docking area can be hidden and recalled by pressing F11. This action will entirely remove the docking windows from view. To restore the docking area to view, press F11 again.

A new feature in docking windows can also be automatically hidden. In the OPTIONS | PREFERENCES | GENERAL dialog, select "Auto hide docking area." This action works the way Windows does when it auto-hides the taskbar while working within the desktop or an application. Just like Windows, moving the cursor to the lower area brings the docking windows back to view. This feature is exceptionally useful for audio editors who prefer to focus on the tracks during recording and only call up plug-ins or editing tools when desired. Video editors will also find this useful during in-depth compositing or when desiring maximum frame view for multiple tracks.

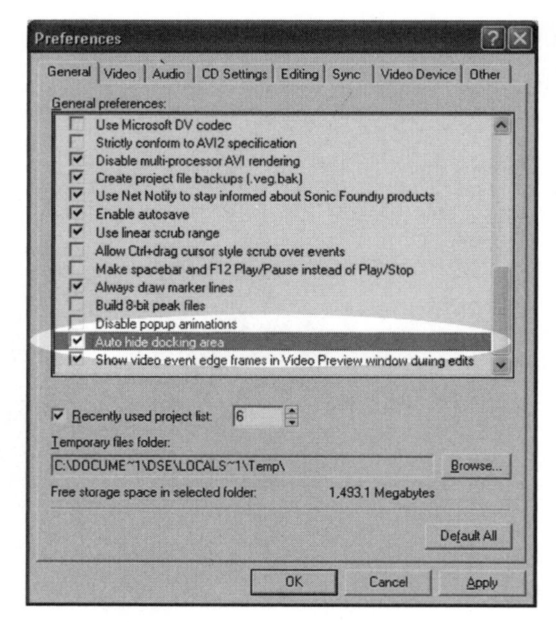

3.18 Hiding the window area makes more timeline space visible. Audio editors will find this feature particularly useful.

File Menu

Several choices under the File menu are available to help create a variety of settings, some of which are duplicates of the toolbar functions.

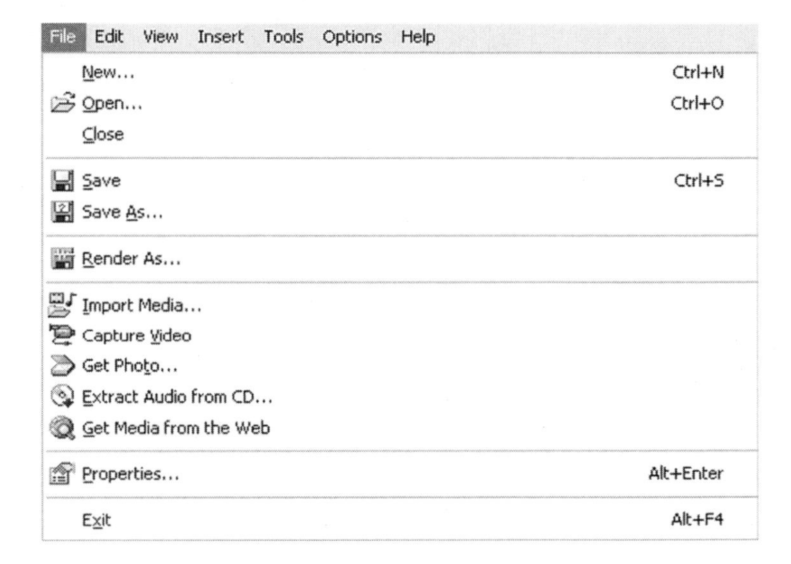

3.19 The File menu.

The Open dialog opens an existing Vegas project. Open can also be accessed by right-clicking the empty track window and selecting Open File or by pressing CTRL+O. This step opens either a VEG file or media to go on the timeline.

The Close dialog closes the current project. Vegas will prompt you to save, not save, or cancel. The Save dialog allows a project to be saved to a new name if it has not been named previously. The Save As dialog allows an existing project to be given a new name, which is useful for saving a number of edits that may need to be recalled at a later date.

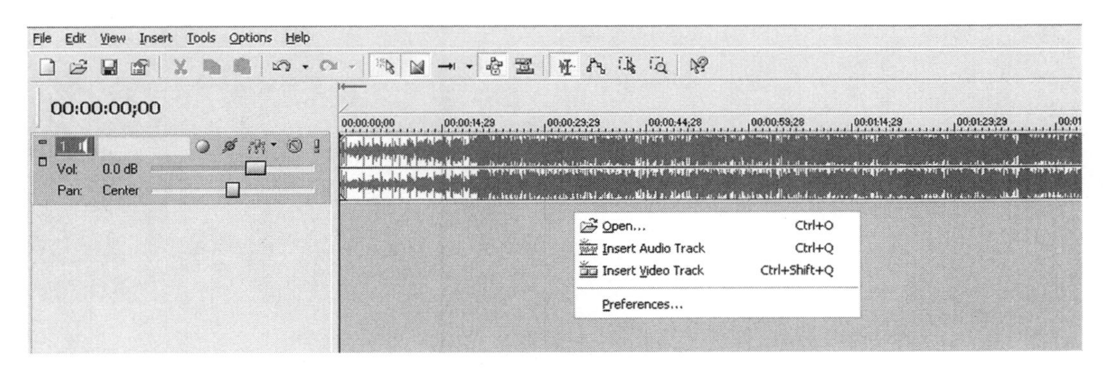

3.20 The timeline/project opens, and any media from the last project saved opens, unless this feature is disabled in the **OPTIONS | PREFERENCES | GENERAL** dialog.

 **Tip**

Saving the same project with different names allows you to save various versions or stages of a project, which lets you instantly recall various mixes and edits. This feature is valuable when a producer wishes to hear/see comparisons between project settings. Save files by time, date, or a mix of time, date, and unique name, for example, `less bass mix-04-15-03`.

The Import Media dialog allows for import of single or multiple files of any nature compatible with Vegas. Files imported are placed in the Media Pool for later use. This feature is useful for setting up entire projects before editing begins.

The Capture Video dialog opens up the Vegas capture utility, allowing for the transfer of DV from a camera or DV deck to a hard drive. This dialog launches a separate application.

The Get Photo dialog communicates with a scanner that is connected to the computer, allowing images to be directly scanned into Vegas' Media Pool, saving users the time it takes to open a separate photo scanning or editing application. Vegas uses the scanning software found on the user's computer.

The Extract Audio from CD dialog is what its name implies: a means of ripping audio from an existing CD. Audio is then placed in the Media Pool. This step only extracts audio from an audio CD, not a CD of WAV or MP3 files. Those file types are opened via the CD/drive selection in the Explorer window.

The Get Media from the Web dialog is a fantastic tool for Vegas users. It's a link to a website provided by Sonic Foundry and their partners that gives users access to video, audio, and other media at no cost as a benefit for owning a registered copy of Vegas 4.0. Bed music and stock video files are regularly changed by Sonic Foundry's partners.

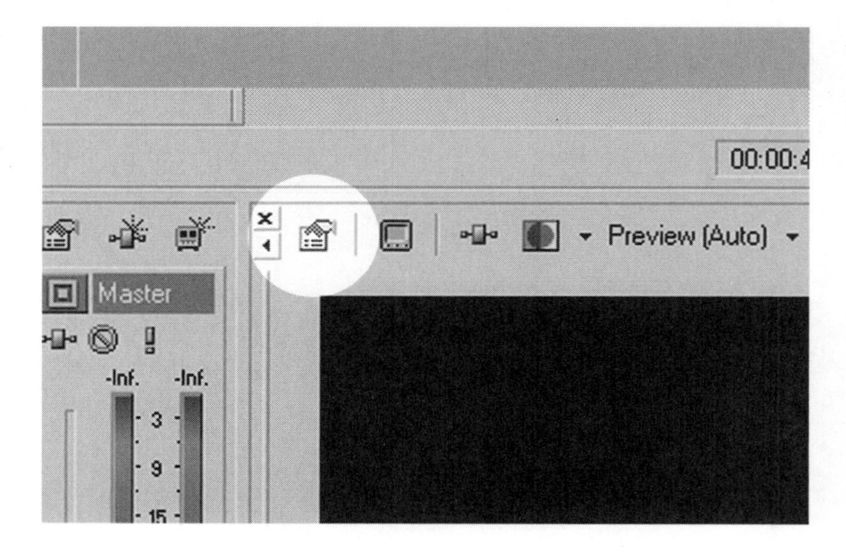

3.21 Access video properties by selecting the Video Properties button found on the Preview Window toolbar.

The Project Properties dialog is a series of choices that allows a user to create project settings related to any number of workflows. Media that is going to be edited just for the web should have project properties that relate to web-oriented video or audio. Recording studio setups might want to adjust properties only for audio and never see any of the video tools. The CD with this book has projects already set up for you to use if you wish.

As with most options in Vegas, the Project Properties dialog can be opened from multiple places. Selecting FILE | PROJECT PROPERTIES is one method of opening the dialog. The Project Properties button found in the Preview pane of the docking Video Preview window also opens the dialog, as does the Audio Properties button found in the Mixer dialog. Pressing ALT+ENTER also opens the Project Properties dialog.

The default setting of the Project Properties is for an NTSC-DV project, which is used for projects captured from a DV camera to be edited and printed back to a DV camera or to another tape machine through a DV camera, DV-to-analog converter, or other DV device. This setting optimizes Vegas' preview and editing behaviors for the DV space. Other presets are there as well, including for PAL-DV, multimedia authoring, web-oriented editing, and MPEG-CD directed media projects. Any of these choices can be used as a starting point and then edited to suit specific needs.

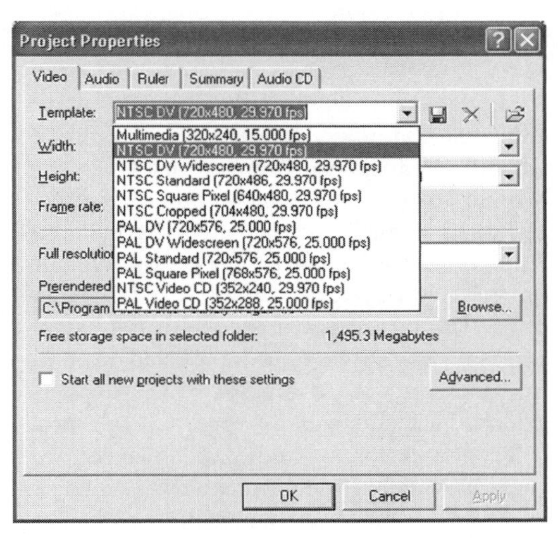

3.22 Using existing templates and creating new templates saves a lot of time when setting up an editing project.

Checking the "Start all new projects with these settings" box ensures that editors will have the same project settings each time they open Vegas for a new session. Having a series of project settings saved as templates is valuable, as it saves time and ensures correct settings in those hurried moments that are common in the studio, whether for audio or video work.

Tip

Save VEG files for specific types of projects, such as a timeline that opens with a station ID or corporate logo. Markers, chapter points, time counters, plug-ins, and other commonly used tools can be saved in a VEG file, ensuring that projects consistently open with all desired tools. This step also sets up a template that includes all consistent media. (See the CD included with this book for VEG files.)

Setting Up Project Properties

Video project types are specified in the Video tab of the Project Properties. As a general rule, it's best to use the template settings to ensure that projects flow with ease and are trouble-free. Advanced users might want to edit the settings for specific results related to third-party tools or applications.

One major consideration with Vegas is to set the Prerendered Files folder to a new default. Vegas defaults to the `C:\Program Files\Sonic Foundry\Vegas 4.0\` folder. Prerendered files will rapidly fill the `C:\` drive and are generally forgotten by most users, which can easily lead to problems. Click the Browse button, locate the second hard drive dedicated to media, and create a Prerendered Files folder on that drive. Point the default prerendered files to that folder on the media drive. This process will speed up prerenders and will ensure that media moved around from computer to computer (if the drive is external) will stay with the project at all times.

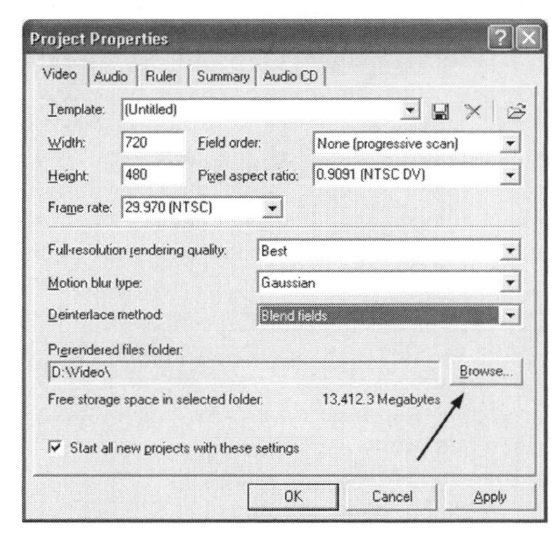

3.23 Video properties can be accessed from the **FILE | PROPERTIES | VIDEO** dialog or from the Properties button on the Preview Window toolbar.

Properties for an audio-only project are also set up here. Engineers wanting to work with high bit rates and sample rates will want to adjust these in Vegas. Default settings are 16 bit/44.1KHz, which are standard CD quality settings. Vegas is capable of working with 24 bit/96KHz files, although for the most part, only professional sound cards can work with these higher sampling rates.

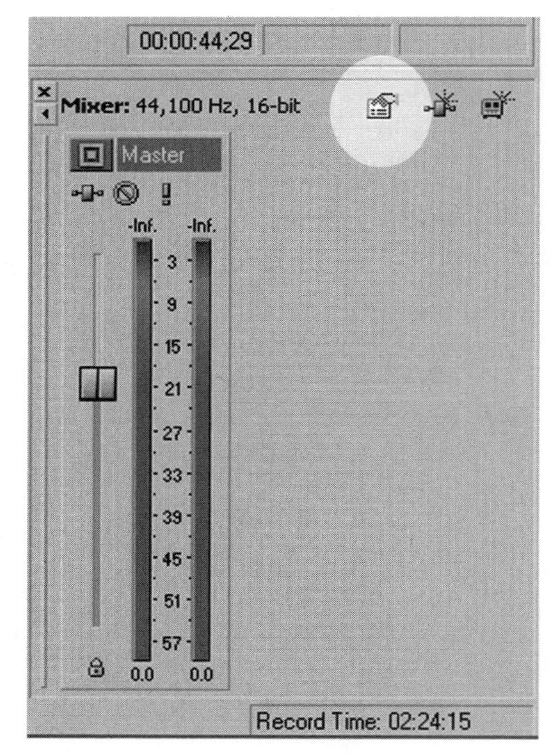

3.24 Audio project properties can be opened by clicking the Audio Properties button above the master volume controls.

Take note of the Master Bus Mode, which defaults to Stereo. This setting is where Vegas 4.0 is adjusted to author 5.1 surround sound files. If the applicable hardware is available to the system, anyone can now author 5.1 surround files in Vegas 4.0. (AC-3 encoding is not included in Vegas 4.0, as it's an additional purchase/plug-in package.) Vegas will output elementary linear pulse code modulation (LPCM) files, which are acceptable by most DVD authoring packages. Users that do not have 5.1-compatible hardware are able to select the 5.1 surround option and can output a 5.1 file. No ability to monitor/master the audio exists, however, which can result in a potentially unbalanced mix.

Up to 26 buses can be assigned in Vegas. (These correspond to the letters of the alphabet.) Buses can be used for inserting effects, such as reverbs, delays, exciters, and other audio plug-ins from Sonic Foundry or other Direct X plug-in developers, such as WAVES, iZotope, and Sonitus. Buses can also be used for submixing.

Resample quality should generally be set to Best in the Project Properties Audio dialog. It defaults to Good.

This location is also where the default recording folder is specified in Vegas. Vegas sets the default audio recording to the `C:\Program Files\Sonic Foundry\Vegas 4.0\` directory. If recording long program material or multiple tracks, or audio is the primary use of Vegas, it is highly recommended that this setting be set to a secondary drive. Vegas will scan the specified drive and indicate how much memory is free on the identified drive.

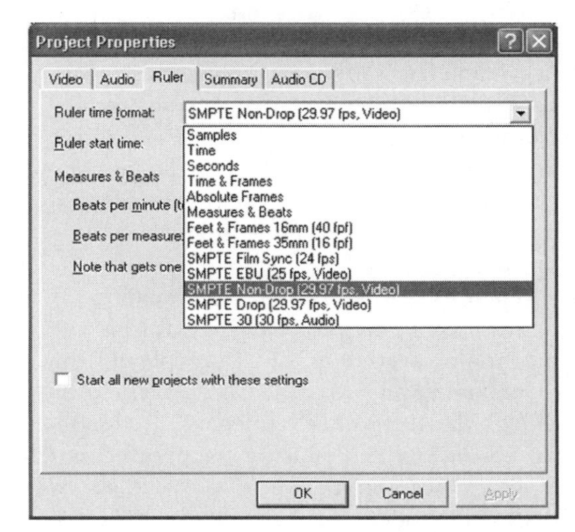

3.25 This dialog sets preferences for how audio-time grids are displayed on the timeline.

Tip

Remember that it's important to have a second drive for media and that the drive controllers should be set to DMA in the Control Panel.

The Project Properties Ruler dialog box allows editors to specify how the track view/timeline appears. Audio editors will want SMPTE time for their house standard; DV editors can set this for 29.97 fps for grids defining their time space, and musicians can set this to beats or tempo.

Tip

Video editors will find the Beats grid valuable if the tempo of bed music is known. This ruler grid can mark the beat of the bed music and assist in placing edits so that they are in time with the beat of the music.

Time display formatting can also be selected by right-clicking the Time Display window and selecting the desired time format, which can be toggled between different time formats. Toggling between beat and SMPTE time won't change the timing of the audio or video; it merely allows for different grid views.

The Summary tab found in the Project Properties dialog allows users to log information about the overall project, such as engineer or copyrights. This information stays with the file in the event that the VEG is transferred to another system, user, or shared file so that the originator of the file is traceable. This data does not become part of the summary information embedded in streaming media, nor does it become part of Scott information embedded for radio play. Those summaries must be filled out separately.

The Audio CD tab in the Project Properties dialog has a form for inputting a Universal Product Code (UPC) or Media Catalog Numbers (MCN) that will be burned in the header of a Redbook-valid CD. Some compact disc players do not display this information, but the information is still embedded in the disc.

There is also an option for changing the first track number on a CD. This feature can potentially prevent a CD from playing on older or fewer-featured CD players even though the disc will be burned as a Redbook-valid CD. This option is only supported as a Disc-at-Once (DAO) burn as specified by Redbook guidelines. Track-at-Once (TAO) authored CDs ignore this setting.

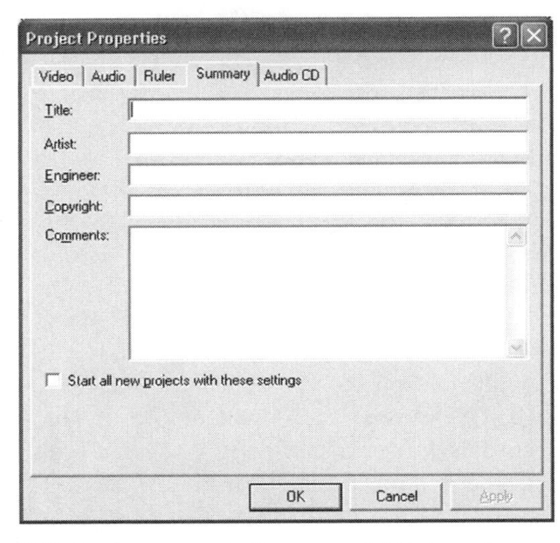

3.26 The Summary tab allows copyright information, engineer information, and other important information about the project to be stored. This feature is particularly important when tracking a project.

Edit Menu

Next on the menu bar is the Edit menu. It contains several options for editing media, some of which are duplicated on the toolbar. First on the list is Undo and Redo. Vegas is unlimited in the number of times that you can undo or redo actions. After a project is saved or closed and reopened, however, the project starts with a new list of undo/redo, so a project is not forever undoable. Be sure of your choices before closing. These tools are also found on the toolbar, identified by the Undo and Redo icons.

The shortcut key combination CTRL+Z triggers the Undo function, and CTRL+SHIFT+Z triggers the Redo function.

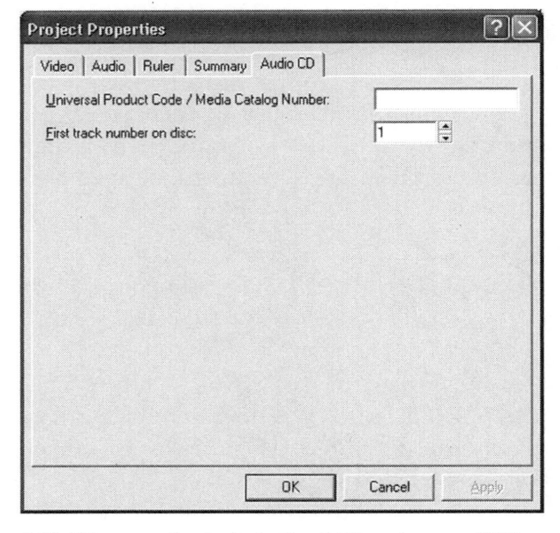

3.27 CD properties include the ability to insert a UPC code onto the disc for media cataloging.

Tip

If you have a Contour ShuttlePro or Space Station, these two shortcuts are handy to have programmed as button strokes.

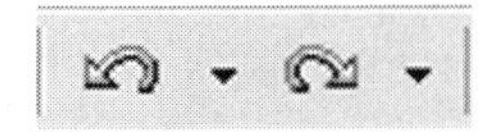

3.28 Undo can be accessed using the button or keyboard shortcut.

Next on the list are the Cut, Copy, and Paste actions. These are the same as nearly all Windows applications. Cut (CTRL+X) removes media from the timeline, and Copy (CTRL+C) copies the media on the timeline. Regardless of whether Cut or Copy is used, the media is copied to the clipboard, so that the Paste (CTRL+V) function can be used to place the clipboard-stored media to another location.

NEW! A new feature found in Vegas 4.0, Paste Attributes, is useful for both audio and video editing. It allows you to copy switches, audio pitch shifting, playback rate, undersample rate, video event effects, and pan/crop settings from one event or media clip to others. As an example, suppose that there are ten tracks of audio, and in those tracks there are 40 events. A blur needs to be applied to half of those events randomly scattered throughout the timeline. Rather than applying the blur to the entire track, click the event containing the correct blur settings and press CTRL+C or go to the Copy button on the toolbar, (or right-click and select Copy or use the Edit/Copy function). Then select the event to which you wish to have the same blur settings applied. From the Edit menu, choose Paste Event Attributes. This step copies the blur and any other unique information contained in the media from which the attributes were copied and can be pasted on the next chosen event(s). Video editors find this method very efficient to pan/crop several still images at once or to copy effects settings to apply to several events at once. Paste Event Attributes can be applied to multiple events of the same type at the same time, so if one event contains the desired settings, those settings can be pasted over hundreds of similar events. Imagine being able to pan/crop 500 still images at once or being able to apply a unique reverb setting to 30 instances of a particular kind of event at once. This result can also be accomplished in the Media Pool via applying a Media FX to all occurrences of video or audio in the Media Pool. Very powerful and useful indeed!

There is no shortcut key for Paste Event Attributes.

Delete is next on the menu and is used to remove an event permanently. Select the event you wish to remove and press the DELETE key or choose it from the menu.

NEW! Post Edit Ripple is the next item in the menu. This feature is a new and long-awaited upgrade in the Vegas series. Editors can now choose the behaviors of Vegas ripple features to suit their own style of editing. *Ripple editing* allows you to insert or delete events on the timeline. At the same time, ripple editing keeps events in front of and behind the deleted or inserted event in the same order and time space that the events occupied before the insertion or deletion of an event.

An example would be a long-form documentary that has been edited, containing many tracks of video information. A piece of the documentary is determined to be incorrect and must be deleted. Deleting the piece, without filling in the exact amount of time occupied by the deleted events, leaves a hole in the timeline. Deleting the hole causes all events occurring after the deletion to slide in place, with nothing lost. The new ripple modes allow Vegas users to decide how they want those inserts/deletions and movements of events to take place. There is a project file called `ripple.veg` contained in the CD with this book.)

The Select All choice on the Edit menu allows for all media on the timeline to be selected. This feature is invaluable for moving an entire project up or down the timeline or for selecting an entire project to be copied and pasted into another instance of Vegas. Selecting either EDIT | SELECT ALL or pressing CTRL+A selects or highlights all media.

The Select Event Start/Select Event End feature jumps the cursor to the beginning or end of a selected event. This feature can also be accessed by using the shortcut keys [(Select Event Start) or] (Select Event End).

From the numeric keypad, press the 7 key for Select Event Start or 9 **for** Select Event End. Use 4 and 6 to move or trim an event edge forward or backward in time, and use 1 or 3 to move the event forward or backward by frames. When not in Event Edge Edit mode, 4 and 6 move the event by one pixel, and 1 and 3 move by one frame. Use 8 and 2 to move the selected event up or down vertically to the next track. Pressing 5 exits the keyboard trimming mode.

3.29 Selecting EDIT | SELECT puts the cursor where it should be with frame accuracy; however, most editors find it faster and easier to use the ten numeric keypad shortcuts.

On a standard laptop, NUM LOCK needs to be enabled to use the numeric keypad. Targus manufactures a numeric keypad that is connected to the laptop via USB for the road warrior editor. This feature allows for rapid location of event edit points, which when coupled with the new ripple features, makes for efficient editing using the keyboard.

3.30 Using the Targus keypad with a laptop makes laptop editing more efficient.

NEW!

The Paste Repeat dialog allows an event or events that have been copied to the clipboard to be pasted multiple times.

To insert events at the cursor position on the timeline and move media down the timeline to accommodate the inserted events, copy (CTRL+C) events to the clipboard, select the location that events are to be pasted, and press CTRL+B or select EDIT | PASTE REPEAT. You will be prompted to input the number of times that the event should be pasted.

Events can be precisely separated in Paste Repeat mode, allowing for many elements, including images and video clips, to be spaced a prespecified distance. Input the length of time between events into this window. The spacing/length of time between pasted images can be determined in milliseconds, frames, seconds, minutes, hours, beats, or musical note values.

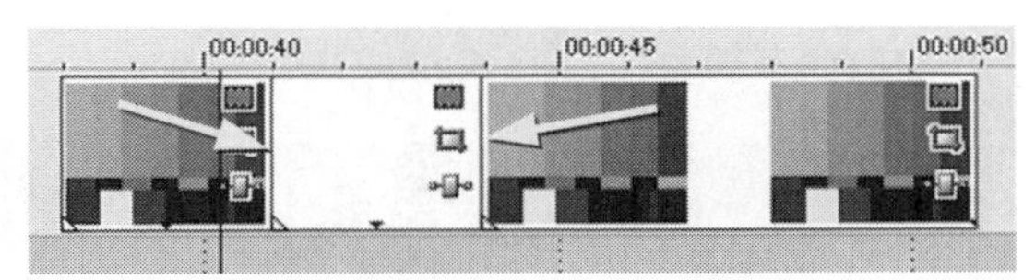

3.31 Use **EDIT | PASTE INSERT** to paste an event on the timeline, while moving everything subsequent to the paste down the timeline.

Tip

Use the Musical Note Values feature when you want a repeating instance of an instrument. For instance, a fast way of building a rhythm would be to insert a sample of a kick drum on every downbeat and third beat and a snare drum on a different track hitting every second and fourth beat. Use the Tempo dialog found in the Project Properties to set the tempo and grid for the project. ACID Loops that match the tempo can also be inserted, making a fast music bed for either editors or musicians to work with. Video editors will find this method a visually and aurally compelling way to edit quickly, if bed music has not been completed but a tempo is known. This process is often called *tempo-mapping* and allows video editors to build rough video comps while a composer finishes a musical cue for the project.

The Trim feature is another excellent feature in Vegas. To delete all portions of the event outside the selected area, select a portion of an event by scrolling the mouse over the event(s), hold the right-click button down, and select EDIT | TRIM or press CTRL+T.

This feature can be used on an individual event or on all events across the timeline.

The Split tool splits a selected event/events at the point where the cursor is parked. If a selection is made on an event, only that event is split. If no selection is made, the split occurs across all events that fall under the cursor. Park the cursor at the desired point and select EDIT | SPLIT or press S. The event on the right side of the split is selected automatically, in the event that it should be deleted.

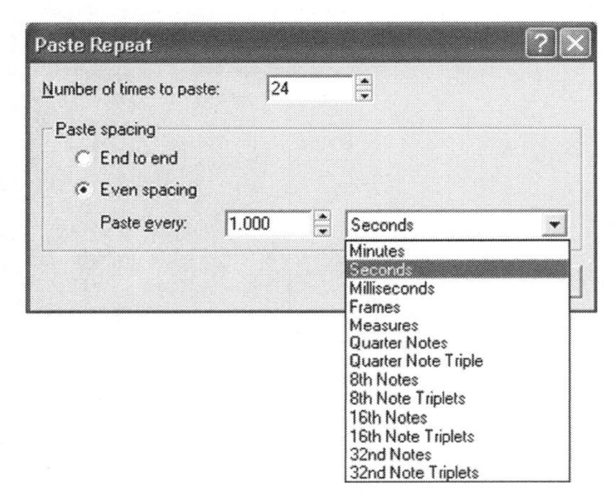

3.32 Repeating a copied image is definable in the **EDIT | PASTE REPEAT** dialog.

The Editing Tool dialog specifies what type of editing behavior the cursor shall have. These same choices are found on the toolbar mentioned earlier. Pressing CTRL+D returns the cursor to the normal Selection Edit cursor mode, which is generally the most commonly used cursor mode. None of the other cursor editing modes have shortcut keys.

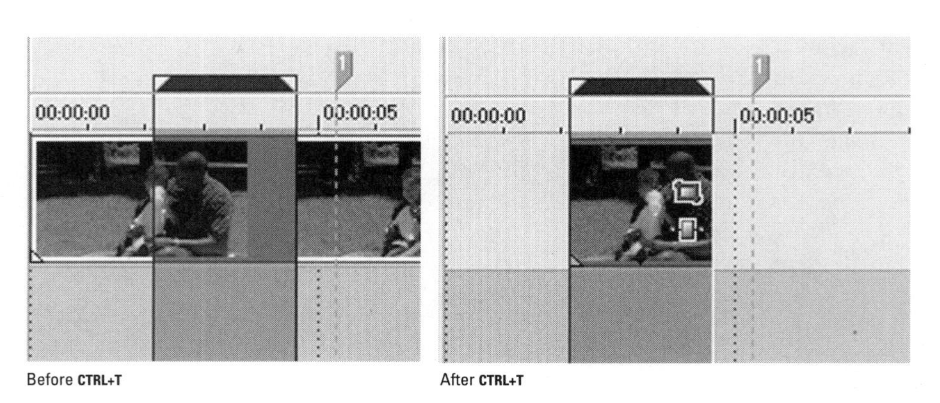

3.33 The **CTRL+T** Edit command deletes all events not inside the selected area. Coupled with the Ripple feature, this feature is very powerful.

NEW!

Using the Switches dialog, users that have been using Vegas in earlier revisions will find that several right-click dialogs have been augmented by shortcut keys and by selecting EDIT | SWITCHES. Mute, Lock, Loop, Invert Phase, Normalize, Aspect Ratio, Reduce Interlace Flicker, and Resample have all been made accessible through the EDIT | SWITCHES menu selection. A Smart Resample, Force Resample, and Disable Resample selection have also been added. Smart Resample instructs Vegas to seek media that has been time-stretched, is a graphic-based event, or does not match the frame rate and the frame rate of the event is greater than 24 fps. Vegas then resamples media meeting those parameters. Force Resample instructs Vegas to resample all events regardless of their format or frame rate. Disable Resample prevents Vegas from resampling any events, regardless of whether they have been selected for resampling or not.

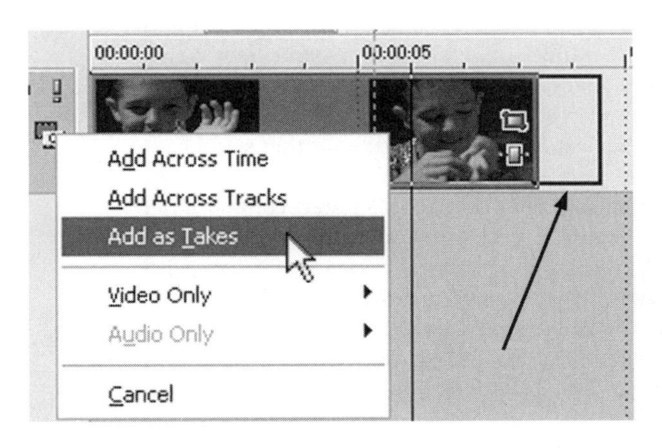

3.34 Although the new take being dragged to the existing event is longer or shorter than the event over which it is being laid, adding as a take trims the new media to fit the existing media. Shuffle through takes during playback by pressing T.

Vegas' Take feature allows you to place several takes on top of each other and allows those takes to be switched to view during playback in real time. For example, perhaps a multicamera shoot occurred, and the director or producer would like to see the scene with the different camera angles that were shot. Either multiple tracks must be used with muting on the fly, or takes are used, having all of the shots living in one event and switching on the fly. This workflow is more efficient in most instances. This process is also useful when a musician wants to perfect a particular part of a song and plays the piece over and over again. Each take can be placed, and different versions can be toggled through in real time by pressing T. Any event that has filters, effects, or other edits made to it can apply those filters, effects, or edits to the take.

Tip

Video editors that work with a formula-based editing situation, such as a wedding video editor, can build a template of empty events marking locations of where media should go, complete with transitions and filters. Video is then dropped onto the empty events as a take, and within minutes, a finished video is ready to print to tape. Studios doing preset-time voice-over work can find this useful as well, fitting 30-second or 60-second radio spots, complete with compression, EQ, and other filtering into an empty event to be replaced by new dialog. For recurring-format editing projects, this method is terrific for cutting repetitive actions to a minimum and keeping projects/productions in a consistent appearance or sound.

The Group menu allows you to create a group of clips that are temporarily locked together so that when they are moved, they can be moved together without having to select multiple events each time. To implement this feature, select a number of events and press G or select EDIT | GROUP | CREATE NEW. This option locks the selected events together, so that when one is moved, all are moved. To remove an event from the group, select the event and press U. The event can also be removed from the group by selecting EDIT | GROUP | REMOVE FROM. All grouped items can be selected at once by using Select All. This option allows for an entire group to be deleted as a whole or to have other processing applied. Grouping is a very powerful tool, and one that is worth the time to get familiar with.

View Menu

The View menu is a series of check boxes that determines what dockable windows are available for immediate view.

Place a check mark in the box next to the name of each view to turn the view on or off.

The Mixer Preview Fader option adds another set of sliders to the menu and allows the previewing volume to be changed regardless of what level the Master Volume is set at. The Mixer Preview Fader option only affects media in the Trimmer and Explorer. This slider set does not affect the level of rendered files, whereas moving the master slider does affect the level of audio mixed from the timeline.

The Show Bus Tracks option is a feature that is new in Vegas 4.0. Buses can now be viewed as tracks, including volume and panning automation control. This feature provides much greater control over mixed tracks that use bus mixes, while also providing a visual reference for mix controls. Buses can now be automated to increase/decrease volume of an effect on the bus or to control panning. Coupled with the Automated Effects now found in Vegas 4.0, Vegas feels more like an actual hardware mixer.

The Show Audio Envelopes option hides or shows volume control views, depending on the current view state of the control. This new feature allows volume controls on all audio tracks to be hidden at once, cleaning up the screen view substantially in the event of a large mix.

The Show Video Envelopes option has the same control over the video envelopes: composite, track fade-to-color, track keyframes, and velocity envelopes.

3.35 Set the Vegas 4.0 workspace to appear exactly as you'd like it to appear.

The Minimize All Tracks option reduces all track views to a minimum size, immediately freeing up space for users to see previews, add full size tracks, or make a large number of tracks fit into a small area.

This option can be accessed via the ~ key. Pressing ~ again restores tracks to the previous view.

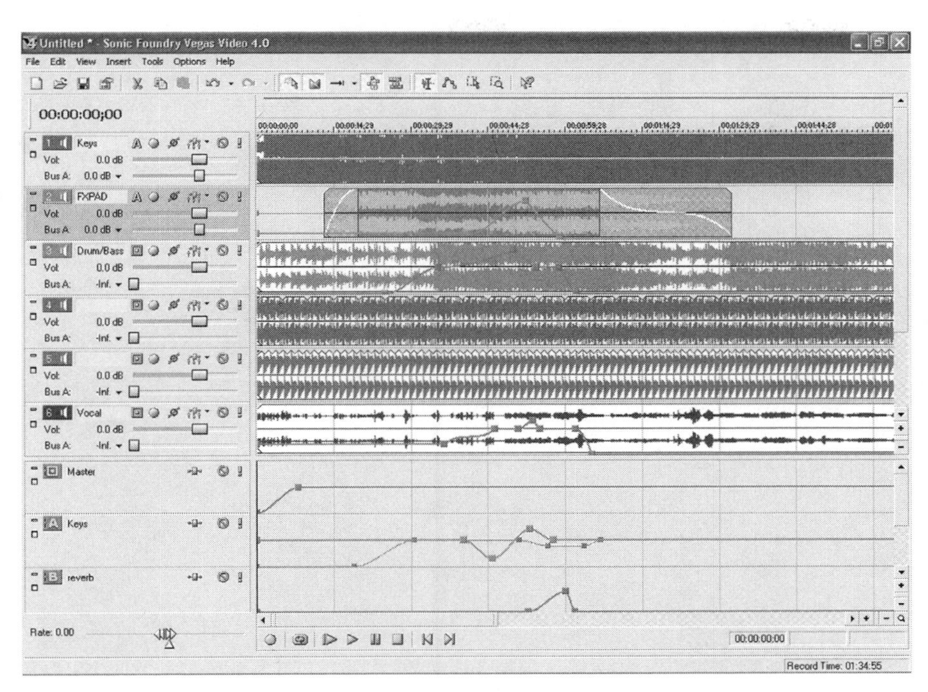

3.36 View bus behavior in Vegas 4.0, allowing full automation attributes in a mix.

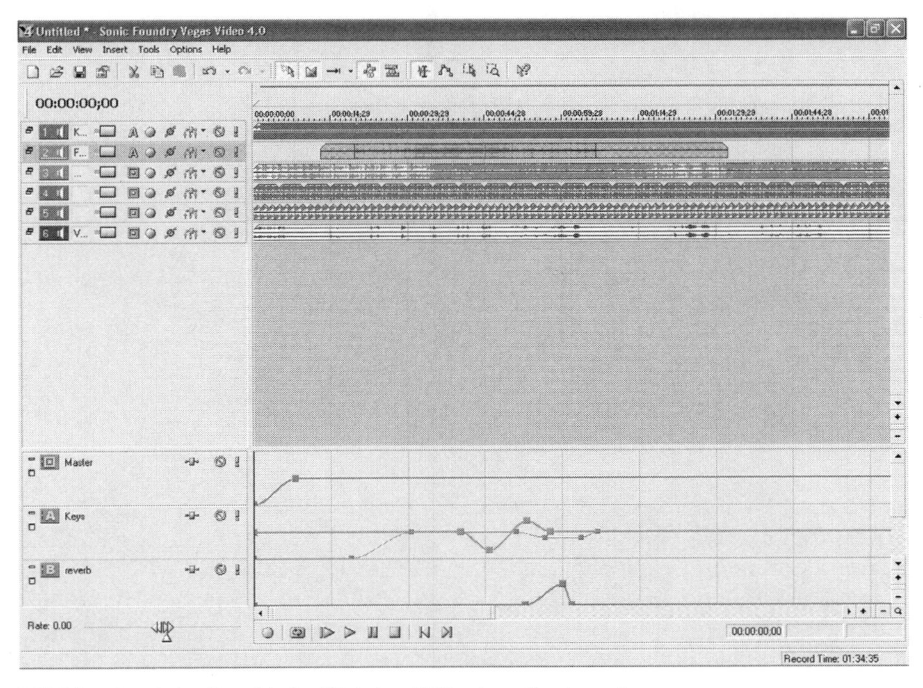

3.37 View bus behavior with the Minimize All Tracks option turned on.

The Rebuild Audio Peaks option redraws/refreshes the wave file views on the timeline. Vegas uses 16-bit peak samples, and on slower machines, you can speed up the system by turning off peak drawing in the OPTIONS | PREFERENCES | GENERAL dialog. This option, also accessed via the F5 key, causes all views to be redrawn.

Tip

If screen views will not be vertically zoomed in-depth, instruct Vegas to draw 8-bit file views. This option allows the process to become faster and increases responsiveness in Vegas. Change the screen draw by going to the OPTIONS | PREFERENCES | GENERAL dialog and scrolling down to the "Build 8-bit peak files."

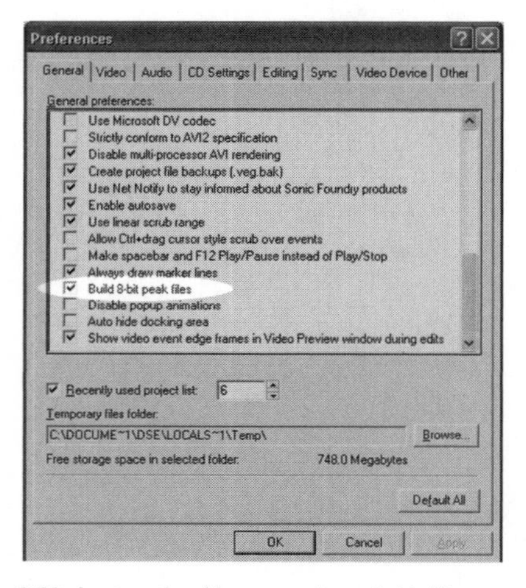

3.38 Instructing Vegas to draw 8-bit file views.

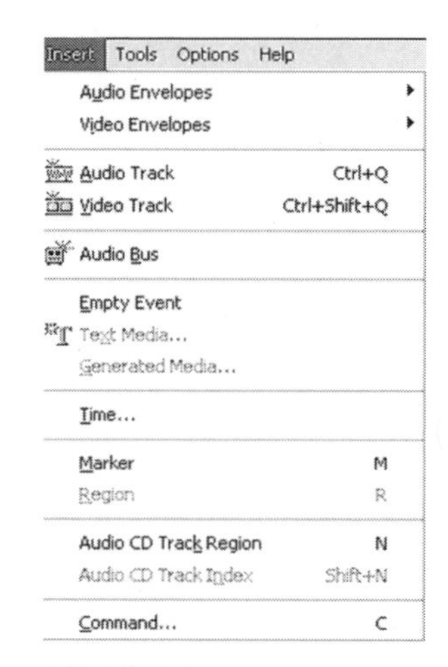

3.39 The Insert menu.

Insert Menu

The Audio Envelopes option in the Insert menu has two submenus: Volume (Shift+V) and Pan (Shift+P). These place volume lines and pan lines on the timeline on selected tracks.

Tip

Select all audio tracks before using SHIFT+V or the menu. This option places volume controls on all audio tracks, saving time.

The Video Envelopes option has three submenus: Composite Level, Fade to Color, and Velocity Envelope. Composite Level allows users to set the opacity of a track over time, by adding nodes or handles to the composite envelope and moving the envelope up or down on the timeline at desired fade points. This feature reduces the opacity of the track, allowing video events below the affected track to become visible. Fade to Color adds an envelope to the selected track, giving control to the color to which a track will fade. The default color is black; however, any color can be selected in the OPTIONS | PREFERENCES | VIDEO dialog.

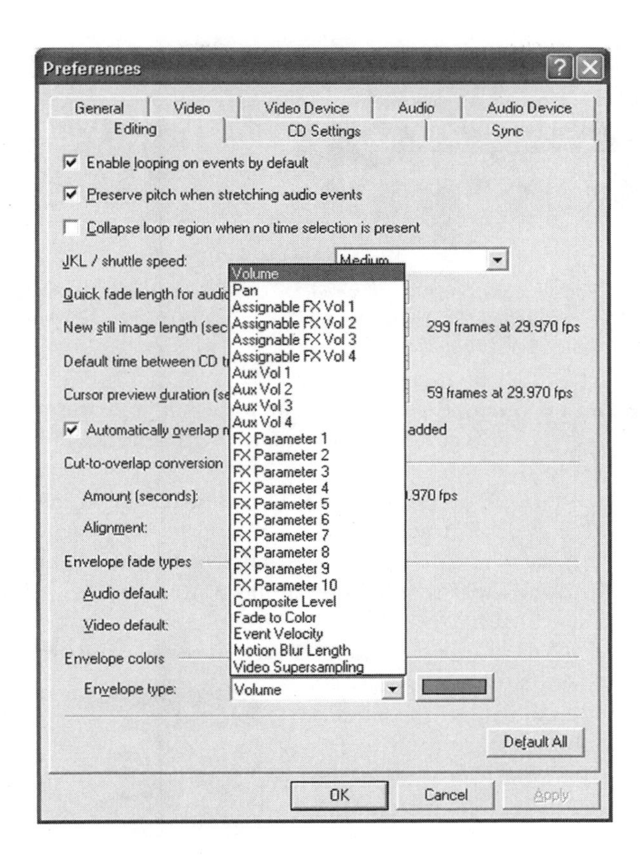

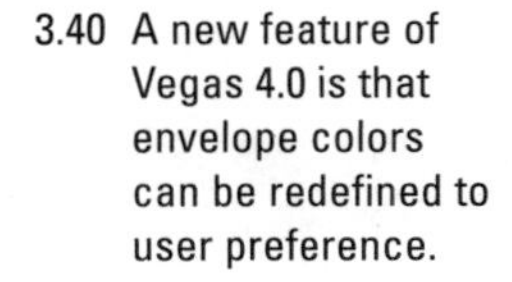

3.40 A new feature of Vegas 4.0 is that envelope colors can be redefined to user preference.

The Velocity Envelopes option allows users to create an envelope that causes a video event to speed up or slow down. Inserting an envelope and creating handles on the envelope, and then either dragging the envelope upward speeds up the event or dragging the envelope downward slows down the event. If an envelope is dragged past the 0 percent speed point, the event reverses itself.

Tip

For the surreal speed effect that looks natural but slightly dreamy, slow files to 80 percent of the original speed. The event won't appear to be slow but will have a natural, dreamy slow motion applied to it. This feature is great for slowing down high-speed events.

An event that has the Velocity Envelope it set to -100 percent will play at normal speed, but in reverse. An event that has it set to +300 percent will play at three times normal speed.

Tip

If you want speeds of greater than three times normal speed, right-click the event and select Properties. Input the additional speed in the Playback Rate window. Insert a Velocity Envelope. Files can be sped up to 12 times their normal speed by using a combination of the Velocity Envelope and the Playback Rate dialog!

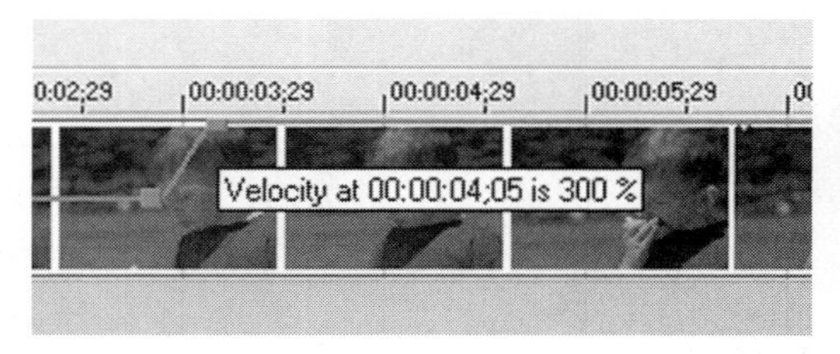

3.41 Velocity can be sped up incrementally and significantly in Vegas via the Velocity Envelope.

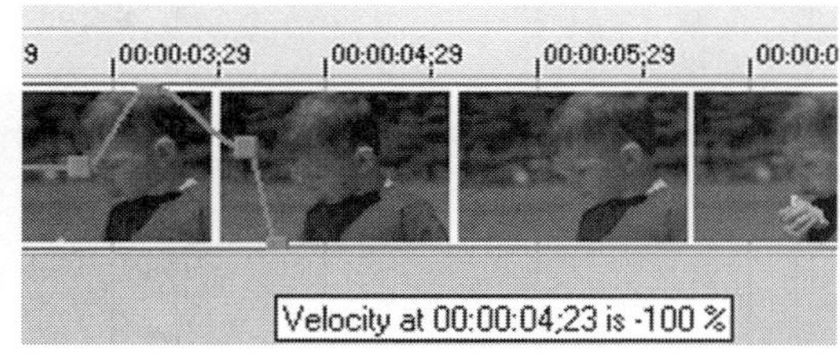

3.42 Velocity can also be slowed down over time. This sort of effect is extremely popular in modern editing.

Always use either the Smart Resample or Force Resample feature with speed-altered events. The Reduce Interlace Flicker feature can be valuable with speed-altered media when flickers are seen in the previewed image.

Selecting the Audio Track option (Ctrl+Q) in the Insert menu inserts a new audio track on the timeline beneath any audio track that is already in place. Dragging audio to a blank space in the timeline workspace where no audio tracks currently exist creates a new audio track. Dragging an audio event to the timeline allows an audio track to be created above or below existing audio or video events depending on where it is dropped.

Selecting the Video Track option (CTRL+SHIFT+Q) in the Insert menu inserts a new video track on the empty timeline or above existing video tracks. Dragging non-audio events to the timeline also accomplishes the same task by creating a new video track for the non-audio event. Dragging a non-audio event to the timeline allows you to create a new video track above or below an existing audio or video track, depending on where it is dropped.

The Audio Bus option (**B**) accesses the audio control or destination tracks. A hardware mixer has assignable buses, allowing the audio signal flow to be routed to a number of places before leaving the mixer. During this routing, the audio signal flow can be processed through a variety of tools, such as delays, reverbs, chorusing, flanging, or other processes. Buses can also be used as routing tools to hardware outputs. If a multichannel sound card is used, bus assignments are required to monitor the audio for multiple outputs beyond a simple stereo monitoring system. Up to 26 buses can be inserted in Vegas. Buses can also be used for submixing, assigning all mixes of a group to one location for final leveling.

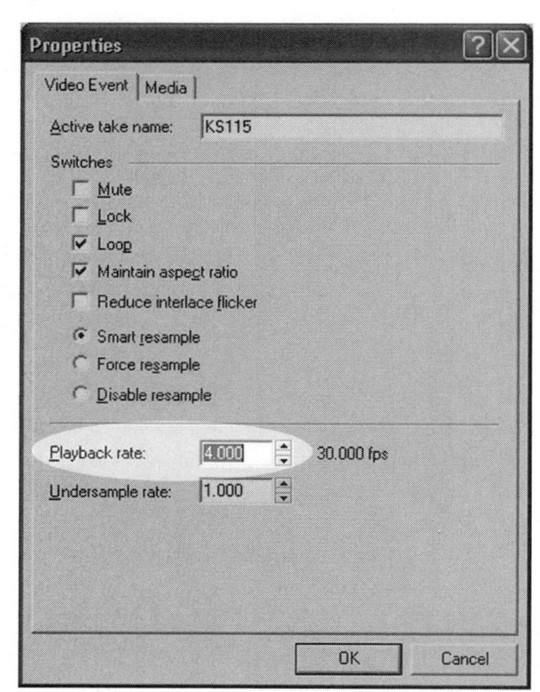

3.43 Adjust the rate of an entire clip with the Playback Rate list box found in the Properties Video Event dialog.

Use bus assignments to route together all harmonies, all guitars, all basses, all drums, and all sound effects for submixing.

The Video Bus Track option (CTRL+SHIFT+B) allows specific effects and processes to be controlled by envelope on the master video bus. Supersampling, MotionBlur, and Master Fade out to Transparent Black are all possible with envelope control in this window.

Selecting the Empty Event option in the Insert menu inserts a blank event on the timeline. Empty events can be used as markers for specified lengths of time in current projects or used as placeholders for takes in projects requiring a consistent edit and workflow, allowing video or audio takes to be dropped on the empty event. Empty events can have filters, effects, or other attributes assigned to them that are maintained when a new event is dropped on the empty event as a take. Empty events can also be inserted by right-clicking an audio or video track and choosing the Empty Event dialog.

Selecting the Text Media option opens the titling dialog.

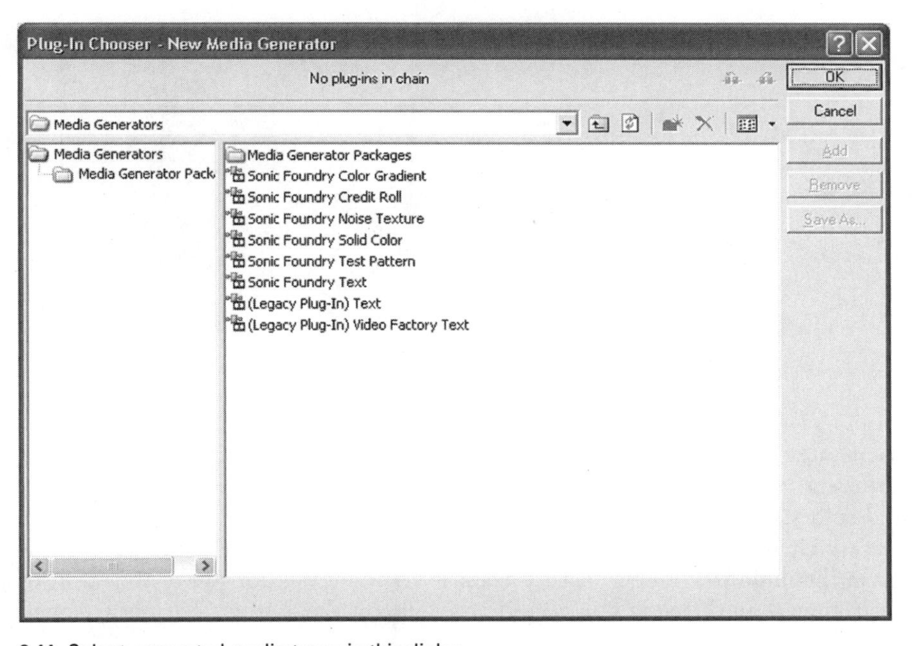

3.44 Select generated media types in this dialog.

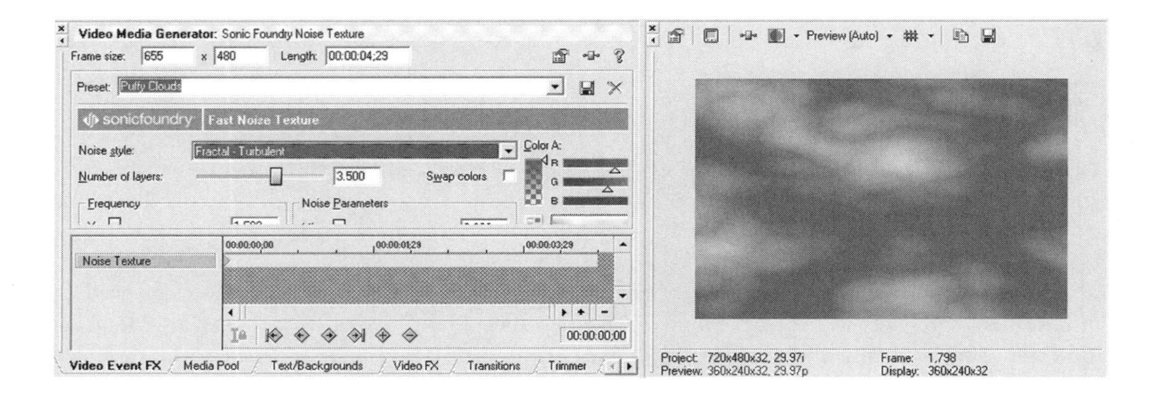

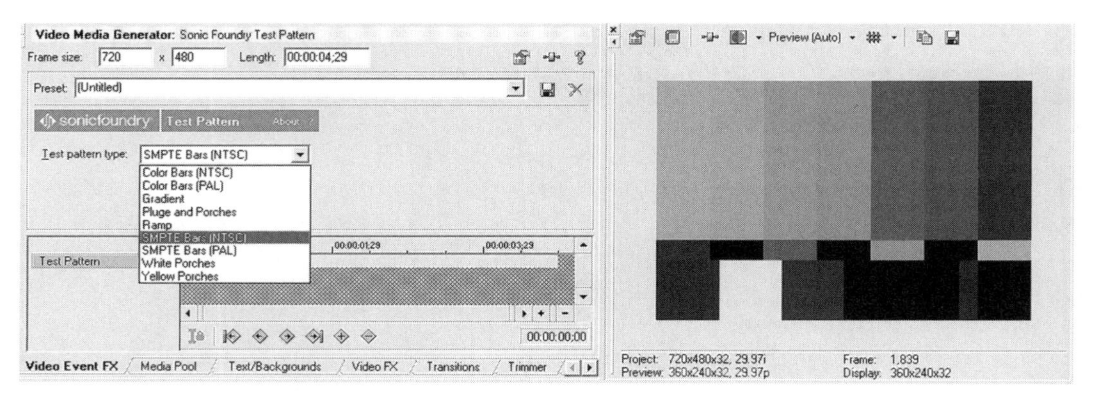

3.45 Creating color bars is one of the functions of the Generated Media tool.

Selecting the Generated Media option in the Insert menu opens a dialog allowing users to create a variety of different forms of media appearances.

Gradients, solid colors, credit rolls, test patterns, and text can all be inserted via this dialog.

Selecting the Insert Time option in the Insert menu inserts time across the entire timeline. If a cursor is placed in the middle of an event, that event is split and the portion of the event to the right, or later in time than the cursor, is moved according to the amount of time specified in the Insert Time dialog box.

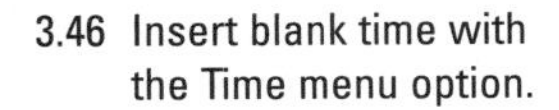
3.46 Insert blank time with the Time menu option.

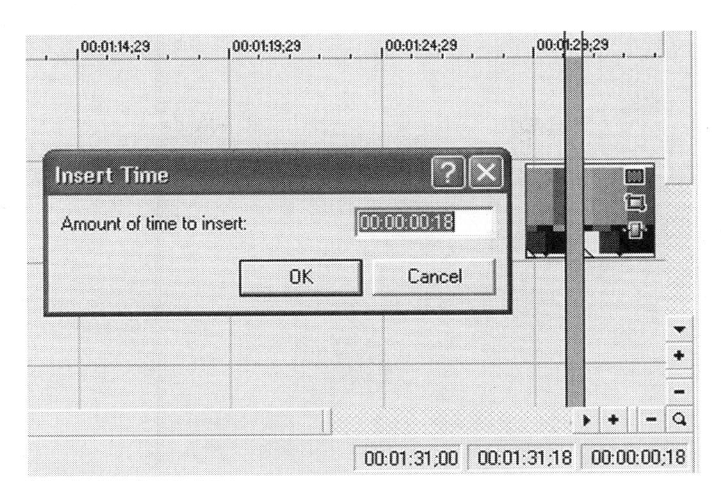

Inserting a marker is possible by selecting Marker from the Insert menu; most users, however, find that inserting a marker is more efficient by pressing M. This option drops a vertical line at whatever point in time the cursor happens to be at when M is pressed or the menu option is selected.

Regions can be defined as a space between two markers or a selection created by dragging the mouse across the timeline. Pressing R also activates the Regions option in the Insert menu.

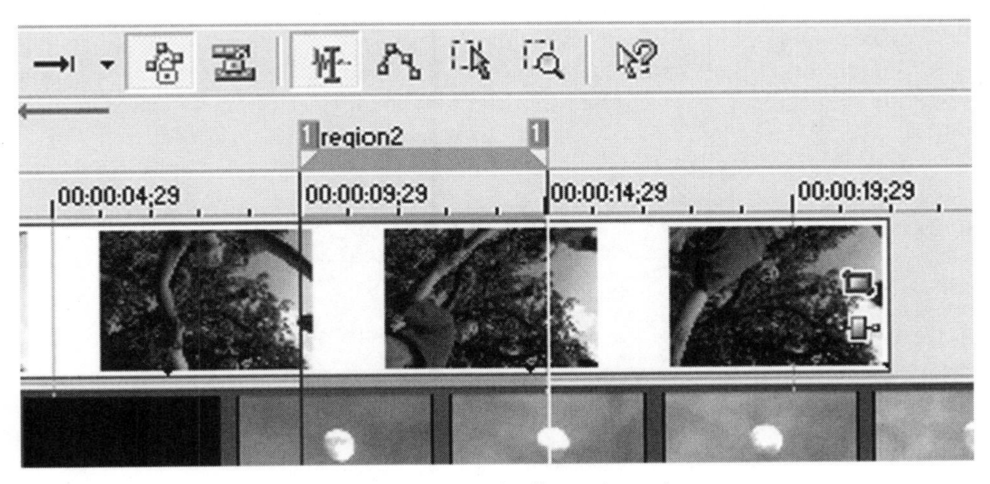

3.47 Regions can be created via a selection on the timeline and pressing R.

CD regions/track numbers can be assigned via the Audio CD Track Region option on the Insert menu. CD regions, however, are also assignable by pressing **N** when a selected event or time frame is on the timeline. This feature is how Vegas creates track numbers for compact audio discs.

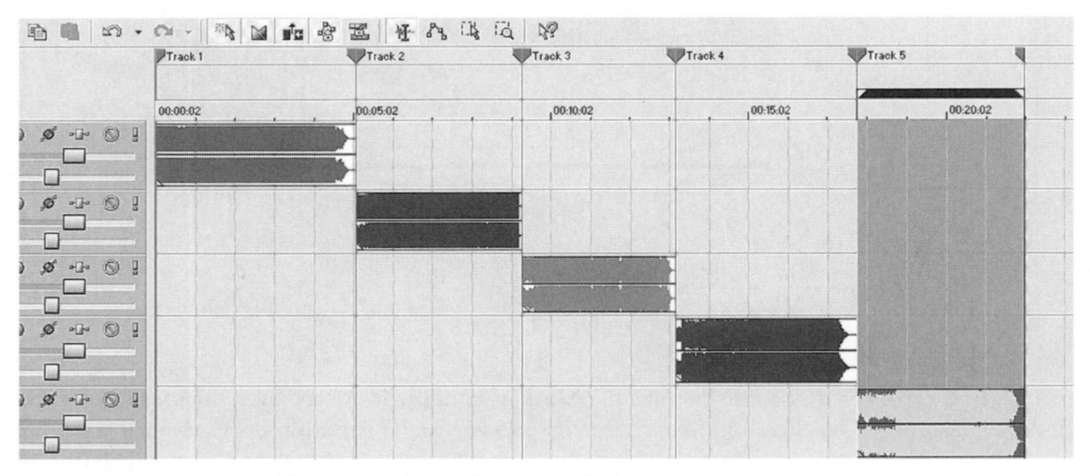

3.48 Using the regions in a CD project makes for fast track indexing.

Selecting the Audio CD Track Index option (**SHIFT+N**) allows for separations between a track number and an index in a track, similar to a chapter and subchapters in a book.

Selecting the Command option in the Insert menu creates metadata commands for streaming media. This menu selection brings up the Command Properties dialog box that gives a selection of different forms of metadata to be inserted into the timeline for authoring of metadata-rich streaming media.

3.49 The Command tools allow for the insertion of metadata that create a more interactive experience for streaming media or distributed disc viewers/listeners.

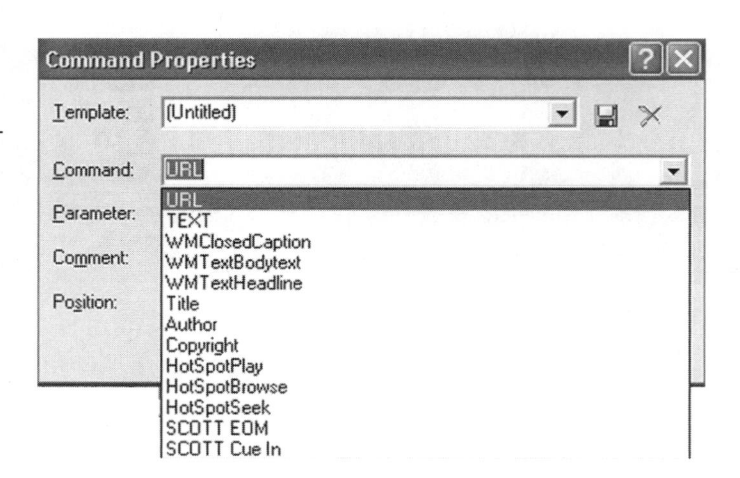

Tools Menu

The Tools menu contains tools for rapidly creating final products, quick renders of edited media, and other highly useful tools. This menu is where the majority of work is done within Vegas outside of the trimming tools and actions.

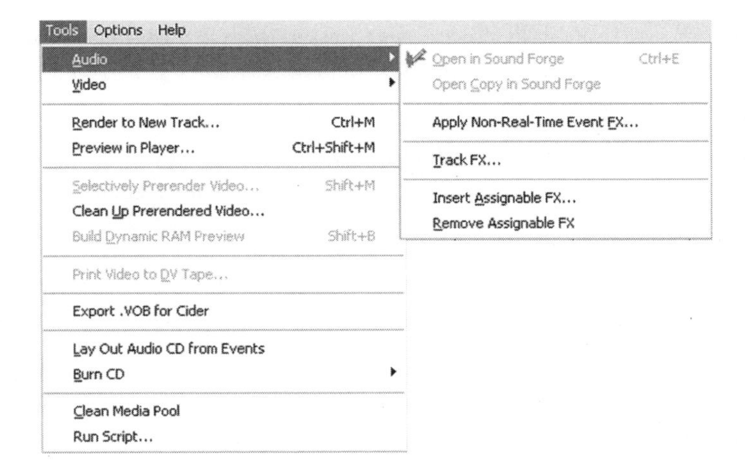

3.50 The Tools menu.

The Audio option in the Tools menu has a number of subchoices within it. These choices are also accessible via a right-click an audio event. The first choice, Open in Sound Forge, opens the actual event—audio, video, or both—in Sound Forge where destructive/permanent edits can be applied to the file. Open Copy in Sound Forge opens a copy of the audio event in Sound Forge, where a take is created that exactly replaces the original event on the timeline. Pressing T lets users toggle back and forth between the original and edited audio. Any audio editor can be chosen as the preferred audio editor; however, the unique relationship between Sonic Foundry products makes Sound Forge a worthwhile application to have in partnership with Vegas 4.0. To set the preferred audio editor, select OPTIONS | PREFERENCES | AUDIO and browse to the .EXE of the preferred audio editor. Audio opened in any audio editing application is replaced in Vegas as a take.

Tip

Always open audio from AVI files as a copy rather than as a destructive file. This option generates a take and does not affect the original. As a result, you can always go back to the original. It also speeds up the save process, as Vegas isn't generating a new video file for the original AVI.

The Apply Non-Real-Time Event FX option is great for slower computers or for applying processes that don't work well in real time, such as Noise Reduction and Acoustic Mirror processes. These processes are so resource-intensive that only the fastest machines can carry the load. They also take resources away from other actions taking place on the timeline.

When using the Apply Non-Real-Time Event FX option, remember that these effects are destructively assigned to the event, allowing the processor to be free of having to deal with the load of real-time effects. Users with slower computers or with little RAM in their machines generally find this feature beneficial as well, as it frees up system resources for processing more audio, video, or mixed events. When working with non–real-time effects, you'll be prompted to provide a new name for the audio file so that the original is preserved. Vegas does not allow you to use the original filename, as the original file is currently open on the timeline. The file with a new name is added as a take to the timeline and can be compared against the original sound by pressing T. Users can toggle back and forth between the original and newly affected sound during playback.

While not readily evident, effects are broken down into various folders, including a folder called `\ALL\`. This folder allows for all Direct X plug-ins to be viewed at once. Other folders, however, show third-party effects, Sonic Foundry–only effects, and FX chains, found in a folder labeled `\FX Packages\`. There is also an `Automatable FX` folder in which access to all FX that can be controlled with the FX automation envelopes are found. Only a limited number of Direct X FX can be automated. The folder `\Track Optimized\` contains a select group of FX, which are optimized for maximum processor performance.

Choosing the Track FX option in the Tools menu opens the effects found on the Track Control pane in Vegas. A Noise Gate, Equalizer, and Compressor are assigned by default to every audio track on the timeline. These can be accessed via the Tools menu (**TOOLS | TRACK FX**) or by clicking the Track Effects button on the Track Control pane. These effects can be deleted, replaced, or augmented, depending on the user's preference.

The Insert Assignable FX option automatically inserts a new bus in the mixer section of Vegas. Selecting this option calls a list of all available plug-ins, where as many as needed can be chosen as a chain of effects.

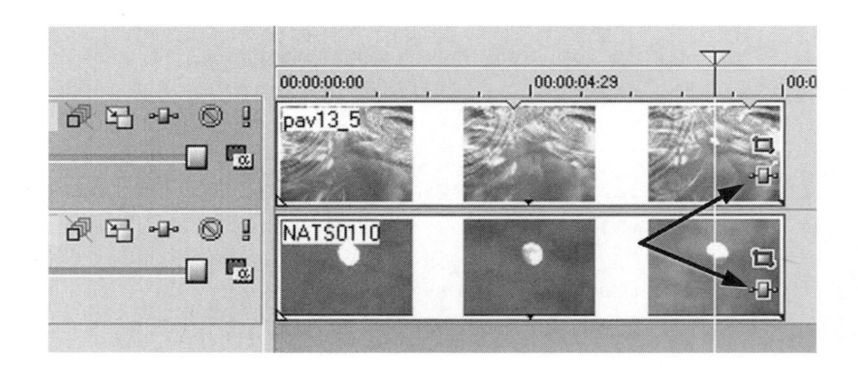

3.51 Each event has an FX button on it unless the button is removed in the **OPTIONS | PREFERENCES** dialog.

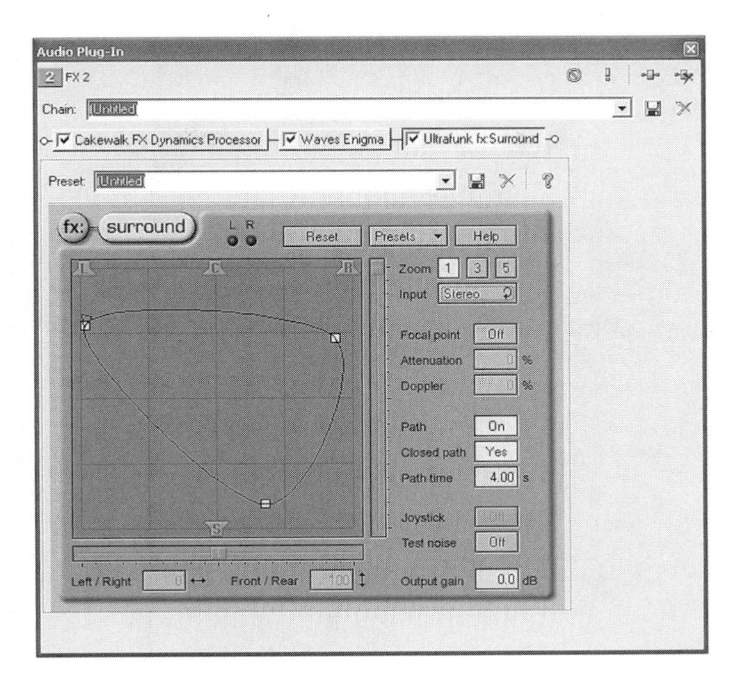

3.52 FX can be sequenced in anyway desired.

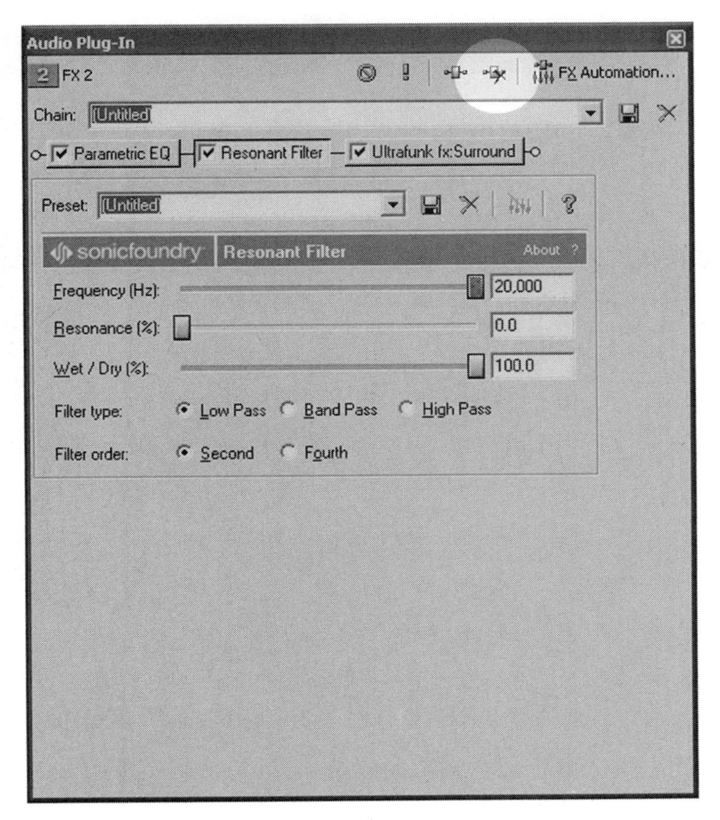

3.53 FX can be removed at any time. FX can also be A/B compared by checking and unchecking the box next to the effect name.

Effects in the chain can be re-ordered by clicking and holding an effect on the chain, dragging it to the desired priority in the chain, and dropping it. Also, right-clicking any effect and selecting the Move Left or Move Right pop-up menu option reorders effects. Effects in the chain can be removed by selecting the Remove Assignable FX option from the Tools menu or by clicking the Remove button in the upper-right corner of the FX dialog box.

Inserting video effects works exactly the same as inserting an audio event effect, except that it is always non-destructive to insert a video event effect.

When the Video Effects option in the Tools menu is selected, all video effects are shown in a dialog box, allowing the user to choose multiple video effects. Effects in the chain can be re-ordered by clicking and holding an effect on the chain, dragging it to the desired priority in the chain, and dropping it. Also, right-clicking any effect and selecting the Move Left or Move Right pop-up menu option reorders effects.

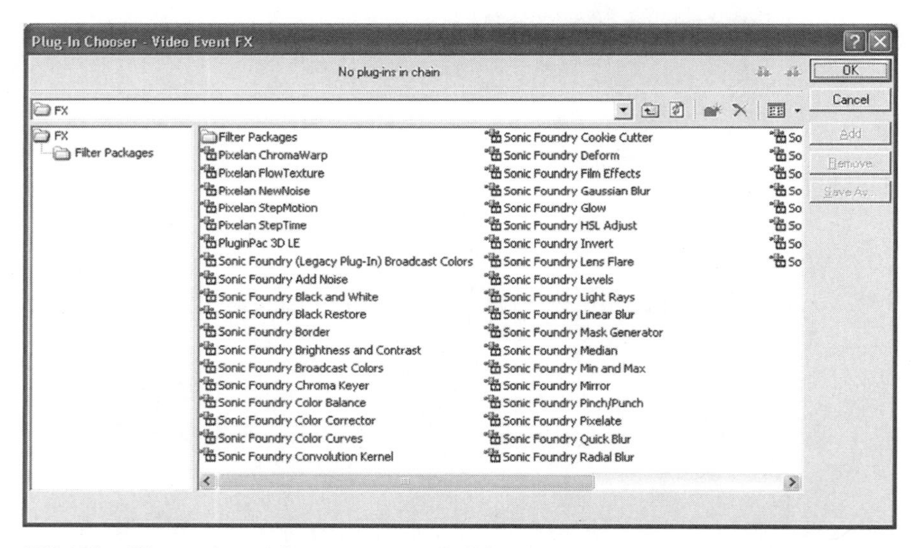

3.54 Video FX are selected the same way audio FX are.

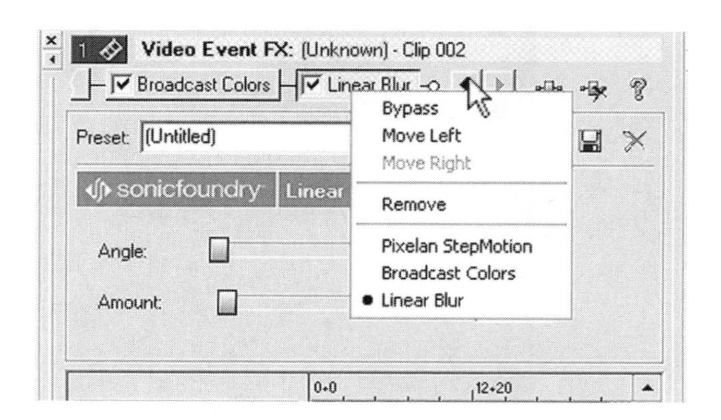

3.55 Reordering video FX is a single button selection.

Clicking an effect on the Video FX chain and then clicking the FX Help button pops up a screen that demonstrates how the selected effect affects the video image. A short explanation of how each slider or button on the effect will behave is also included.

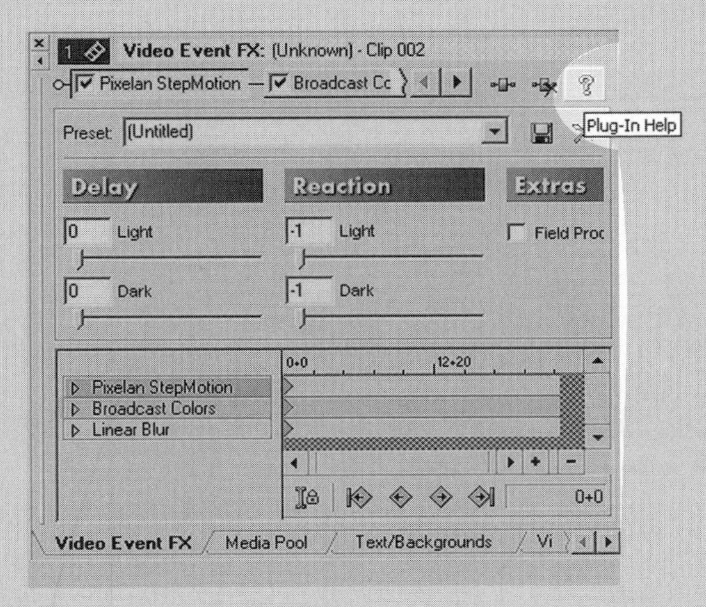

3.56 All aspects of Vegas 4.0 have a Help icon or button associated with them.

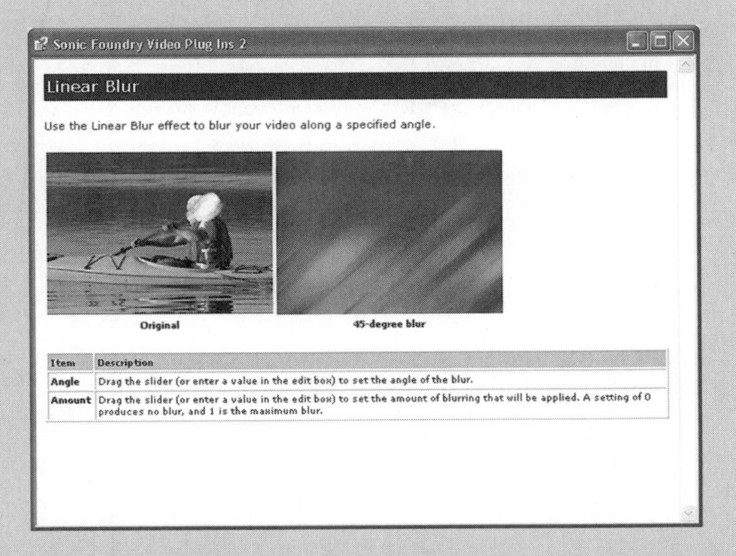

Item	Description
Angle	Drag the slider (or enter a value in the edit box) to set the angle of the blur.
Amount	Drag the slider (or enter a value in the edit box) to set the amount of blurring that will be applied. A setting of 0 produces no blur, and 1 is the maximum blur.

3.57 Calling up help by selecting the FX Help button opens a brief tutorial explaining how effects work.

This feature allows users to preview the event effect as a static image before it is applied to the event on the timeline. It's also a great way to learn the attributes of each video effect.

Effects in the chain can be removed by selecting the Remove pop-up menu option or by clicking the Remove button in the upper-right corner of the FX dialog box.

The Video FX dialog can also be accessed by right-clicking the event on the timeline or by pressing the Video FX button found on each non-audio event.

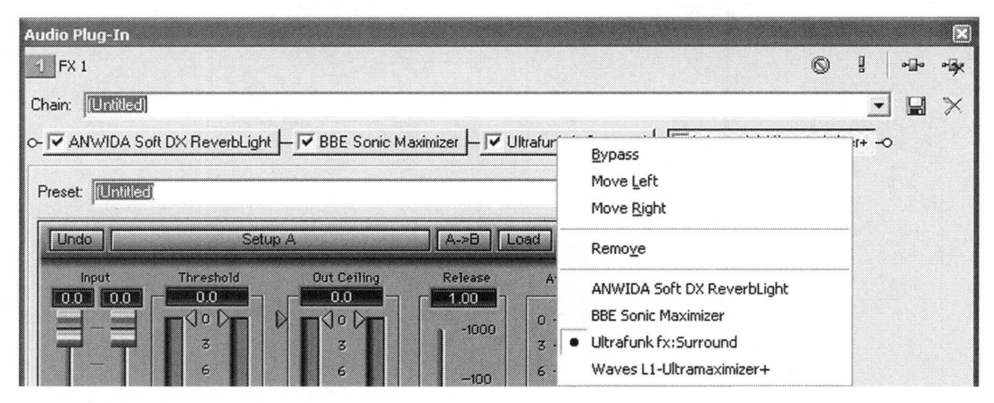

3.58 Select the effect requiring removal or adjustment.

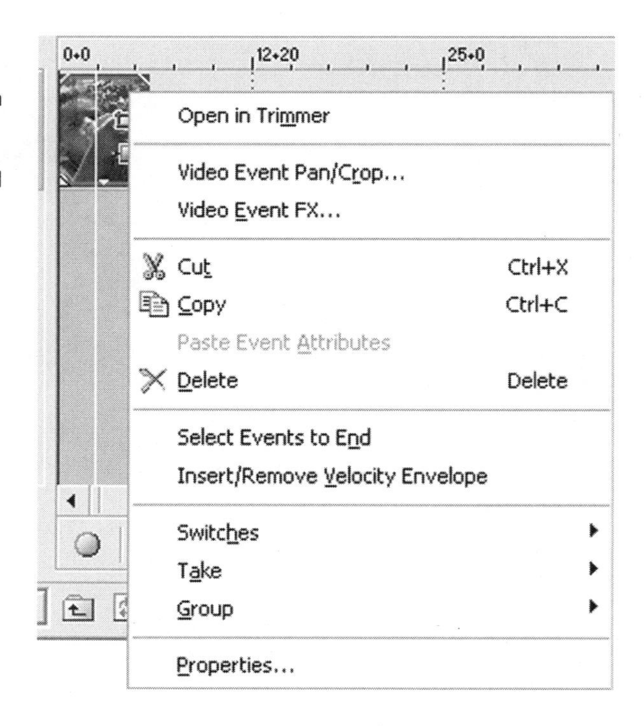

3.59 Right-clicking an event accesses a pop-up menu in which you can open the FX dialogs if the FX button has been removed from the event.

The Pan/Crop tool can be opened by selecting TOOLS | PAN/CROP. This tool is one of the most powerful and oft-used tools found in Vegas, particularly by users of still imagery. Selecting the Pan/Crop option opens a dialog on the timeline, in which the event image can be cropped to meet the screen size, forced to match the aspect ratio of the project settings, or have zooms, pans, and other movements applied to otherwise static images. Images facing right or left, up or down can be reversed in view easily by right-clicking the Pan/Crop tool.

3.60 FX button found by default on all events.

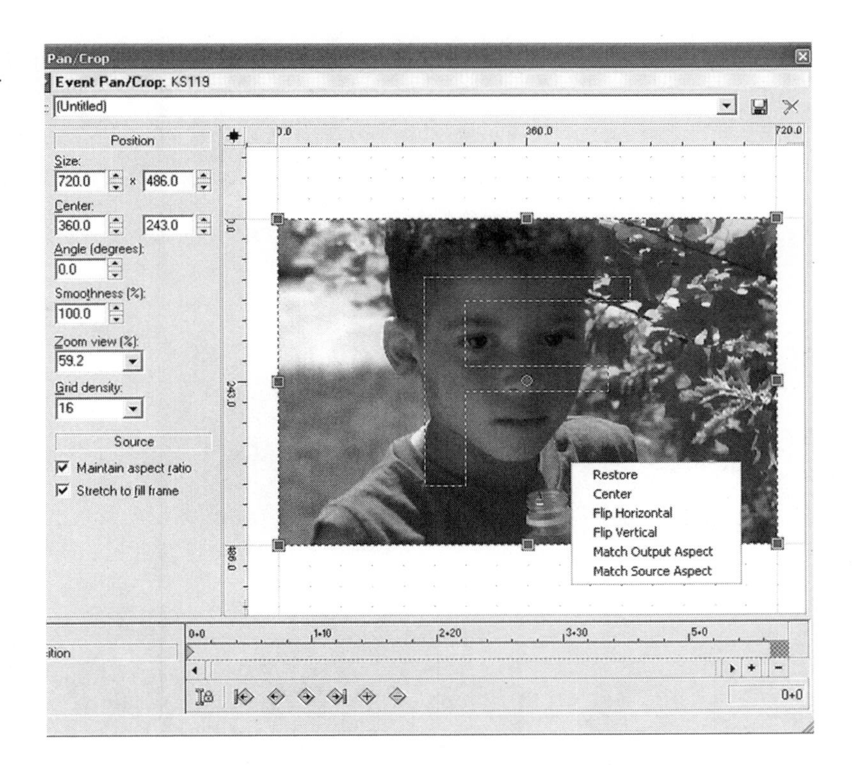

3.61 Open the Pan/Crop dialog either by right-clicking the video event or by selecting the Pan/Crop button.

Right-clicking inside the Pan/Crop tool brings up a secondary dialog. Choosing Restore restores an event image to its prepanned/cropped status. Choosing Center centers the image around the small dot or center indicator found in the middle of the Pan/Crop tool. Moving the dot to another location in a cropped image, right-clicking, and choosing the Center option moves the cropped selection to a point where the dot will re-center itself.

Choosing the Flip Horizontal option in the Pan/Crop pop-up menu reverses the horizontal image. This option is indicated by the "F" (focus) in the middle of the event image in the Pan/Crop dialog. The "F" is backward if this choice is made.

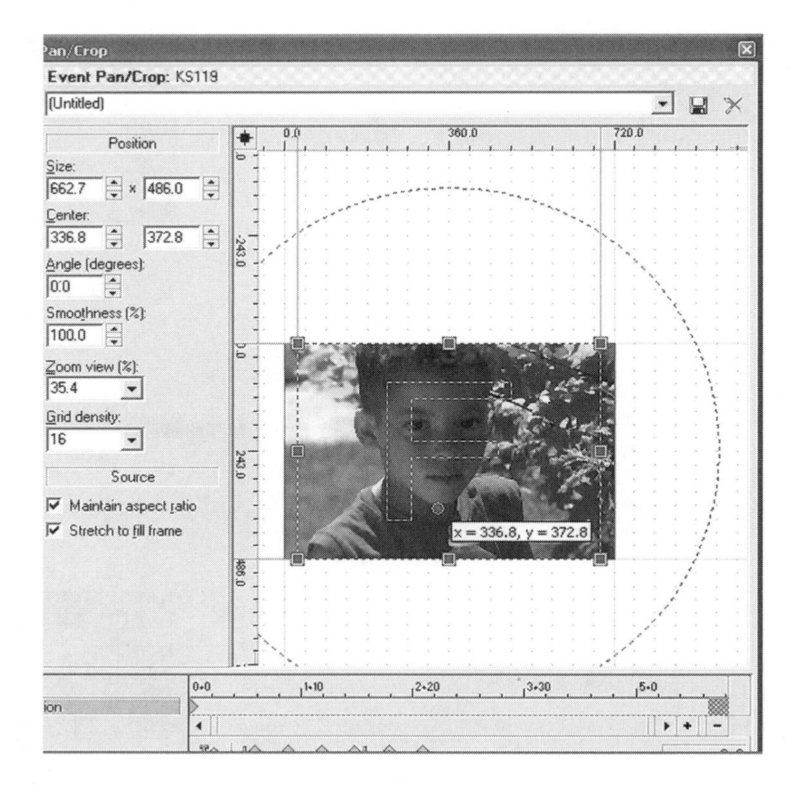

3.62 Before cropping.

3.63 After cropping.

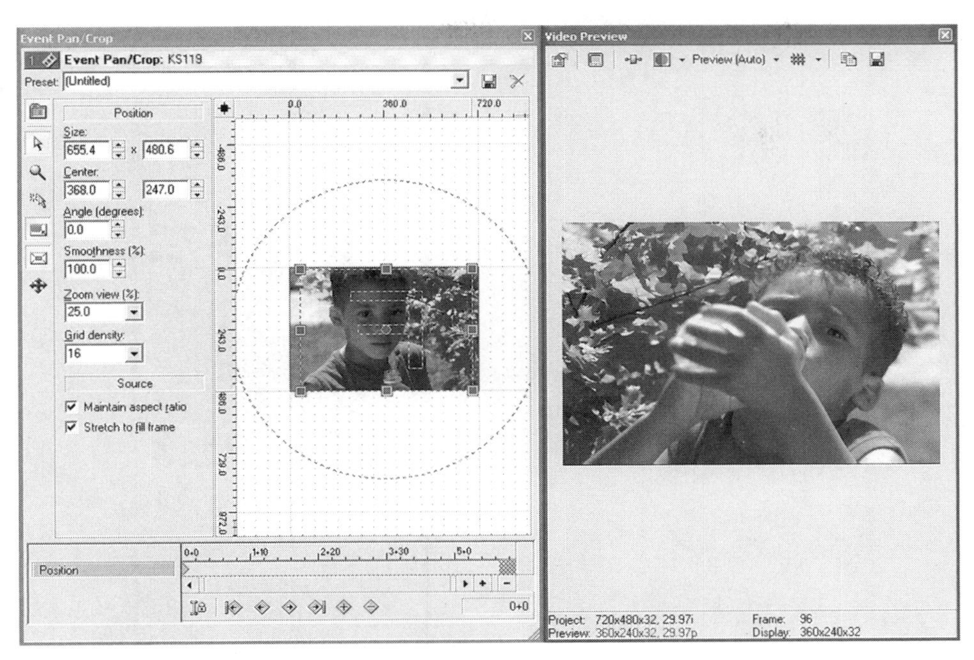

3.64 In Preview window, preview image is reversed in horizontally.

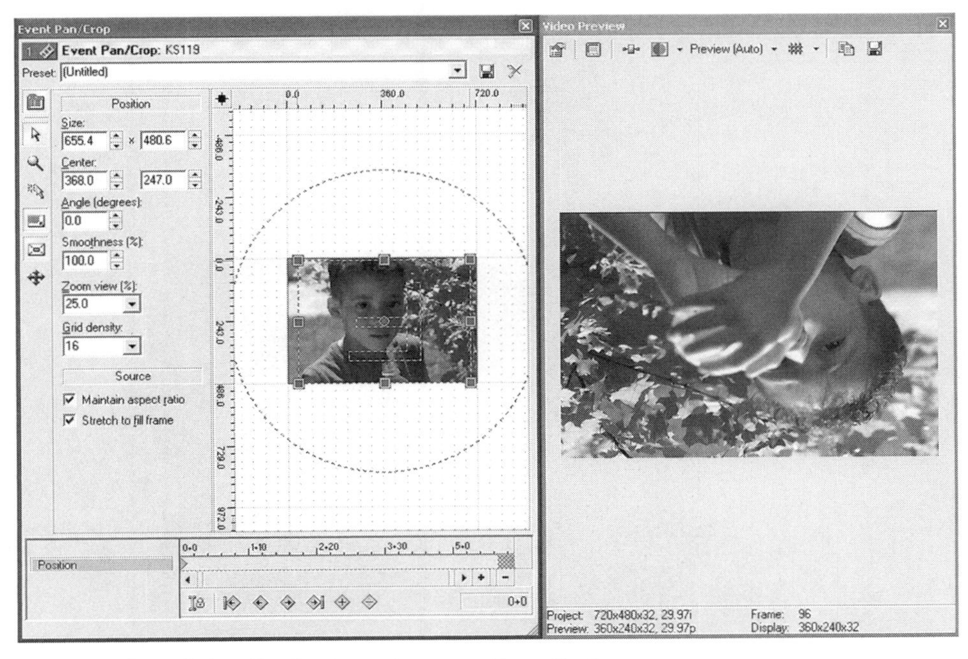

3.65 In Preview window, Preview image is reversed in vertically.

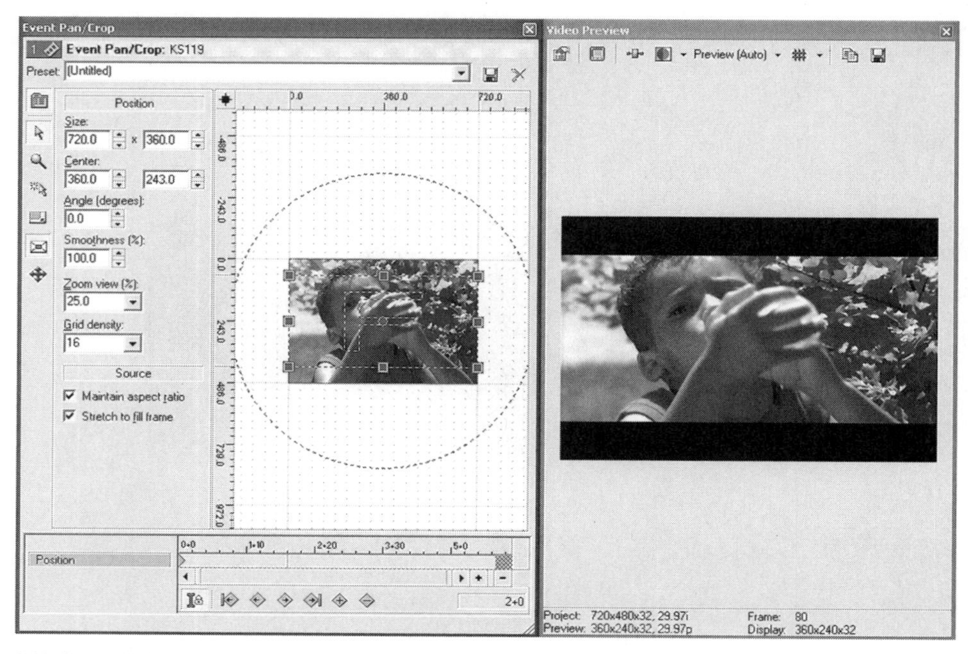

3.66 Image before matched aspect ratio is applied.

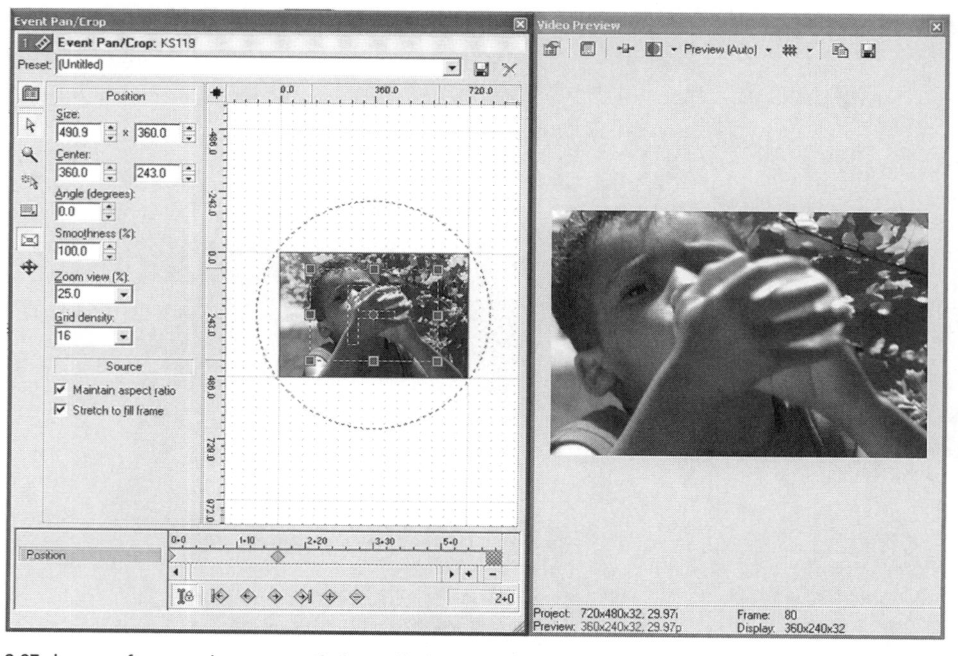

3.67 Image after match aspect ratio is applied.

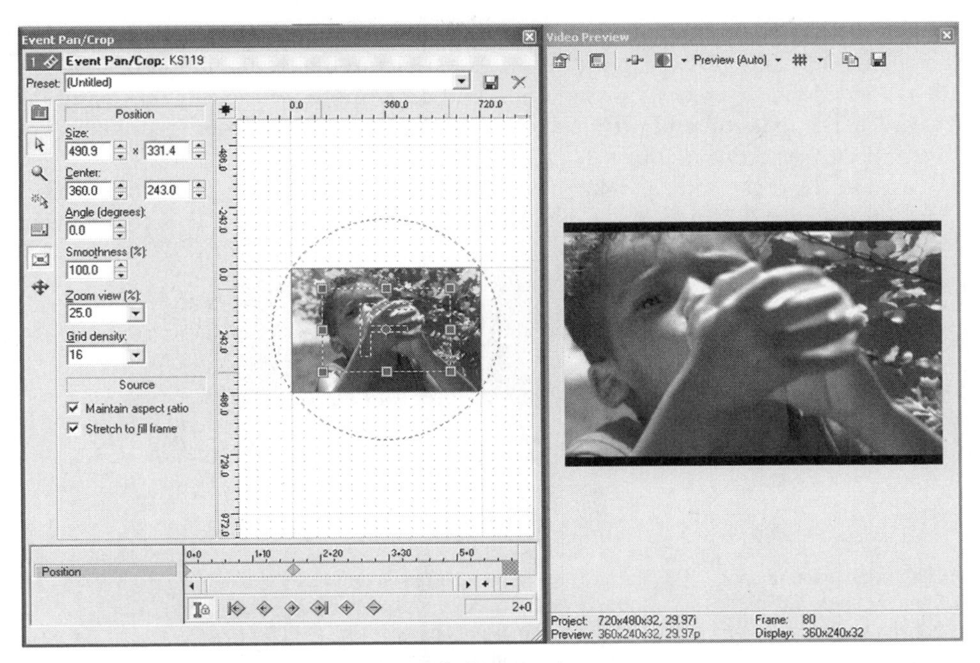

3.68 The source aspect ratio can also be applied to a pan/crop.

Choosing the Flip Vertical option turns the image upside down in the Preview window.

Choosing the Match Output Aspect option assigns the aspect ratio of the Project Properties to the image, regardless of the original aspect ratio of the event before panning or cropping the image.

The original aspect ratio of the image in the figure is 16:9 and does not match the project properties of this project. Choosing Match Output Aspect crops the image to match the 4:3 aspect ratio of the DV-NTSC Project Properties defined in the FILE | PROJECT PROPERTIES dialog.

Choosing the Match Source Aspect option matches the aspect ratio of the source, while maintaining any cropped status applied to the event image.

The Pan/Crop tool can be accessed by right-clicking the media or by choosing the Pan/Crop icon found on each non-audio event on the timeline. (See the advanced editing and still image in Chapter 5 for more information about the Pan/Crop tool.)

Selecting TOOLS | VIDEO | TRACK FX opens the Video FX dialog box and allows the assignment of FX to an entire track of video on the timeline, rather than just a video event on the timeline. This action can also be chosen on the Track Control pane of the desired track.

3.69 The Pan/Crop button is found on all video/graphic events.

Choosing the Track Motion option in the Video Tools submenu opens the Track Motion dialog box. This dialog box allows users to create picture-in-picture imagery, motion of multiple layers of video, divided video screens, moving titles, and other motion imagery. This feature is another of Vegas' most powerful and oft-used tools for the creation of composited media, in-depth motion, and other creative motion.

This tool is used to create the popular *Brady Bunch* screen format with multiple motion images applied to grids on a screen. A *Brady Bunch*-like VEG file is included on the CD accompanying this book.

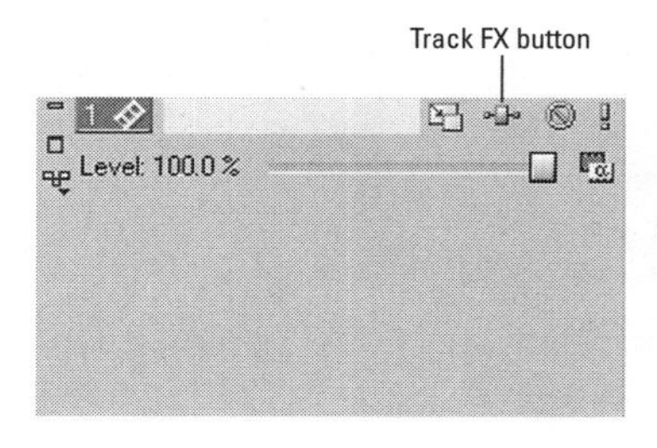

3.70 The Track FX button is found on all Track Control panes. If this button is not shown, drag the Track Control pane to the right into the timeline. The button will become visible.

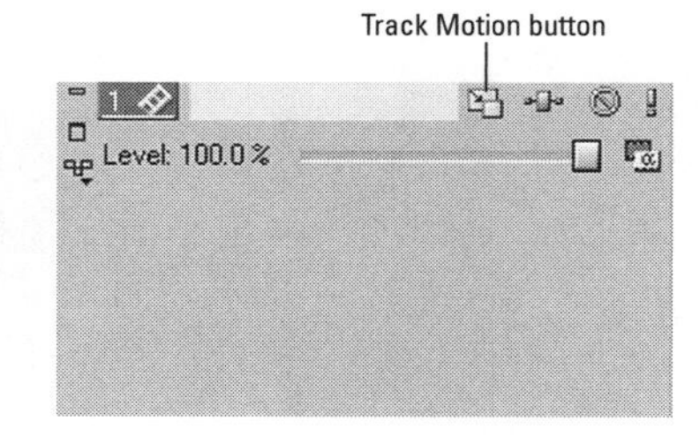

3.71 The Track Motion button is found on all video Track Control panes. If this button is not shown, drag the Track Control pane to the right into the timeline. The button will become visible.

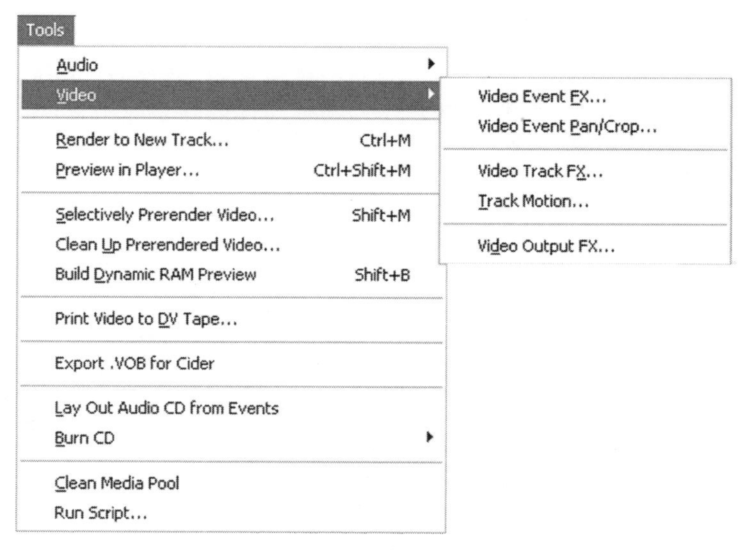

3.72 Selecting **TOOLS | VIDEO** also opens the Video Tools submenu.

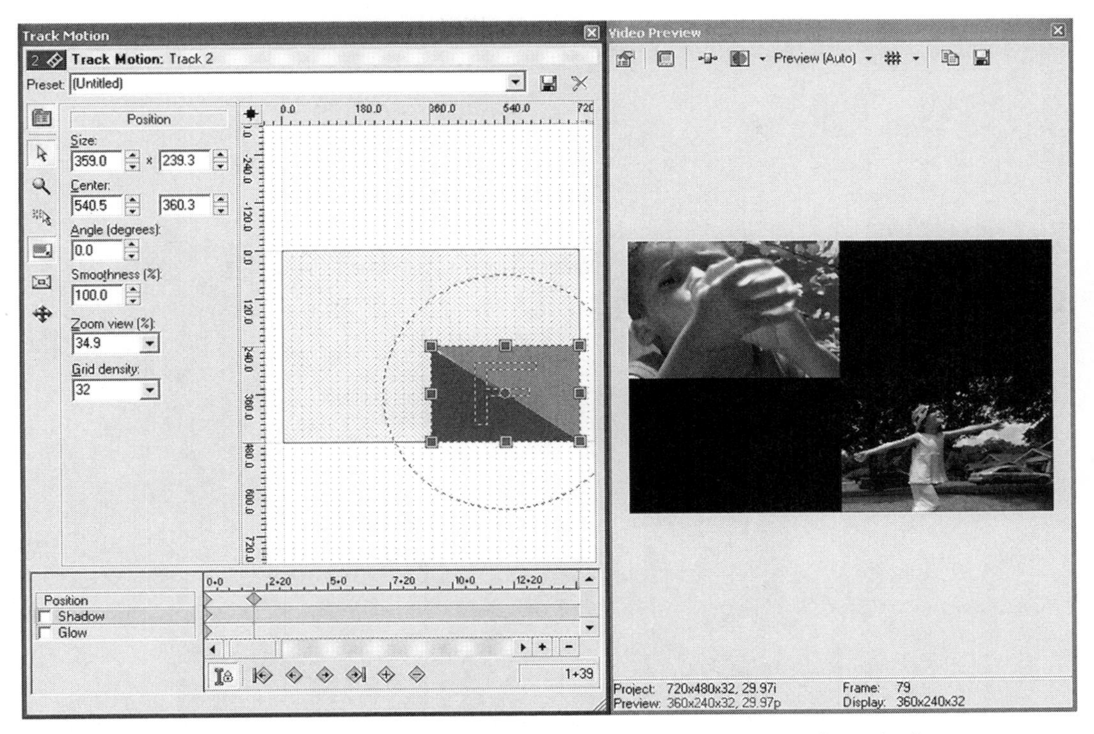

3.73 Track motion is a fast way to split screens, build multiple video views, or move a picture in picture.

Two layers of video are placed in the Preview window with the Track Motion tool. By moving event images to grid points, exactly one-quarter of the screen can be filled with a video image. Shadows or glows can be applied to the event image. The Track Motion tool affects all media on the track to which it is applied. Motion can be keyframed for motion/movement. Shadows and glows can also be keyframed. (See Chapter 5 for more information on keyframing.)

Right-clicking in the Track Motion dialog opens several menus. Selecting Restore restores the Track Motion settings to a null position, removing any track motion applied, and thus restoring any event image to the original settings.

Choosing the Center option centers event imaging on the dot found in the middle of the Track Motion dialog with the same behavior as the Pan/Crop dialog. Moving the dot to the point that is desired as the center focus and then choosing the Center option moves the track motion box to so that it is centered over the dot.

Choosing the Flip Horizontal option reverses the horizontal image. This option is indicated by the "F" in the middle of the event image in the Track Motion dialog. The "F" is shown backward if this choice is made.

Choosing the Flip Vertical option turns an image upside down in the Preview window. This option is indicated in the Track Motion window by the direction of the "F" shown in the window.

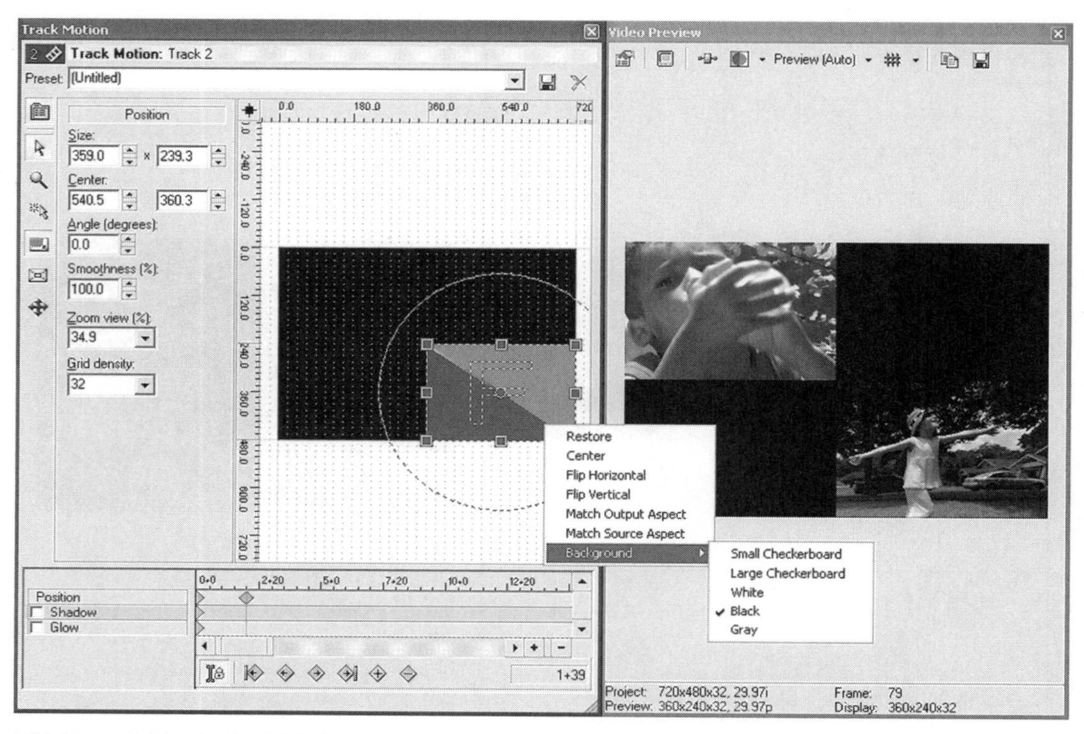

3.74 Right-clicking the Track Motion dialog opens several pop-up menus.

Choosing the Match Output Aspect option assigns the aspect ratio of the Project Properties to the image, regardless of the original aspect ratio of the event image.

Choosing the Match Source Aspect option matches the aspect ratio of the source, while maintaining any cropped or reduced-size status applied to the event image.

NEW! Selecting the Background option opens a submenu in the Track Motion dialog. This Background menu allows users to specify the color of the Track Motion control box. Grids become more visible with a black background/white dots or with a gray background with white dots.

Selecting TOOLS | VIDEO OUTPUT FX applies effects to the entire project. This step is the final one in the event/track/project opportunity to apply effects to the video imagery. Regardless of how many effects are placed on an event or track, applying effects at the project level causes effects to be applied over the top of existing effects. Effects can also be applied at the project level by dragging an effect to the Preview window or by clicking the Video Output FX button in the Preview window.

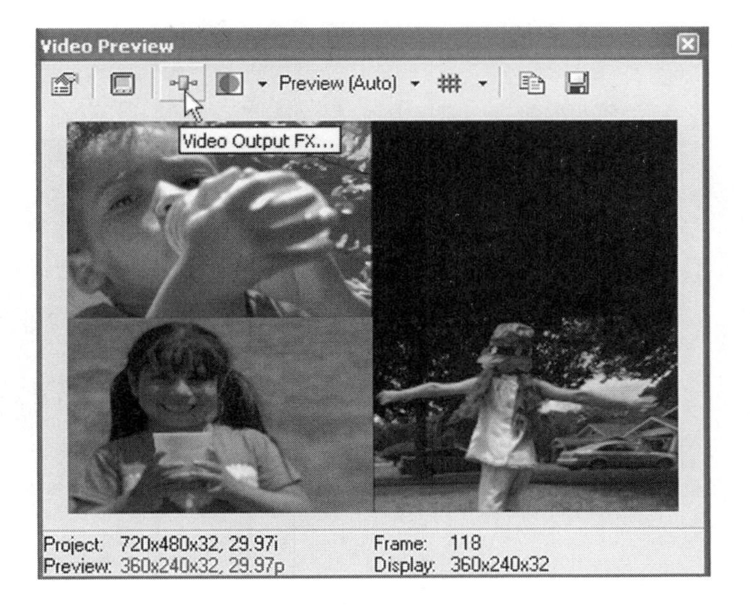

3.75 FX can be applied at the project level, affecting all media in the project.

The Render to New Track option (CTRL+M) in the Tools menu is a valuable feature in Vegas. This feature allows edited media, regardless of the number of tracks, to be rendered to a new file that is placed on a new track at the top of the timeline. Because the media is rendered, it plays back at full frame rate, full resolution at all times until it is filtered, edited, or has new media placed above it. This feature is useful when large composites are made and playback is slow, yet the events are finished in the editing process and the user doesn't want to have it consistently slowing down the rest of the project. At the same time, if edits to the composited or filtered media beneath the new track might be needed in the future, the events are still in place beneath the newly rendered track, making it accessible at all times.

The Preview in Player option in the Tools menu is a necessary tool for streaming media authors. This feature allows authors to preview selectable sections on the timeline in a Windows Media Player, a REAL Media Player, or a QuickTime player complete with metadata embedded in the media stream. Files can be previewed at any speed, using templates or custom settings. In the past, authors of streaming media needed to render out entire files, and if dissatisfied with file quality, re-encode the file. Sonic Foundry set the pace with this tool in their 2.0 version of Vegas, and the encode quality has only gotten better with time. This feature saves enormous amounts of time and effort as encoders can see exactly how the stream will appear in a player from any place in the timeline.

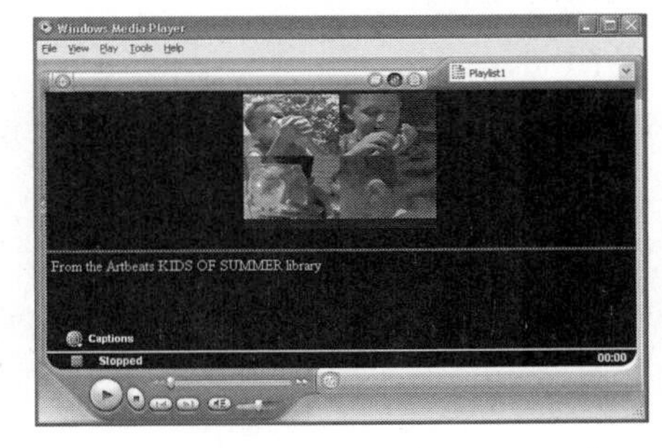

3.76 View streaming media (or other media formats) in their related players, complete with metadata.

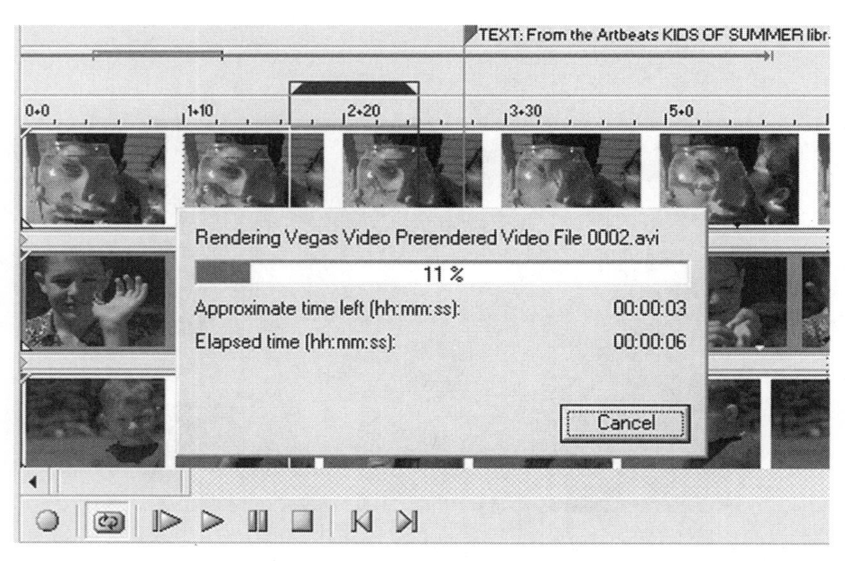

3.77 Prerendering video.

The Preview in Player tool isn't limited to streaming media. AVI, MPEG, MP3, Ogg Vorbis, and other file formats can be previewed in any of the major players complete with any applicable embedded metadata. Synchronized multimedia integration language (SMIL) embedded in REAL media is viewable in the REAL player, along with URL flips, text information, or other call-outs.

The Selectively Prerender Video option (SHIFT+M) in the Tools menu is used to prerender video sections so that the user can see exactly what the rendered video will look like. When a section of video is prerendered, a new temp file is written to the hard drive/folder specified in the FILE | PROJECT PROPERTIES dialog. This drive should be the second hard drive on the system for maximum efficiency.

After a file has been prerendered, a small blue bar is displayed above the timeline, indicating that the time section covered by the bar has a temporary prerendered file associated with it.

After an event time selection has been prerendered, moving, editing, removing, or any other edit applied to that time section invalidates the temporary file/prerender, as the temp file is no longer accurate as to what is contained in the time/event selection.

Be sure prerendered video is going to the folder you specified in Project Properties. Otherwise temp files are being written to the default `C:\Program Files\Sonic Foundry\Vegas 4.0\`, and your boot drive will rapidly become filled with temp files that aren't readily apparent. This issue significantly affects performance in a generally undesirable manner, slowing down the entire application and system.

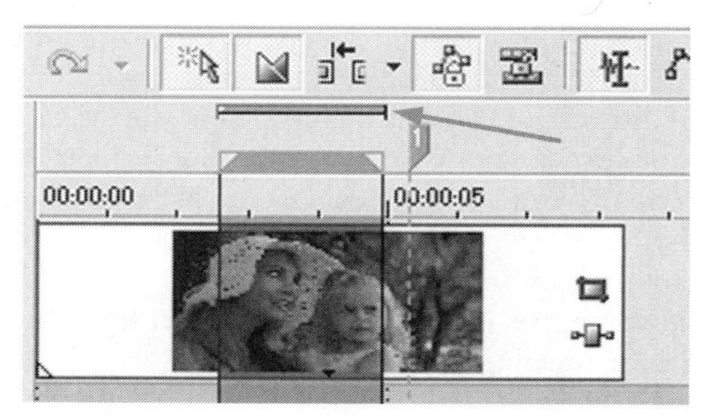

3.78 Prerendered section.

This tip is a great shortcut to program on a Contour Shuttle Pro or other hardware device, as it is typically used fairly often.

Selecting the Clean Up Prerendered Video option presents a dialog box asking which prerendered files should be cleaned up or deleted.

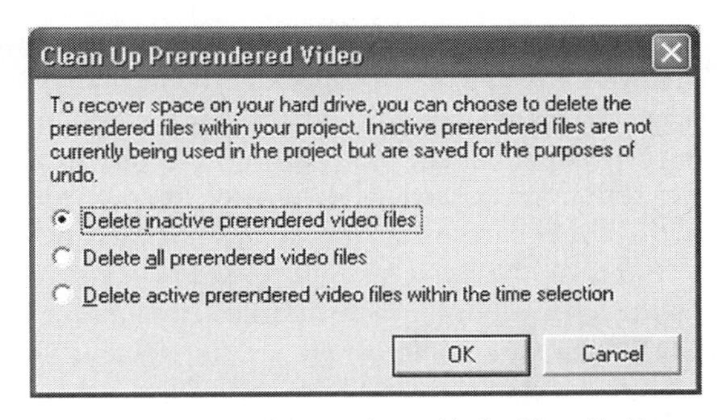

3.79 Remove unused prerenders with the Clean Up Prerendered Video dialog.

The default of "Delete inactive prerendered video files" is the most often required choice, but in the event that all prerendered files should be deleted or if a section has been prerendered several times, it might be best to clean up files related to a specific time selection only. After files are cleaned up/deleted, they cannot be recalled.

The Build Dynamic RAM Preview option (SHIFT+B) in the Tools menu works much the same as Selectively Prerender Video option, except that it does not write a temp file to a hard drive. Instead, it uses available RAM for the temp file. This render is faster and is deleted as soon as another prerender is called on or when an event is moved or edited. The amount of RAM available for a prerender is specified in the OPTIONS | PREFERENCES | VIDEO dialog. Approximately 60 percent of RAM installed in the machine can be made available for Vegas to prerender files with. The entire RAM space cannot be used, as the operating system, applications, and services require a minimal amount. This issue is why Vegas does not show all RAM installed in the machine as available.

The Print Video to DV Tape option is just as it implies. This option prints the project on the timeline to videotape without rendering the entire project as a new AVI. When Print Video to DV Tape is selected, Vegas renders all edited portions of the file to a temp prerendered track. It can give a warning that over 80 percent of the timeline needs to be rendered. This issue is expected in longer or more intense projects with lots of filters, edits, transitions, or color correction.

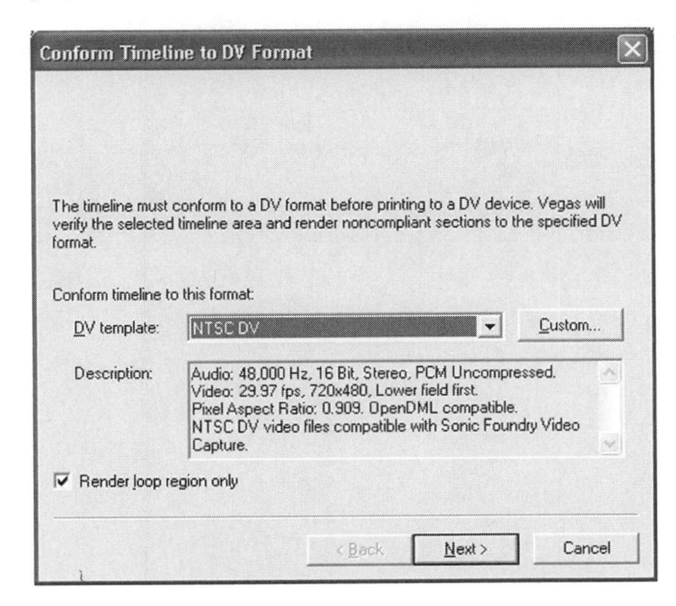

3.80 Print-to-tape presents faster options for placing media on a DV tape.

The dialog box asks for preferred black leader and tail at the end of the print-to-tape. A minimum of three seconds of black at the beginning and end of the print are recommended to compensate for any camera compatibility issues. It also provides an option to print an SMPTE color bar pattern and test tone at the head of the file and provides options for length of color bar and test tones. Forty seconds or greater of color bar and test tone are considered standard. Check with your replicator or broadcaster for their requirements.

Selecting the Lay Out Audio CD From Events option instructs Vegas to look at audio events on the timeline, insert index markers, and create necessary information to burn a successful audio CD. Make sure audio events are not touching on the timeline, or Vegas will see the touching events as a single event.

Vegas auto-inserts two seconds at the head of the audio and two-second index spaces in order to create a Redbook-compliant CD master. This issue only applies to disc at once (DAO).

Selecting the Burn CD option opens a submenu that lays out either audio or video CDs for burning, displaying the following options:

- **Track at Once Audio CD**—burns one song at a time to an audio disc.
- **Disc at Once Audio CD**—burns an entire project at a time, including index markers. This option is the Redbook standard.
- **Video CD**—burns an MPEG-1 video CD. MPEG-1 is the VCD standard.
- **Multimedia CD**—burns an MPEG, REAL, WMV, QuickTime, or other multi-media form of CD. MPEG-1 files are not compliant.

Both of the video formats can be burned from a template, existing file, or timeline-based file. Files not previously rendered require rendering before burning to disc, so Vegas prompts you through the required steps.

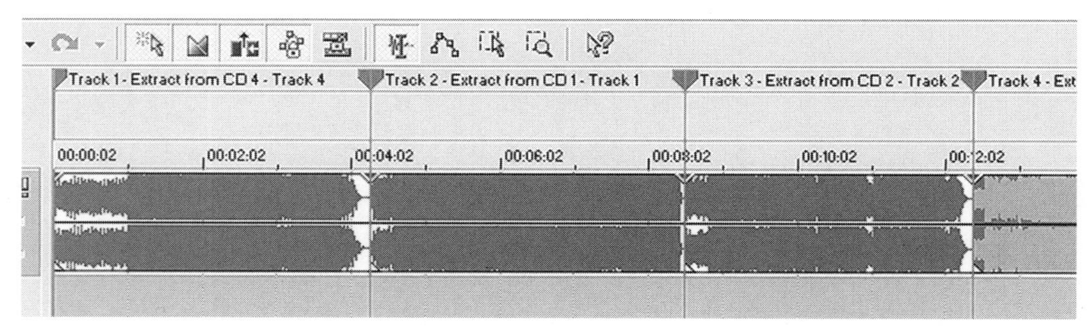

3.81 Vegas can automatically lay out a CD for burning.

Selecting the Clean Media Pool option removes from the Media Pool media references that are not active on the timeline. This step should be done before saving a project with media. If files have been placed on the timeline, they do not appear in the Media Pool whether they were kept on the timeline or not.

NEW! The Run Script option is a new feature in Vegas 4.0 that allows users to write custom macro scripts for batch processing events in nearly any way that a user can imagine or require. Need a script that tells Vegas to render an AVI, use the AVI as a master to render a QuickTime, Windows Media Video, and REAL file, beep twice, then render an MPEG-2 file with AC-3 audio? The Run Script option can make this happen if the scripting code is written correctly. An example of a script in Vegas 4.0 looks like the following listing:

```
**
 * This script will remove all effects of a particular type from items
 * in the project's media pool.
 *
 * Revision Date: Jan. 30, 2003
 **/
import System.Windows.Forms;
import SonicFoundry.Vegas;

// This is the full name of the plug-in associated with the effects
// you want to remove.
var plugInName = "Sonic Foundry Timecode";

try {
    var mediaEnum = new Enumerator(Vegas.Project.MediaPool);
    while (!mediaEnum.atEnd()) {
        var media = mediaEnum.item();
        var effectsEnum = new Enumerator(media.Effects);
        while (!effectsEnum.atEnd()) {
            var effect = effectsEnum.item();
            if (plugInName == effect.PlugIn.Name) {
                media.Effects.Remove(effect);
            }
            effectsEnum.moveNext();
        }
        mediaEnum.moveNext();
    }
} catch (e) {
    MessageBox.Show(e);
}
```

Custom scripts are found on the disc included with this book.

A button can be created on the toolbar making scripts immediately accessible. To insert a button, select TOOLS | SCRIPTS and locate the script for assignment to a button. Assign the script to one of the ten script location points.

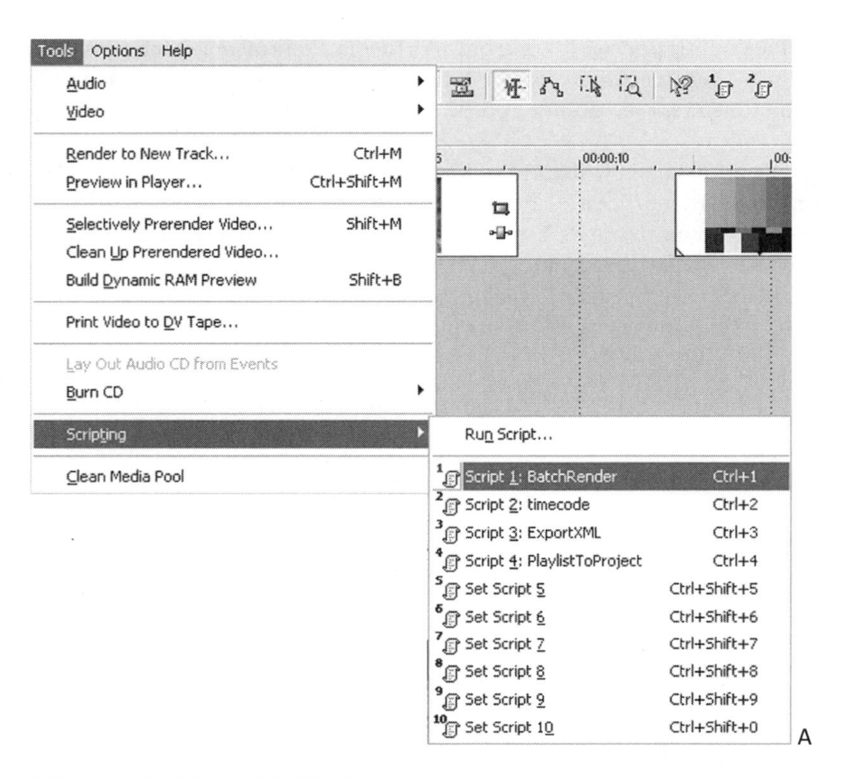

3.82 Assign a script to a shortcut (a), which can be assigned to a button on the toolbar (b).

Using the Customize toolbar dialog, assign a key shortcut to a button on the toolbar. The script is now a one-button operation. Up to ten scripts can be loaded on the timeline at one moment.

Excalibur is a scripting interface that may be purchased from a third party. It enables as many as 19 scripts to be found on the toolbar at one time. Excalibur may execute multiple scripts at one time, such as multicamera edit functions, audio normalization, gap removal, and more. This tool is available at http://www.excalibur.com.

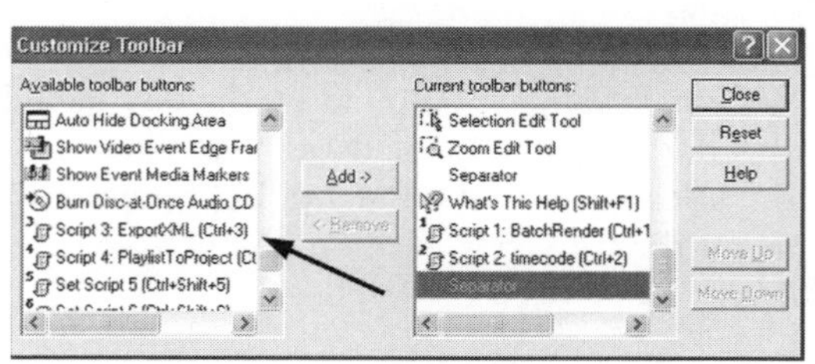

3.83 Loading scripts to a toolbar button makes running scripts fast and easy.

Options Menu

This series of menus are predominantly for setting behaviors and functions of Vegas to specific needs. Many are related to video-centric, and others are not.

The Quantize to Frames option generally should be checked for most video editing situations; it should be unchecked if Vegas is being used as an audio-only tool. This option forces video in an event to snap to a frame boundary, preventing a trim from occurring in the middle of a frame, regardless of the grid or time settings. Pressing ALT+F8 turns the Quantize to Frames action on or off.

Selecting the Enable Snapping option causes events to line up or snap to grid markers, other events, cursor positions, or inserted markers. This feature is useful when a user wants to ensure that timing is correct in the recording studio environment, that no frames are missed in the video editing environment, and that timing is accurate in placing events on the timeline. Snapping can be enabled via the menu or by pressing F8 to turn this feature on or off. Turning Snapping off allows media to be slid around in time if necessary. Snapping is typically left on, except in instances where sound effects or video events should not fall on a marker, grid line, or inaccurate cursor placement.

The Snap to Grid (CTRL+F8) and Snap to Markers (SHIFT+F8) choices are dependent on whether Enable Snapping (F8) is on or off. If snapping is turned on and either of these selections are enabled, events will snap to the marker or grid line to which they are dragged closest. If no grid line or marker is present, an event will snap to the closest event to its release point when dropped or moved on the timeline.

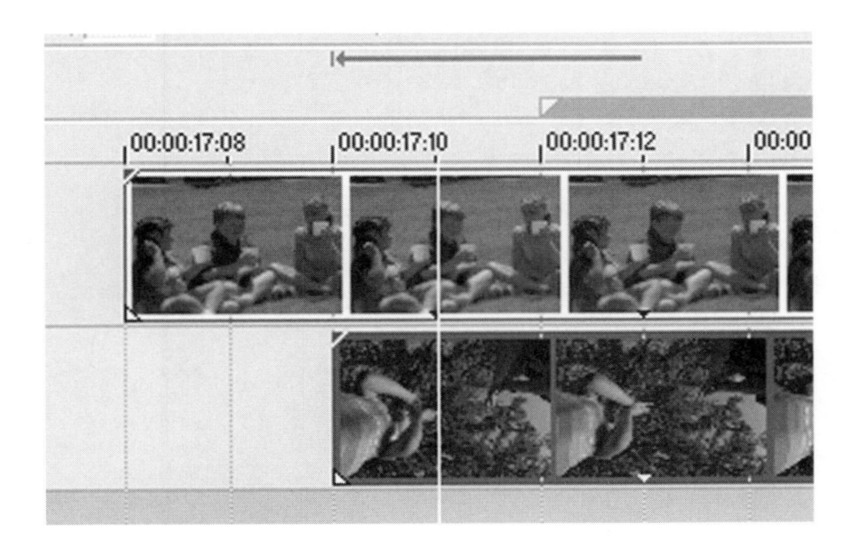

3.84 Notice the events snapped to frame-level grid lines.

Snapping can be enabled to music tempo, video frames at selectable frame rates, or absolute times.

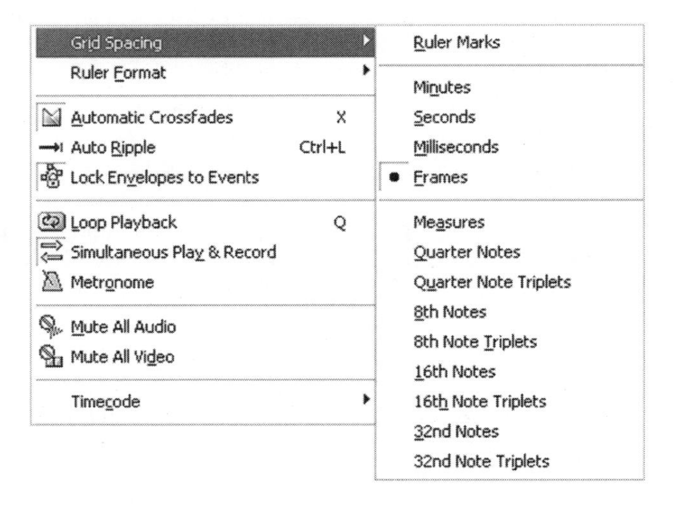

3.85 Assign grid spacing to personal preference or to a house standard.

Grid Spacing options allow for a variety of gridlines to be selected, which is dependent on the user's working methods. This selection controls how tightly or closely spaced gridlines are placed on the timeline.

Recording studios or musicians might want to use musically based timings, such as tempos and beats, to lay out a grid. Video editors might desire the grid to be frame-based at whatever frame rate they can be editing at, such as 25 fps for PAL or 29.97 drop-frame for NTSC DV. Musical grids can be laid out as tightly as 32nd note triplets, while video grids can be laid out as tightly as single frames.

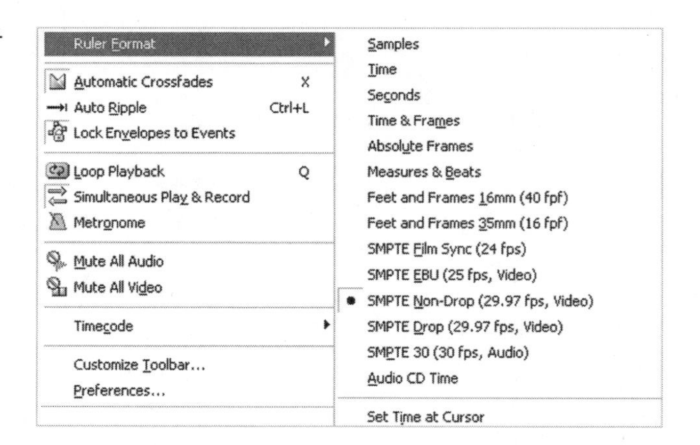

3.86 Set rulers for the manner in which the grid is viewed for snapping, alignment, or time selection.

Different from the Grid Spacing options, the Ruler Format options determine how grids are laid out. If different forms of ruler formats are chosen from the selection of grid spacing, such as choosing a frame-based ruler while a beat-mapped grid is selected, the grid lines disappear from the timeline. Vegas cannot make a distinction between incompatible ruler and grid formats, leaving it to the user to define these settings. Studios recording music to picture will probably want to use the more common beat maps but can switch back and forth between grid forms, helping calculate tempo to time for musical cues.

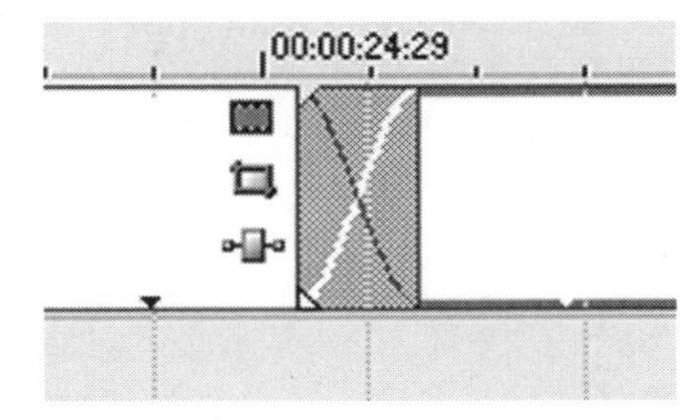

3.87 Crossfades are indicated by the "X" between events.

The Automatic Crossfades option, selectable by menu or the X key, defaults to on. When two events cross over each other, Vegas creates a cross-fade, indicated by an "X" in the middle of the space in which they are crossed.

Automatic cross-fades are generally left on. Events crossed over each other with automatic cross-fades turned off become hard cuts.

NEW! Auto Ripple (CTRL+L) is a new feature found in Vegas 4.0. This feature allows the trimming of an event edge. Subsequent events snap into place to fill the hole left by the trim, or if the event is pulled longer, Vegas auto-moves all events down the timeline for the same duration as the pulled event's time addition. If media is pasted in place and auto ripple is turned on, Vegas moves all subsequent events down the timeline to accommodate the inserted event. By the same behavior, if an event is deleted, auto ripple slides all events forward in time to fill the space left by the deleted media, ensuring that whatever amount of time is added or deleted, it ripples in or out events following the changes.

This new feature also exhibits other behaviors, in that it can be track-dependent or project-dependent. Further, events can be shuffled on the timeline, with all events moving with shuffled events.

Notice that when "A" is moved down the timeline with auto-ripple turned on, "C" stays snapped to "A" and moves down the timeline for the same duration as "A" was moved down the timeline.

Tip

Be aware that if inserting media underneath existing events with auto ripple enabled, all events are shifted in time for the length of the media event insert. If video events are on the timeline, disable auto ripple before inserting media. Press CTRL+L to enable/disable auto ripple.

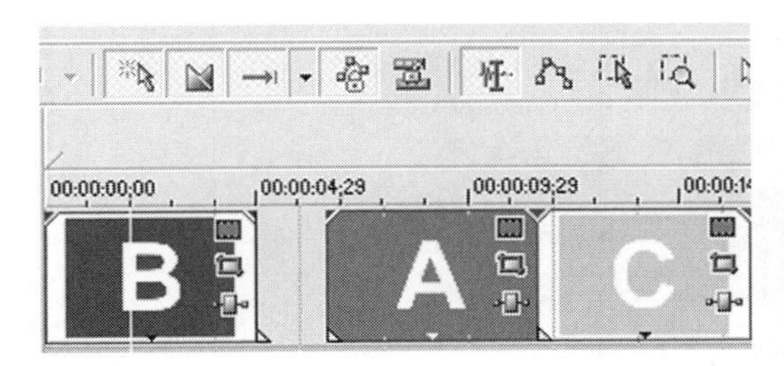

3.88 Layout of media before a ripple is executed.

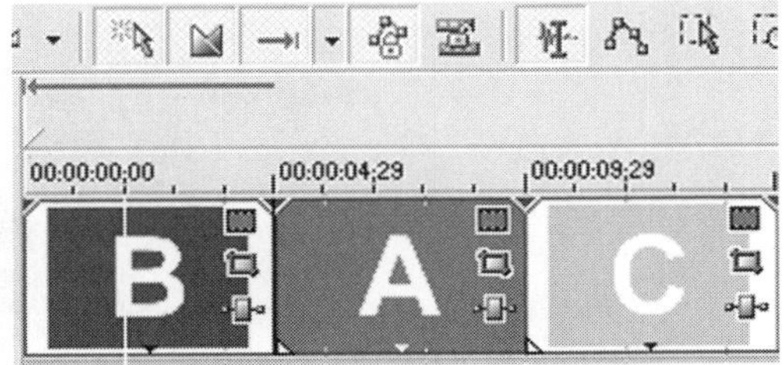

3.89 Layout of media after a ripple is executed.

Selecting the Lock Envelopes to Event option causes volume controls, panning envelopes, bus envelopes, compositing envelopes, and velocity envelopes to move on the timeline with their corresponding events. If 80 audio events are on the timeline with 200 envelope changes, if this feature is disabled, and if any one of the events is moved, the audio fades will be incorrectly timed.

The Loop tool is another one of Vegas' most-used tools and can be accessed through the menu or pressing Q. This feature allows users to draw a selection on the Marker bar and have the cursor loop over that selection over and over, while edits are being performed. Using this tool is an excellent way to tweak transitions, color correction, panning, track motion, and other edit effects and tools, while seeing the events change in real time on either the internal or external monitor.

Tip

Use the Loop tool during editing of critical spaces while watching video on the computer monitor rather than the external monitor. As Vegas loops through the selection, it draws a RAM render automatically, increasing the resolution and frame rate as it loops. Let a section loop while having a cup of coffee or on a phone call, and within a few short moments, the entire section will be playing smoothly and clearly. Be sure to have enough RAM specified in the OPTIONS | PREFERENCES | VIDEO dialog.

The Simultaneous Play & Record option, if enabled, allows audio to be recorded at the same time as audio is being monitored. This feature is critical for a recording studio. The majority of sound cards are full-duplex, which means that they allow audio to pass in both directions at the same time. Most laptop sound cards are half-duplex, not allowing audio to be bi-directional. In this case, the feature requires disabling to avoid error messages. If a laptop is used for monitoring and recording, an external sound card/device is usually required via FireWire, USB, or CardBus.

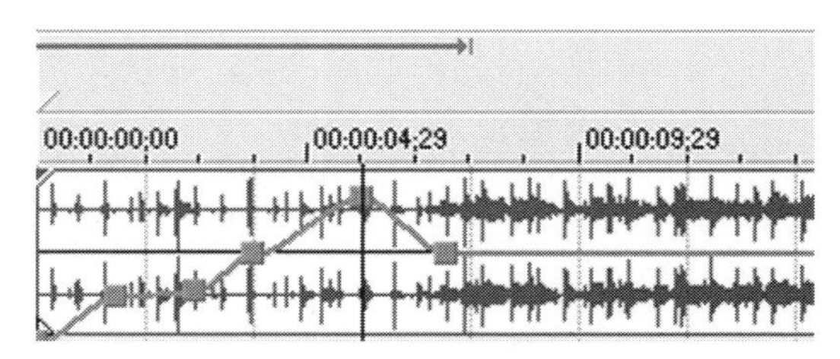

3.90 Original envelope.

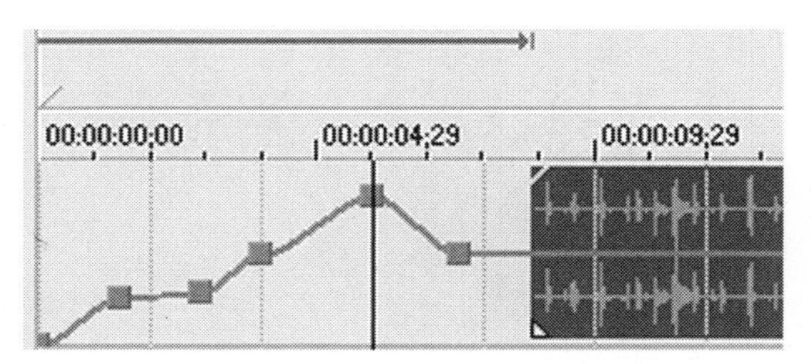

3.91 An event that is not locked to envelopes moves without affecting the position of volume/pan/FX.

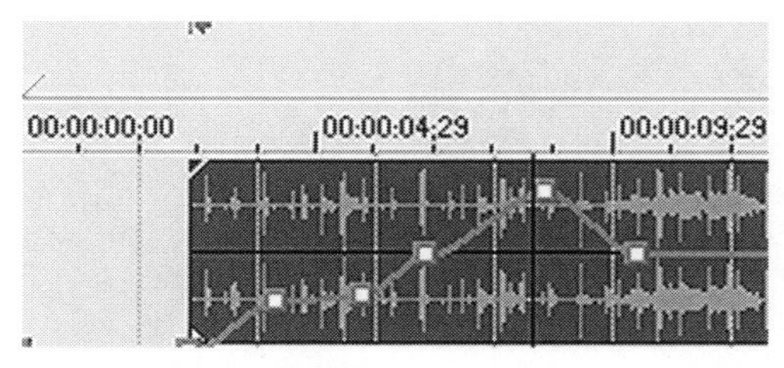

3.92 A locked envelope moves all markers; volume/pan/FX changes with it.

The Metronome feature is an audio metronome that can count time at the tempo and beat settings specified in the FILE | PROPERTIES | AUDIO dialog. This tool is handy for both musicians and video editors. The metronome is not recorded to hard disk if used during the recording process.

The Mute All Audio option is exactly as its name implies. All audio on the timeline is muted/silenced. Deselect this feature or select a track directly to unmute audio.

Selecting the Mute All Video option has the same effect, with the exception that it causes all video tracks on the timeline to go dark. Individual tracks can be unmuted and viewed or deselect this menu to bring video back to the screen.

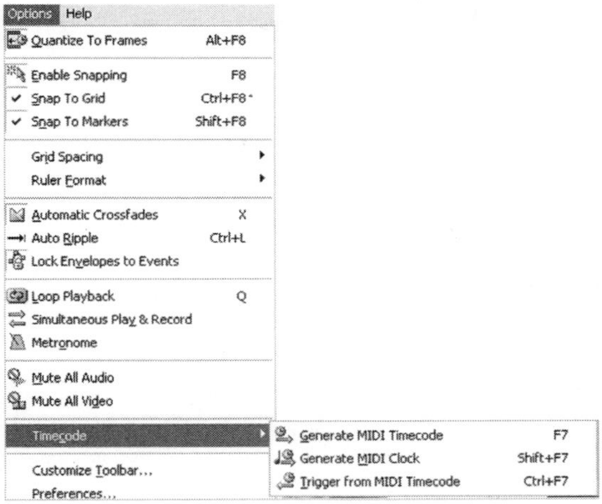

3.93 Select timecode options to fit hardware or house requirements.

Selecting the Timecode option calls up a submenu, asking what and where timecode is to be read/generated.

Timecode can be generated for MIDI devices. MIDI time code (MTC) is read by nearly all software applications and many hardware tools. MTC is used for synchronizing software applications together or synchronizing hardware and software together.

Vegas generates MTC each time playback is started.

Vegas also has the ability to generate Midi clock. The MIDI clock feature is different from MTC in that it also contains song pointer position (SPP) and carries tempo information as well. Musicians will want to use this feature, as it allows Vegas to carry increasing or decreasing tempos and the slaved application or hardware will then increase or decrease in tempo according to the tempo map laid out in Vegas.

Vegas can also be started/stopped via the Triggering option from timecode. An external device that generates timecode, or that feeds a generating device that can accept SMPTE timecode and convert it to MTC, can be a master device, and Vegas can be its slave.

Tip

Registered users of Vegas can access Sonic Foundry's website and download the virtual MIDI router (VMR) utility at no cost or get it from the Vegas installation disk. The VMR utility allows Vegas to drive other timecode-capable applications such as ACID or triggering Sound Forge. This option is great for syncing a music bed that isn't completed in ACID to the Vegas video timeline.

Vegas users can customize the toolbar to suit their specific needs and desires for Vegas' appearance with this toolbar. Video editors might wish to hide audio-only tools, just as audio-only studios might desire to hide the video tools found on the toolbar in Vegas. In any event, Vegas can be set up to look and feel the way any user chooses.

The OPTIONS | PREFERENCES dialog is the primary place in which the look and feel of Vegas can be set to individual preference.

3.94 Adding a custom tool to the toolbar is the same for all tools as shown in the Scripting section.

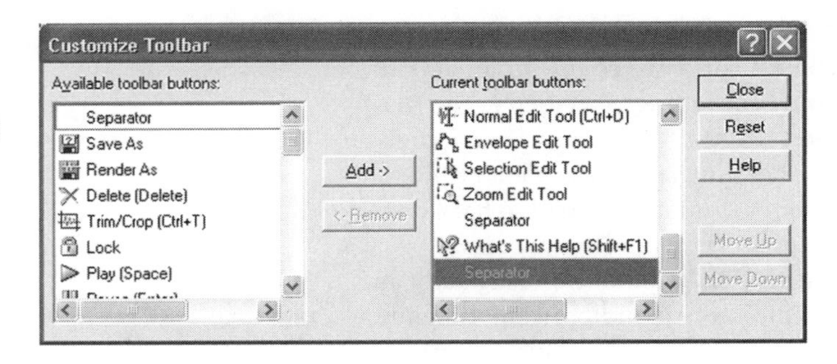

Select preferences that suit your workflow. Experiment with settings that work best for you. Each tab has several options relating to General, Video, Audio, CD, Editing, Sync, Video Device, and a new tab in Vegas 4.0 for Audio Hardware.

Tip

Make sure that volume and pan envelopes are of greatly differing colors if the default colors are not used, so that they can be differentiated at a glance when mixing in Vegas. Applying color schemes that are similar in style, such as making all EQ gain envelopes varying colors of green, bandwidth envelopes varying colors of blue, and so on help in rapidly assessing a mix and determining what envelopes are performing various functions.

3.95 Set Vegas 4.0 to feel the way you'd like it to feel in the Preferences dialog box.

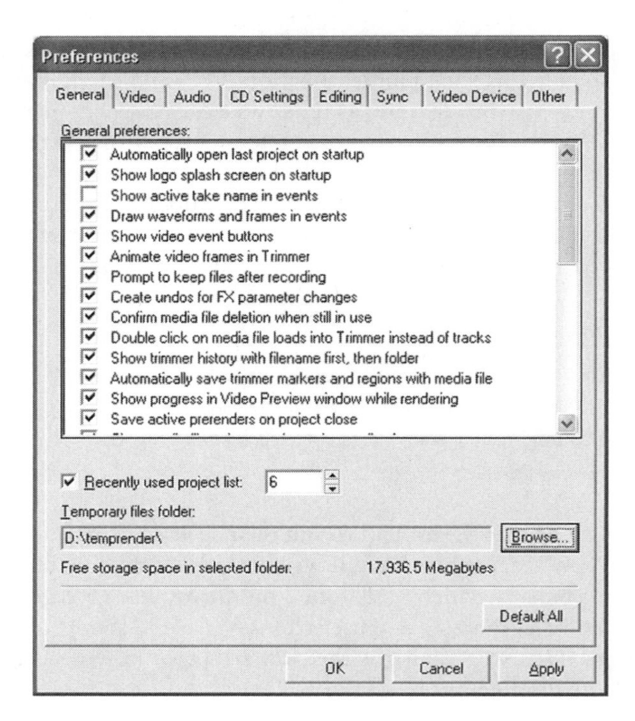

Capturing Video

Getting Media from the Camera to the Hard Drive

Capturing media from a camera is the first step in the video editing process. Vegas has a unique application that functions from within Vegas but can also function outside of Vegas as a separately accessible tool. A separate capture icon can be placed on the desktop if so desired. Browse for the `Video Capture40.exe` application or find it directly at `C:\Program Files\Sonic Foundry\Vegas 4.0\` for most installations. (You might need to search for the `Vegas40.exe` file if Vegas was not installed to the `C:\` drive.) It is highly recommended that you launch the capture tool from within Vegas. If launched from Vegas, media captured in Video Capture is added automatically to the Media Pool in Vegas.

Transferring video from tape to a hard drive is called *capturing*. In the OHCI world, this process is different from the hardware-assisted capture card world as no compression or altering of the video takes place in the transfer. With M-JPEG cards or hardware-assisted cards, media is compressed or has header information that some editing applications can't read. Fortunately, while Vegas can't read every form of video file captured in the Microsoft Windows environments, it can read nearly every form of video file captured in the Windows environment if the proper video for Windows (VfW) codec is installed on the machine. For M-JPEG AVIs, you might need to download and install an M-JPEG codec. Matrox and Main Concept are good sources for these codecs. However, the Video Capture utility only controls cameras through the OHCI cards. Vegas cannot take advantage of hardware-assisted codecs, such as those offered by Matrox, Pinnacle, and Canopus, although Vegas can edit media captured with the capture utilities that come with those cards.

Video capture enables media from the camera to be placed on the hard drive. This process works with live or pre-recorded media. Vegas uses a standard called open host controller interface (OHCI) to control cameras and to facilitate the transfer of video from the camera to the hard drive. Vegas functions with some analog capture cards but consult your owners' manual for more information. To control the start, stop, rewind, fast-forward, and record functions of a digital camera, however, the only card Vegas functions with is the OHCI standard. Cards manufactured by ADS, SIIG, Unibrain, as well as the new OHCI-compliant card from Canopus, the ADVC1394 card, are all compliant examples. Analog or non-OHCI cards do not have the ability to control a DV camera from within Vegas.

Vegas automatically senses the frame rate, image size, and any other information that might be part of the DV stream. No special settings are required for capturing DV from a DV camera via an OHCI card. Whether the media is PAL, NTSC, 24p, 16:9 anamorphic, or other supported format, it is captured, or transferred, from the camera with those attributes recognized.

Manual Capturing/Capturing Entire Tape

Connect the camera to the OHCI/FireWire card installed in the back of the computer. If a shortcut is not on the desktop, open Vegas and select FILE | CAPTURE VIDEO, which opens the Vegas Video Capture utility. The capture utility can also be opened from the Media Pool by clicking the Capture Video button.

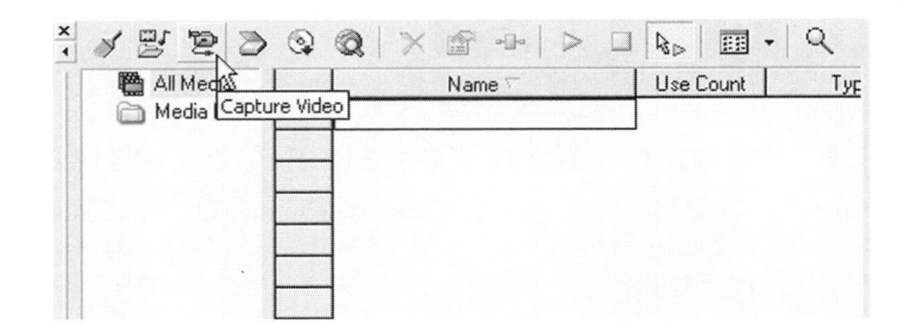

4.1 Opening the Video Capture utility from the Media Pool.

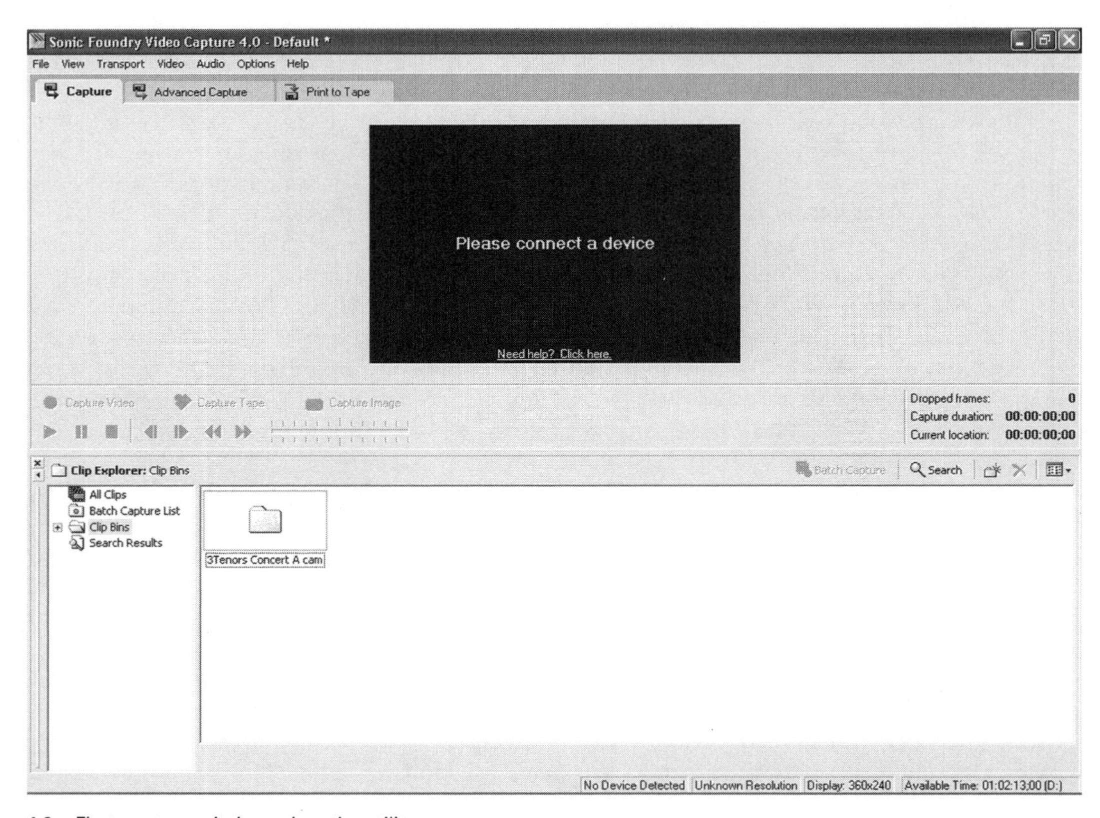

4.2 First capture window when the utility opens.

In some systems, the camera is not auto-detected. If this happens to you, go to CONTROL PANEL | SYSTEM | HARDWARE and look for the 1394 Bus Host controllers device line. Double-click the line, and the properties of the bus will open, showing the installed drivers. In some instances, a Microsoft driver and another driver, such as a TI, Lucent, or VIA driver will be present. Right-click any driver that is not a Microsoft driver and uninstall it. Repeat for all non-Microsoft drivers.

Some 1394 cards are installed as network cards, which can cause conflicts with the camera and bus operations. If this becomes an issue, open the network adapter hardware and disable the 1394 NIC/Network card. This process should not affect other networking tasks or abilities.

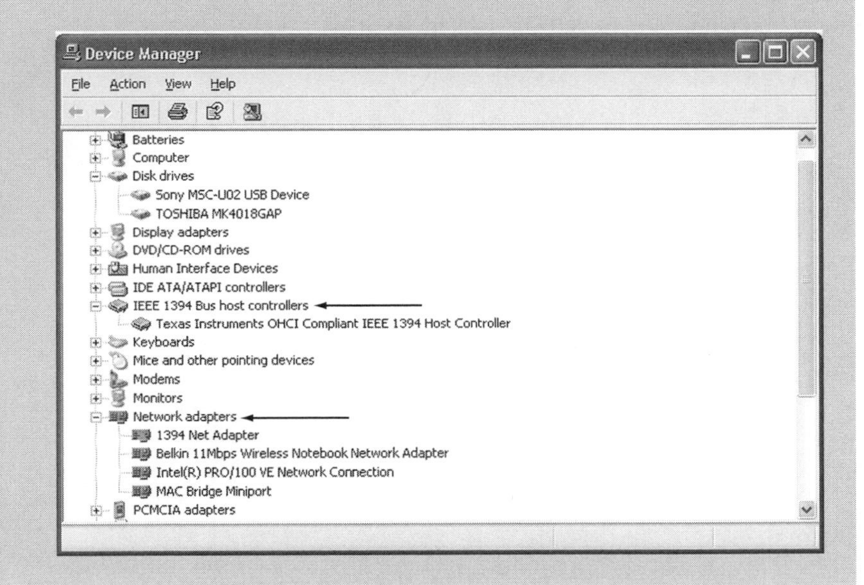

4.3 Open the CONTROL PANEL | SYSTEM | HARDWARE tab to adjust 1394 and Network properties.

When Video Capture launches, select **OPTIONS | PREFERENCES.** Select the Disk Management tab. The default capture location for captured files is `C:\Documents and Settings\username\My Documents\`.

If the camera is turned off, Video Capture will prompt you to turn it on.

When the camera is turned on, Windows will automatically open screens asking what it should do with the camera. Cancel the Windows screens, as Video Capture is already open. If you select anything else, it may override the Video Capture drivers, and then you'll need to start again by shutting down the newly launched application and recycling the camera power. If Video Capture and/or Windows doesn't auto-sense the camera when it is turned on, be sure the camera is in videotape recorder (VTR) mode.

4.4 Windows automatically senses most DV cameras.

If a capture drive has not been specified, Video Capture will ask for a drive to be specified. Unless absolutely necessary, never capture media to the same hard drive as the operating system. Specify a hard drive and specify a disk overflow size that is at least three percent less than the overall drive size. This option will allow room for defragging drives and will prevent a drive from filling to total capacity, possibly rendering the drive incapable of being recovered in the event of a drive or system crash. Default overflow is 360MB. Larger drives can require a slightly larger overflow, such as 500MB.

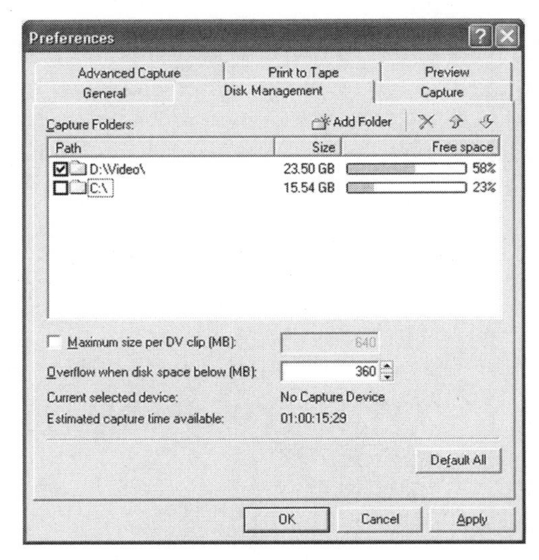

4.5 Specify capture drive. Notice that Vegas indicates capture time available based on selected hard space.

Video Capture will prompt for a tape name with the Verify Tape Name dialog. This step is the first in proficient media asset management (MAM) or digital asset management (DAM). Enter the name of the tape you wish to capture. This tape name will be appended to filenames as they are captured, so be sure it's correct. You will have one more opportunity to correct this easily later on. This process is known as *logging*.

Tip

Enter the name of the tape exactly as it's written on the tape spine in the event that you want go back and recapture the media. This process avoids errors in logging and locating tape.

The Verify Tape Name dialog also provides choices as follows:

- "Don't capture any clips right now"
- "Start capturing all clips from the current tape position"
- "Start capturing all clips from the beginning of the tape"

4.6 Be sure to label tapes correctly at capture to assure quick location of clips during editing.

For the moment, select "Don't capture any clips right now" and click OK. Video Capture is now ready to capture video.

Clicking the Play button or pressing the SPACEBAR causes the camera to begin playing. Video is not captured until you click the Capture Video button or press CTRL+R.

Video Capture can capture video from the camera/tape in several ways, including:

- Manual capture by starting and stopping (labor/time intensive)
- Capture from a specific location and detect or not detect scenes, starting and stopping with the control (users' preference, can be labor/time intensive)
- Capture entire tape with scene detection (little effort)
- Batch capture logged clips (fair amount of effort to log clips; time saving/disk space saving in end view)

Tip

For a rapid-start capture:

1. Connect camera to OHCI card.
2. Turn camera on.
3. Launch Video Capture.
4. Choose Capture Entire Tape.

Scene detection is a feature in which Video Capture sees breaks in the date and time stamps created when the camera is started and stopped during the recording process. Each time the camera is stopped, Video Capture sees the change in the date and time stamp and starts a new file. So if you've started and stopped the camera ten times while videotaping at an event and then connected the camera to the computer and opened Video Capture, Video Capture creates ten files in the media pool when it is activated. (Files are stored in the folder/drive specified in Disk Management. Media Pool creates a pointer to that file; files are not actually stored in Media Pool.)

Tip

When using DV tape, be certain that timecode isn't broken on the tape. Broken timecode often happens when reusing old videotape or when cameras aren't allowed to run for a moment following shooting. This problem also happens when tape is viewed in the field and allowed to play past the end of timecode. Avoid this problem by:

- Always recording blank time completely over previously used videotape by recording with the lens cap on for the length of the tape.
- Allowing 5–10 seconds of tape to roll by when recording in the field, if you know you will be reviewing tape in the field.

Video Capture is capable of batch capturing an entire tape while left unattended. Click the Capture Entire Tape button. Video Capture will rewind the tape to its beginning/top and then start to capture the tape to the folders/drives specified in the Disk Management preferences. If the specified hard drive does not have enough space to store an entire tape, be certain to specify more than one drive. Video Capture will automatically roll files over to the next specified drive.

Tip

Hard drives formatted as FAT 32 drives may not have files larger than 4GB, or about 18 minutes in tape time. Video Capture will automatically find the best point to divide files in the event of FAT 32 drives. NTFS drives have a file size limitation of 4TB, which is roughly 330 hours of DV.

Begin capturing tape by either pressing CTRL+R or clicking the Capture button in the capture tool. Video Capture will start capturing from the moment you click the button. If the camera or tape deck is OHCI/1394 standard, the device will start playing and transferring media from the tape to the hard drive. Clicking the Capture Entire Tape button instructs Video Capture to rewind to the beginning of the tape and start recording. Video Capture will auto-detect scenes in the tape and create new files for each scene. (Enable Scene Detection is enabled by default in the OPTIONS | PREFERENCES | CAPTURE dialog.)

In the Video Capture OPTIONS | PREFERENCES | CAPTURE tab, set the minimum clip length to 5 or 10 seconds. This instructs Video Capture to ignore the OOPS!-type of files, in which a camera was inadvertently started and stopped, leaving small video clips on the tape. Of course, if working with animations or other projects that require short shots, you might wish to leave this set unchecked, which is the default setting.

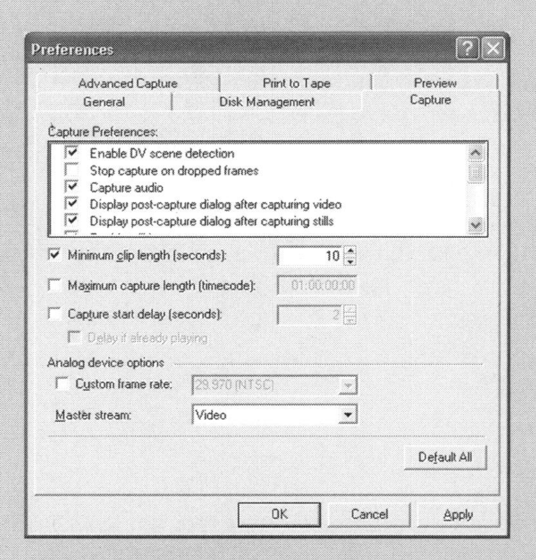

4.3 Setting the minimum clip length.

Animation-type capture is also possible by selecting a maximum capture length of only one or two frames. Experiment with clay models or action figures to get the hang of editing this type of captured media.

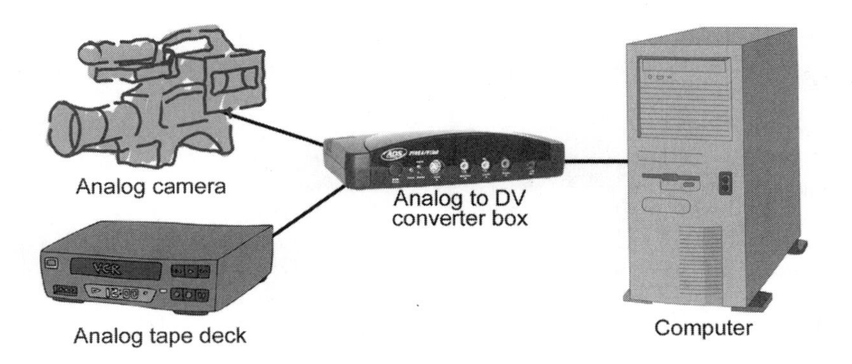

4.8 An analog camera or video deck, such as a VHS or Beta SP machine, uses composite, S-Video, or component video output to a DVC, which converts the analog video to digital video so that the computer can see the video signal as data.

In the analog tape world, such as a VHS or Beta SP, a digital video converter (DVC) is required to convert analog to DV. Several converters are available in the market today, from the very high-end Laird Blue Flame to the Canopus ADVC 1394 card. YUV input, composite input, and S-Video input are all common input features, making it fairly simple to find a converter that meets your specifications.

If an analog machine is used with a converter or analog card, the machine must be turned on and play enabled before capturing video. Scene detection does not take place with analog tape machines passed through a converter. Machine control is not possible with analog machines, with the exception of some high-end converters that read DV control signals and translate them to RS-422 control signals to which some analog decks will respond. DV control can be disabled in the OPTIONS | PREFERENCES | GENERAL dialog. Typically, leaving DV control enabled does not affect a DV capture but can create some confusion if shifting between DV and non-DV sources.

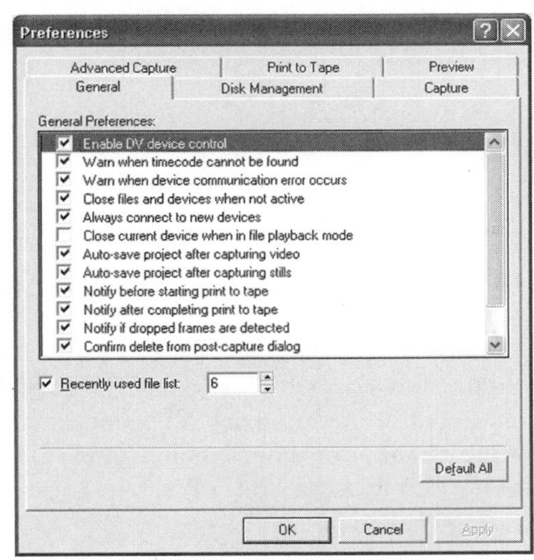

4.9 Disable the DV control when working with non-DV decks or analog capture cards.

When Video Capture has captured the entire tape or when you have manually captured all clips desired from the tape, Video Capture will prompt you to save the capture session. This feature is useful if there is any possibility that you will be going back to the tape to recapture at some future point. Lost video, accidentally deleted video, and failed hard drives are all reasons to recapture tape, so it is generally good practice to save the capture session.

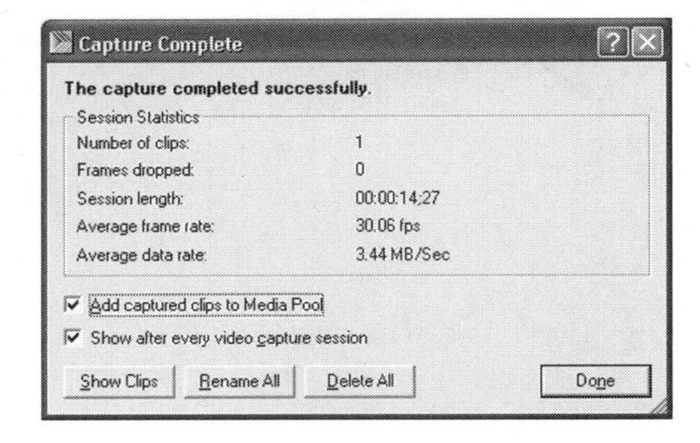

4.10 The post-capture dialog provides information about clip length, dropped frames (if any), location of media storage, frame and data rate, and check boxes to place media in the Vegas Media Pool. Unchecking the "Show after every video capture session" option will prevent this dialog from popping up after each capture.

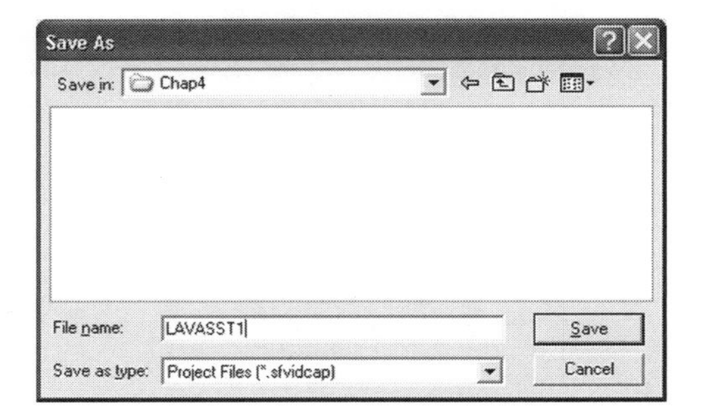

4.11 Store the capture session if any likelihood exists that the same tape will need to be recaptured later or if the session needs to be shared on a network.

After capturing video and closing the Video Capture application, the captured media will appear in the Media Pool in Vegas 4.0.

Video Capture has other tools available for logging and capturing in Vegas. Click the Advanced Capture tab. This tab opens a different view of Video Capture. Comments, ratings, length of clip, and in/out information can all be logged and added in this view. Use this feature to keep track of media and how it appears for rapid editing decisions later in the editing process. These comments appear in the Media Pool and can assist in making editing choices.

Advanced Capture Tools

The Advanced Capture tab has dialogs that allow for more advanced file management. Open the Advanced Capture tab at the upper left of the capture utility. Detailed capture logging tools are located on the right side of the capture screen.

Using Batch Capture

Batch capture only functions with OHCI devices. To use the Batch Capture tools in Video Capture, click the Advanced Capture tab. Begin playing tape via the control seen on screen.

To mark the point on the tape where you want capture to begin, click the Mark In button.

To mark the place on the tape where capture should end, click the Mark Out button.

Video Capture will note the in/out times in the windows next to the Mark In/Out icons and will show the clip length in the Clip Length field.

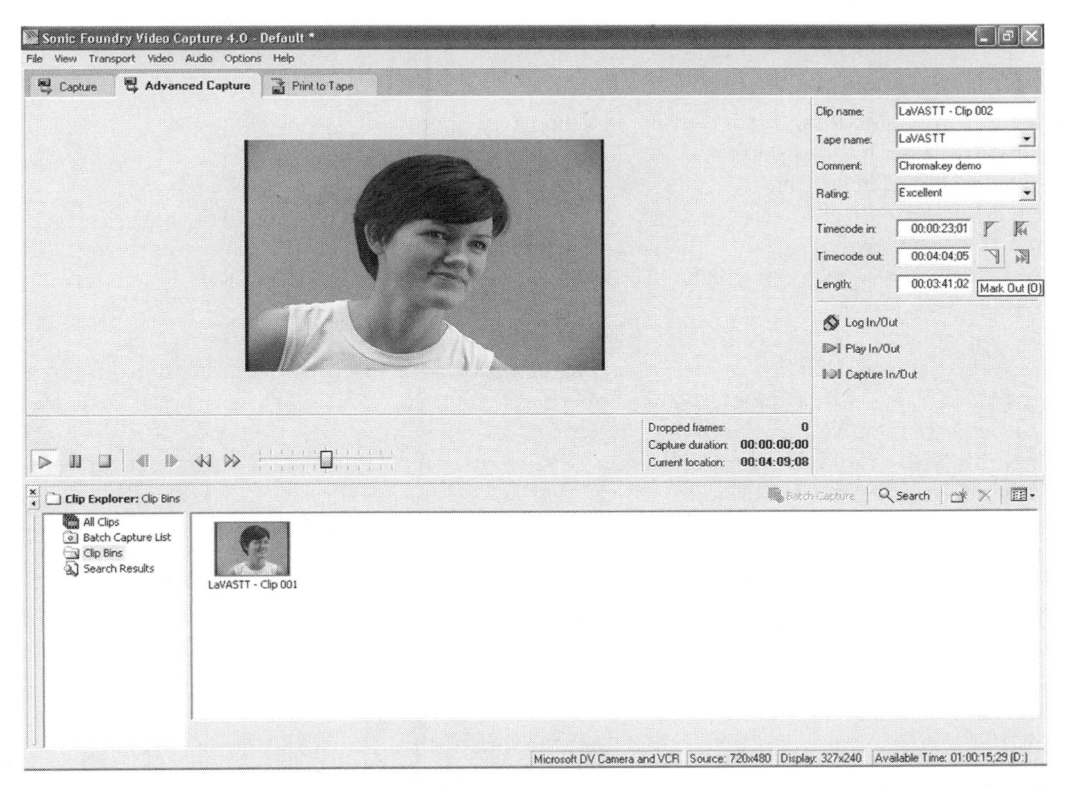

4.12 Advanced capture tools offer efficient capture logging and automated capture tools, saving tremendous time in the editing/capturing process, in addition to saving disk space.

The Mark In/Mark Out dialogs contain two buttons for returning the tape to the in or out point. Selecting the "Cue To In" (SHIFT+I) or "Cue To Out" (SHIFT+O) buttons will cause the tape deck or camera (T/C) to auto-locate to the selected in or out points indicated in the In/Out dialog boxes.

To log the clip, click the Log In/Out button under the Length field. Video Capture creates a note based on the tape deck or camera in/out markers and displays the clip in the current bin in the right pane of the Clip Explorer. Change in/out markers as many times as necessary to correctly mark the clip. The clip is not logged until you click the Log In/Out button. Only the displayed time in the T/C in and T/C out fields are used to mark the clip for batching later on.

Mark as many clips as necessary. Multiple tapes can be logged at once. Be sure that tape names are changed in the logging dialog so that Video Capture knows when to prompt for a new tape to be put in the deck.

As clips are marked for batch capture, an icon of the capture will appear in the list view/bin window. These icons have a red dot in the center, indicating that they are marked for capture. When all clips are logged, click the Batch Capture button found above the list view/bin window. Clips can be automatically marked for batch capture by selecting the OPTIONS | ADVANCED CAPTURE check box.

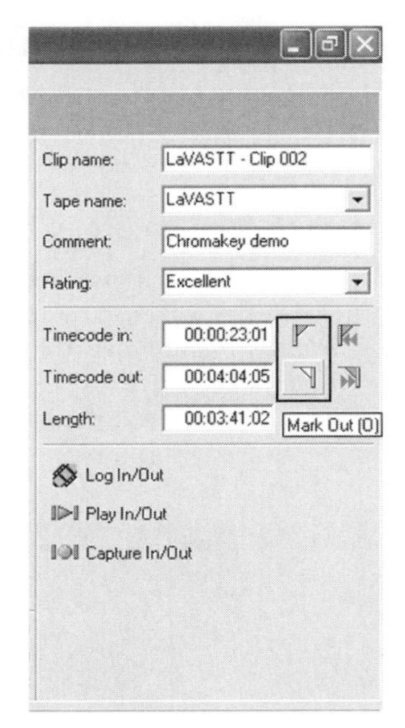

4.13 Use the Mark In (I) and Mark Out (O) buttons to mark timecode points for a batch capture of selective clips.

In this same dialog box, clips can be commented on and rated. Comments assigned to an individual clip can be viewed in the media pool after the clip is identified in Vegas 4.0. The new Media Pool search tool allows Vegas to search for keywords based on these comments, so while this feature is convenient, it's fairly indispensable for long-form work where rapid searches for specific clips are needed. It's also good media management, even if it takes a bit of time to do.

Video can be graded on the quality of shot, helping to assign the priority clips for later identification and location.

The Capture and Advanced Capture dialogs provide the ability to create and use Clip Bins. Each master tape or B-Roll might have its own bin, and clips can be sorted in any fashion that makes sense to the project editor or producer. Remember how Radar O'Reilly stored the maps to the minefields on *M.A.S.H.* under "B" for "BOOM"? Each person has their own method of sorting and storing files. Media bins accommodate this. Right-click the Clip Bins folder, select Create New Bin, and name the bin with the filename information that relates to the part that the clips will play in the project. If the "Add Clips to Media Bin" check box is selected in the post-capture dialog, clips are auto-added to the media pool bins in Vegas 4.0, and bins from Video Capture can be dragged to the Media Pool in Vegas as well.

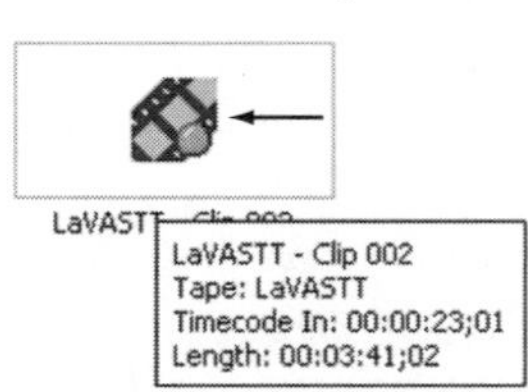

4.14 The red dot on the clip indicates that a clip is marked for batch capture.

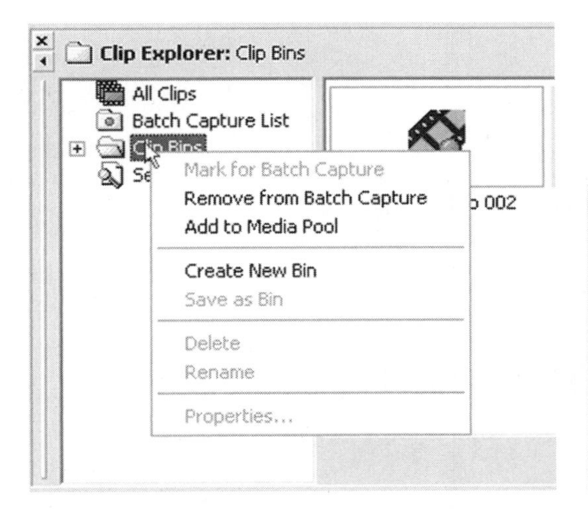

4.15 Right-click the Media Bin to create a new bin to aid in locating, managing, and logging media.

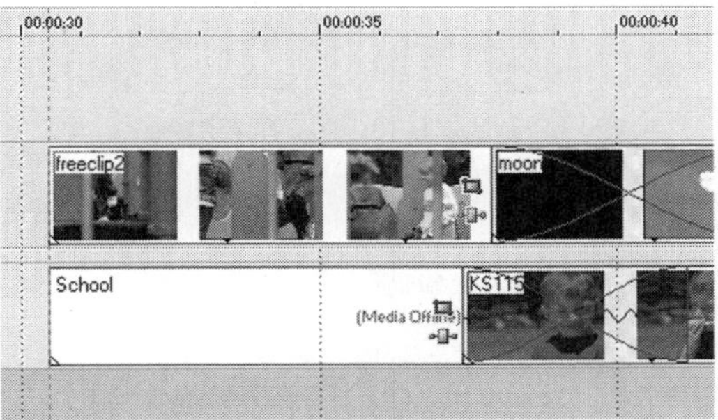

4.16 When video is lost, is moved to a different folder, or is corrupted, but is associated with an event on the timeline, the event will provide a placeholder on the timeline that can be replaced with a take or recaptured from the DV tape.

Bins from the Video Capture utility can be imported to the Media Pool in Vegas by creating the bin in Vegas before capturing media.

4.17 To recapture DV media with Video Capture, right-click in the Media Pool in Vegas and select "Recapture All Offline Media."

In the NLE world, if a clip is on the hard drive, it is considered to be online. If the clip disappears from the hard drive but is part of a project, it is considered to be offline, and Vegas shows the file as offline in the Media Pool, shown as grayed out or missing media with the text "Offline Media" in the lower-left corner of the Media Pool. Vegas provides the opportunity to recapture offline media by right-clicking the offline media and selecting "Recapture Offline DV Media."

Recapturing media from the Media Pool launches the Video Capture application automatically, and you will be prompted to place the correct tape in the deck/camera if you have not already done so. This point is where careful logging/labeling is invaluable. Video Capture will seek out the timecode related to the clip that is marked for recapture and start the capture process with no assistance. The recaptured media will need to be directed to a folder on one of the drives.

Capturing Stills with Video Capture

The Video Capture tool allows stills to be captured and cataloged during the capture process. To capture stills, click the Capture Still button. Still images by default are stored as JPEG files unless otherwise specified in the OPTIONS | PREFERENCES | CAPTURE dialog. In this same dialog, Video Capture can be instructed to de-interlace video images and can be instructed to apply correct aspect ratio to still images. (Refer to Chapter 2 for aspect ratio information.) If the "Saved Captured Stills as JPEG" box is left unchecked, images are stored as uncompressed Bitmaps.

Stills captured to the clipboard can be opened in any photo editor for color correction, image correction, resizing, or other manipulation.

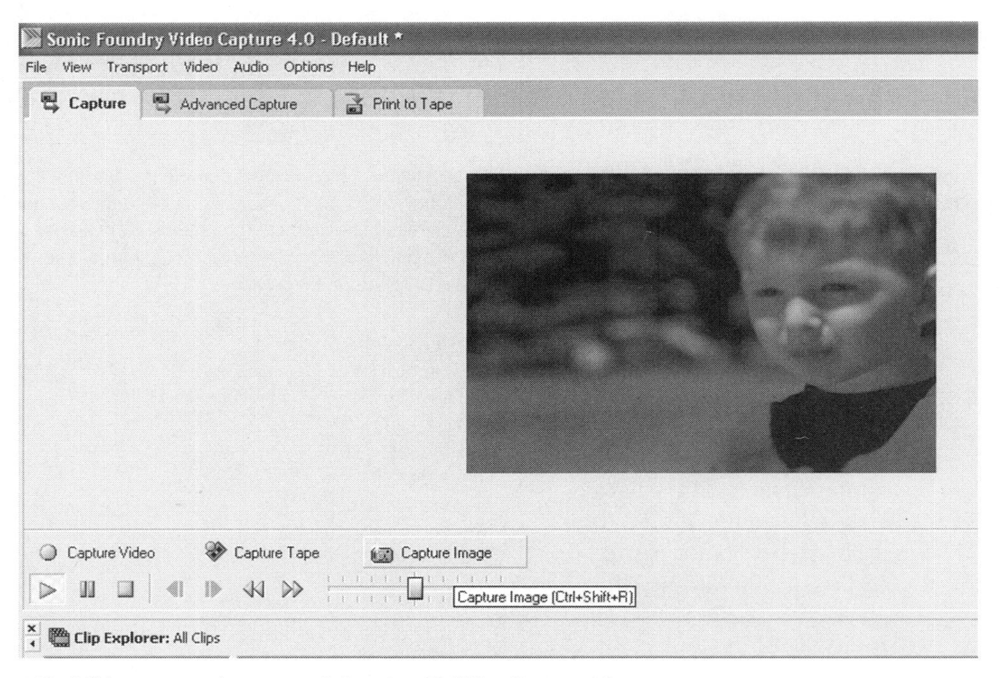

4.18 Still images can be captured directly with Video Capture 4.0.

Many video cameras have the ability to capture stills as well. Still images are generally captured to videotape and can be narrated during the still capture on the video camera. Video Capture treats these still images as video because the tape is moving and the still image generation is created by the video camera. Default still-image length on most DV cameras is five seconds. If Scene Detection is enabled, Video Capture creates a new file for each still photo taken with the camera.

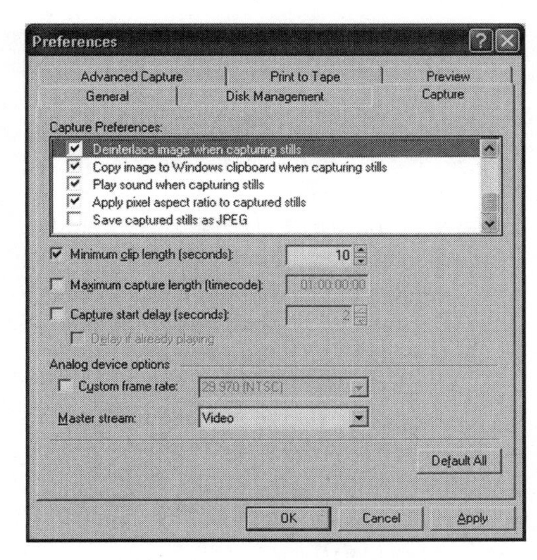

4.19 Specify preferences for managing still images in Video Capture and about viewing formats.

Changes in Video Capture

In the very near future, most DV cameras in the consumer and semi-professional markets will sport not only IEEE-1394 and S-Video connectors but will also have USB mini connectors (Type B) as well. This option will occur because of increased manufacturer support of USB2 as a data transfer protocol. All DV cameras currently have a FireWire/IEEE-1394 connector on them. Sony cameras label this output as an i.Link connection. Having both USB2 (Type B) connectors and 1394 connectors means that cameras will be able to connect to any type of computer, regardless of hardware configuration. Both formats operate reasonably the same as the other.

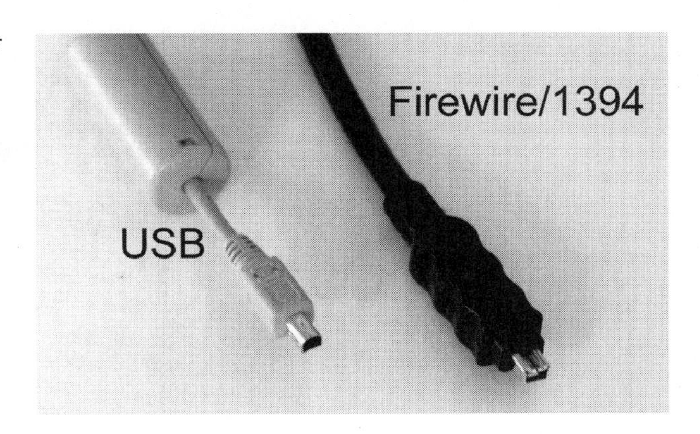

4.20 USB Type B connector and FireWire/1394 four-pin connector. Soon, all cameras will have both types of connectors.

Another major change coming in digital video is the advent of portable, battery-powered hard drives that will eventually be built directly into cameras. For the time being, these drives are belt-worn and have rechargeable batteries. Many of the units use the IBM TravelStar drives found in laptops. As drive speeds and sizes increase, these drives or similar storage mechanisms will be stored in the cameras themselves. The ADS Technology belt-clip drive is capable of holding up to five hours of DV media, with a built-in rechargeable battery. The drive housing has a built-in six-pin 1394 connector, which is connected to the FireWire output of the DV camera.

4.21 The ADS 2.5 drive kit has a self-powered housing with a six-pin FireWire connection.

4.22 The four-pin FireWire output from the DV camera plugs directly into the FireWire input on the drive kit. Video captured directly to the drive kit will not require capture, as it is already in digital format on the drive and merely needs to be plugged into the computer/editing system.

Editing Tools, Transitions, Filters, and Other Basic Video Tools

Importing Media to the Timeline

Aside from capturing media, many methods can be used to bring media into Vegas. Vegas is format agnostic and resolution independent, meaning that nearly any form of media recognized by the Microsoft Windows operating system can be imported to the Vegas timeline as an event.

Vegas has an Explorer that by default opens in the lower-left corner of the dockable tool space. This Explorer can be dismissed from view and recalled at any time by pressing ALT+1. The Explorer is a dockable tool that can be sent to a second monitor to preserve space on the timeline.

Open and view the Explorer. Any form of media that can be imported is shown in the Explorer. Media files, such as GIF, TGA, JPEG, BMP, PNG, TIFF, MOV, AVI, WMV, and MPG are all media file formats that can be opened in Vegas.

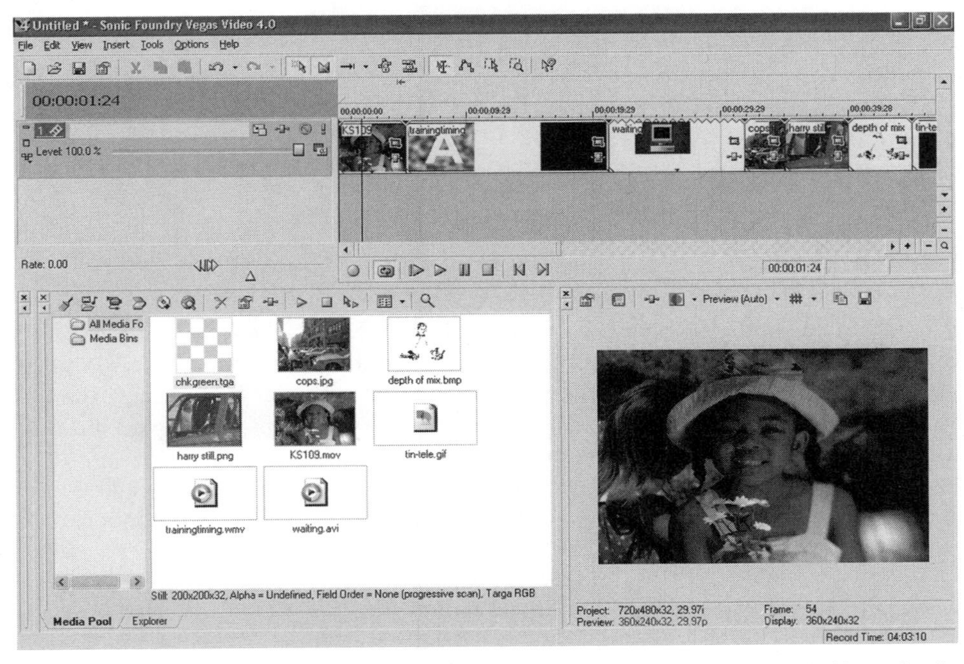

5.1 In this picture of the Media Pool, different media formats are placed as events on the Vegas timeline.

Tip

If QuickTime files are to be viewed and edited within Vegas, the QuickTime full install is required. Visit `www.quicktime.com` and download the current player. When installing, three choices are presented: Full Install (recommended), Custom Install, and Minimal Install. Choose full installation so that Vegas is able to access authoring **tools** for the QuickTime MOV file type. Without this installation, Vegas won't be able to access or edit QuickTime files or TIFF files.

All media formats may be placed on the same track, which means that an AVI file might be next to a JPEG file, which in turn might be next to a QuickTime file, which might be next to a WMV file, and so on. About the only Windows-recognized media format that cannot be opened inside of Vegas is the REAL (`rm.ram`) file format, as this format is protected by Real's protection algorithms.

To place media on the timeline, drag the media from the Explorer to the timeline, where the media becomes an event. If a track is not present, Vegas automatically creates one for the media that has been dragged and dropped. Double-clicking a file in the Explorer window automatically adds the file to the timeline as an event and creates a track for the file if a video track isn't already present.

To manually insert a new video track, right-click the track control pane and select INSERT | VIDEO TRACK or press CTRL+SHIFT+Q.

Media can also be opened in the Trimmer window. For those familiar with other NLE systems, the Trimmer window is much like a source window where media can be previewed and trimmed to the desired length/information before being placed on the timeline. To accomplish this task, right-click the media and select Open in Trimmer. Vegas can be set to open files directly in Trimmer, by going to the OPTIONS | PREFERENCES | GENERAL dialog and checking the "Double-click on media file loads into Trimmer instead of tracks" check box.

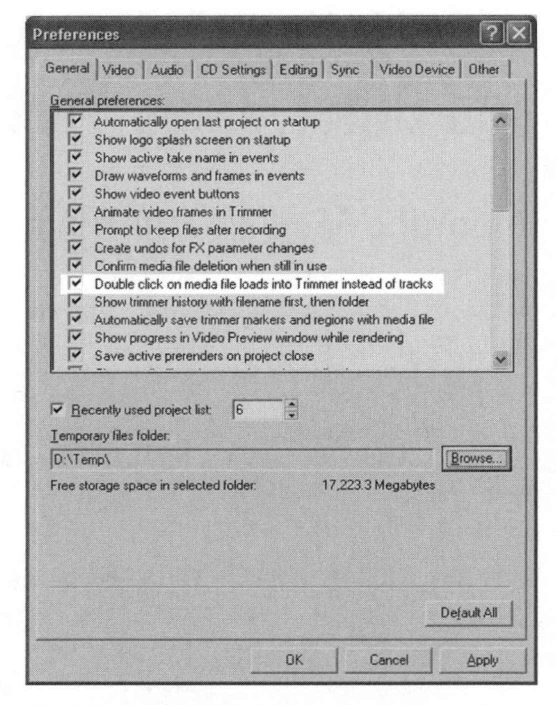

5.2 Selecting preferences for loading media in Trimmer.

Media can also be dragged directly into the Media Pool, which is where Vegas stores location information for all media used in a project. Media can be stored in the Media Pool regardless of whether the media is found on the timeline or not. Any event or piece of media placed on the timeline, however, is automatically added to the Media Pool.

NEW! Vegas 4.0 has new media management features in the Media Pool. The addition of media bins allows users to preset where media will be pointed to and how media is located. A master project bin resides in the Media Pool where all media is pointed to by default. This master bin is labeled "All Media Folder." Beneath it is a folder labeled "Media Bins." Right-clicking the Media Bins folder opens a menu/dialog that allows for new bins to be created and named.

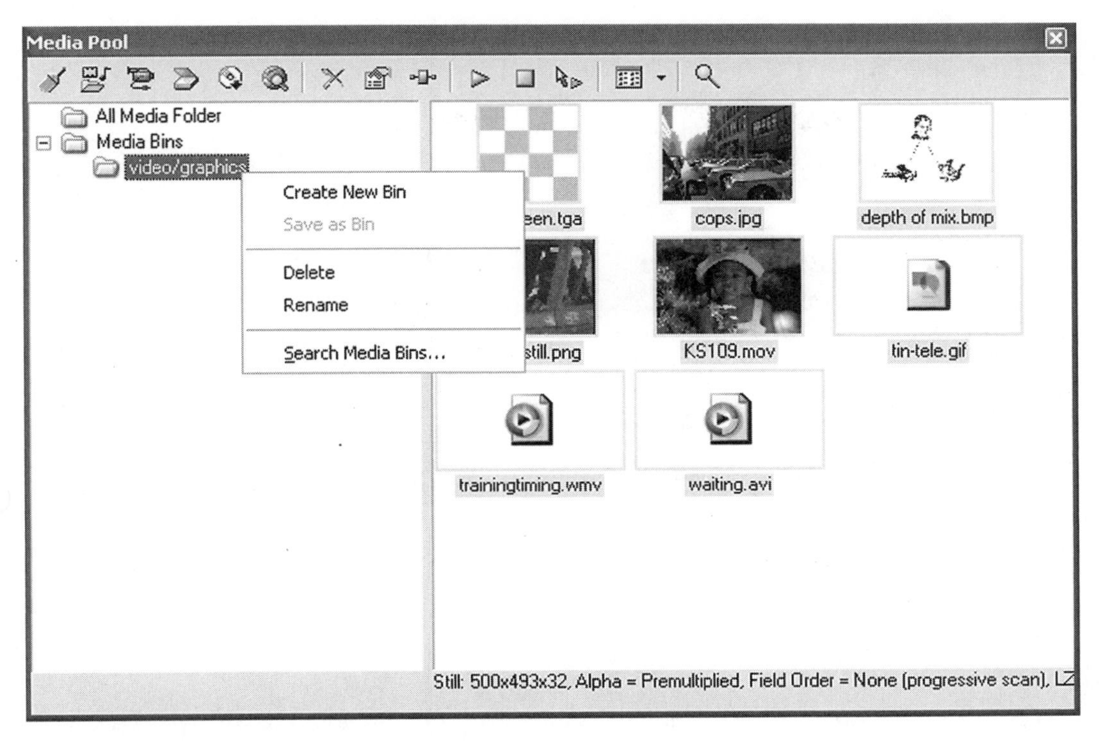

5.3 Creating and naming the media bins.

Be aware that Vegas does not copy media and place it in bins. There is no point nor value in having two copies of the same piece of media on the hard drive simply for the point of filling bins as some NLEs demand. On the contrary, Vegas points to a file from within the bin. While the media is always accessed from the bin, it truly resides wherever it was stored in the acquisition of the media. Bins are simply a means of collecting the storage points of all media and putting those indexes in one place.

Bins can be dragged from project to project in Vegas or stored as project bins so that they'll always be found within the preferences of Vegas. Audio, video, graphics, and anything else for which you would like to specify bins can be created, with sub-bins for every bin format. This way, you'll always know where media is stored.

Bins can be dragged from one open instance of Vegas to another open instance of Vegas.

Media can also be imported from other projects by selecting FILE | OPEN, choosing the VEG file from which media is to be imported, and checking the "Merge media from Vegas project files into current project" check box, as shown in Figure 5.5. Media is then automatically added to the Media Pool in the same folders from which the imported project media came. If the imported project has no bins specified in its Media Pool, Vegas will add media to the All Media Folder, where it can be dragged to specific bins within Media Pool for consistent file management.

Use bins to sort media by the name of the tape that media is captured from, a bin for bed music, a bin for various takes, or a bin for JPEGs, BMPs, or PNGs. Good project management starts with locating and knowing where all media is stored at all times. This process also assists in clearing out a project when it is finished.

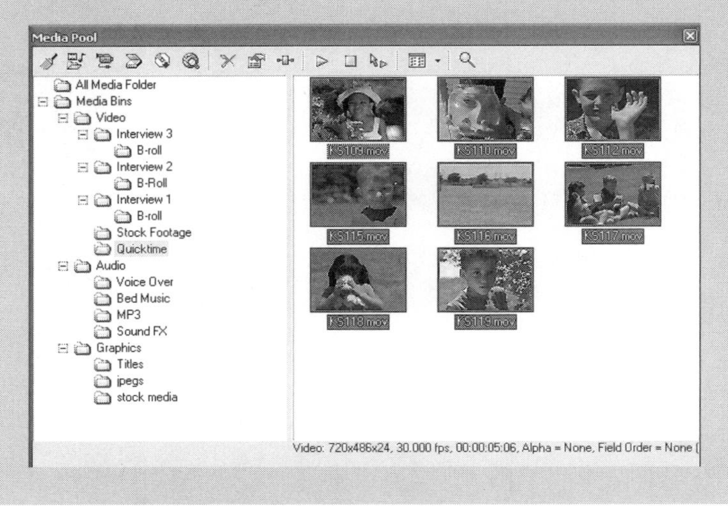

5.4 Bins separating various media types for efficient file location.

The Media Pool has several additional options for file import. Audio from CDs can be extracted directly from the Media Pool toolbar.

These options include the following:

- **Sweep Media Pool**—removes all media from Media Pool that is not in use on the timeline of current project.
- **Import Media**—imports media from other VEG files/Vegas projects to the Media Pool.
- **Capture Video**—opens the Video Capture application to transfer video from a camera to Media Pool.
- **Import Photo**—opens a scanner or camera attached to computer for importing photos to the Media Pool.
- **Extract Audio**—extracts, or rips, audio from a CD to the Media Pool.
- **Get Media from the Web**—provides a direct link to Sonic Foundry's partners, some of whom provide free media for your use.

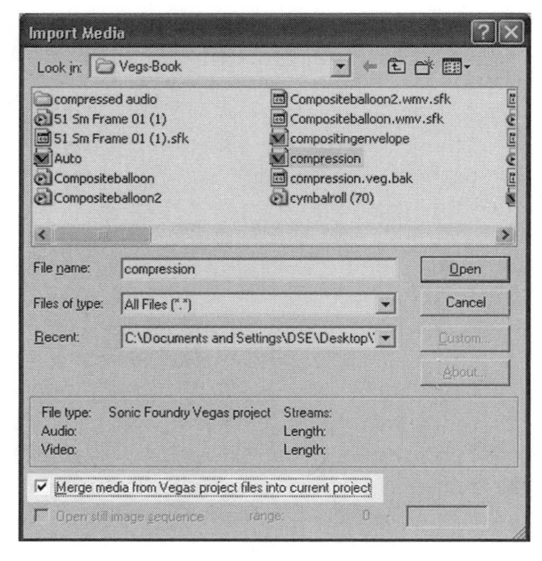

5.5 Importing media from other Vegas projects, which are either open or closed.

Editors coming from other NLE applications often ask about having subclips or selections appear in the Media Bins. Vegas does not have this function unless subclips are rendered as new files. Rendering straight cuts to a new file is faster than real time, because nothing is actually done to the clip. However, some people won't want to render subclips. For those, here is a workaround.

Open a clip in Trimmer. Make regional selections and name the regions. After selecting/naming all regions on a clip in Trimmer, right-click in the open area of the Trimmer and press S or select Save Markers/Regions, which will embed the markers/regions in the file.

In the Explorer window, enable the Regions view as shown here. Notice the regions created in the Trimmer tool are now visible in the Regions view of the Explorer. These regions will not show things such as effects or envelopes, but are a means of quickly locating selected regions especially if the regions have been named with a recognizable name and then sorted by name.

5.6 Enabling the Region view.

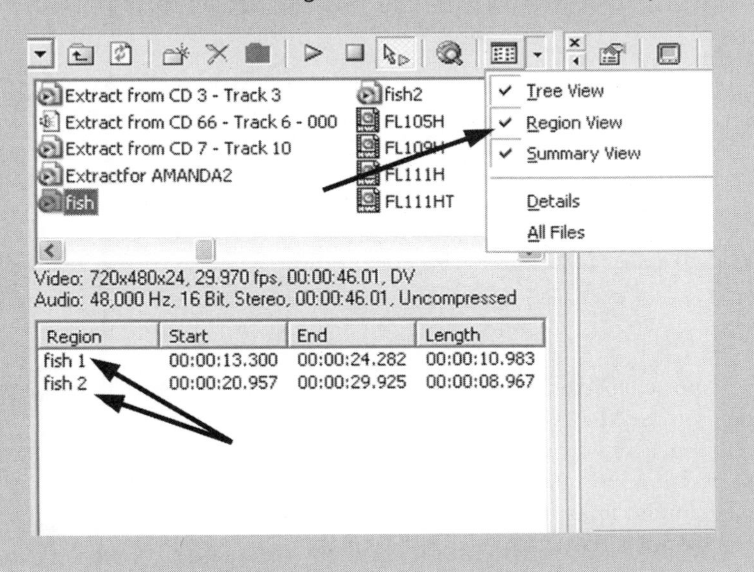

5.7 Import buttons from the Media Pool.

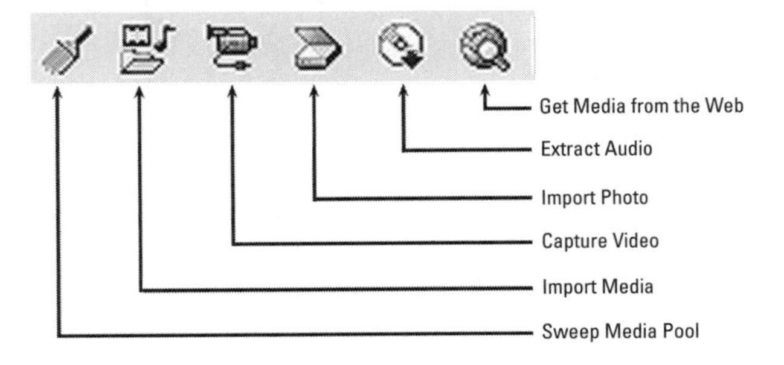

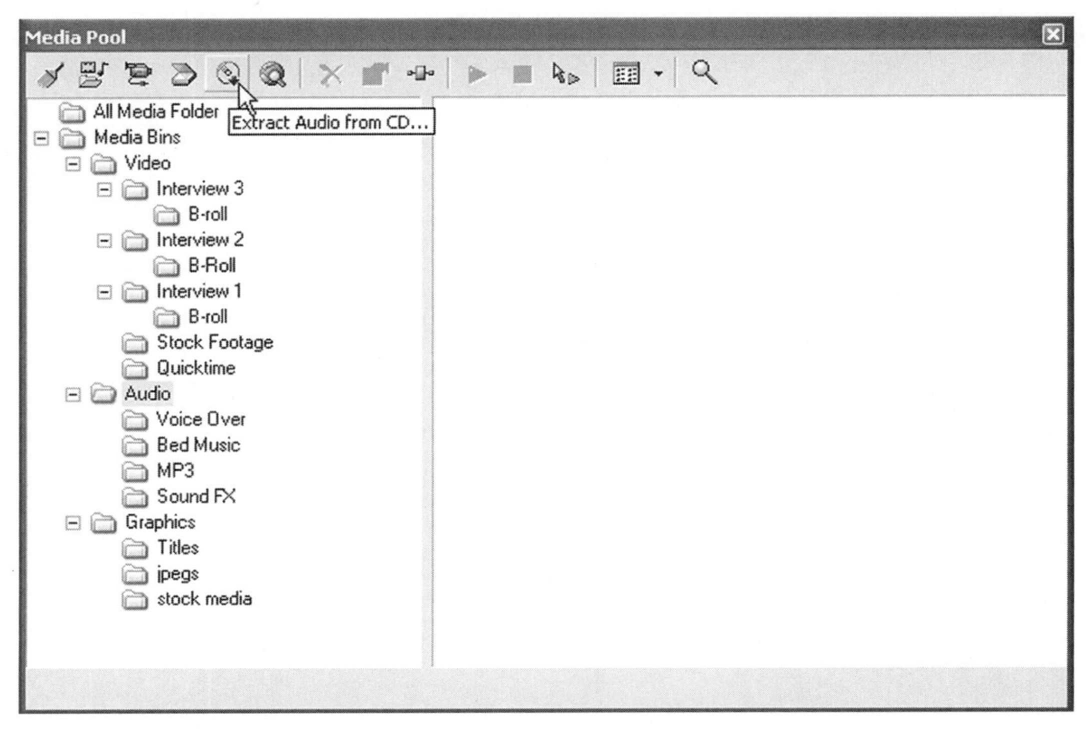

5.8 Choosing this button auto-extracts audio from a CD inserted in the CD/DVD player.

Imported media is placed in the All Media Files folder in the Media Pool and, once imported, can be dragged and dropped into a specific bin. Vegas supports import to the currently selected bin; bins only contain links or pointers to existing media.

Bins are also searchable, which makes cataloging files very valuable. Comments inserted in the Batch Capture/Advanced Capture dialog can be searched/located in the bin search operation.

Located media can be dragged and dropped into any bin in the Media Pool.

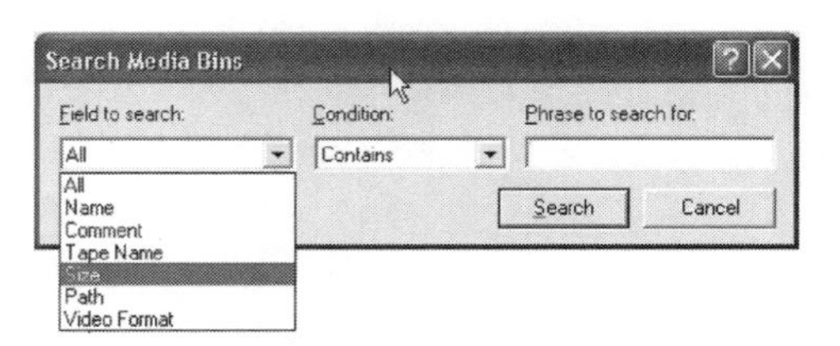

5.9 Search media bins to find elusive files that might not have been properly named or cataloged.

The Media Pool toolbar contains other useful tools for managing the Media Pool contents, including the following:

- **Remove Media**—removes media from the Media Pool or bins but does not delete media from the storage drive.
- **Media Properties**—opens a dialog indicating the type of media, storage location, length, size, and other properties of the media file.

- **Media FX**—allows FX to be placed on media before it's placed on the timeline. This option is very useful for single, long-length files that need the same format, such as of color correction or special FX, but will be trimmed in the editing process. Select the file to be affected, then press the Media FX button, select the desired FX, and adjust accordingly. This process does not alter the original file.
- **Start Preview**—plays/previews selected media file in the Preview window.
- **Stop Preview**—stops playback of selected media in the Preview window.
- **Auto-Preview**—when pressed, any time a media file is selected, it will be displayed in the Preview window automatically. This process is exceptionally useful when quickly going through media, saving time.
- **Views**—determines how media is shown in the Media Pool, whether as icons, detailed information, or a list.
- **Search Media Bins**—uses keywords or attributes to locate media whether captured or imported and does not work over non-mapped network drives.

Vegas 4.0 and the new media management features are helpful, powerful, and useful. Take the time to use them correctly, and projects will be much easier to manage.

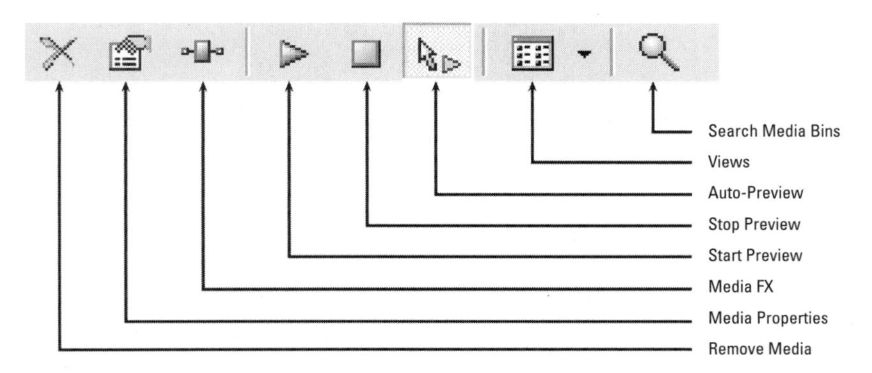

5.10 File management viewing attributes in the Media Pool.

Placing Media on the Timeline

The Media Pool is a great place to start building a project and makes for a much more efficient workflow. Getting media to the timeline from the Media Pool is well organized and easily traceable. The Media Pool is not the only way to work within Vegas, however.

Media can be dragged to the timeline directly from the Explorer within Vegas or from the Windows Explorer. Notice that the Explorer window also has an Auto Preview button on the Explorer toolbar, just as the Media Pool does. Leaving this checked allows rapid preview of various media formats shown in the Preview window. A search can be done in the Windows Find feature (START | FIND or START | SEARCH in Microsoft Windows XP) and dragged directly to the Vegas timeline.

What if you happen to have a long AVI file and only want selected portions of that file?

You can open the file in the Trimmer in one of two ways. First, you can right-click the file and select Open in Trimmer. Second, you can set the general preferences to automatically open a file in the Trimmer by double-clicking the file.

Notice that the file takes a moment to build an audio preview. This graphic represents the audio. This process can be sped up by checking the "Build 8-bit peak files" in the OPTIONS | PREFERENCES | GENERAL dialog.

Tip

The graphic drawing reference files are stored in Vegas (and all other Sonic Foundry products) as SFK files. These files may be deleted at any time. If a file is loaded that does not have a linked SFK file, Vegas will redraw the file and save as a new SFK file.

After the file is drawn in the Trimmer, select the portions of the media that are to be used. Place the cursor at the point that the video clip should begin and press I for In Point. Place the cursor at the point that the video clip should end and press O for Out Point. This process automatically creates a time selection in the Trimmer. Media can either be scrubbed over or the Play (SPACEBAR) button can be clicked in the Trimmer.

An alternative to pressing the I and O keys to select the in/out points is to draw the selection directly in the Trimmer. Place the cursor where the in point should be and draw by holding down the cursor and dragging the cursor to the out point. You can also work in reverse by placing the cursor where the out point is desired and dragging the cursor to where the in point should be. Notice that as the cursor is dragged the video plays in the Preview window. Dragging the cursor with the left-click button held down behaves like a scrubbing tool in the Trimmer window.

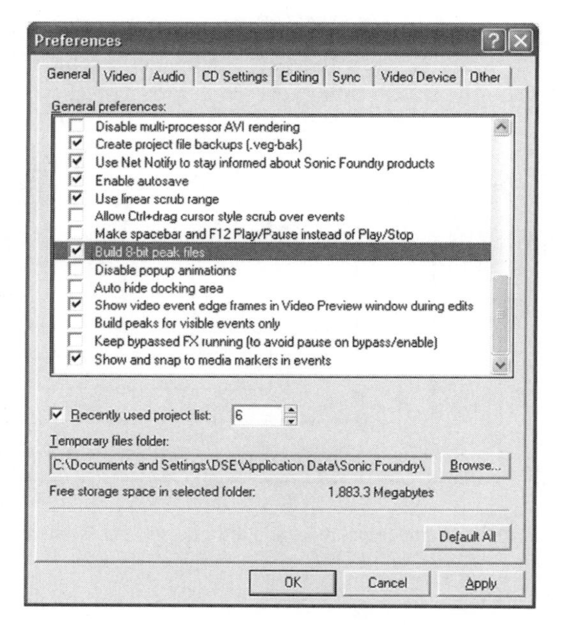

5.11 Reducing the size of screen peak draws will speed up screen draws in Vegas.

After a time selection is defined in the Trimmer, a few choices can be made as to how the media can be handled. Media can be placed directly on the timeline by pressing A for Add, it can be dragged directly from the Trimmer to the timeline, or a region can be created. If using the Add feature, place the cursor on the timeline where the added media should be placed. After adding the media to the timeline, the media becomes an event.

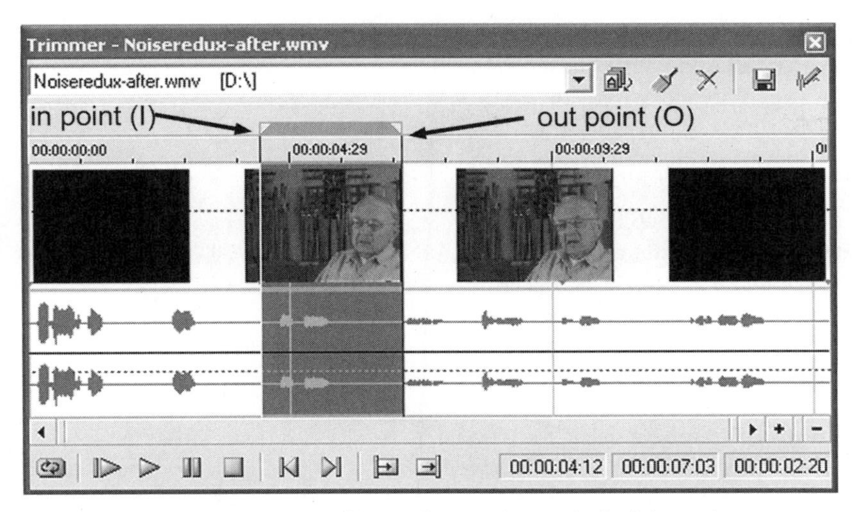

5.12 Time selection created in the Trimmer by pressing **I** and **O** for in/out points.

If no time selection is defined as a drawn area, as an in/out point or as a region, media can be selected from the start point to where the cursor is parked. By pressing **A**, media in the Trimmer is added from the start to the moment that the cursor is set. Alternatively, pressing **SHIFT+A** adds media up to the cursor to the end of the file.

When placing events on the timeline, keep a fast, visual reference to them by using the scribble strip on the Track Control pane. This feature also helps others quickly know what you are intending with the track. For example, a master track might be named as such, while a picture-in-picture track might be named "PIP track." Good track management is critical, especially in the digital realm where file names on a hard drive can be extremely varied.

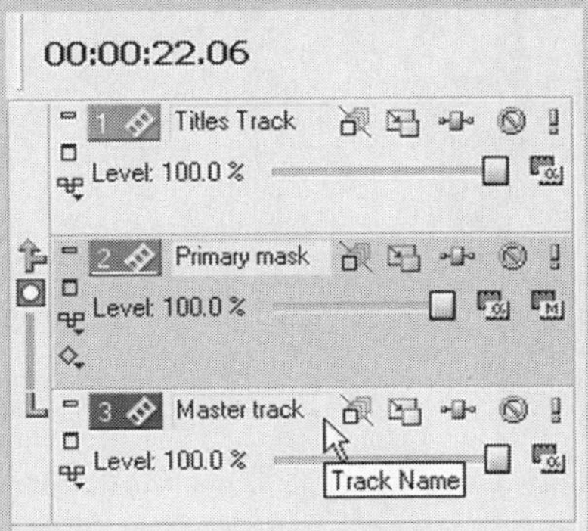

Regions are great, as a defined region will stay with the media even after Vegas is closed. To define a region, press R, and Vegas will open a field where the region name can be defined. Multiple regions can be defined, and region in/out points may overlap. When new media is loaded into the Trimmer and replaces the media containing regions, the defined regions will stay with the media file and will be shown any time a file is loaded into the Trimmer. (By default, this option is turned on in the OPTIONS | PREFERENCES | GENERAL dialog. It can be unchecked if for some reason this feature is not desired.)

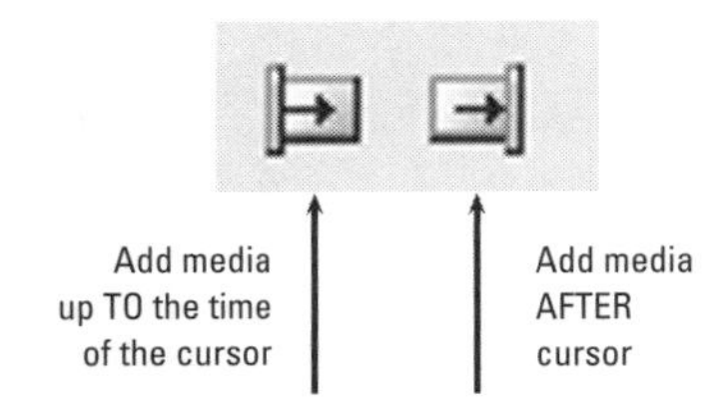

5.13 Buttons found in the Trimmer tool act as shortcuts for the **A** or **SHIFT+A** commands.

Markers can also be used in the trimmer to serve as reminders for in or out points. Pressing M will drop a marker. Markers can be named if a name is desired. Should an unnamed marker need to be named later, right-click the marker and choose Rename from the menu. Double-click between two markers to create a time selection from which a region can be named. A region may contain many markers.

NEW! Markers and regions are exceptionally important for many editors, especially those that are creating keyframes by the frame or in very tight visual spaces. Vegas 4.0 adds the ability to see markers or regions inserted in the Trimmer on the timeline for greater consistency in editing. If a region is created in the Trimmer and several markers are placed at desired points inside that region, these markers (and regions) are visible in events on the timeline. Markers and regions cannot be edited on the timeline if added in the Trimmer. Events can be reopened in the Trimmer, however, and edited there if necessary.

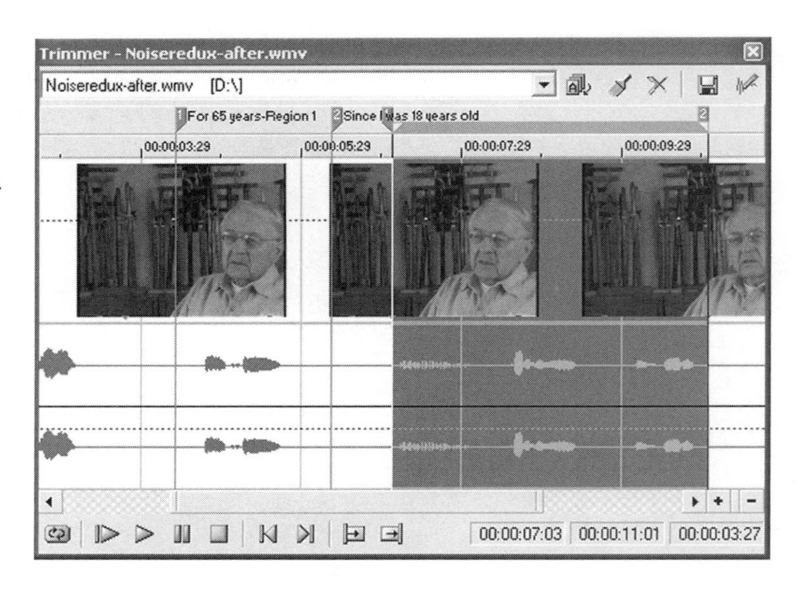

5.14 Regions may overlap and may also be named. No limit exists on the number of regions that can be created in the Trimmer.

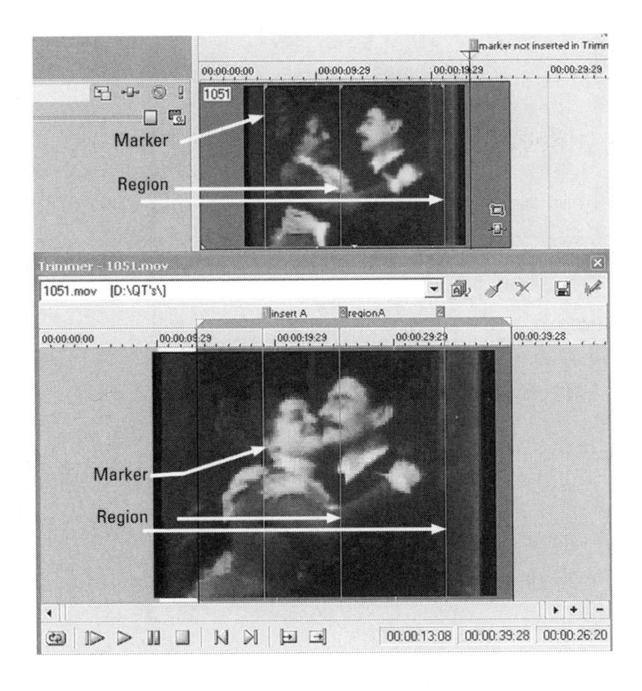

5.15 Markers/Regions inserted in Trimmer appear in events on the timeline.

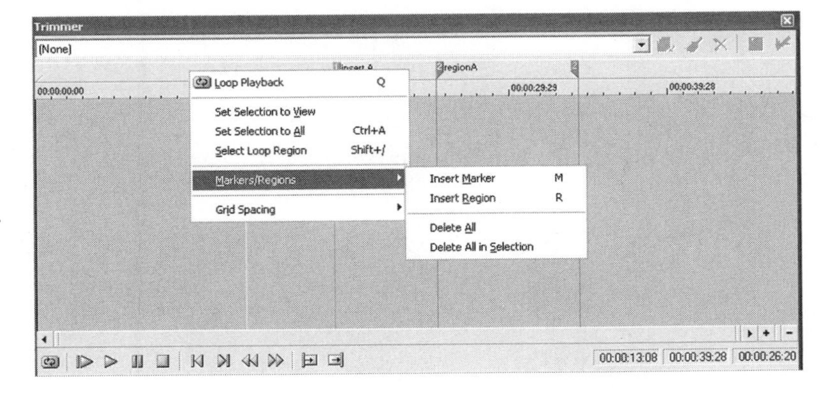

5.16 To clear markers/regions on events, open the event in the Trimmer, right-click the Marker bar, select Markers/Regions, and select Delete All or Delete All In Selection.

Placing Still Images or Graphics on the Timeline

Photos play a large role in creating video, regardless of the format. Vegas is capable of reading nearly every photo or image format that the Windows environment will allow. Some formats, however, are more optimal than others. Whether scanning via the Vegas Get Photo option that uses your scanner utilities or scanning outside of Vegas, the process is essentially the same.

Scan photos at a resolution of not higher than 300 dots per inch (dpi). Because video has a resolution of 72 dpi, extremely high resolutions are not only extra effort, they can also cause your image to look poor if the resolution is set too high. If no pan or crop will be done with the photo, a resolution of 72 dpi is sufficient. Generally, a resolution of 150 dpi is recommended. This resolution is a comfortable size for Vegas to manage quickly and yet provides the ability to pan or crop the image without fear of pixilation. If the Pan/Crop tool will be used to zoom in very tightly on the image, a higher resolution of 300 dpi may be called for.

Earlier in the book, we looked at pixels in the computer world as being square and pixels in the DV world as being rectangular. This issue affects stills and graphics that will be inserted into the Vegas timeline, as Vegas treats graphics and stills as DV when they are square-pixel images from the digital still camera, image editing application, or other source.

With most DV editors, still images and graphics must be created with the correct aspect ratio in mind when planning on dropping the stills or graphics into the DV timeline. Vegas isn't quite as rigid when it comes to this issue as the Pan/Crop tool allows the correct aspect to be maintained. When working with stills in an image editor, however, the correct size (NTSC) for a still image is 655 × 480. This size compensates for the square pixel being stretched in the non-square pixel world of DV. The math to derive this pixel aspect ratio (par) is as follows:

720 × .909 (par of DV) = 654.48

For those who like to round down, 654 works equally well. The correct aspect for PAL images is 704 × 576.

One experiment that demonstrates square versus non-square pixels is creating an image in your favorite image editor, such as Adobe Photoshop or Paint Shop Pro that is nothing but a circle at a project size of 720 × 480. Then drop the image straight into Vegas and don't use the Pan/Crop tool. Create the same image at 655 × 480, place that image in Vegas, and notice the difference. Users of other NLE systems might be familiar with saving files at sizes of 720 × 534/NTSC. For example,

534 × .909 = 485.4

This format works equally well, and either format is usable.

Tip

For best results, save still images, graphics, or titles created outside of Vegas at 655 × 480/150 dpi (NTSC) or 704 × 576/150 dpi (PAL).

Files should be saved as portable network graphics (PNG) files for optimum use in Vegas. Vegas reads JPEG, BMP, GIF, TGA, and most other formats, including the Adobe PSD file format, as well as the TIFF format. The Adobe Photoshop files, however, will not be broken down into individual layers; they will be shown as one layer. If an alpha channel is present, Vegas will recognize the alpha channel, and those areas will be transparent. TIFF files can be used in Vegas only if QuickTime is installed on the host system (`www.quicktime.com`). Because Vegas uses QuickTime to read the TIFF file format, TIFF files will slow down the rendering process somewhat. TIFF files also show problematic symptoms in automatically displaying alpha channels. PNG files are a lossless format, so the image is not compressed when saved in the image editing application.

Files may be scanned in at their normal aspect or size and then cropped within Vegas at any time to match aspect ratio, regardless of the original aspect ratio of the photo. This feature is a major timesaver. Sonic Foundry has created a script (see Chapter 12 for more information on scripts) that will match aspect ratio on all images on the timeline.

In Chapter 8, the pan/cropping section demonstrates how to match aspect ratio of images when they are not of the correct aspect in order to fill the screen.

Editing Events on the Timeline

Each piece of media placed on the timeline becomes an event. Events are then edited to form the final production. Events may be on single or multiple tracks.

Events containing video may be edited in any number of methods. One issue that exists with some NLE systems, however, is that a frame might be split. Vegas gets around this issue for videographers by offering a Quantize to Frames option, ensuring that frames are not split and forcing edits to occur on frame boundaries. When working with video, Quantize to Frames should nearly always be enabled by selecting OPTIONS | QUANTIZE TO FRAMES or pressing ALT+F8. When working with audio only, this feature may, and generally should be, disabled.

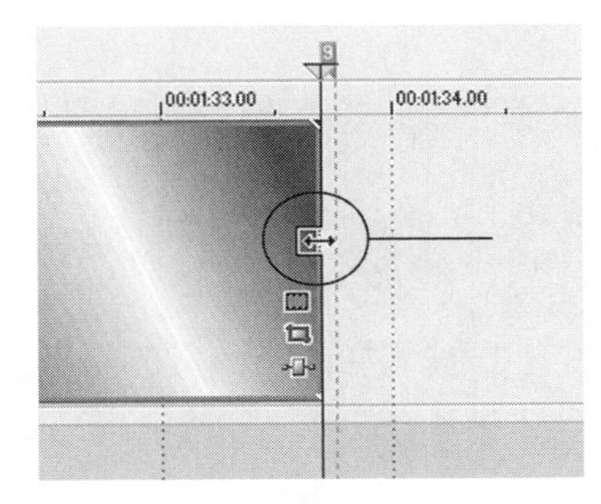

5.17 When the cursor is placed at the head or tail of an event, the cursor changes to a trim-editing cursor. This allows the event to be dragged out or dragged in, thereby lengthening or reducing the event.

The first editing behavior is to edit on the timeline by extending/reducing the length of an event on the timeline. This type of editing is called *trim-editing*.

When a trim-edit occurs with the Ripple tool enabled, media subsequent to the edited event will slide to the left to fill the hole left by the shortened event or slide to the right to make room for the lengthened event. Trim-edits can be done with only a numeric keypad. Selecting keys on the keypad moves the cursor to the beginning or end of an event, and the keypad can be used to edit an event by frames or pixels.

Table 5.1 Keyboard Shortcuts Used in Trim-Editing

Shortcut	Description
Press 7 or 9	Moves the cursor to the next event edge, either moving backward or forward.
Press 1	Trims the event edge to the left by one frame.
Press 3	Trims the event edge to the right by one frame.
Press 4	Trims the event by one pixel to the left.
Press 6	Trims the event by one pixel to the right.
Press 5	Exits the numeric keypad trimming mode.
Press 0	Creates a region/loop around the cursor. The length of the region is specified in OPTIONS | PREFERENCES | EDITING.

Pressing and holding CTRL+ALT while the cursor is placed between two events will lengthen one event while the other event is shortened, depending on whether the cursor is dragged to the right or left. This feature allows both events to continue to occupy their current total length, while editing the point at which one event ends and the other event begins.

Slip-editing is another basic editing format that allows an event to remain a static length while changing the in/out points of the event. You can slip-edit by holding the ALT key down while moving the mouse inside an event. During slip-edits, the Preview screen shows the first frame and the last frame of an event based on the slip-edit point.

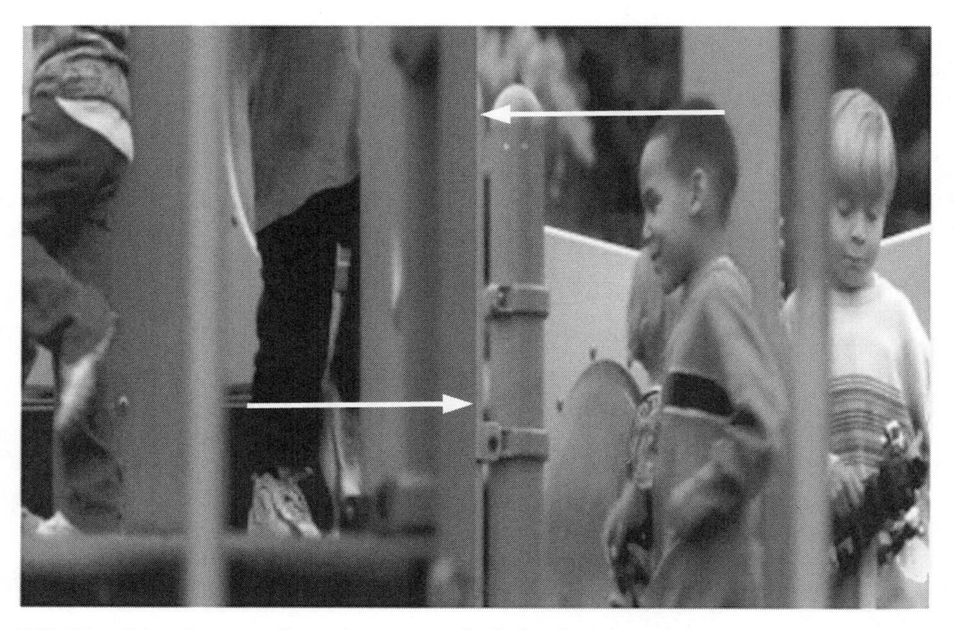

5.18 Slip-editing shows a split preview screen, displaying the in/out points of a slip-edited event.

The second edit function is to perform a transition. The most basic transition is the *crossfade*, otherwise known as a *dissolve* in the rest of the video world. A crossfade occurs when two events overlap, and one blends into the other as one event fades out and the subsequent event fades in. By default, Vegas crossfades events together any time they are overlapped. Automatic crossfades can be turned off at any time using the Automatic Crossfades button found on the timeline or by pressing X.

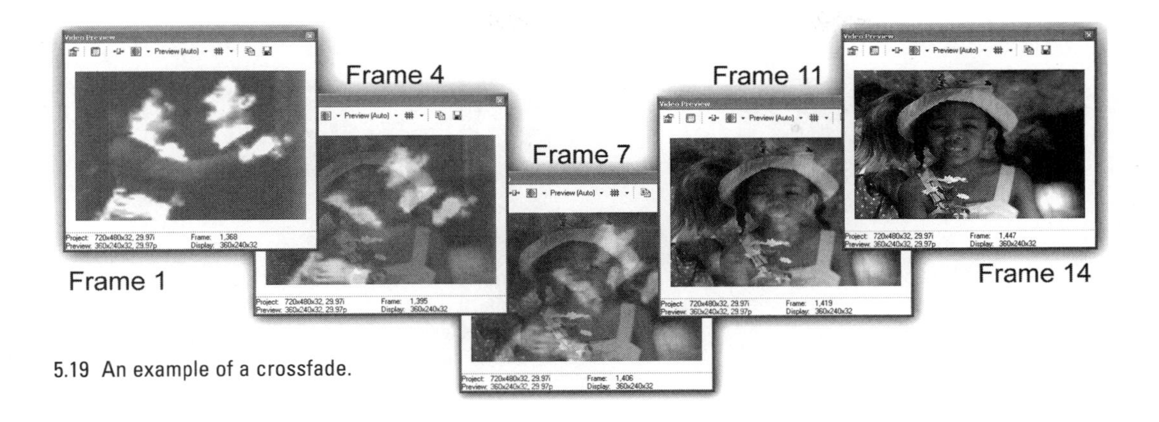

5.19 An example of a crossfade.

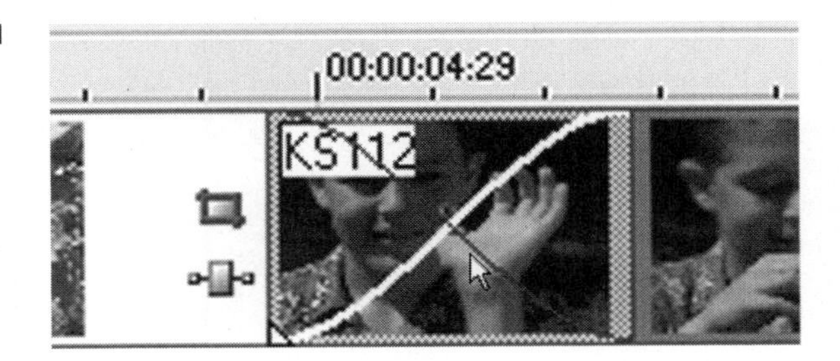

5.20 Crossfade is indicated by an "X" between the two events.

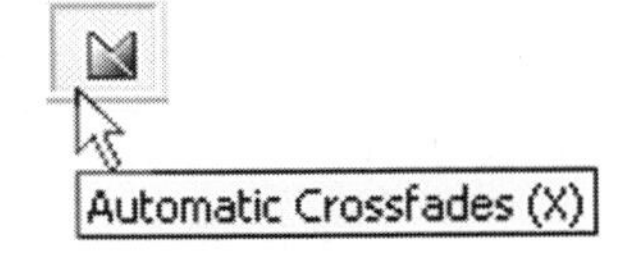

Place two events butted against each other on the timeline. Now drag the front of the second event partially over the end of the first event. An "X" appears in the space in which the two events overlap.

Now play the timeline by pressing the SPACEBAR or clicking the Play button found on the control bar of the timeline. The two events will fade from one to the other in equal time. If a faster crossfade is desired, drag the second event farther away from the first event, or trim either the end of the first event or the beginning of the second event directly on the timeline. This process is accomplished by placing the cursor in the middle edge of either of the events on the timeline and by clicking and dragging in either direction. This process shortens either the outgoing or incoming event and causes the crossfade to be shorter in duration.

The shape of the crossfade can also be controlled. Right-clicking a crossfade brings up a Crossfade menu, which has several choices regarding the way the crossfade should occur.

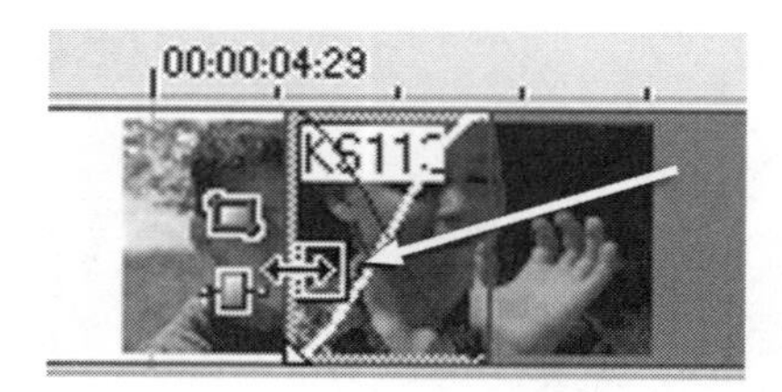

5.21 Grabbing and dragging the end of the first event is shown by a small box-like icon indicating that the end of the outgoing event is being dragged.

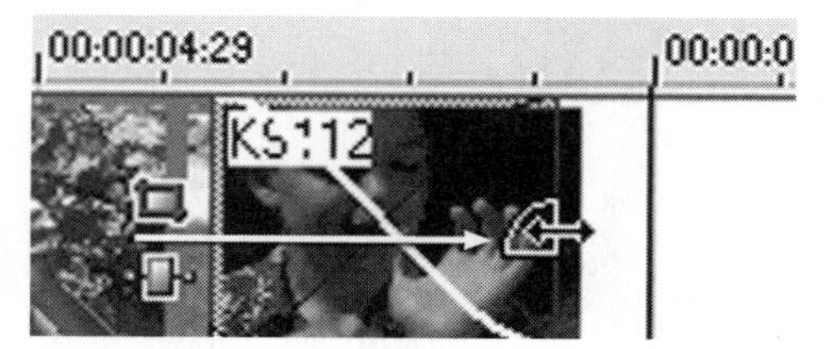

5.22 Grabbing and dragging the beginning of the incoming event is demonstrated by a small quarter-circle-shaped icon. This feature helps eliminate confusion as to which event is being edited.

Create a time selection over a crossfade, beginning slightly before the crossfade and ending slightly after the crossfade. Then press Q, which instructs Vegas to play continuously and loop over the crossfaded area. While playing over the area, all edits and adjustments are seen in real time in the Preview window as Vegas plays.

After right-clicking the crossfade, choose the Fade Type menu. A series of crossfade attributes opens. Choosing one causes this menu to close. Surprisingly, the attributes of a crossfade have a great impact on how some visual events mix.

Crossfade behaviors can also be applied to the head or tail of an event. Positioning the cursor at the top of the beginning or end of an event will change the cursor to the quarter circle form. The head or tail of the event can then be dragged inward to create a fade. This process creates a fade in or out from black.

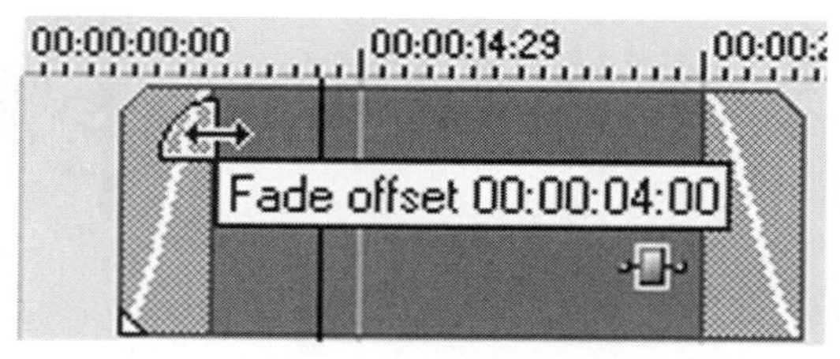

5.23 Fading in from black is a common editing necessity.

5.24 Vegas offers many crossfade styles.

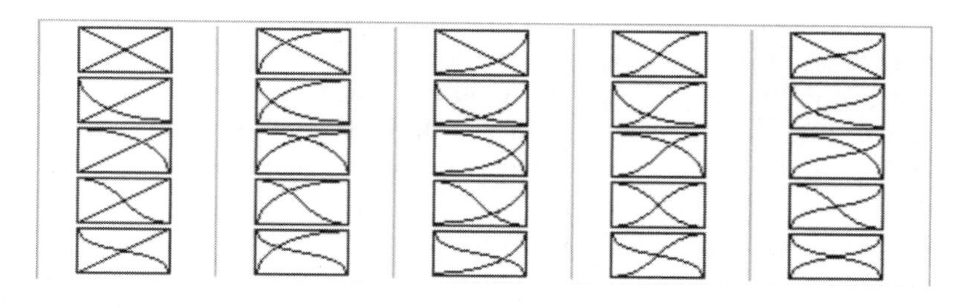

The default fade-in color can be changed by selecting OPTIONS | PREFERENCES | VIDEO and clicking the default colors.

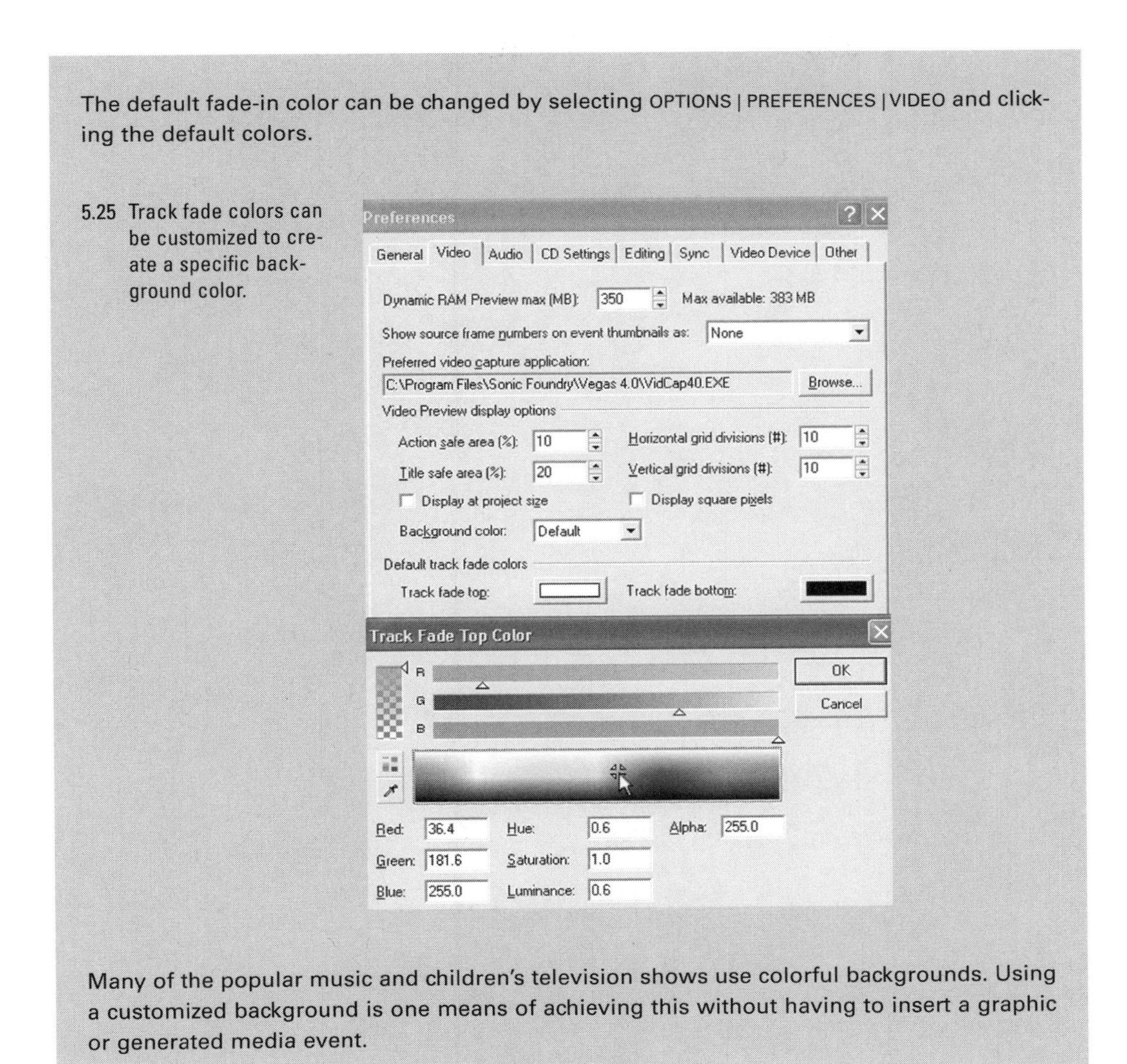

5.25 Track fade colors can be customized to create a specific background color.

Many of the popular music and children's television shows use colorful backgrounds. Using a customized background is one means of achieving this without having to insert a graphic or generated media event.

A crossfade is a form of a transition. Vegas has hundreds of built-in transition possibilities, and several thousand more transitions can be installed from third-party vendors, such as Pixelan, Boris, and DebugMode.

To change the transition from the default crossfade to another transition, select the Transitions tab in the lower-left corner of the dockable windows area to open the Transitions window. If you can't see a tab, press ALT+7, and the Transitions window is displayed. This window is dockable and may be placed anywhere on the screen that is convenient. To change a transition from any style back to a crossfade, right-click the transition and select TRANSITION | CHANGE TO CROSSFADE or press the / key on the numeric keypad. Other transition choices are also available if you right-click the Transition menu.

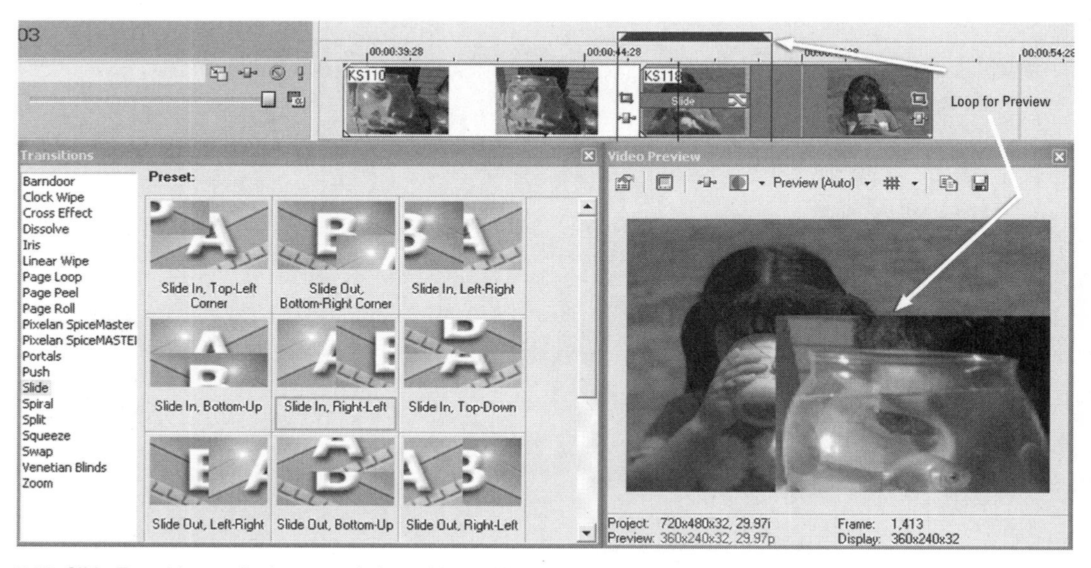

5.26 Slide Transition replacing crossfade and looped for real-time preview.

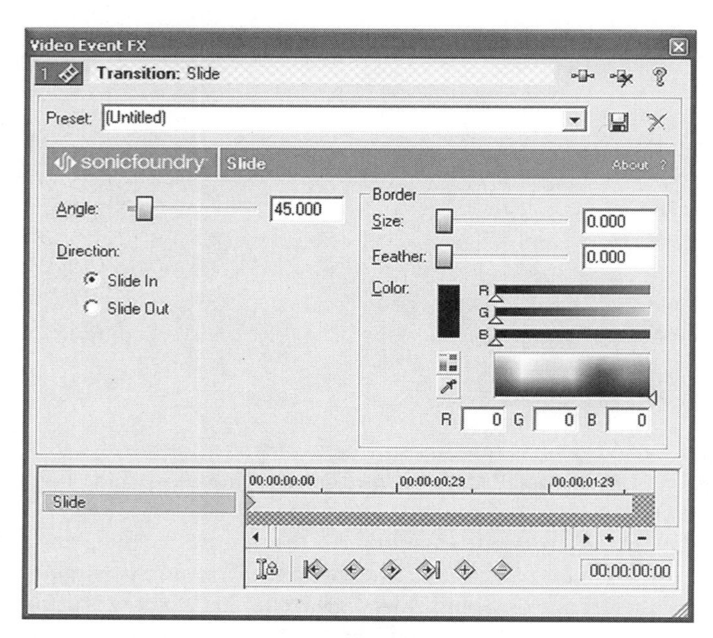

5.27 All transitions have dialog boxes to edit transition attributes. The Transition dialog box can be called at any time by right-clicking a transition and choosing the Transition Properties menu.

To preview transitions, click the different transition styles and pass the cursor over the choices. Each transition preset becomes active as the cursor moves over it. When you find a desired transition effect, click and drag the transition over the top of the "X" that defines the crossfade point. The transition dialog window will open, allowing for you to fine-tune the transition, adjusting colors, edges, softness, speed, and other parameters that vary with the different transitions.

Transition properties can be placed on fades as well. This feature is great for bringing in or sending out video for the beginning or end of a presentation.

By creating a fade in or out on a video event and then dragging a transitional element to the fade, the fade creates a transition from transparent black.

When the Transition properties dialog opens, at the bottom of every dialog is a keyframe timeline. This timeline allows for unique and defined behaviors of a transition in time by adjusting keyframes and the sliders around them. (See the "Using Keyframes in Vegas 4.0" section for more information on how to use keyframes.) Feathering edges, rates of transition, and color of borders all can be controlled with the available attributes of the specific transitions.

Right-clicking a crossfade or transition opens a dialog that allows for more transition choices. Vegas will also keep the last five types of transitions in memory, so that when you right-click, the last five transitions used are found for rapid access.

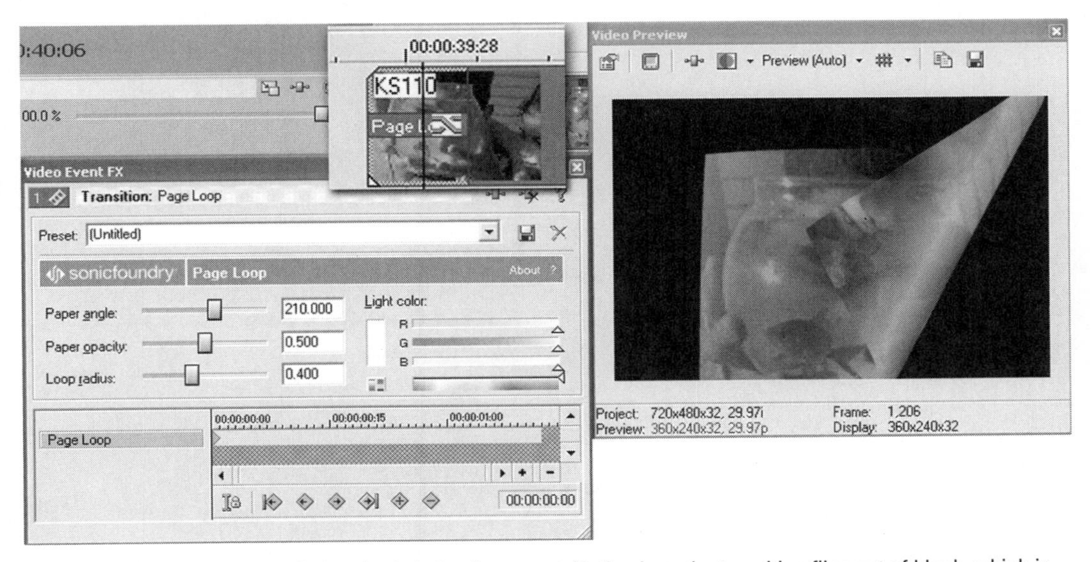

5.28 Transitional element applied to the fade in of an event. Notice how the transition flies out of black, which is the default track fade color.

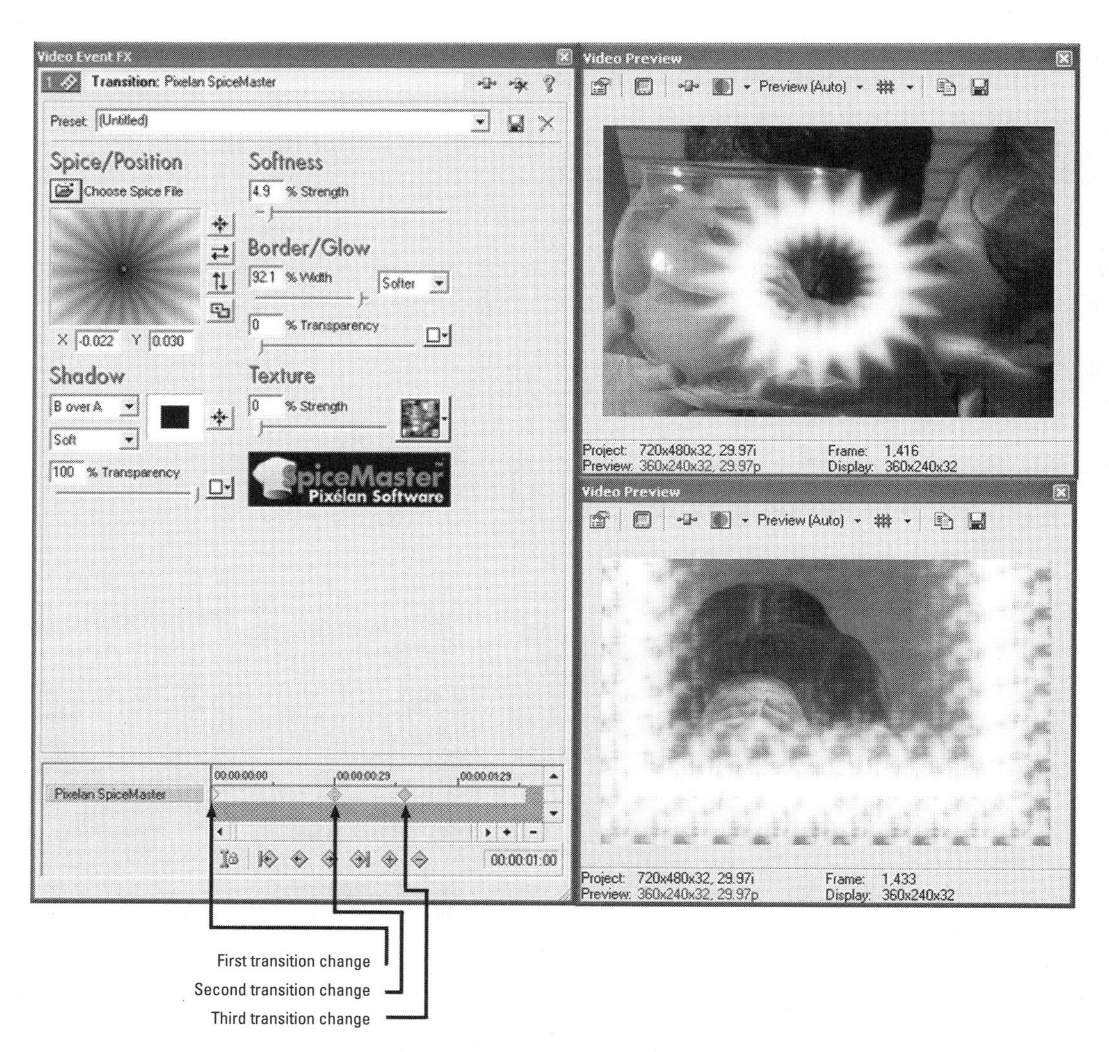

5.29 Selecting other transition types with shortcut keys or by selecting in the menu.

Edited sequences can be quickly put together in Vegas using Cuts Only, in which events are butted against each other. Position the cursor at the point at which two events are butted up and press one of the following shortcut keys related to an automatic transition:

Press / to crossfade events

Press * to create a dissolve

Press - to create a linear wipe

The two events are pulled together to create a crossfade, dissolve, or linear wipe.

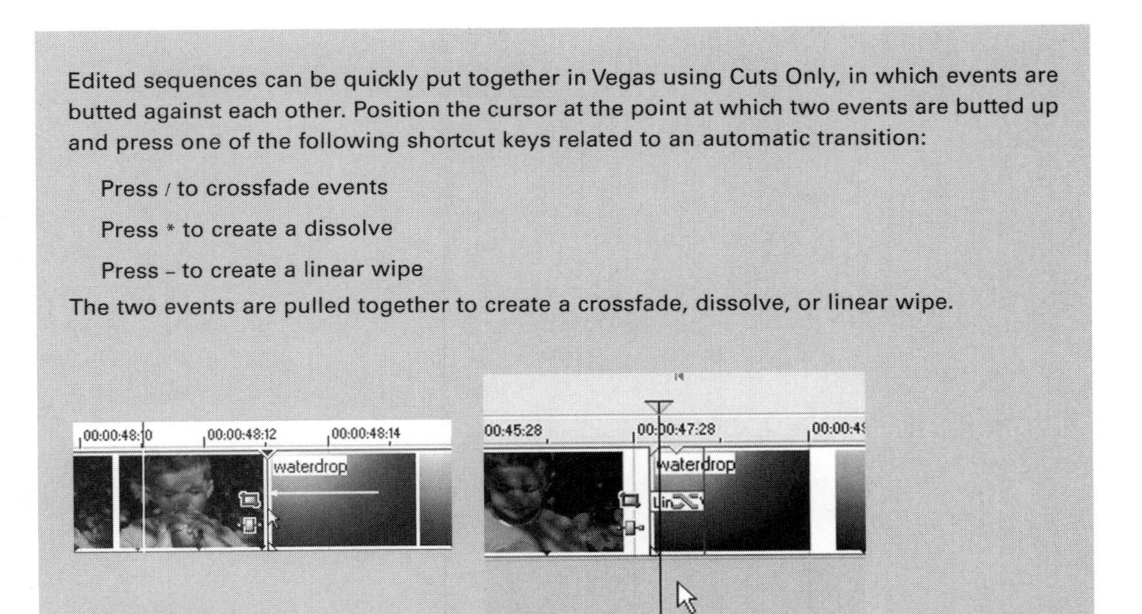

One unique feature in Vegas is that when an event is slid out of the area in which a transition occurs, Vegas will remember where the transitional element took place and recall all features of the transition. Whether the original event is placed back in the area in which the transition occurred or a new event is placed there, Vegas will still hold the transition location and information. This issue results from transitions living within events.

NEW! Users of older versions of Vegas will find new transitions in the Transition Selection menu, including Portals, Page Rolls, Page Loops, and more, as shown in Figure 5.32.

For experienced NLE users, Vegas' method of displaying crossfades and transitions is definitely unorthodox. Vegas can display the traditional A/B roll as well, should this be the desired look in the workspace. Click the Expand Track Layers button found on each Video track to show the standard A/B roll.

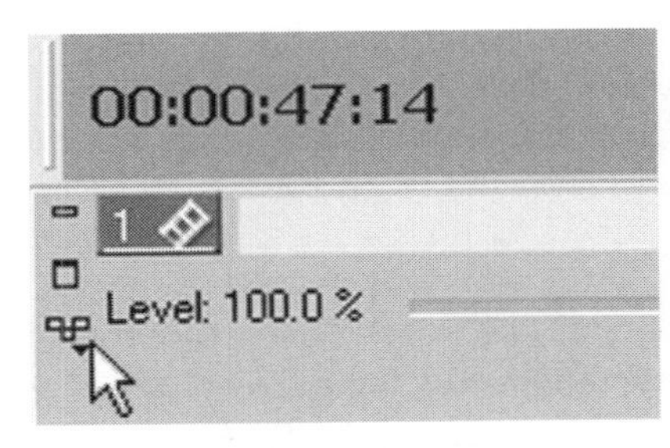

5.30 Expand Track Layers button.

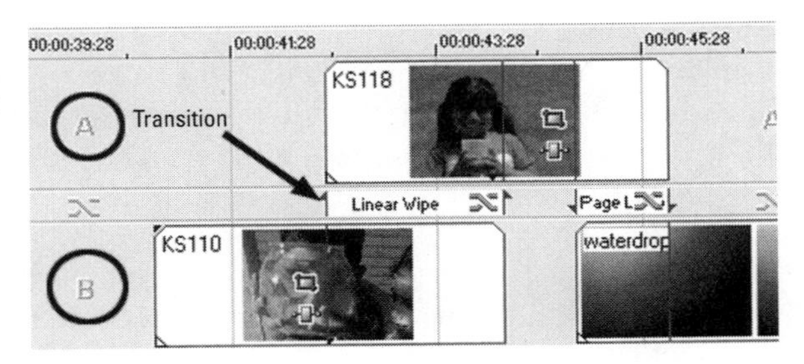

5.31 Vegas displays A/B roll views, although A/B roll views take up twice the amount of screen space as a single-track view.

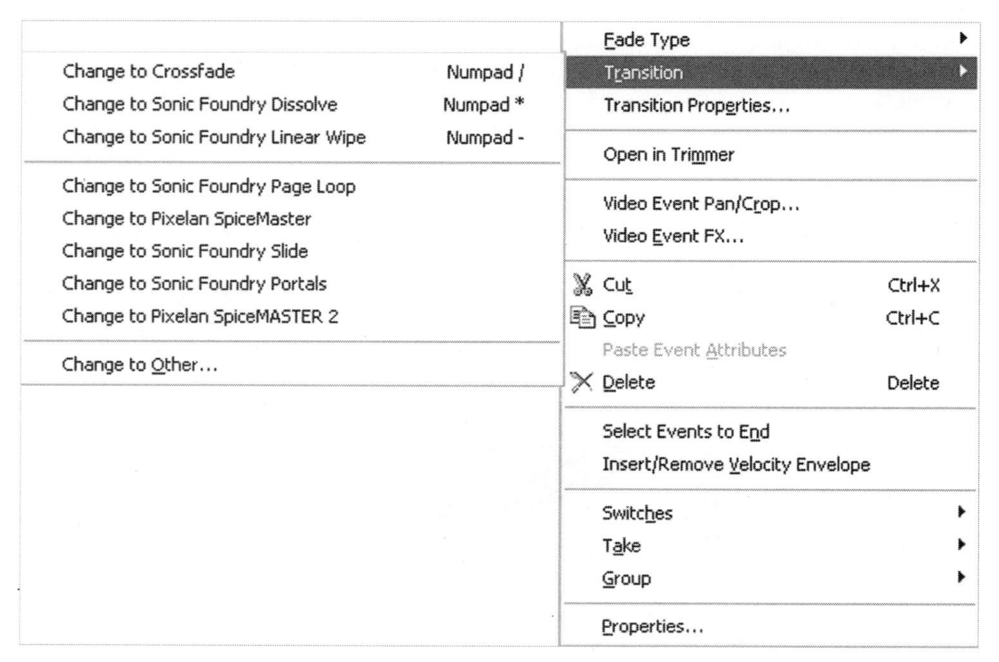

5.32 Keyframes controlling transitional changes over time.

The concept behind hiding the A/B roll view and showing transitions as crossfades stems from Vegas' origin as an audio editing tool, in which crossfades for audio are always viewed as single tracks. Further, it makes more sense in a real-estate–challenged environment, such as a computer screen, to bring as many elements together as possible.

Using Keyframes in Vegas 4.0

Vegas has always had a unique method of keyframing, and 4.0 is no exception. Keyframes are often misunderstood and are sometimes difficult for newcomers to the world of video editing.

Keyframes are nothing more than indicators in time, defined by how an event should behave at a specific moment and an instruction sent to a process at a specific moment during that time. They animate sequences of images or control effect parameters over time.

We live our lives by keyframes. All of us have specific moments in time when we do certain things, such as going to bed, eating a meal, traveling to work, or making a phone call. Appointment books are nothing more than analog keyframes.

In the video editing world, animation world, and audio editing world, keyframes are used to instruct an event to color-correct, change speed, delay, move onscreen, pan, change volume, and much more at specific points in time. The tool that defines these instructions are keyframes. Keyframe windows are nothing more than standard clocks, but instead of being round, they are linear.

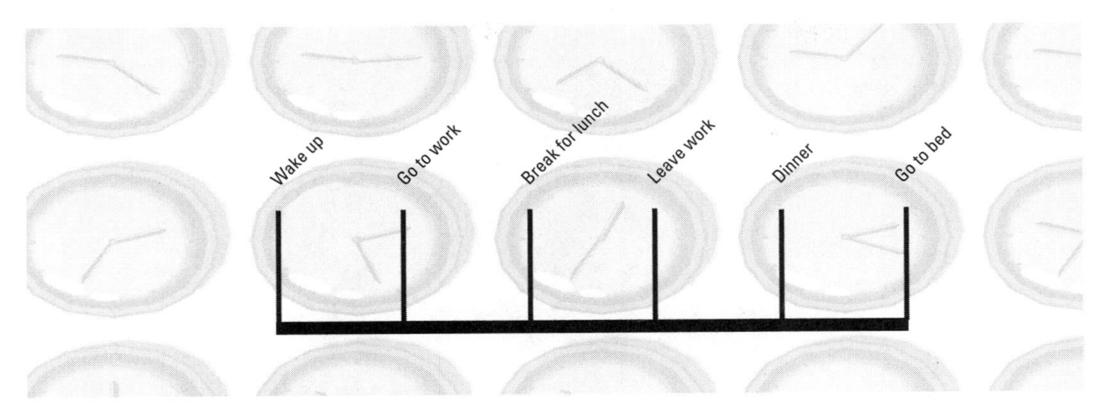

5.33 An average daily keyframe.

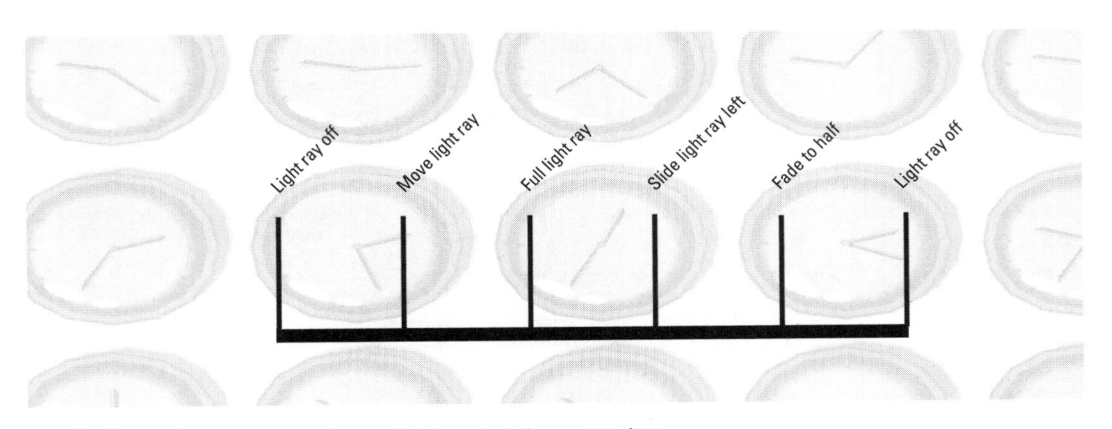

5.34 Keyframes instruct an event regarding how to behave over time.

Vegas departs from most applications in regard to keyframes in that Vegas is event-, track-, or project-based, and separate events can have their own sets of processing and keyframing within those processes, independent of the timeline. What this means is that an event can be moved on the timeline, and the keyframes will stay with it!

Imagine doing a 52-minute project for television, and, after the entire editing process is complete, a three-frame error is found in the first two minutes of the program. In most applications, if an event is moved even three frames, the entire project will be affected. With Vegas, this is simply not so. In fact, the entire series of events that exists in the hypothetical three seconds can be moved to the end of the timeline, with keyframes and processing intact.

5.35 The selected event length is displayed in the lower-right side of the timeline.

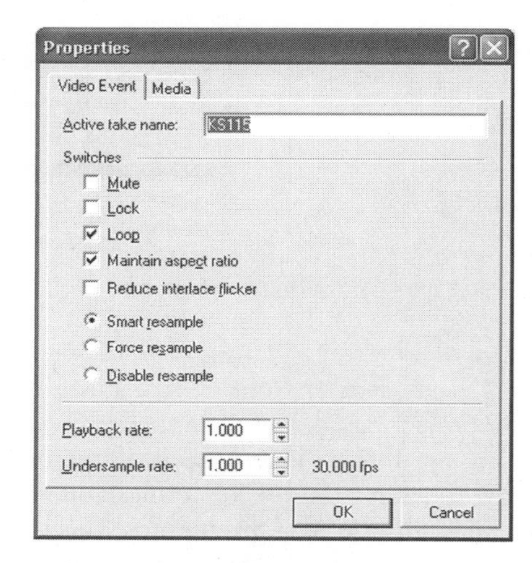

5.36 Event properties are seen when the event is right-clicked and Properties is selected.

Notice that the length of the event illustrated is ten seconds and six frames in length. Double-click the event, and, in the lower-right corner of the editing timeline, Vegas displays the length of the event selection. You can also right-click the event and look at its properties, where the media length is displayed.

Insert the CD found in this book and browse to the \veg\ folder. Locate and open the VEG file titled Keyframes. The timeline will open with the video file shown in the example.

Now, click the Event FX icon found on the event. (If you don't see the icon, go to the **OPTIONS | PREFERENCES | GENERAL** dialog and check the Show Video Event Buttons.)

This process opens the Video FX plug-in folder. In this example and CD project, the Wave plug-in has been selected.

Note the first keyframe in the keyframe timeline, which is indicated by a small diamond.

Now click the five-second indicator on the timeline in the keyframe timeline window and select the Horizontal Only preset in the Preset dialog box. Notice that a new keyframe or diamond appears on the keyframe timeline. This diamond indicates that over time the slider will begin to move, causing the event image to gradually change over time, getting more and more wavy until at the five-second point, it reaches the full effect as dictated by the slider position.

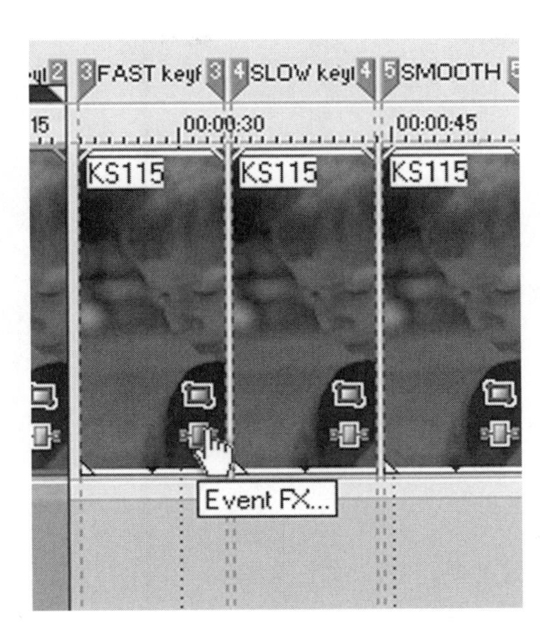

5.37 The Event FX button is found on every event by default. This feature can be turned off in the **OPTIONS | PREFERENCES | GENERAL** dialog.

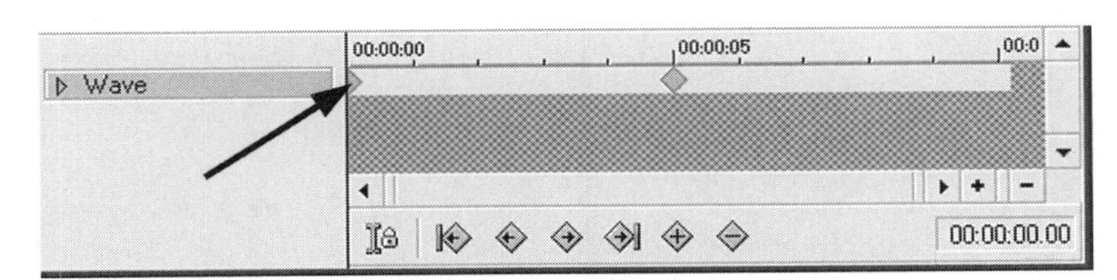

5.38 The first keyframe may be moved anywhere on the keyframe timeline. Attributes of the first keyframe are shown regardless of the position of the first keyframe, as it is the first indicator of time-based event/track/project behavior.

What if you don't want the event to become more wavy over time but instead want the effect to come in only at a determined time? Click and hold the first keyframe and slide it to the right. Now right-click the keyframe and select one of the keyframe behaviors. Attributes that can be assigned to keyframes include Hold, Linear, Fast, Slow, and Smooth.

Hold instructs the keyframe to freeze the settings for the FX in place until a new keyframe is shown, at which point the process jumps to the new settings. Linear creates a gradual change with no curve in the manner in which the sliders move. Fast causes the instructed behavior to slope in quickly and then gradually slow to meet the setting dictated by the slider. Slow does just the opposite; it slowly slopes toward the settings for the second keyframe and speeds up as it reaches its indicated position. Smooth provides temporal and spatial behavior, appearing to start slowly, speed up, and slow down again in time as the controls reach their keyframed positions. All of these attributes are demonstrated on the CD.

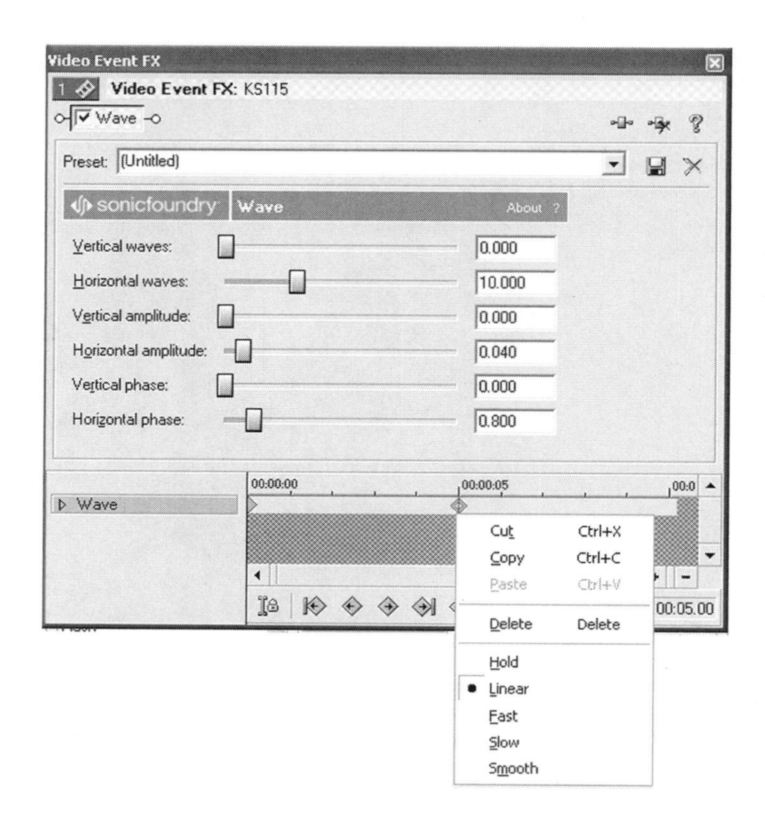

5.39 Keyframe velocities/attributes can be changed by right-clicking a keyframe.

Vegas is efficient in how it syncs the locations of keyframes to a cursor position. This feature can save hours of time trying to place keyframes correctly.

Move to Region 6 on the editing timeline of the Keyframes project. Place the cursor anywhere on the editing timeline inside the event contained in Region 6. Notice that inside the keyframe timeline, the cursor will move in sync to wherever the cursor is located on the editing timeline. Now move the horizontal amplitude slider. Notice that Vegas automatically inserts a keyframe at that moment in time.

Now double-click the time indicator inside the plug-in.

Enter a new time of 00:00:06:10, which will move the cursor inside the keyframe timeline and will move the cursor to the same position on the editing timeline. Now slide the vertical amplitude slider all the way to the right. Notice how the preview of the event image changes to match the setting, while at the same time Vegas inserts a new keyframe at the position of 00:00:06:10. Either move the cursor tool with arrow keys or mouse or double-click the time indicator inside the keyframe tool or the main editing window time indicator to place the cursor with frame accuracy for keyframing effects or other processing.

Nearly everything in Vegas is keyframeable. Pan/Crop settings can be keyframed, as can Track Motion behaviors. Track Motion will show up on the timeline of the main editing window. Region 7 of the Keyframes project contains track motion keyframes. Notice that they are displayed on

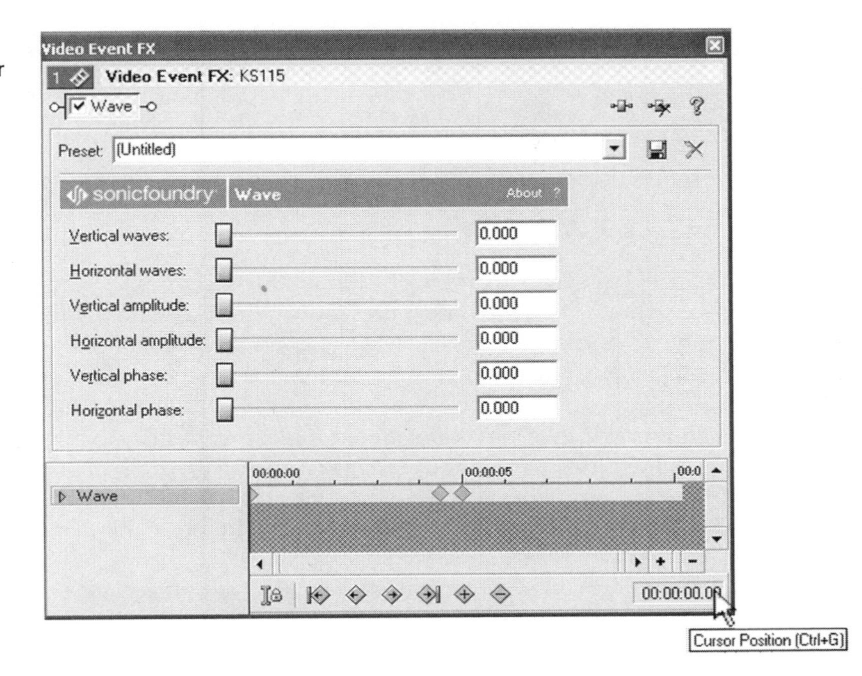

5.40 Double-click the time indicator inside the plug-in to place the cursor at the exact time.

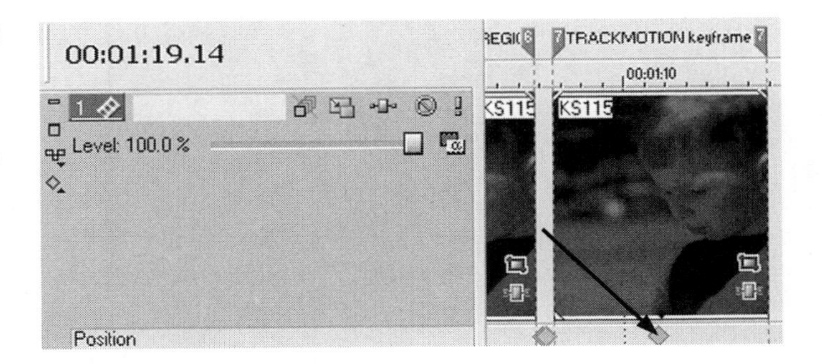

5.41 Track motion keyframes can be viewed directly on the timeline by double-clicking a track keyframe and expanding the Track Motion Keyframes view.

the edit timeline. Track motion can dramatically affect how events are shown, and, to prevent confusion about oddly sized events on a track, these are displayed prominently.

Keyframes are unlimited as to their application. They can be applied to as many FX as necessary to make an event meet the desired appearance. Some effects seen on television today contain hundreds of keyframes in one-to-two seconds of video.

Keyframes can be copied and pasted inside the keyframe timeline. Move the cursor to the Region 9 event on the editing timeline. Click the Edit Generated Media button in the upper-right corner of the event.

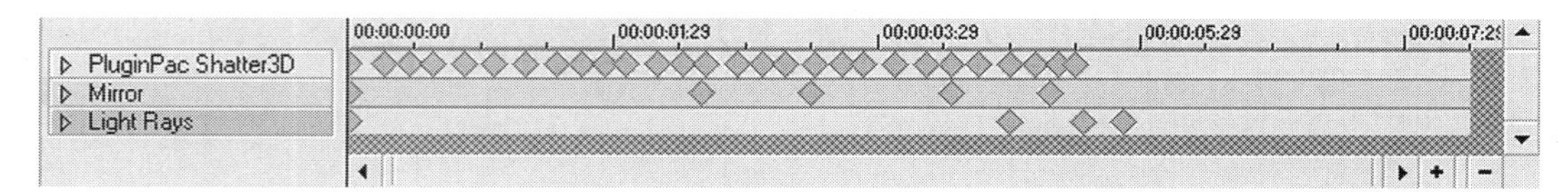

5.42 This event keyframes page contains 28 keyframes in less than three seconds of video. Multiple keyframes allow for powerful uses of FX.

The Edit Generated Media dialog will open, and five keyframes will show on the keyframe timeline. While holding the left mouse button down, draw around the five keyframes. When all keyframes are selected, they'll have a white diamond in the middle of each keyframe and will become white in color. Copy the keyframes by either right-clicking and selecting Copy or pressing CTRL+C. Move the cursor inside the keyframe timeline to 00:00:01:15 by either clicking the cursor to position or by double-clicking the keyframe time display. Now paste the keyframes by right-clicking and selecting Paste or by pressing CTRL+V. Repeat the action again at 00:00:03:00. All five keyframes are now repeated three times. Play the media. Notice how it loops the keyframe attributes. These keyframes can also be copied and pasted into another keyframe timeline of a similar nature.

To complete the explanation of Vegas' keyframing attributes and their incredible power, go to the editing timeline of the Keyframes project. Click and hold Region 1, Linear Keyframe setting. Slide it to the end of the project, past Region 8 Multiple keyframes-FX. Place the cursor at the front of Region 1 and play. Notice that the keyframes stayed with the event. Therefore, events can be copied and pasted, slid on the timeline, or moved to completely different tracks. If a keyframed event is copied in one instance of Vegas, it can be pasted into another instance of Vegas with keyframes intact. This feature is unique in the NLE world. Keyframes placed on Events stay with the events. Keyframes placed on a track will stay with the track. Tracks, however, may not be moved horizontally in time; they may only be moved vertically in priority.

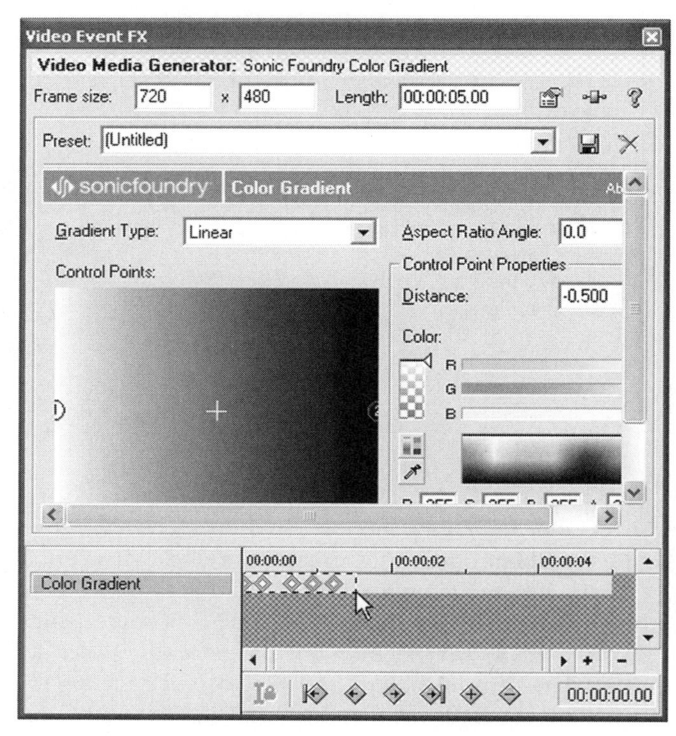

5.43 Keyframes can be copied and pasted within their own keyframe timeline or pasted as event attributes to another event.

Keyframes may also be applied at the project setting. Any video plug-in that shows up in Vegas is keyframeable. Audio plug-ins are automatable, which is similar to keyframing, yet are managed and viewed differently. (Vegas uses keyframes to control the surround sound movement, which is discussed in Chapter 7.)

Learning to understand and manage keyframes is perhaps the most important and valuable lesson to be learned and understood in the video editing world, whether working with Vegas or other NLE systems. Various NLEs have unique keyframing attributes, but regardless of the system, all NLEs use keyframes in one form or another.

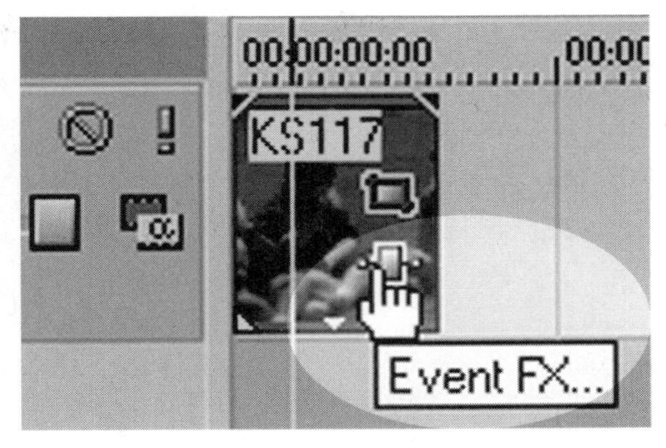

5.44 Click the Event Video FX button found on each event by default. It is seen on every visual event, regardless of whether it's a still photo, graphic, or video event.

Filters in Vegas 4.0

Vegas 4.0 is jam-packed with filters to enhance, correct, and design graphic and video events, either bringing substandard images to an acceptable level or allowing editors the expression of artistic freedom.

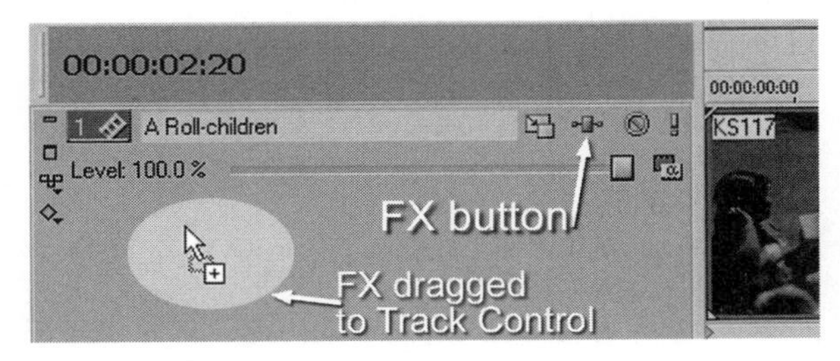

5.45 Dragging FX or clicking an FX button on the Track Control pane has the same effect.

Filters, also known as plug-ins, are found in the plug-ins folder and are accessible from many points in Vegas. Opening the VIEW | PLUGINS menu will show the effects found in Vegas. Vegas comes with more than 40 plug-ins, ranging from color correction tools to swirls, waves, and pixilate filters.

While all filters/plug-ins have presets, learning the settings and behaviors of the plug-ins makes for very creative and powerful image processing in Vegas.

As with nearly all other aspects of Vegas, plug-ins can be inserted at the event-, track-, or project-level. Plug-ins can be dragged and dropped or selected from menus selected at the event-, track-, or project-levels.

Plug-ins dropped on

- An event affect only that specific event on the timeline
- A track affect all events on the track that are affected by the plug-in's attributes
- Project/video output level affect every track, event, and final output in the project

To add plug-ins to an event, click the FX button on the event. These buttons are visible by default, but if for some reason they are not shown, the OPTIONS | PREFERENCES | GENERAL dialog has a check box for Show Video Event Buttons that should be checked. If having the buttons visible is not desired, uncheck this box. Events can also be right-clicked and Video Event Effects chosen from the menu.

FX can also be dragged directly onto events from the FX menus. To view all FX at one time, press ALT+8 or select VIEW | VIDEO FX to open the FX selection box.

To have a thumbnail preview of some FX properties, keep the FX window open and in the dockable windows section or drag it onto a second monitor. It also makes for a great space from which to drag and drop FX.

Dragging FX to a Track Control pane places FX on an entire track. Each track also has an FX button on it.

Any FX dragged to the Track Control pane will result in all events having FX applied to them. The only way to control FX on individual events is to use keyframes.

FX can be applied at the project-level by dragging FX to the Preview window or by clicking the Video Output FX button on the toolbar in the Preview window. All video, graphics, and still events on every track will be processed at the project- or output-level.

The primary use for FX at the project- or output-level is for plug-ins along the lines of the

- Timecode plug-in
- NTSC/broadcast clamp
- Black restore

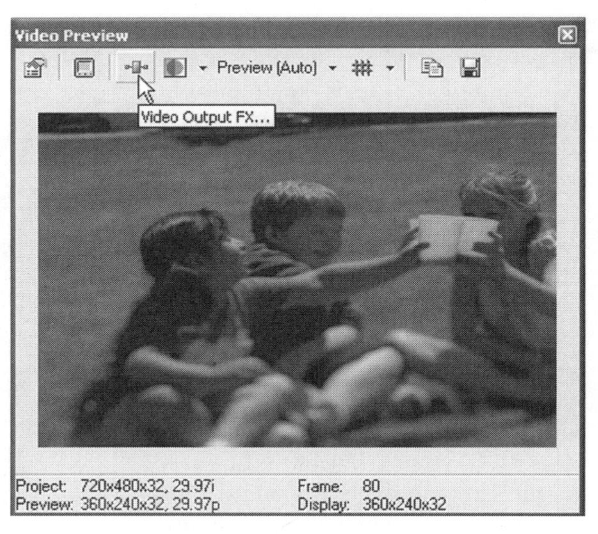

5.46 Inserting FX at the project-level.

5.47 The timecode plug-in places a timecode burn on the project window.

Any FX, however, can be placed on the project.

As many FX as necessary can be applied to events, tracks, or the project. FX are arranged in the order in which they are applied; this order is easily changed by either dragging one plug-in to its place on the plug-in bar or by clicking the "Shift plug-in left/shift plug-in right" button found on the FX Chooser plug-in bar.

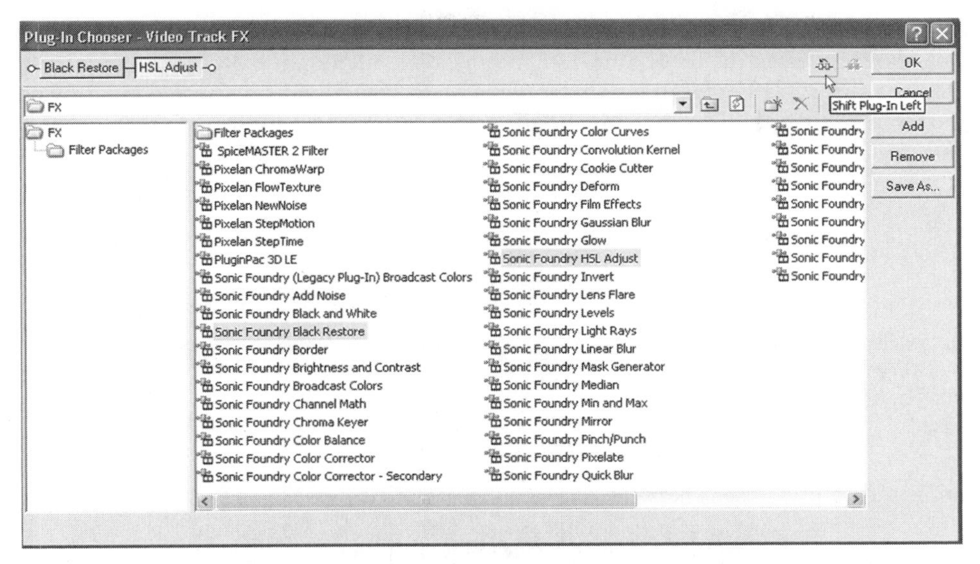

5.48 Arranging the order of plug-ins can be done via button press or by dragging plug-ins to the desired priority.

You can save *chains* of FX in Vegas. Many times, a project requires repetitive use of the same series of FX. Creating presets and inserting FX in sequence can be time-consuming. For example, a project might have poorly shot video and requires color correction, brightness adjustment, and broadcast color clamp. Rather than inserting these FX time after time, the series of FX, complete with presets, can be recalled after it is saved as a chain. This process saves tremendous time, especially in an editing house where a set flow of work exists. Each editor can create their own chains to suit their own workflow.

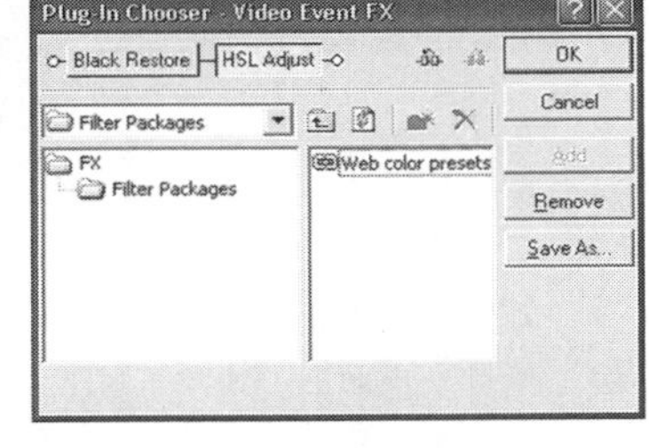

5.49 Find preset chains in the FX packages folder, which are shown each time FX are to be applied to events, tracks, or projects.

To save the chain of plug-ins as a preset:

1. Select the plug-ins and arrange their priority.
2. Click the Save-As button. The Chain dialog is displayed.
3. Name the chain/preset.
4. Press ENTER to accept the changes you've made to save the preset.

The chain preset is now displayed in the plug-in chooser and is available from the Filter Packages view. (The same process applies to audio presets.)

NEW! Vegas 4.0 has a tremendously powerful new feature that allows video output to be viewed at both pre- and post-process settings. Often times, editors find themselves wanting to see original video next to the processed video. This option has not been possible in the past without complex hardware. Now it is possible with Vegas 4.0's split-screen preview.

The Split Screen View Selection Tool button is shown on the Preview Window toolbar.

Selecting this button in the Preview window makes Vegas 4.0 automatically split the preview screen, which shows affected media on one half and original video on the other half. However, Vegas 4.0 allows users to define the areas where preview or original video is viewed by drawing a square or rectangle over the screen. Drawing the shape on the preview screen creates a hole where original video is seen.

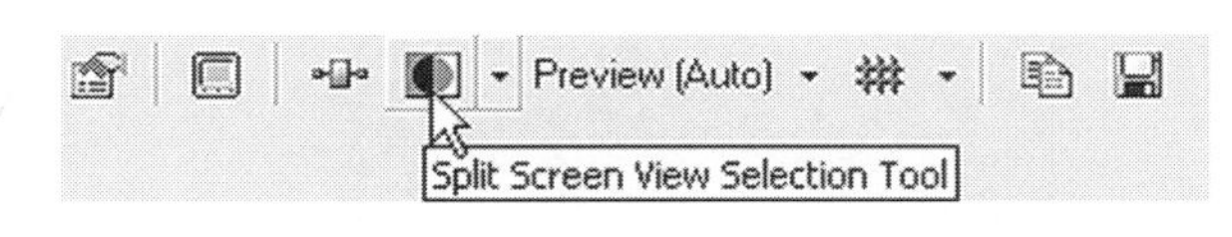

5.50 Split Screen View Selection Tool button.

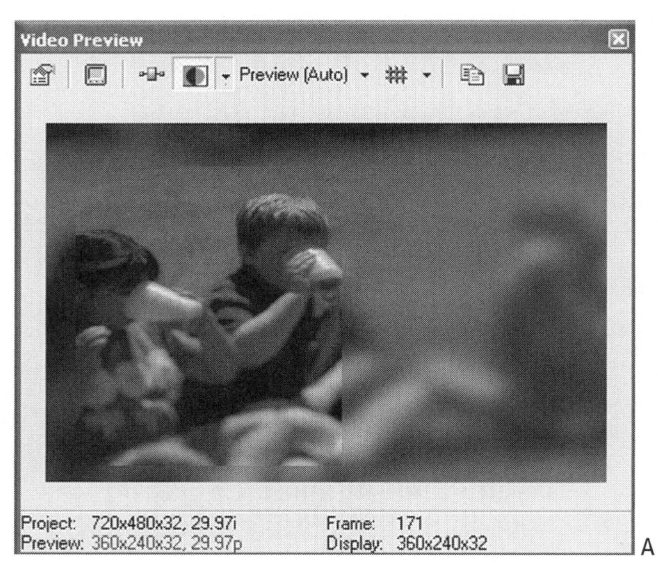

5.51 The area to view the original (a) versus processed video (b) is user-defined.

The split preview of processed and original video is also viewable on an external monitor, which makes color correction in Vegas advanced compared to even exceptionally high-end NLE systems. The capability to compare processed video to original video on an external monitor without rendering a file provides editors with a newfound speed, workflow, and freedom from concerns about accuracy in the editing workflow. This feature coupled with RAM renders allow editors to immediately see full motion/full frame-rate accurate views of edited media, limited only by RAM and processor speed.

The Split-Screen Preview window can be used to show contents of the clipboard during the preview of a timeline section. Position the cursor on the timeline and click the Copy Snapshot button in the Video Preview window to copy a frame as a still image to the clipboard. Move the cursor to another point on the timeline.

On the Split Screen View Selection Tool button, click the drop-down menu arrow and select the Clipboard. The contents of the clipboard are displayed in the Preview window or external monitor. The area in which the clipboard contents are displayed is defined by the area drawn with the cursor inside the Preview window area. If the entire clipboard clip is to be shown, redraw the cursor over the entire Preview window area. The underlying clip indicating the location of the cursor is no longer shown. Turn off the split-screen preview or redraw the viewing area to view underlying events. Double-clicking the Preview window returns the Preview window to its default state.

How Filters/Plug-ins Function

While far too many plug-ins are available in Vegas to describe the attributes of each, some common plug-ins are of great benefit to understand. Each plug-in has a series of presets but experimenting with the settings increases the power of the plug-in tremendously. A number of third-party filters are available, some of which are described in the following list:

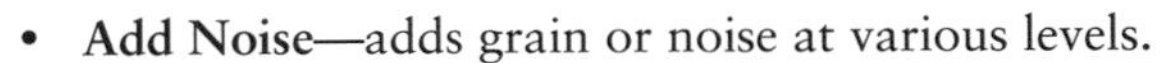

- **Add Noise**—adds grain or noise at various levels.

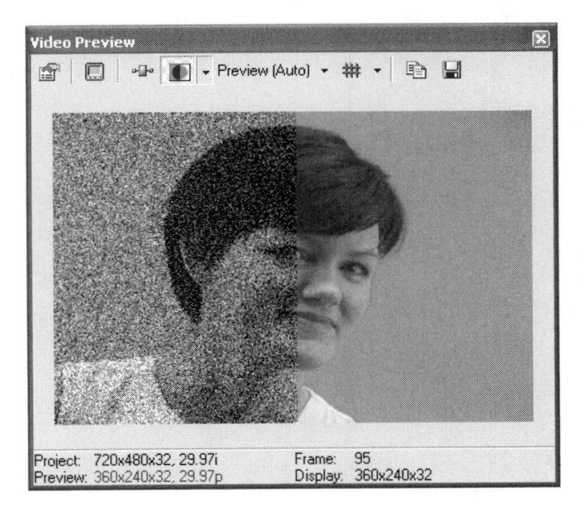

- **Chromakey**—removes colors from an image and replaces with other information, similar to the weatherperson on the television news.

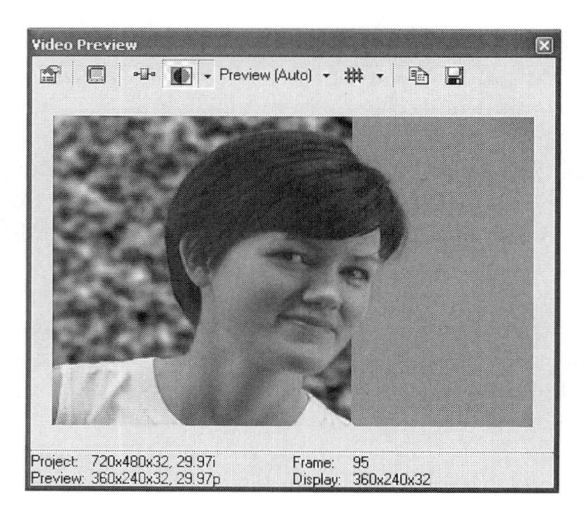

- **Convolution Kernel**—embosses, finds lines, or creates a motion photographic image from video.

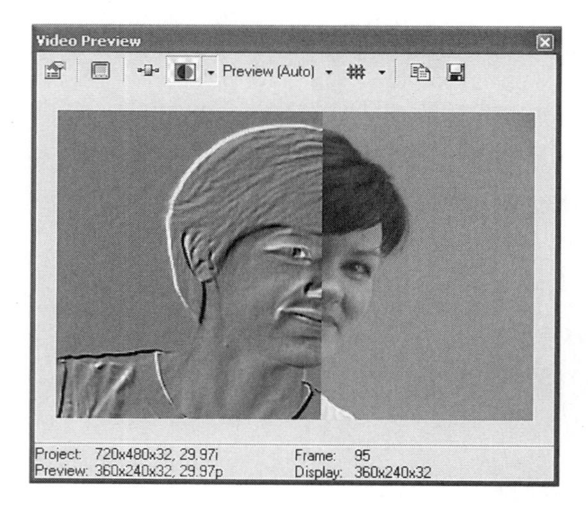

- **Lens Flares**—hide anomalies, make still images look more like video, and give flair to wide-video shots.

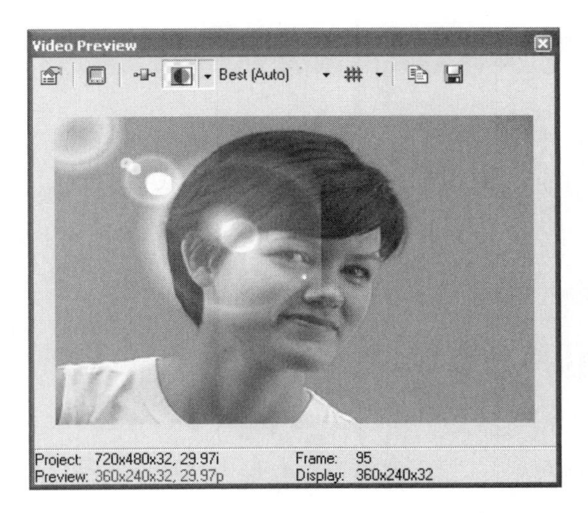

- **Mirror**—creates mirrored images of any portion of the screen. When used correctly with generated media, the plug-in can emulate a kaleidoscope.

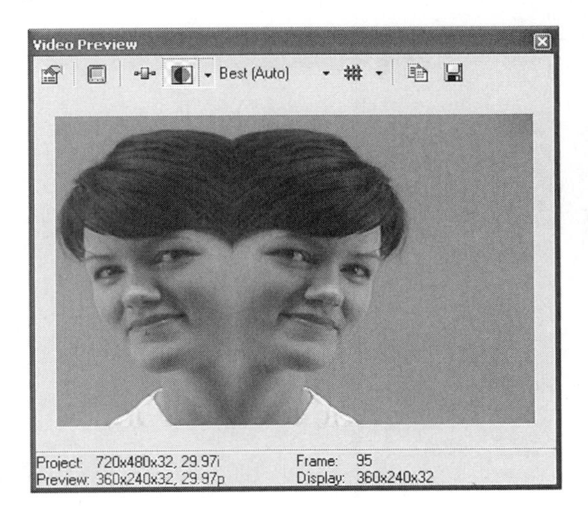

- **Pixelan's ChromaWarp**—provides infinite artistic choices for the beginning and experienced editor.

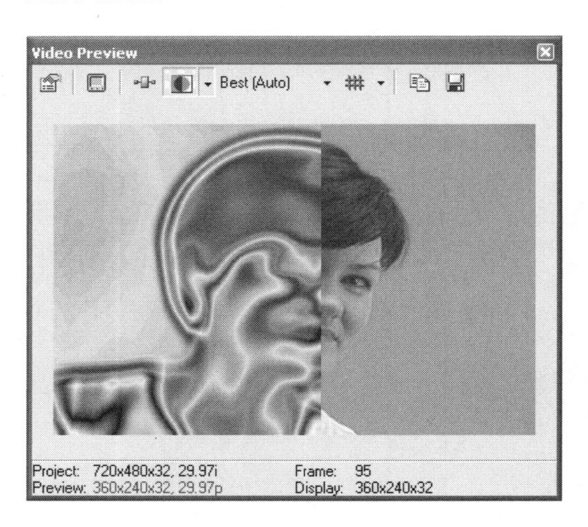

- **DebugMode's 3D Pak**—adds some wonderful 3-D elements to the Vegas editing toolbox.

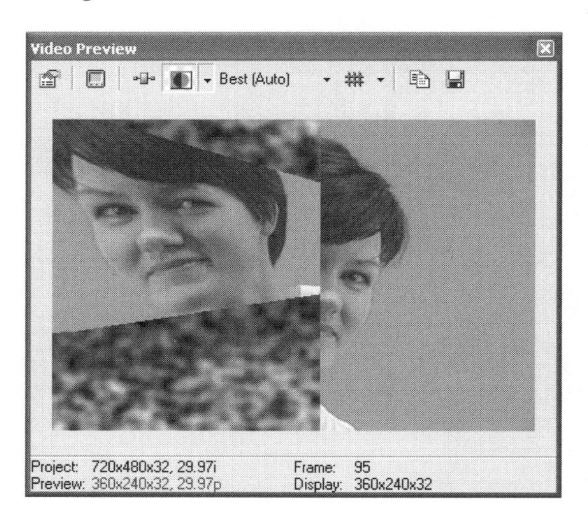

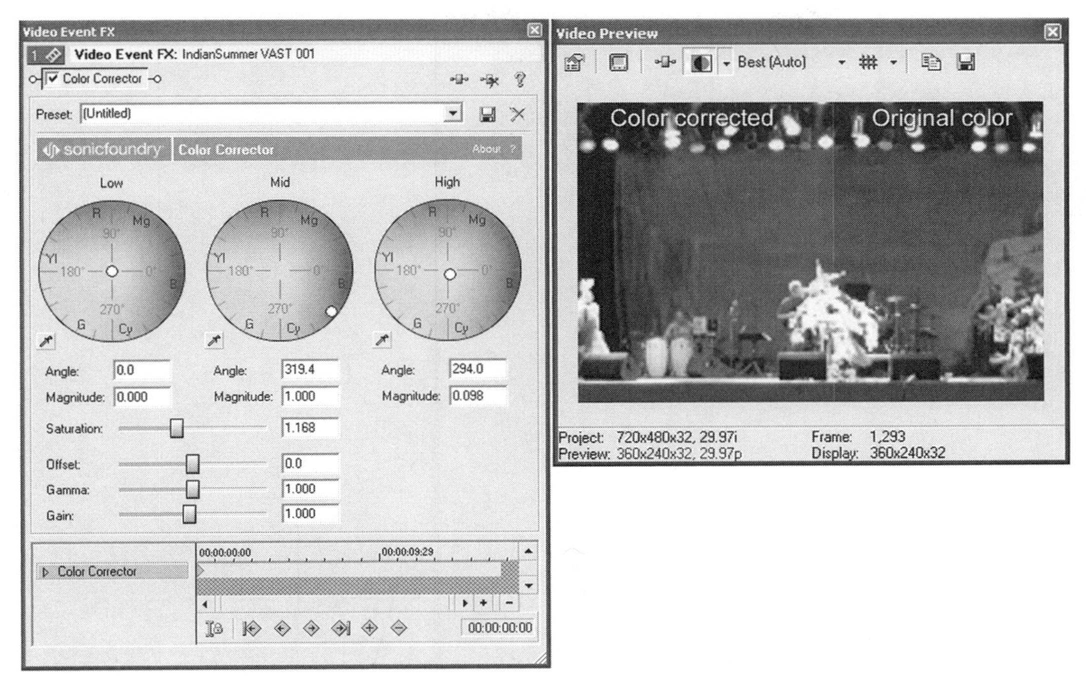

5.52 Color correction can be viewed in real time on an external or internal monitor.

These plug-ins are free and are included on the disk in this book.

Vegas has a feature that is often overlooked yet should not be. Vegas has the ability to build a RAM preview of the final appearance of an event so that it can be seen full-motion/full frame rate. The RAM preview is important to be able to see, particularly in situations where keyframes are inserted in tight time spaces. Slower processors or FX-heavy, multiple events can tax the processor so much that some keyframes simply can't be drawn due to lack of processor power. Using a RAM render causes these sections to be drawn without creating a temporary file on a hard drive.

To initiate a RAM render, make a selection on the timeline. Pressing SHIFT+B instructs Vegas to begin rendering the time selection to RAM. The amount of RAM installed in the computer determines the length of time that RAM can render. This is why as much RAM as possible is best in most editing systems. By default, Vegas allocates 16MB of RAM to the RAM preview. Opening the OPTIONS | PREFERENCES | VIDEO dialog gives users the ability to adjust the amount of RAM to be used for RAM preview renders. Vegas subtracts 128MB of RAM from all available RAM for preview purposes. Even if only 128MB are available, Vegas makes allowances for up to 16MB of RAM. However, any available RAM in excess of 128MB can be used for RAM renders. (100 percent of RAM may not be used, as the operating system and Vegas require 128MB of RAM space to operate properly.)

While RAM renders benefit external previews, previews on external monitors are not full frame rate except in the faster machines. If full resolution/full frame rate are required for an external preview, a standard pre-render is required on edited/recompressed media. Pressing SHIFT+M opens the pre-render dialog.

In Vegas 4.0, new plug-ins have been added. One of the more in-demand and powerful plug-ins added is the new Color Corrector tool. This palette-based tool offers extremely fast color correction, and, when combined with the Split-Pane Preview window, it's simple to view changes in color on either an external monitor or on a gamma-corrected computer monitor.

The Color Corrector plug-in can be dragged and dropped on an event, track, or entire project. It can be used to simulate color effects as well, such as those seen in the movies *Payback* or *Traffic*.

Double-clicking the dot in the color correction palettes resets the palette to zero. Most controls in Vegas act in this manner; double-clicking resets a control to default or null.

Colors can be sampled using the Eyedropper tool found in the low/mid/highlight palettes to assist in selecting the color curve.

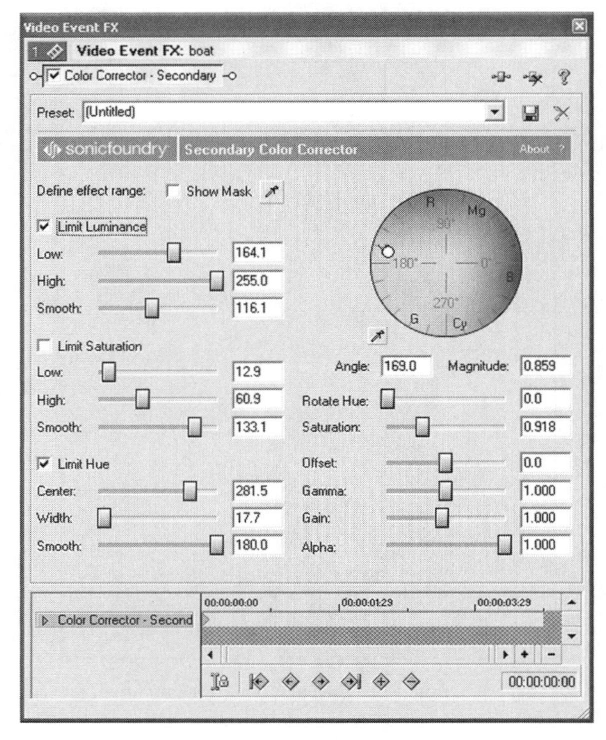

5.53 The Secondary Color Corrector makes replacing colors easy and intuitive.

The Secondary Color Corrector is another new feature in Vegas 4.0. It allows for a color or series of colors to be replaced or used as masks through which other media can show. Start by selecting a color to be replaced. Shut off external monitor preview while selecting colors. A single color or gradient of color can be selected. Best results are obtained by working with a single color and using more than one instance of the Secondary Color Corrector.

After the color for replacement is defined, choose a color with which to replace it. External monitor preview may now be turned back on if this is how video is being monitored.

Tip

Any time video has a final destination where it might be viewed on a television monitor of any kind, it is strongly advised that a good quality television or broadcast monitor be used to preview video; otherwise it is far too easy to have incorrect colors, borders, or margins. If video will only be seen on a computer monitor as streaming media or intranet media, an external monitor is unnecessary.

Opacity Envelopes

Each video track in Vegas has an opacity control, although no shortcut key is available. *Opacity* is the level at which an event is visible. At zero percent opacity, an event is completely transparent, and at 100 percent opacity, the event is completely visible with zero transparency, which is the opposite of how Photoshop and Ulead control opacity.

The opacity of a track can be altered in one of two methods. The first method is in the Track Control window, where level/opacity for the entire track can be set. This is not controllable over time but rather sets the level or opacity for every event on the track.

The second method is to set an Opacity Envelope in the track, which is done by right-clicking the Track Control pane and selecting Insert/Remove Envelope and then selecting Composite Level from the submenu that will open. This process will insert a blue line (default color) at the top of the video track. Double-clicking this blue line inserts handles or nodes. These handles/nodes allow the compositing envelope to be pulled down and adjusted for opacity level at any point on the video track. To delete a node, right-click it and select Delete.

5.54 Track Control pane–level adjustment.

5.55 Adjusting opacity over time. No limit exists as to how many handles/nodes can be placed on the track timeline.

The following are the envelope modes you have to choose from:

- **Linear**—Straight line fades with no velocity or curve to the actions.

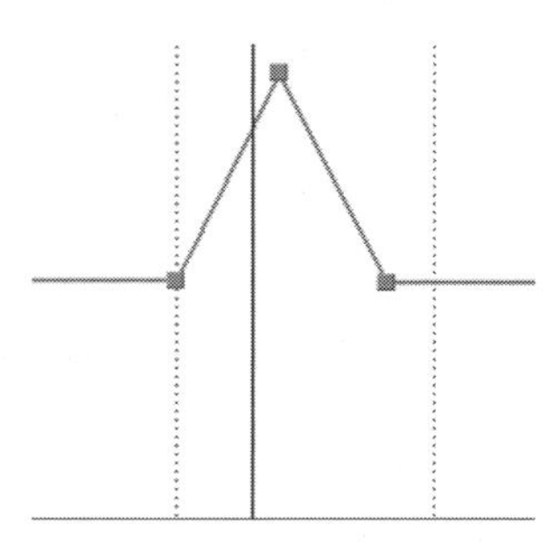

- **Smooth**—Smooth fade ramps slightly into the event before ramping back down.

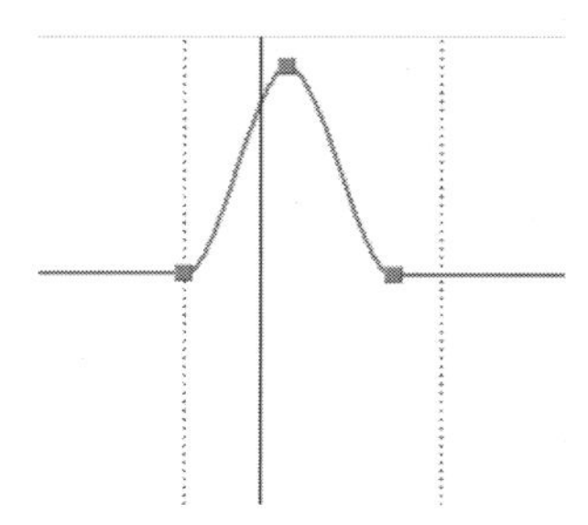

- **Fast**—Fast fade quickly raises or quickly lowers settings.

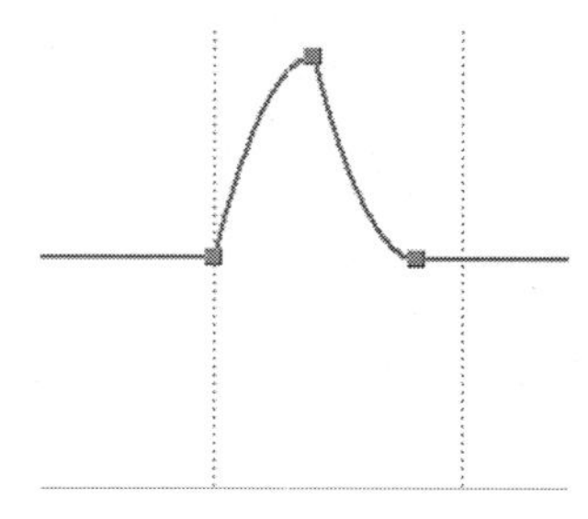

- **Slow**—Slow fade gradually moves in or out.

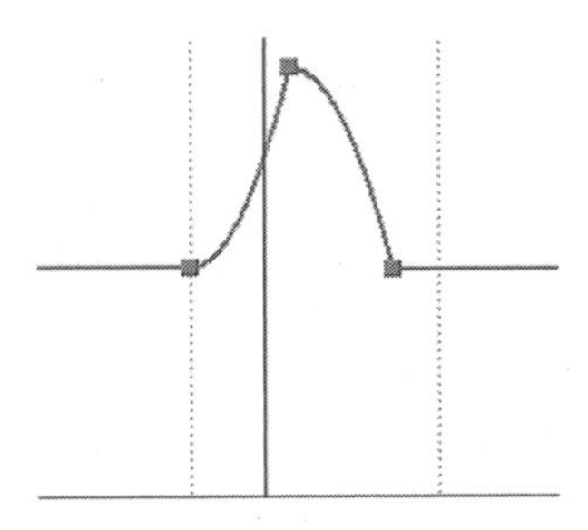

- **Sharp**—Sharp fade moves quickly to the peak movement.

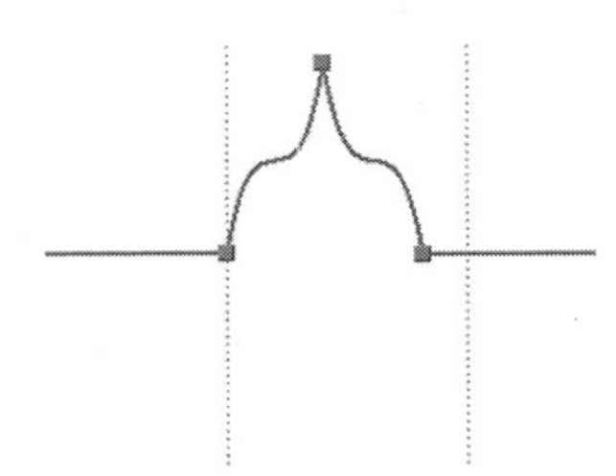

- **Hold**—Hold position has no subtle movement at all and is fairly harsh and obvious.

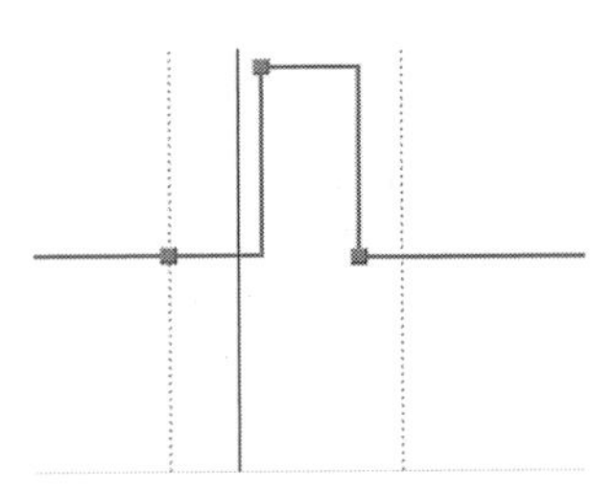

Using the opacity envelope or envelopes, events can be faded to another event on a different track or to a transparent black.

Open the project called `Compositing Envelope.veg`, which is on the tutorial CD. Two video tracks are on the timeline. Track 1 already has a compositing envelope on it. On the control pane for track 2, right-click and select Insert/Remove Envelope, and then select Composite Level. A blue line, similar to the line found on track 1, will appear. Double-click at the first marker on this line to insert handles/nodes. Pull the handle down to 50 and notice how the gradient found on track 3 becomes visible.

By using compositing envelopes, you can create beautiful fades over multiple layers of video. Compositing envelopes also allow for rapid transition from one event to another with overlay or multiple images. With compositing envelopes in place, as many tracks as the imagination might create can be visible.

If multiple cameras are used on the same shoot, the opacity envelope is an indispensable tool. Place video from multiple cameras on different tracks on the timeline, lining up their sync with the audio on each track, so that all video tracks are lined up with sync. Delete the audio tracks that are unused, keeping only the audio master track.

For this example, we'll assume three tracks of video and one track of audio are being used. On video tracks 1 and 2, insert compositing envelopes. Leave video track 3 without any adjustment at this time. Play the video until you'd like a fade or cut to occur and drop a marker at that point, pressing M. Stop playback and double-click the compositing envelope at the top of video track 1 to insert a handle. Double-click again slightly after the first handle. Now pull the opacity envelope all the way down to the bottom of video track 1. Now the events in video track 2 are visible. Play the timeline again, until another fade or cut is desired. Drop another marker and repeat the actions performed above at the new edit point. Now video track 3 is visible. Again play the video until a new fade or cut should occur. Now events from track 2 or track 1 can be brought in by double-clicking the bottom of either video track on the compositing envelope, inserting two points, and pulling the compositing envelope up. In this manner, an entire show can be roughed out quickly.

After all fades and cuts have been made using the opacity envelope, each time a track has been removed from view with the opacity envelope, you can place the cursor on the event, making sure the event is selected. Press S to split the event. Repeat the split at the point in the event just before it fades in again. Now the middle portions of the event can be deleted.

Open the `compositingenvelope#s2.veg` project from the CD. (This project also includes the use of the Pan/Crop tools, discussed earlier in this chapter.) This project demonstrates the use of the same video track, panned and cropped, to create the false image of a three-camera shoot from one camera.

The compositing envelope is used to create the original edit points. In this project, you'll need to split out the unused video. Removing the unused portions of the event is not necessary, merely helpful in viewing the overall project.

Velocity Filters

Velocity filtering allows for media to speed up, slow down, reverse, or freeze-frame, all without cutting clips of media into individual pieces as previously required in most applications. Velocity filtering, or envelopes, as they are known in Vegas, can be applied to any number of clips. The speed at which the envelope interacts with the media can be controlled as well via keyframes. Every effect, envelope, or filter can be keyframe-controlled in Vegas 4.0, with a variety of shapes for the envelope properties.

Open the `velocitytraining.veg` file found on the CD.

Notice that on the timeline a green (default color) line is in the middle of the event. This line demonstrates how the velocity envelope is adjusted. The second event on the timeline is for you to adjust your own velocity envelope.

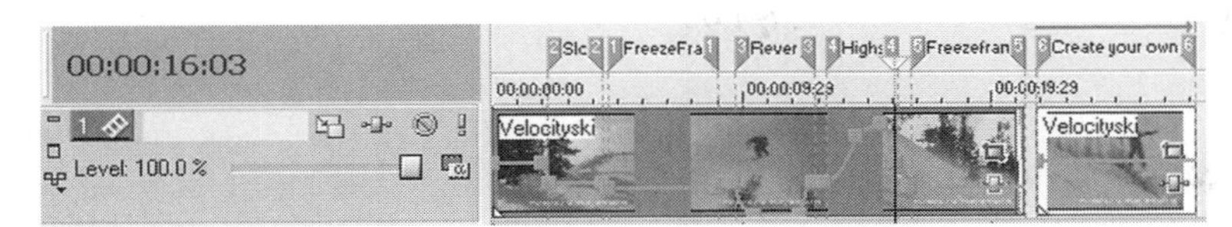

5.56 Velocity envelope inserted, with regions expressing each change in velocity. Handles have various envelope settings on them.

Start by previewing the first clip. Notice how it slows down during the first moments of the event, slowly climbs back to regular speed, and then freezes frame at the end? (You might need to perform a RAM render [SHIFT+B] if the processor cannot draw the video at normal speed.)

In the second half of the event, the video actually reverses to give an instant replay type of feel. Again, the event freezes frame at the end, allowing time for a blur, added title, special effect, or just a simple dissolve to the subject shot. This option is especially effective with video of high motion, such as sports, chases, or fight scenes. Notice how during the freeze frame, the Pan/Crop tool is used to create an artificial camera zoom. Another great space to use the velocity envelope is when you have a series of camera zooms that maybe weren't as smooth as they could have been. The envelope can slow down the jerkiness of a poor zoom or speed up a zoom that was too slow. A velocity envelope applied to speed up a fast zoom makes it faster, more frantic, and draws immediate attention to the subject on the screen.

To insert a velocity envelope into a media clip, right-click the clip and select Insert/Remove Velocity Envelope. A green line will appear. (This line is green by default in v4 and blue by default in v3.) The line will be set by default to normal speed, or 100%. Double-click the timeline, and a node or handle will appear. Move down the timeline a few seconds and double-click again, inserting a second handle. These handles are similar to keyframes, allowing the setting of velocity or speed on the individual event. To delete a node, right-click it and select Delete.

Now, move the handle up or down to set the desired speed. If the desired percentage of speed change can't be precisely achieved by moving the handle up and down, right-click the handle. The dialog box shows the fade choices and speed settings. Select Set T and type in the desired setting. Slow the envelope down just as the zoom reaches its end and make sure the handle on the keyframe is set to Smooth if the media is being sped up or slowed down over a short number of frames. By setting the envelope handle to Smooth, the media won't appear to jerk or jump from slow to fast or from fast to slow.

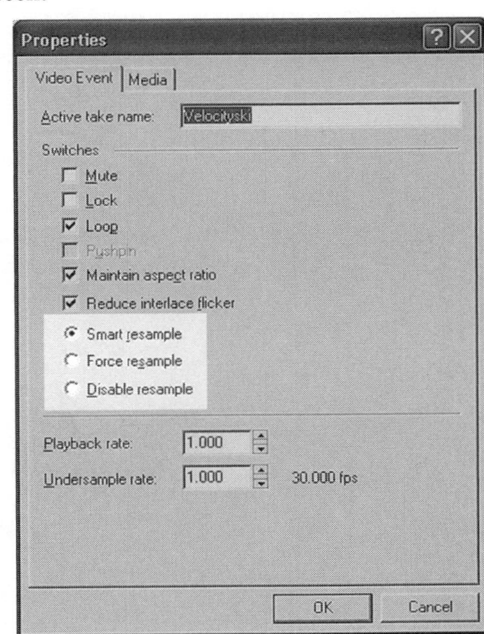

5.57 New resampling features found in Vegas 4.0 provide for smooth slow motion and velocity events.

As the event is sped up or slowed down, artifacts can occur, particularly in Vegas 2.0. Avoid this problem by right-clicking the clip, selecting Properties, and checking the Resample box. This process causes the render to slow down somewhat, yet carefully resamples the video to assure smooth picture and motion in almost every instance.

In Vegas 4.0, a new feature for resampling has been created, which allows for smart, forced, or disabled resampling, as shown in Figure 5.57. If Smart Resample is enabled, which is the default option, Vegas determines whether the event requires resampling or not. Force Resample is similar to the resampling found in older versions of Vegas, in which Vegas is instructed to resample the event whether it requires it or not. Disable Resample turns resampling off.

Velocity envelopes have been used in nearly every form of special effect there is and can be used for titles, movement control, color keying, and other forms of effects. One favorite is to use an envelope during a title shot, slowing or freezing the subject matter, panning/cropping the media as though the camera zoomed in on the subject, and dropping the opacity of the media so that the title can punch through. This method is exceptionally effective when coupled with a mask and moving media behind the text. Velocity envelopes can be applied to all moving media, including GIF files.

Tip

Here's a tip for using velocity envelopes on long files: use the Split tool (S key) to split the clip before the actual change/handle of the velocity envelope. Otherwise, the envelope affects the entire file and slows down the render process if the rate is accidentally set to any level but 100%, when only a small portion of the event needs to be affected and perhaps resampled.

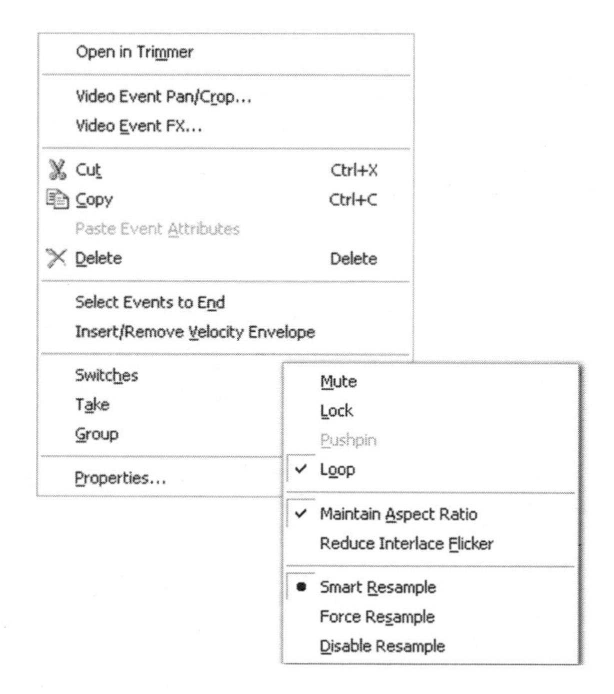

5.58 Right-clicking an event calls up the Switches dialog.

Event Switches in Vegas

Right-clicking any event in Vegas 4.0 calls up a Switches dialog. These switches dictate how selected events are treated. Events can be muted, locked, looped, video events maintained to aspect ratio, interlace flicker reduced, and audio events normalized. As mentioned previously, Smart Resample, Force Resample, and Disable Resample are also options in the Switches dialog.

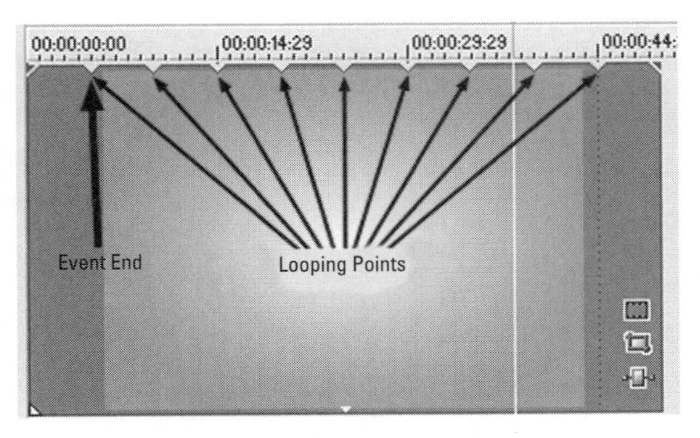

5.59 Looping events is helpful for extending titles and generated media.

Muting a selected audio event or video event prevents the event from being heard or seen on the timeline. Muted events are dark on the timeline, demonstrating the muted status.

Locking an event prevents it from being moved or edited. Locking events is good practice when a section is complete in a long-form video. Locking events causes those events to be grayed out, indicating that they are locked.

Looping events causes events to repeat if they are dragged past their actual end. This option allows events to be extended on the timeline.

If FX have been applied to an event that is being looped, the FX holds at the last keyframe. This option can be a valuable artistic editing tool with titles, generated media, and backgrounds. If a loop is applied to a generated media event, keyframes loops with the generated media length.

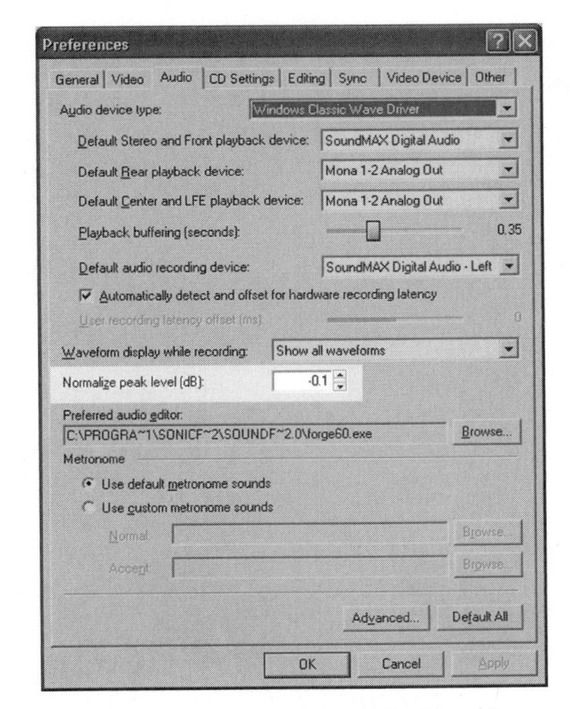

5.60 Setting the level at which normalization peaks.

Normalizing an audio event is typically for bringing audio to its maximum volume, based on the highest peak found in the audio. When normalizing is used, Vegas looks for the loudest point of the audio and adjusts the entire file to the loudest possible level without clipping. The maximum level for normalization can also be set manually, if necessary.

Aspect ratio lock can be applied to an event as well. This option prevents Vegas from stretching the aspect of the events' contents should it not match the project settings.

Sometimes graphics or stills that are converted to video do not display well on an interlaced monitor, depending on the colors and resolution. Reducing interlace flicker generally repairs this condition. Without applying the reduce interlace flicker, graphics and stills often show fringing on video edges or show crawling lines on the screen.

When editing long-form video and some short-form videos, it's helpful to be able to lock various events together in a group. By doing this, cut, copy/paste, and delete can be executed simultaneously, and multiple events can be moved simultaneously. It's also an efficient means of tracking sectionalized video.

To create a group:

1. Select all events to be included in a group.
2. Press G or right-click any event.
3. Select GROUP | CREATE NEW in the submenu that opens.

No limit exists as to how many groups may be created; however, no two groups may share the same event.

Groups may be shifted, copied and pasted, or deleted together. After a group is created, right-click any event in the group to select all events in the group. Select GROUP | SELECT ALL from the submenu.

Audio and video files are grouped by default when placed on the timeline. To remove audio or video from a group or to separate them, select the event to be removed from the group and press U to ungroup that event. The event will be removed from the group.

To remove all events from a group, right-click any event in the group and select Clear.

Grouping can be temporarily ignored by clicking the Ignore Event Grouping button on the toolbar.

This option instructs Vegas to ignore the groups temporarily, while allowing events to be moved. After this button is released/selected again, however, the group is restored with all event members.

Groups or individual events can be locked on the timeline to prevent them from being moved.

The Shuffle feature, which is part of the new Ripple Editing toolset, is also a powerful and efficient means of creating a storyboard and rough cut on a single Vegas timeline.

At Sundance Media Group, we commonly use one instance to rough cut, one instance to edit/finish, and a third instance rendering in the background.

In a facility that uses multiple editing stations, one editor creates the rough cuts and saves them as a rough-cut VEG file. A second editor grabs the VEG file without ever copying media and does the color correction and finish work on the project. With a network, this workflow is exceptionally fast, particularly in a small company setting.

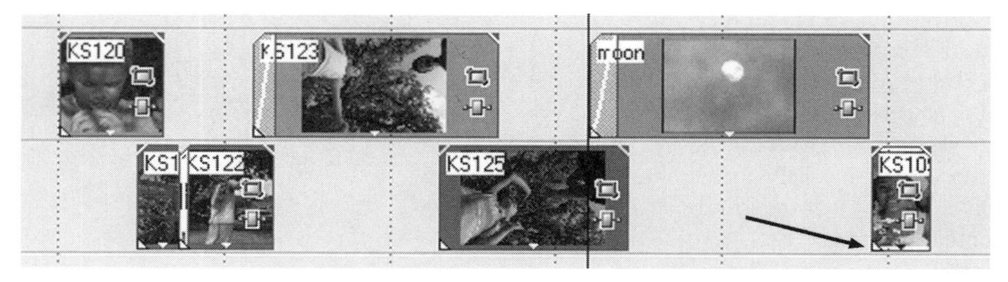

5.61 Six events are grouped. Notice the dark color at the top and bottom of the events. The seventh event is not grouped. Notice the white at the top and bottom, which indicates that this event is not part of the group.

Some NLE systems use a layout concept called *storyboarding,* which uses thumbnails that can be dragged into a specific order and placed on the timeline for fine editing. Vegas does not have this feature. The following two methods, however, can be used for those desiring to edit with a storyboard workflow.

AVI clips in the Windows Explorer (not the Vegas Explorer) show thumbnails if the thumbnail view is enabled. Clips can be shuffled from position to position in the Explorer, creating a sort of storyboard that works for most general purposes.

A more powerful and useful method of storyboarding is to have two instances of Vegas open at one time. The first instance is the active editing workspace, and the second instance is used for creating a storyboard/rough cut that is then copied and pasted into the first instance. What makes this option the more powerful storyboarding tool is that comments, markers, regions, and trims, in addition to storyboarding capabilities, are retained in the second instance. With either a single or dual monitor setup, this is a fast and powerful way to storyboard and edit.

Create a rough cut in the second instance, copy the timeline by selecting a region (press CTRL+A) or the entire timeline (press CTRL+C) and then paste into the first instance (press CTRL+V). Be certain to have "Automatically save markers and regions" enabled in the OPTIONS | PREFERENCES | GENERAL dialog. Otherwise, markers and regions created in the copy-from instance will not be carried over into the copy-to instance.

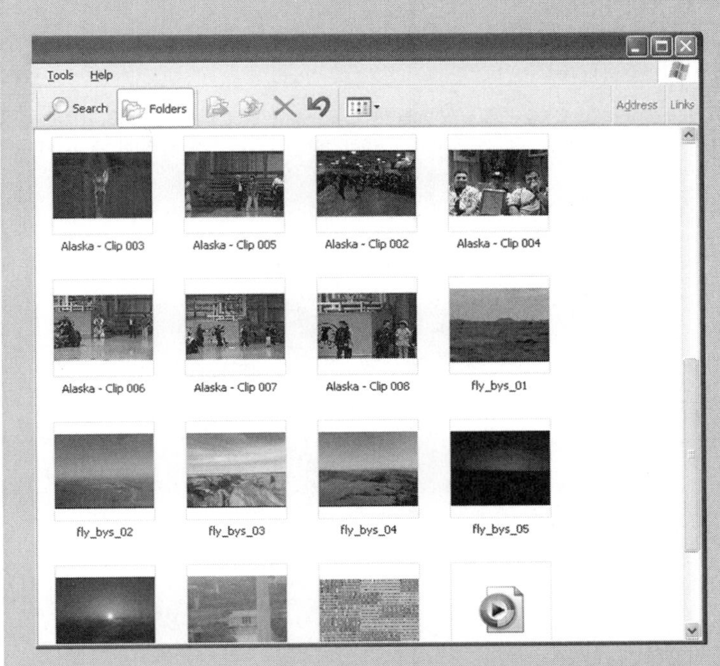

5.62 Notice that clips are not in numeric order, rather they have been shuffled to create a storyboard.

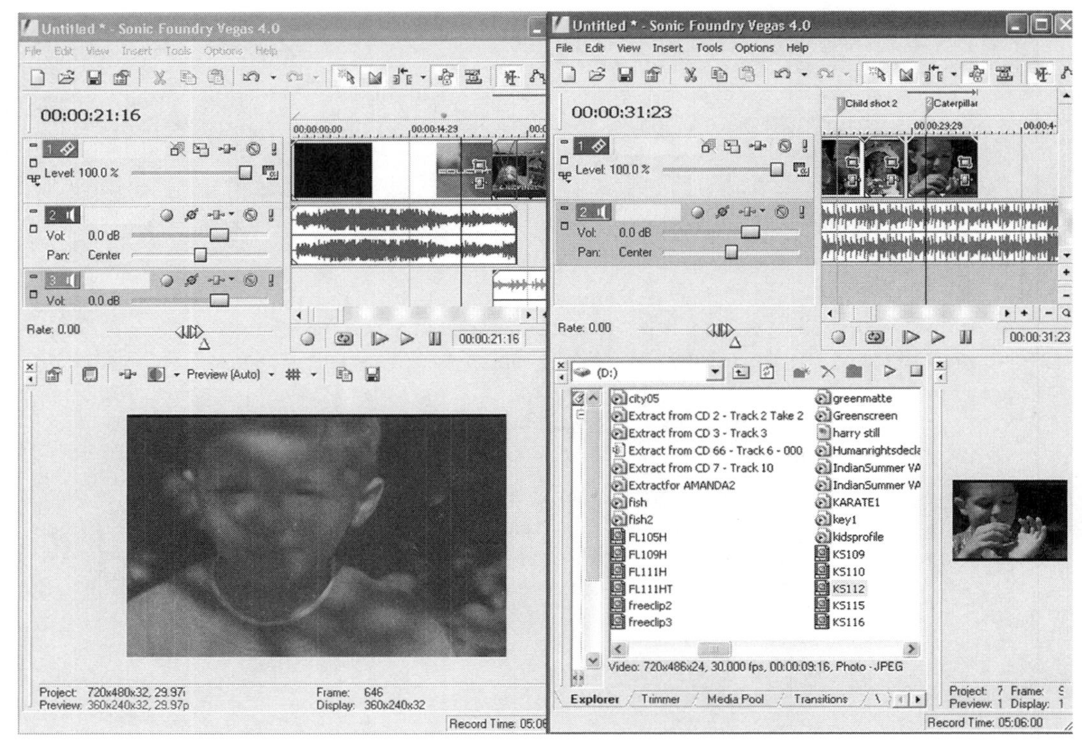

5.63 Two open windows of Vegas 4.0. The first instance, or copy-to instance, doesn't need the Explorer, Trimmer, or other tools open. The second instance, or copy-from timeline, is where media location, sequencing, markers, and other finishing aids are created.

Filters and Add-ons

Working with Plug-ins in Vegas

Several third-party manufacturers have stepped up to Vegas and have written plug-ins for the product. The tools that come with Vegas provide a wide variety of additional expressions and emotions for editors in Vegas; however, these plug-ins give editors far more arrows for their bows, as well as a feel that the Vegas plug-ins don't come with. Pixelan Software, BorisFX, and Debug-Mode have released powerful new tools for Vegas, and demos of both of these products are found on the CD in the back of this book.

Pixelan SpiceMASTER 2

Pixelan, long famous for their plug-ins for Avid, Adobe, Pinnacle, Media 100, Ulead, and other fine editing products, created the SpiceMASTER plug-in several years ago. Often mistakenly thought to be just a transition tool, this plug-in provides tremendous power to users. This tool has 500 BMPs alone that can be used as backgrounds, foregrounds, masks, and more. Combine this feature with other tools, and all of a sudden, there are millions of possibilities.

6.1 Spice explosion with Bump Map applied over Generated Media makes a very realistic solar flare.

Installing SpiceMASTER takes only a few minutes. Five hundred Spices provide the backbone of how images are manipulated, and the Spices are automatically installed to a folder that you can find quickly and easily. When working with Vegas and DV, load only the DV/720 × 480 Spices (for NTSC) and 720 × 576 (for PAL), as the other sizes will be of little use in Vegas. (If other applications are used, loading other Spice sizes might be of benefit.)

After installing the SpiceMASTER application, open Vegas and insert a video track, placing two events on a track with a crossfade.

In the Transitions tab, select the Pixelan SpiceMASTER 2 plug-in. Drag it to the transition between the two events, or right-click the crossfade and choose Pixelan SpiceMASTER 2, and a dialog box opens. This dialog can be confusing to first-time users. The dialog box instructs users to disable the Sync Cursor button in the Vegas keyframe timeline.

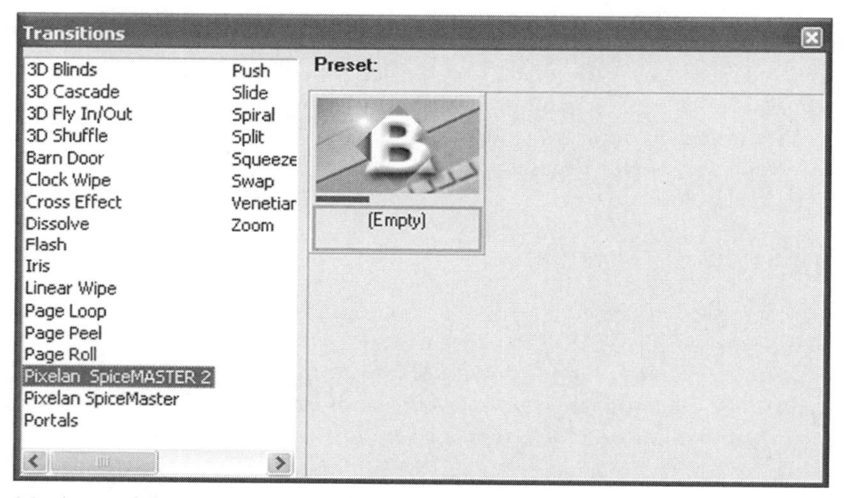

6.2 Locate SpiceMASTER 2 in the Transitions tab.

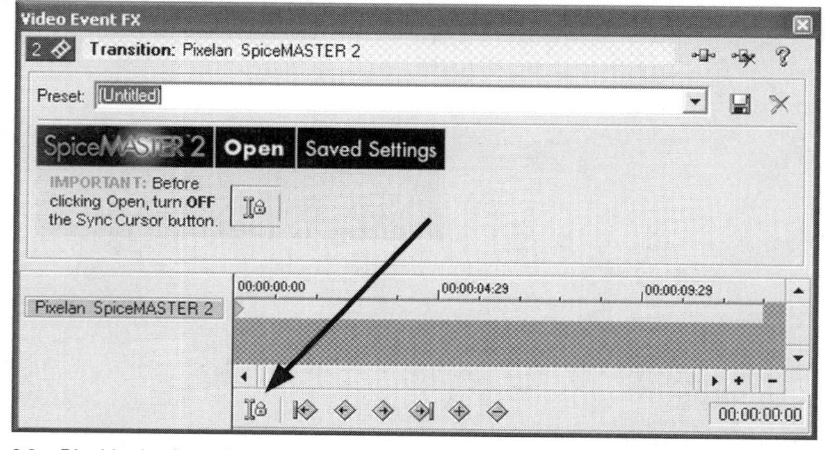

6.3 Disable the Sync Cursor button in the Vegas 4.0 keyframe timeline.

The Sync Cursor button must be disabled in order to use some of the more powerful features of SpiceMASTER. SpiceMASTER 2 has its own keyframe timeline that allows for a level of control that the Vegas keyframe timeline cannot allow on its own.

The first window that opens shows the keyframing area, Preview window, the Choose Spice File button, and controls for how the Spices behave and appear.

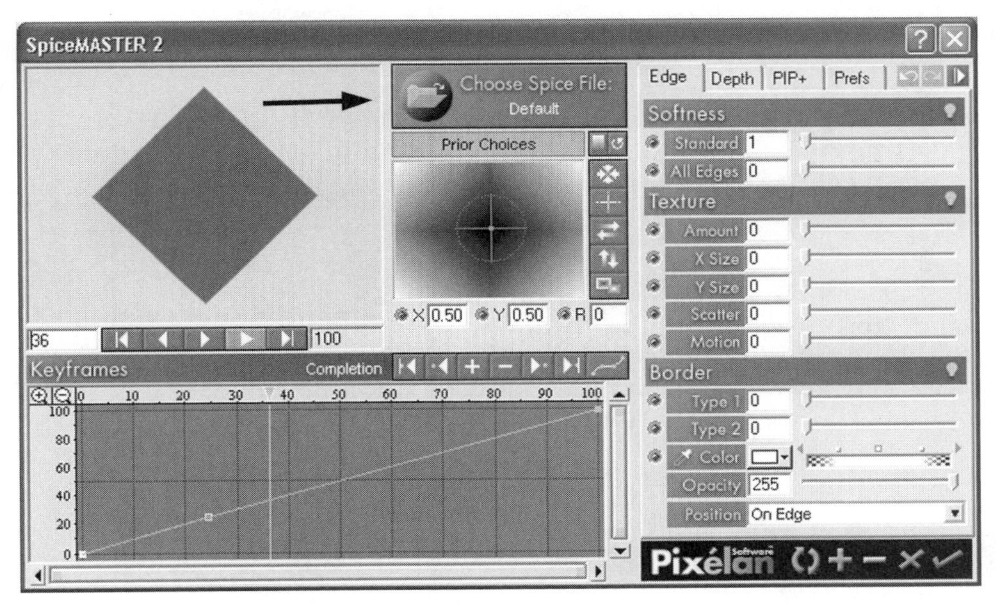

6.4 Main SpiceMASTER 2 window with the keyframing area to the lower left, the Preview window in the upper left, and the control panel to the right.

The Keyframe window is where the velocity or attack/release of the Spices are controlled. The keyframer controls every aspect of how a Spice behaves, based on user input. SpiceMASTER has added a Bezier curve control that allows the user to smooth or ramp keyframes. The keyframer allows images to be brought in and out with a Spice, to be frozen, and even to be reversed. The line in the keyframer is a linear indicator of how the Spice will flow from start to finish. In the case of a transition, the keyframe controls the outgoing event, with Spice behavior on the left and the incoming event controlled on the right. Where the Spice falls in the center and how it behaves is entirely up to the user, who can slide the curve to satisfaction. A line from 0 on the left to 100 percent on the right and without a curve represents a traditional, straight transition. A line shaped like the line shown in Figure 6.5 creates a slow outgoing transition with a slight hold in the middle of the transition and a rapid curve to the new event at the end.

With the two events on the timeline and a SpiceMASTER 2 transition inserted, click the Choose Spice File button, which opens the menu of Spices from which you can choose. Spices are animated in the Library, demonstrating how the motion of the effect will appear in the transition. Select a Spice. The effect is shown in outline in the SpiceMASTER's Preview window.

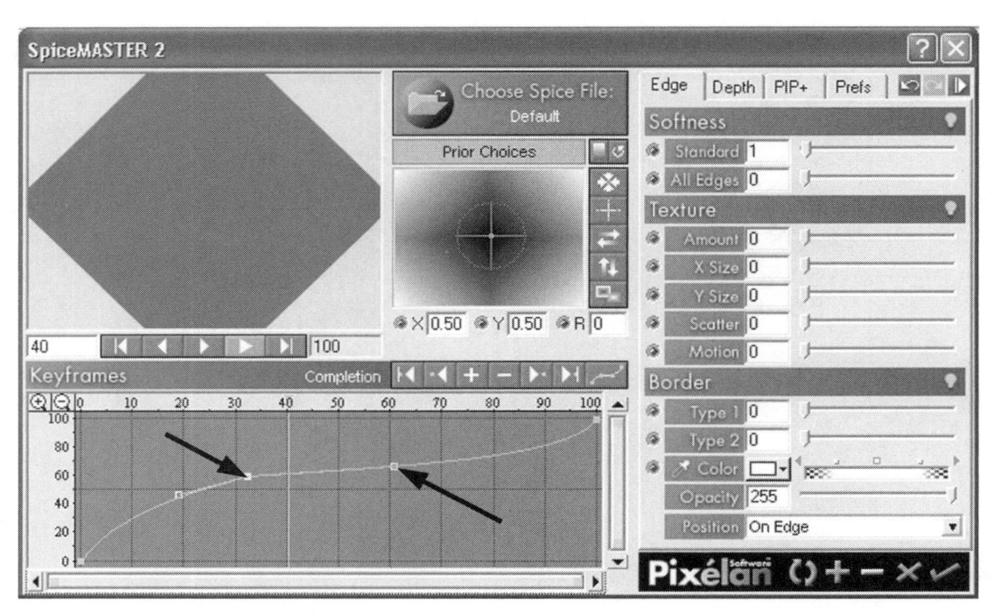

6.5 Notice the keyframes holding a curve, controlling how the Spice flows between two events.

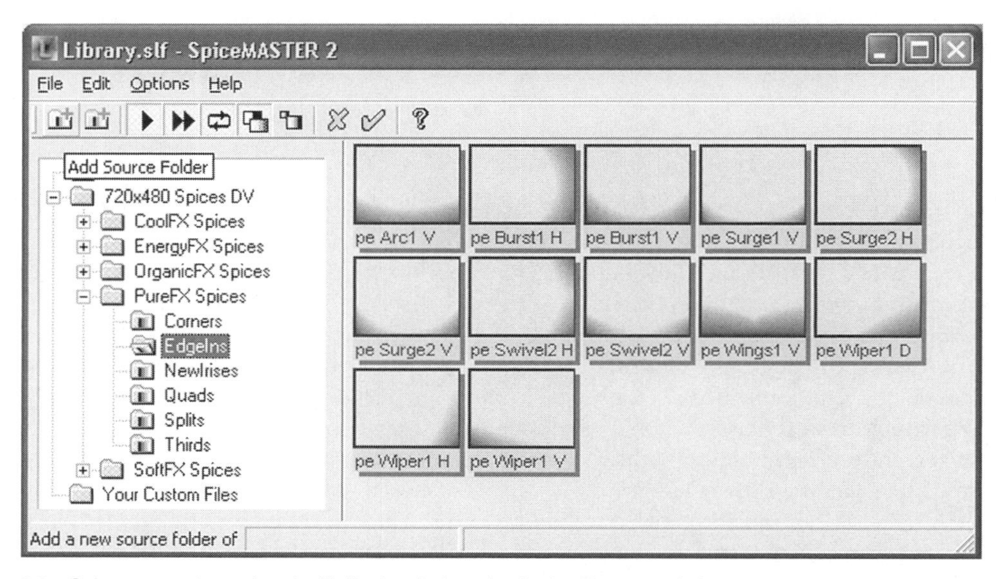

6.6 Spices are animated and will display their action in the Chooser window.

Just below the preview image is a Transport bar. This bar has a Play button that can be used to play the built-in preview or to show how the keyframe points are controlling the effect. This option saves time by allowing users to see how a Spice transition will flow in real time. At the same time as adjustments are being made to the keyframes in the keyframing tool, the Vegas Preview window can also be active, showing in real time what adjustments are being made to the transition and Spice behavior.

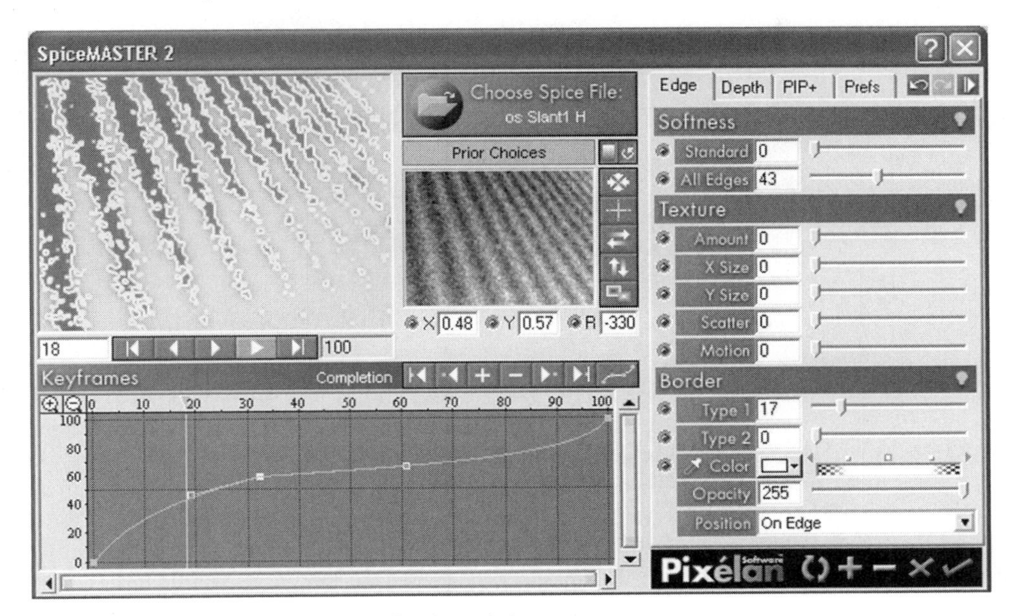

6.7 A selected Spice displayed in the Preview window.

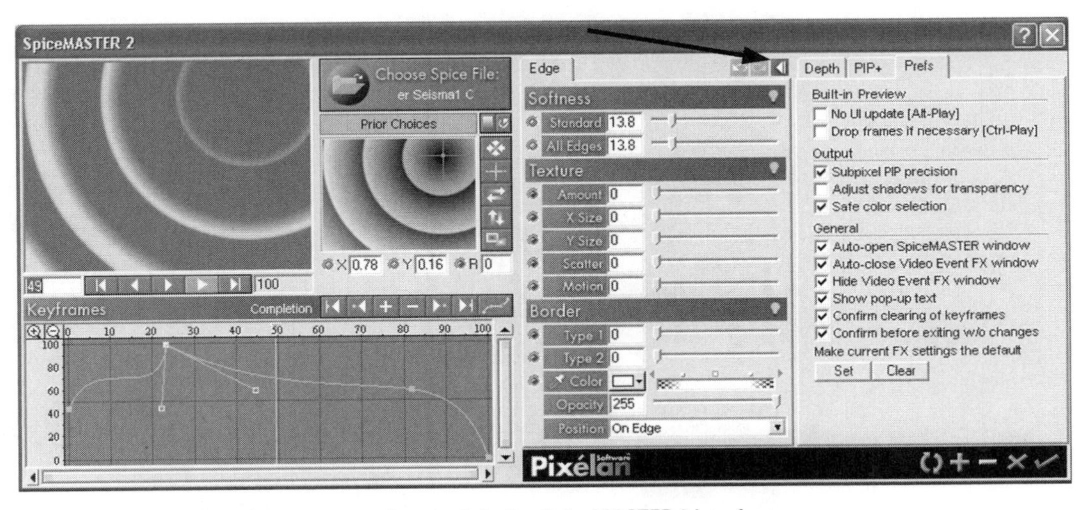

6.8 Click the Dual Pane button to see all controls in the SpiceMASTER 2 interface.

The keyframer takes a bit of adjustment for Vegas users familiar only with Vegas' keyframing techniques. The concept of keyframing in SpiceMASTER, however, is exactly the same as Vegas, minus the sync cursor capability. Users will find the keyframer in SpiceMASTER more capable and very fast to use. The cursor bar in the keyframer allows scrubbing in the keyframe window in order to see exact points in the behavior of the Spice file.

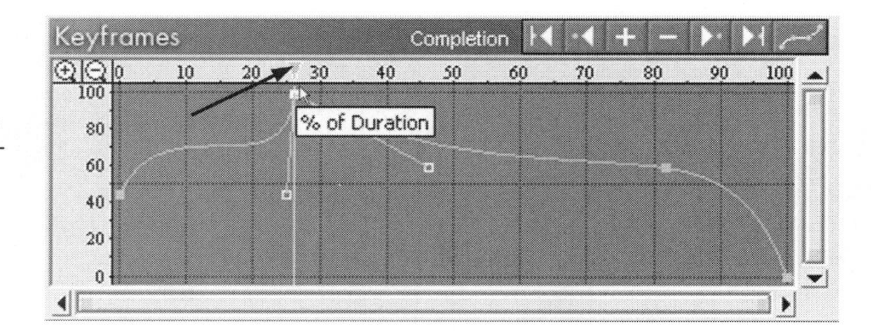

6.9 Above the keyframe tool is a cursor that can be scrubbed across the keyframer, allowing for frame-by-frame adjustment of keyframes.

Using the mouse or cursor, grab the top of the cursor bar in the keyframer and scrub across the timeline. Notice how in both the Vegas Preview window and in the SpiceMASTER Preview window, changes are seen as the cursor bar rolls over time. This ability allows keyframe points to be set to exact moments in time. Stop the cursor in the keyframer where a desired change should be made. Adjust for softness, adjust for color or transparency, or just adjust the curve. A new keyframe is inserted. When the new keyframe is inserted, observe the small Bezier curve indicator that appears to the right and left of the keyframe. This small line is what controls the Bezier curve properties of the keyframe. Raising the line upward causes the keyframe to adjust the image rapidly, while lowering the curve slows down the keyframe's adjustments to the image, depending on whether the event is the incoming or outgoing event.

6.10 Changing the Bezier points in a file, coupled with the use of a Climactic Spice, gives a sense of foreboding in this image overlay. Using a standard overlay without a Spice would be too sharp or clear and would not have the "mysterious" nature of this image.

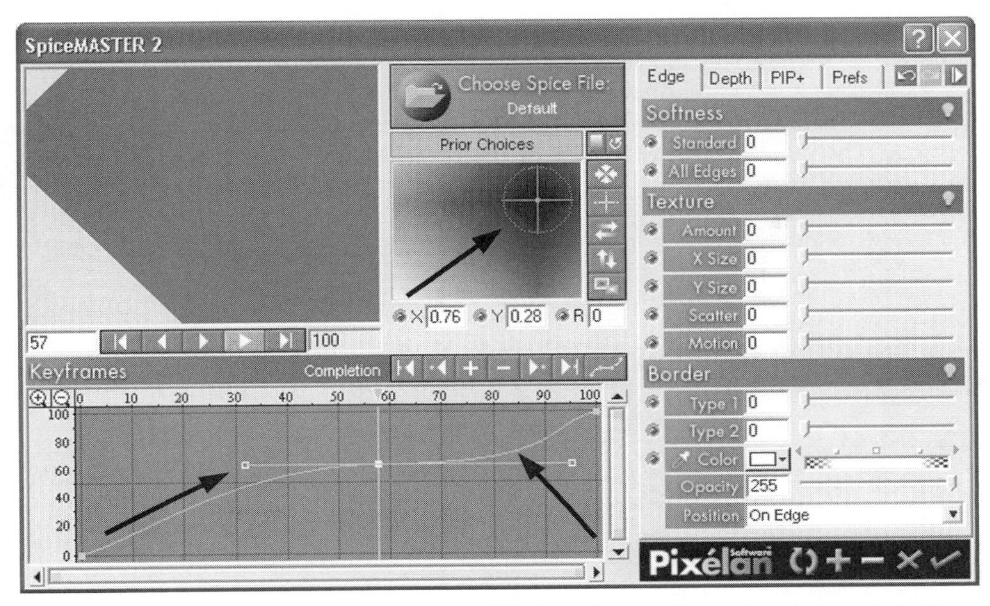

6.11 Bezier curves control the velocity and depth of the incoming event's transparency, coupled with the moving position of the transition. Transition appears slowly in the upper-right corner, moving to the center, with a rapid slide to fully opaque in the last frames of the transition.

6.12 Image appearing from the upper-right corner as displayed in the previous example.

Nearly every aspect of the SpiceMASTER interface has a built-in preset that can be mixed with other presets to give deep functionality and a unique transition between events. Using a softness preset coupled with a texture preset presents more than 1,000 possibilities alone, and all five of the control panes contain dozens of presets in addition to any random manual adjustments you might wish to make. To reach the presets, select the Light Bulb icon found on each control pane.

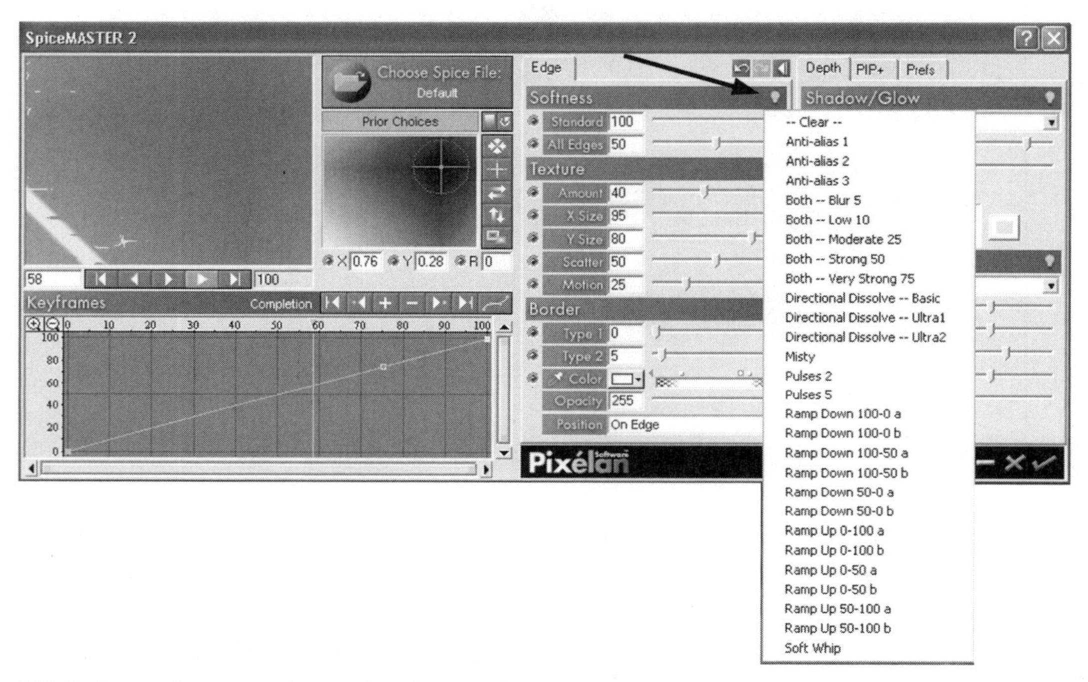

6.13 Each control pane contains a series of presets for users starting out. Personal presets can also be saved.

The true magic behind how Spices function is the ability to make Spices more organic and rich in motion and depth, in addition to allowing editors more creative inspiration. SpiceMASTER and accompanying Spices are not simply transitions. The Spices can be used as filters or as associates to filters in Vegas that create extremely deep and creative images and effects over text, still photos, video, or generated media. Combined with bump or height mapping, the new compositing tools found in Vegas 4.0 take on even greater power and opportunity for creative flow.

Inserting a Spice as an effect or filter is as simple as dragging the SpiceMASTER effect from the Video FX tab and dropping it on an upper event. (You'll need two tracks; otherwise, the Spice will be flowing over the default solid color background.) This process is useful as a means of showing an event via a created mask, for creating a virtual transition, or for moving a Spice over an event as if it were an event on a separate track. The angle, opacity, softness, and so much more can be controlled in the SpiceMASTER tool, but possibilities are virtually without limit. Colors can be applied to borders with one color on the outside border and a second color on the inside border, with varying opacities for each border setting. Keyframes can be applied to change the opacity or color of the borders, or simple presets can be applied.

6.14 Same transition as in Figure 6.12, with a texture and softness preset applied.

Spices can be added to an unlimited number of tracks. One creative opportunity is to place three or more tracks of video on the timeline, with different Spices flowing over each track. Shift different Spices around to find what best displays all tracks as a deep composite. Spices make incredible masks that can be inverted, softened, position-shifted, and more.

6.15 Combining a Spice with a Cookie Cutter Mask and a Spherize filter creates the appearance of a super fish-eye lens coming out of a fog-like mist.

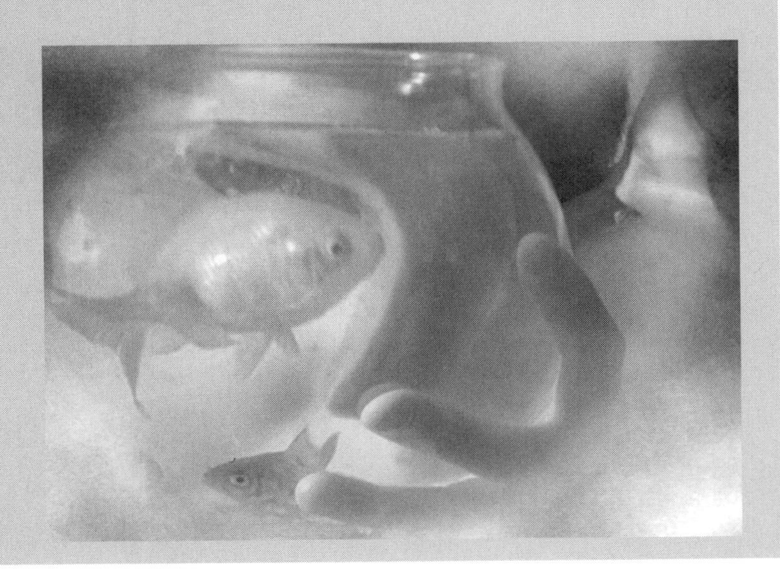

Spices can also be used directly on masks, creating beautiful flowing titles or other masked information. Using multiple Spices on multiple masks and events allows for even greater creativity. In the next example, the image flows from one Spice to another without one Spice interfering with the flow of another. One Spice is applied to the mask/letters, while a second Spice is applied to the generated events, acting as a child to the parent. The background event also contains a Spice masking a still of camera lenses in a very subtle background motion.

Extremely soft, subtle transitions or filter flows can be created; in fact, far too many to list in this book. The tips and tricks presented here were intended to whet your appetite.

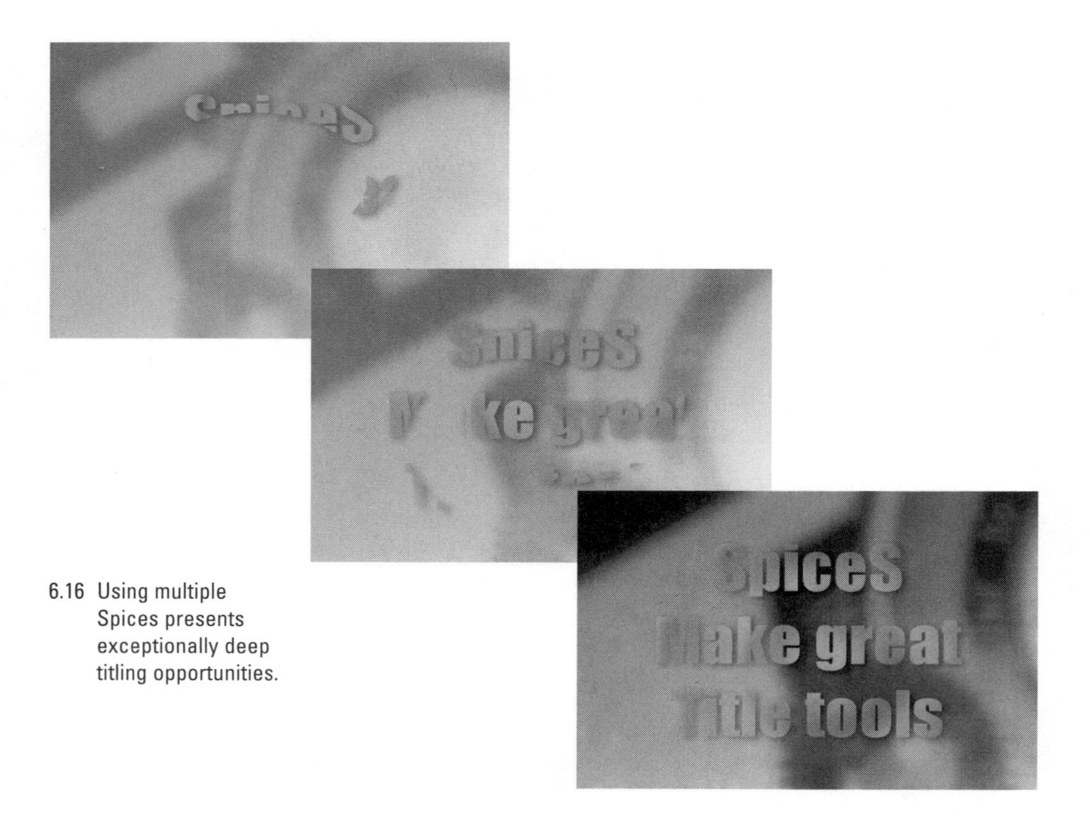

6.16 Using multiple Spices presents exceptionally deep titling opportunities.

Tip

The SpiceMASTER 2 projects seen in the previous examples can be found on the disc in the back of this book, in `spices.veg`.

SpiceFILTERS

SpiceFILTERS, also from Pixelan, open many, many doors in Vegas. Five primary filter tools are available: ChromaWarp, FlowTexture, NewNoise, StepMotion, and StepTime.

The ChromaWarp plug-in is a color-shifting plug-in that displaces pixels over time or over color space. Psychedelics, inversions of colors, additive or subtractive tints, or even complex color correction can be performed with the ChromaWarp plug-in.

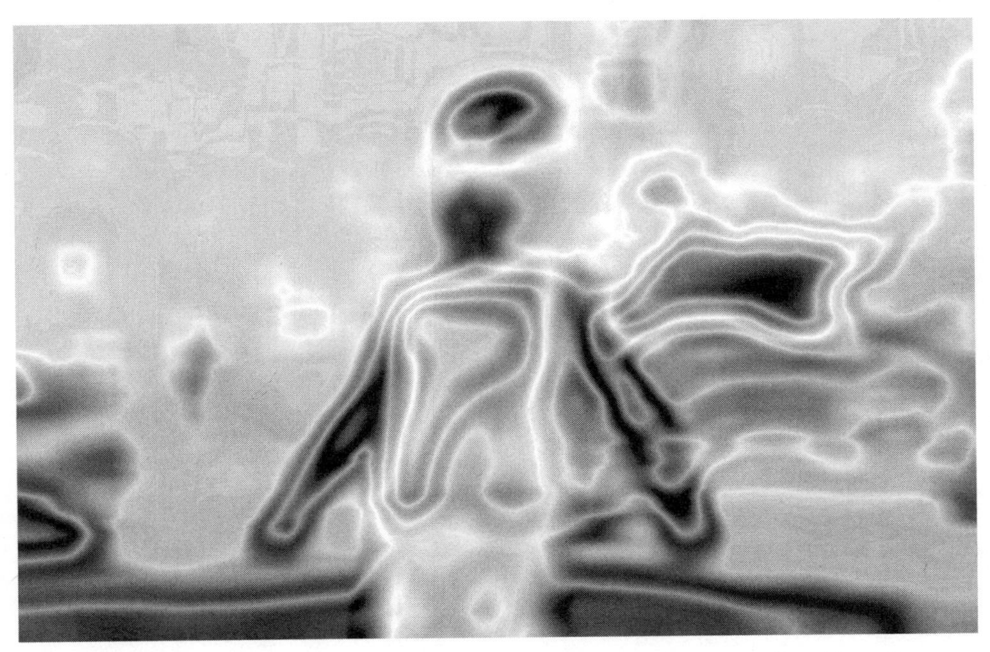

6.17 ChromaWarp can be used to create a psychedelic among many other user-adjustable settings. This plug-in can be used to quickly simulate an infrared heat signature, a negative film image, or other color-based effects.

The FlowTexture plug-in uses Spices from SpiceMASTER or other graphic files to create textures with displacement mapping. Extremely complex textures, bump maps, displacement maps, and rippling effects can be created with this effect. Coupled with other effects, the possibilities are essentially endless. This plug-in is great for flowing over titles, creating motion backgrounds, or for creating unique visual images. The next example provides the appearance of a Monet-style painting over video using the Moderate Displacement preset and some minor adjustment using the Seep1 Spice.

The NewNoise plug-in is used for creating film noise, color streaks, random color patterns, and grain effects among other effects that are noise or granular based. The streaking on the screen in Figure 6.19 is just one of the many presets in the NewNoise plug-in.

6.18 A stippling-style painting is simple to achieve with the FlowTexture plug-in.

6.19 Streaks preview in real time with the NewNoise plug-in. Reduced opacity and a slight blur create the illusion of high-speed motion while an event is playing at normal speed.

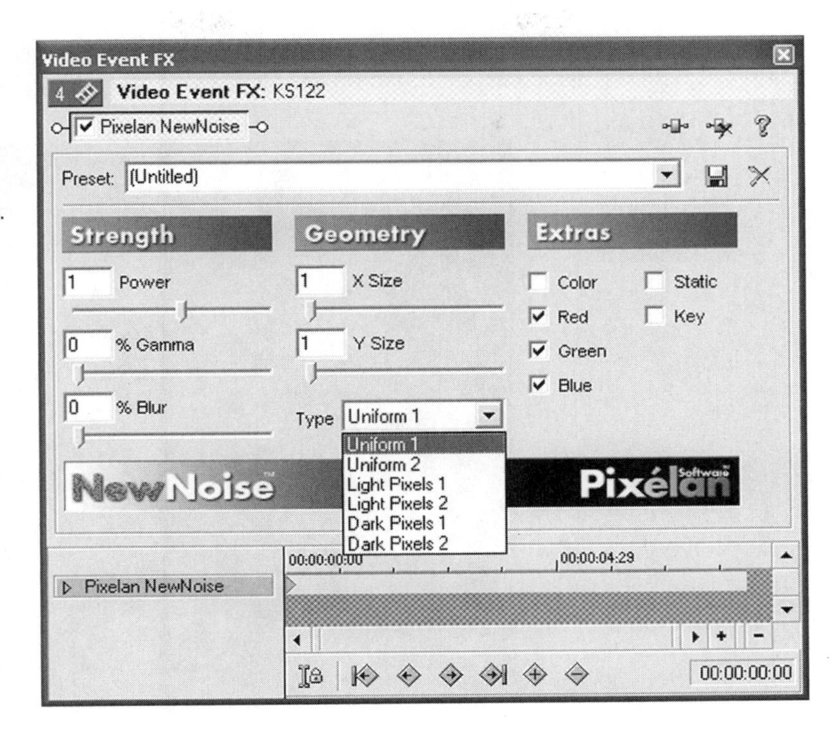

6.20 Properties of the SpiceFILTERS are adjustable and have tremendous options. All filters have several presets in addition to adjustable parameters.

The StepMotion plug-in is similar to an audio time delay or reverb, creating visual echoes of the event. Dark or light pixels can be selected for delay, either via preset or manually adjusted parameters. No wedding videographer should be without this filter; when applied to a bride with a flowing wedding dress, the filter creates a beautiful effect of dream-like motion. Applied to motion graphics, such as the echo rings found on the CD in this book, applied to the sweep hand of a clock, or to another motion over the top of reasonably still backgrounds, the filter creates a stunning effect. Try using this filter on bullets or other fast-moving images on a screen to give a new look to an old image. Images can be frozen in time, ghosted, and echoed.

The StepTime plug-in can also provide ghosting, but more importantly, low-frame rates can be simulated, in addition to strobing, stop action, and more. Using keyframes, various strobe and stop motion effects can be blended or keyed to other clips, music beats, or other effects. By using the key parameter and a strobe, events underlying the event with a strobe on it can be revealed, creating a timed alpha revealing any underlying events without needing to composite them together. This feature is powerful for titles or for creating a dream sequence in which two different images are required to set up the feeling of alternating perspectives. One powerful effect is to take the same image on two tracks, reverse the lower track, and apply a strobe with a key. At the mid-point, the two images come together and then fly apart again. This same technique can be used to create a stunning effect by copying a track and desaturating the lower track so it is black and white, while the strobe on the event above is in color. This process creates a strobe between the black and white event and the color event.

SpiceFILTERS open many doors to creativity, color enhancement/correction, and special effects. Have fun with the demo version found on the CD!

6.21 The StepMotion plug-in acts like a video version of an audio delay. Added to high motion, it provides amazing effects.

6.22 Using a key and a strobe, events can be composited without using a parent/child, with opacity creating a unique good against evil effect. To add some softness to one or the other, add a StepMotion filter to one or the other event in the adjust, or overlay, mode.

6.23 Blend the strobe with a black-and-white image beneath to create color strobing over a monochrome image. The Spice-FILTERS are powerful, effective, and inexpensive.

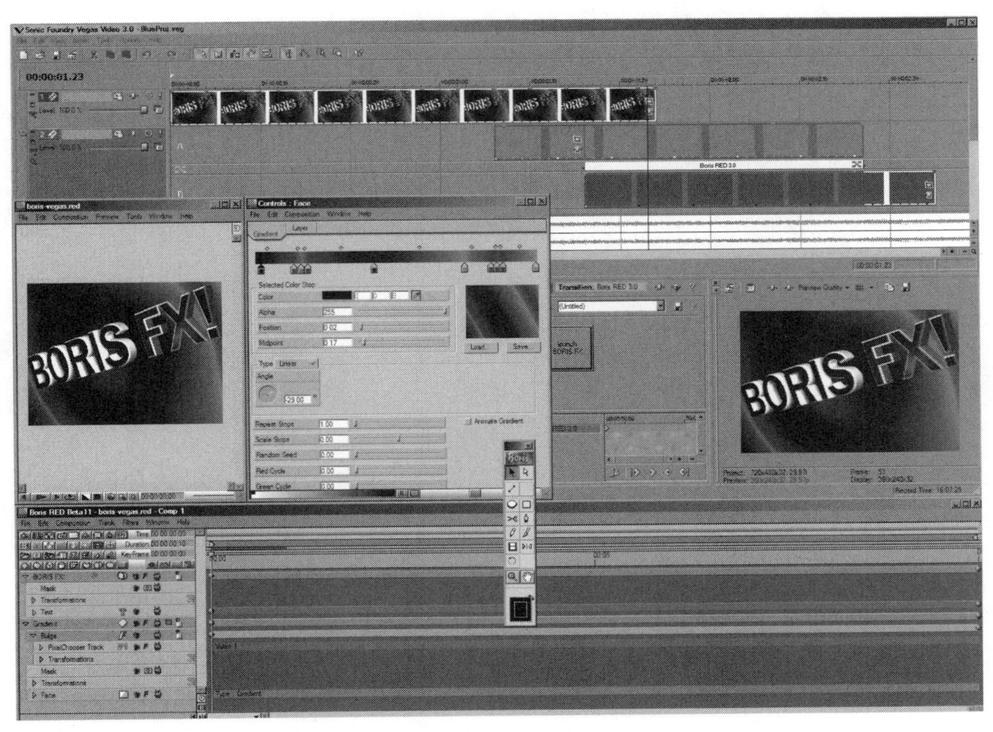

6.24 Boris RED interface working within Vegas 4.0.

BorisFX

BorisFX makes a series of compositing, titling, and special effects tools for a wide variety of NLEs, in addition to providing a stand-alone product. At the time of this writing, BorisFX and Sonic Foundry are developing Boris products for users of Vegas 3.0 and 4.0. Figure 6.24 shows the Boris RED interface working within Vegas 4.0. Boris RED is both a Vegas transition and filter, with a surface generator (or what Boris terms a *synthetic importer*) in the works. BorisFX is very aware of the workflow that Vegas offers users and molds their development around this. Within Vegas, users can both apply a Boris effect to the timeline and preview it in the same manner they are accustomed to with other Vegas filters. After the effect is applied, users can launch the Boris user interface from the native Vegas FX Property page, create their effect, and upon closing, automatically import the Boris effect into the Vegas timeline.

From this point, whenever the Vegas cursor is scrubbed across the Boris effect in the Vegas timeline, users can view the Boris frames natively in the Vegas Preview window. Unfortunately, the plug-in was not complete at this book's press time, but will be available to Vegas users very soon.

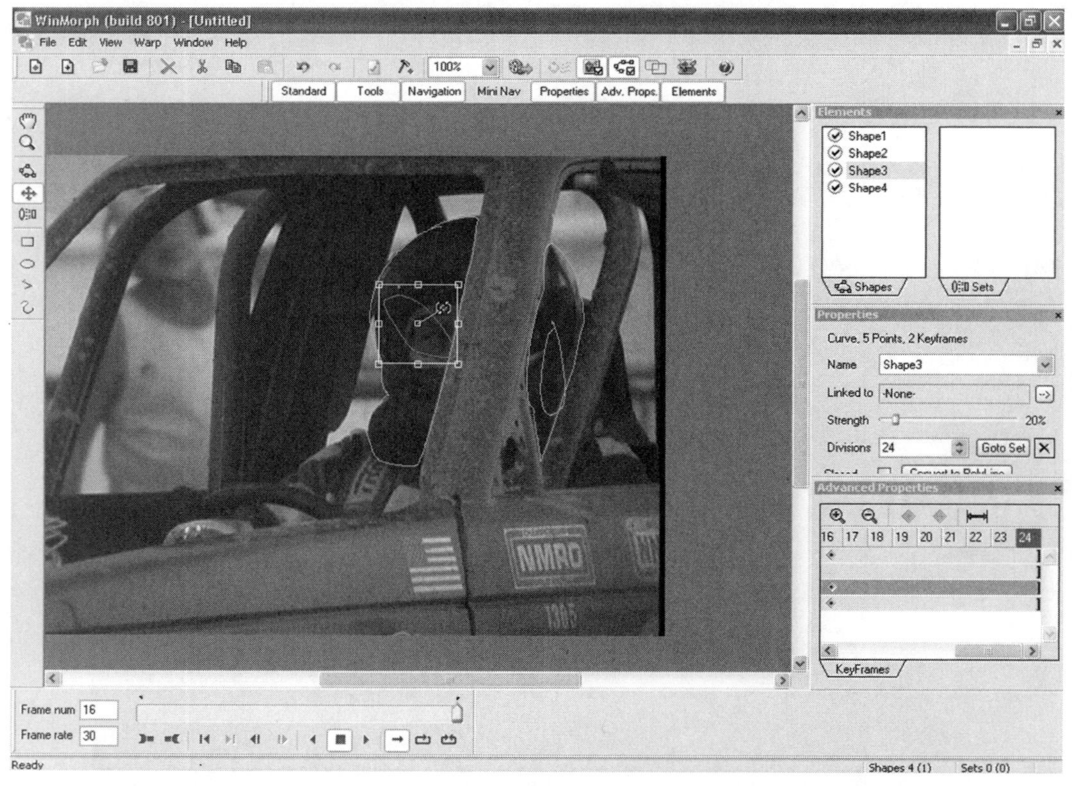

6.25 Drawing morph and warp characteristics can be done with built-in shapes or freehand drawings in the WinMorph workspace.

WinMorph/PluginPac

The DebugMode products that work under Vegas are the result of the work of Satish Kumar, a software programmer who also enjoys Vegas. These plug-ins bring new capabilities to Vegas not otherwise possible with X-Y-Z plane adjustment to create multidimensional imagery.

The WinMorph plug-in allows images similar to those seen in the smash hit movie *The Matrix,* wherein the subject freezes and the background spins around the subject, or similar to the Michael Jackson music video *Black and White,* wherein one face morphs into another with different parts of the image changing at different times. Warps of specific sections in the event, whether a still image or video image, can be created with shifting space based on shapes drawn in the WinMorph main workspace. WinMorph is included on the CD in this book, complete with tutorials on how to use it to create some of these effects.

The Plug-inPac filters offer similar attributes to WinMorph in that shapes can be rapidly shifted in three dimensions and moved on various X-Y-Z planes and aspect ratios. Three Plug-inPac filters are available: PixelStretch, Shatter3D, and 3D LE.

PixelStretch allows pixels at a defined space to be shifted, stretched, and pulled into unbelievable shapes. Laser beams can be created from letters, faces distorted, and, when combined with other plug-ins, amazing images can be quickly created. Like the Pixelan plug-in tools, the Plug-inPac tools can unleash incredible creativity.

As its name implies, the Shatter3D plug-in creates the image of shattered glass or other fragile material. The number and size of the shards or shattered pieces are controlled by the parameters of the plug-in.

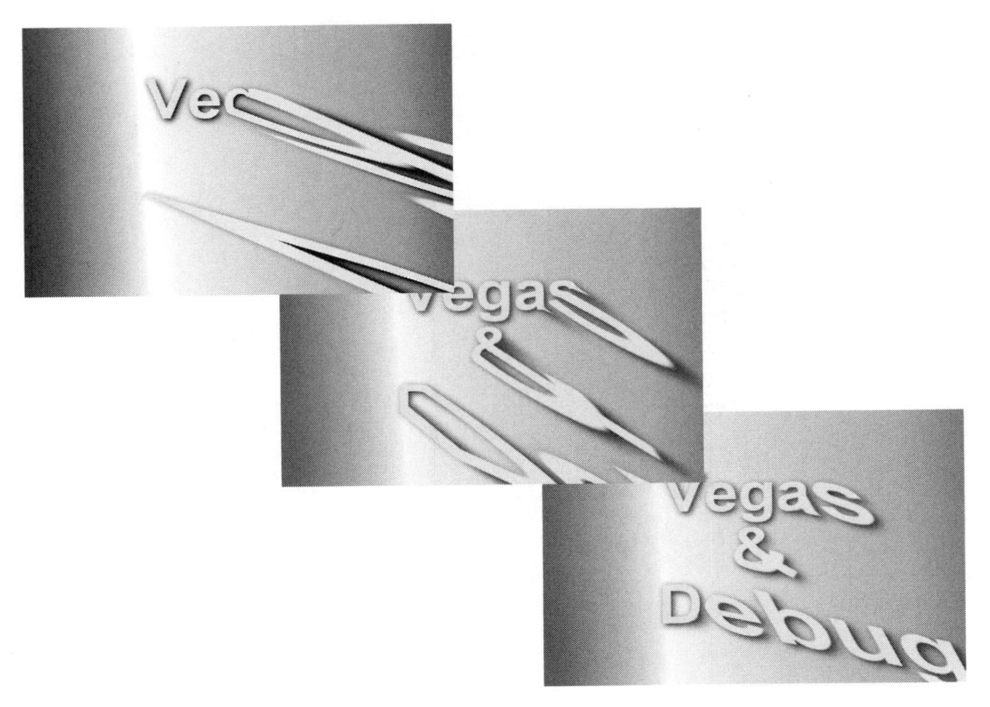

6.26 PixelStretch stretches a defined area of pixels, keyframable over time.

6.27 Shattered images can be very complex with the Shatter3D plug-in.

6.28 The Shatter3D plug-in controls.

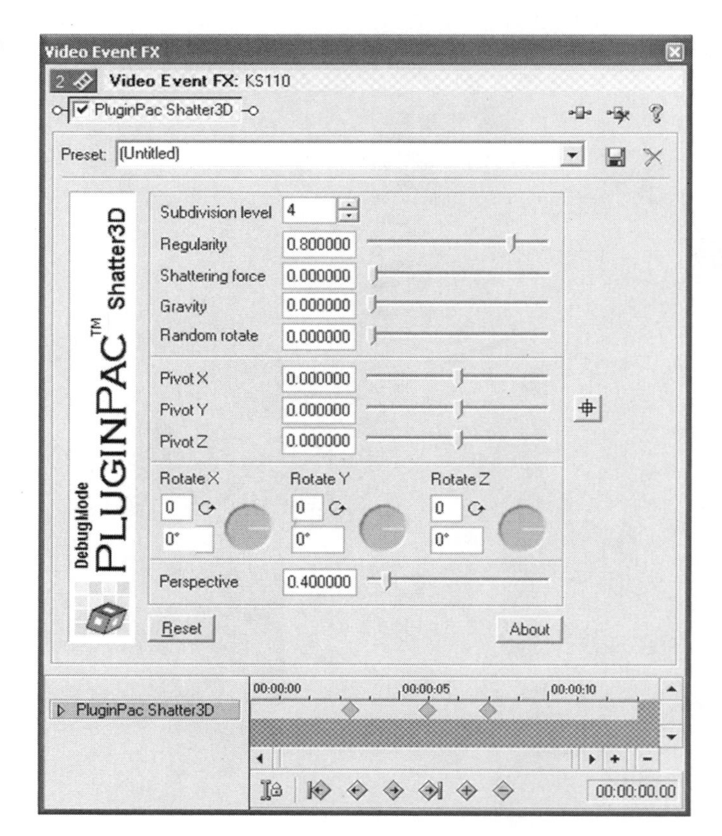

Tip

The images from the PluginPac examples can be found in the `debugfiles.veg` files on the disc in the back of this book.

The parameters of the Shatter3D plug-in allow images to be simple or deep, with keyframe ability over time. Shattering images can be reversed in the middle of an event, allowing an image to be shattered and restored during an event. The plug-in has controls to simulate the amount of force with which the image is struck, the angles of the shards as they fly across the screen, the angle at which the shards are viewed, the rotation of the shards, and the perspective/depth of focus.

The 3D LE plug-in is fantastic for creating 3-D images that fold together and move apart with varying angles and depths and for creating trapezoids and multiplane angles. These can be combined with shadow and glow from the Track Motion tool and used as transitions or event fly-bys. This tool is also great for creating the ubiquitous *Star Wars* titling effect.

Tip

More tutorials and information on the Pixelan and DebugMode plug-ins are available at `www.pixelan.com` and `www.DebugMode.com`. The tutorials for WinMorph on the disc in this book are courtesy of Satish.

6.29 A 3-D transition flies in to lock to one side of the screen before closing like a door.

6.30 Trapezoidal screens created with the 3D LE PluginPac are fast and simple to do.

Audio Tools in Vegas 4.0

Recording Audio in Vegas 4.0

Vegas was one of the early multitrack editors available for the PC before the advent of Vegas Video 2.0, Sonic Foundry's first venture into the video field. Having set the standard for audio on the PC with Sound Forge several years previously, a multitrack tool was a logical step, and like Forge, it became a standard for the Windows environment.

Vegas can be used for a basic voice-over-only setup or as the heart of a full-blown multitrack studio setup for recording live bands, film scoring, television production work, or other professional audio requisite.

For those doing audio for video, it's often thought that video is more difficult to edit than audio. Perhaps so, but audio is the greater importance in the two media forms. Those who would argue the point might want to take any movie, turn the sound off, and see how enjoyable the film is. Even in the days of silent film, a musical score accompanied the film, and most theaters had a piano or organ player who performed the score along with the silent film. In current times, audio and video artists and engineers are crossing formerly strongly drawn lines in the authoring of various forms of media.

Vegas manages media at three levels:

- Event
- Track
- Project

Events contain graphic, video, text, or audio information. Events can be individually edited, and have effects added, processed, and controlled. Events live on tracks.

Tracks contain multiple events. Tracks can control, process, and add effects to all events on the individual track.

Projects contain a single track or multiple tracks. All events contained in all tracks can be controlled, processed, and finished.

Basic Setup

To record audio from a microphone plugged into the computer, a microphone and sound card are needed. Plug the microphone into the soundcard's microphone input and open Vegas. The timeline/workspace will open with no audio or video control panes. An audio control track will need to be inserted, which is done by pressing CTRL+Q or by selecting INSERT | AUDIO TRACK. Click the Record button on the control pane of the new track. Vegas displays a dialog box asking for the location to which audio files are to be recorded.

7.1 Clicking the Record button will call up a dialog asking for the destination of audio.

Tip

Recording to a second hard drive is highly recommended for purposes of keeping audio files off the system drive. This process allows input/output (I/O) to run more efficiently and helps ensure flawless recording.

Audio is now ready to be recorded. Click the Record button on the transport control to begin recording. Click the Stop button on the transport control to stop recording. You won't believe how many people will not think of that! This is all there is to recording basic audio in Vegas. From here, much, much more can be done.

When the Record button is armed (clicked), an icon that looks like two speakers is located in the Track Control pane next to the Record button. If a multichannel card is installed on the system, a number will appear instead of the dual speaker icon.

Clicking the speaker or number icon allows audio input to be converted from stereo to mono or a unique input channel to be selected on the multichannel device. If voice or a single instrument is all that needs to be recorded, it's good practice to use a mono input, as it saves disk space.

NEW! The Input Monitor option also appears in the same menu as the stereo/mono inputs. Vegas 4.0 now allows inputs to be monitored with effects/processing as it's recording. For musicians, this upgrade is fantastic, as a reverb can now be monitored with compression added during the performance. The reverb/compression/other processing is not recorded to the track—it is only there to assist the vocalist in getting the best performance possible, without using additional external hardware. This upgrade gives Vegas the ability to be a multitrack recording system with no additional hardware whatsoever. The ability to monitor inputs with processing is entirely dependent on processing speed and system ability. FX Automation envelopes are bypassed during monitored recording.

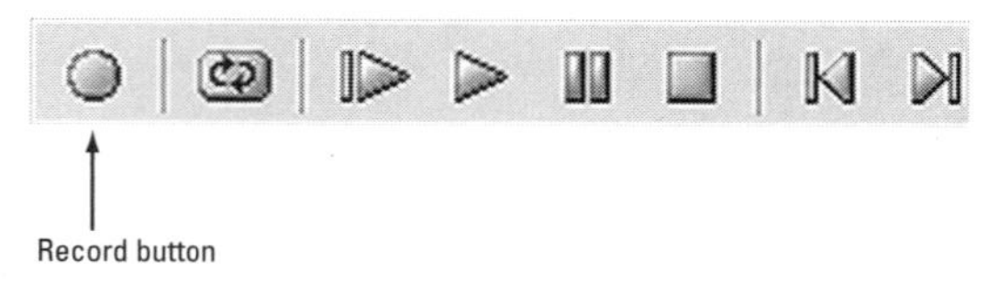

7.2 Click the Record button on the transport control to begin recording.

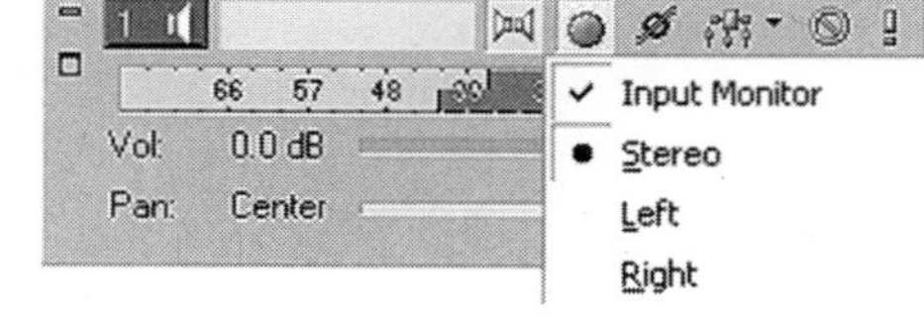

7.3 Audio may be recorded in stereo or mono and may be live-monitored in Vegas 4.0.

Checking the Input Monitor box enables processing to be heard through the output specified in the output/bus assignment dialog. To assign outputs other than the default sound card outputs, select OPTIONS | PREFERENCES | AUDIO and specify input/output settings. If only a two-channel sound card/Blaster-type card is present, nothing more can be set, as there is nothing for audio to be assigned to.

If a multichannel sound card is available, such as an Echo Audio Layla, M-Audio Delta, or other similar card, up to 26 input/outputs can be specified. Multichannel cards are required for mixing 4.1, 5.1, or 7.1 surround sound.

7.4 Echo Audio Layla.

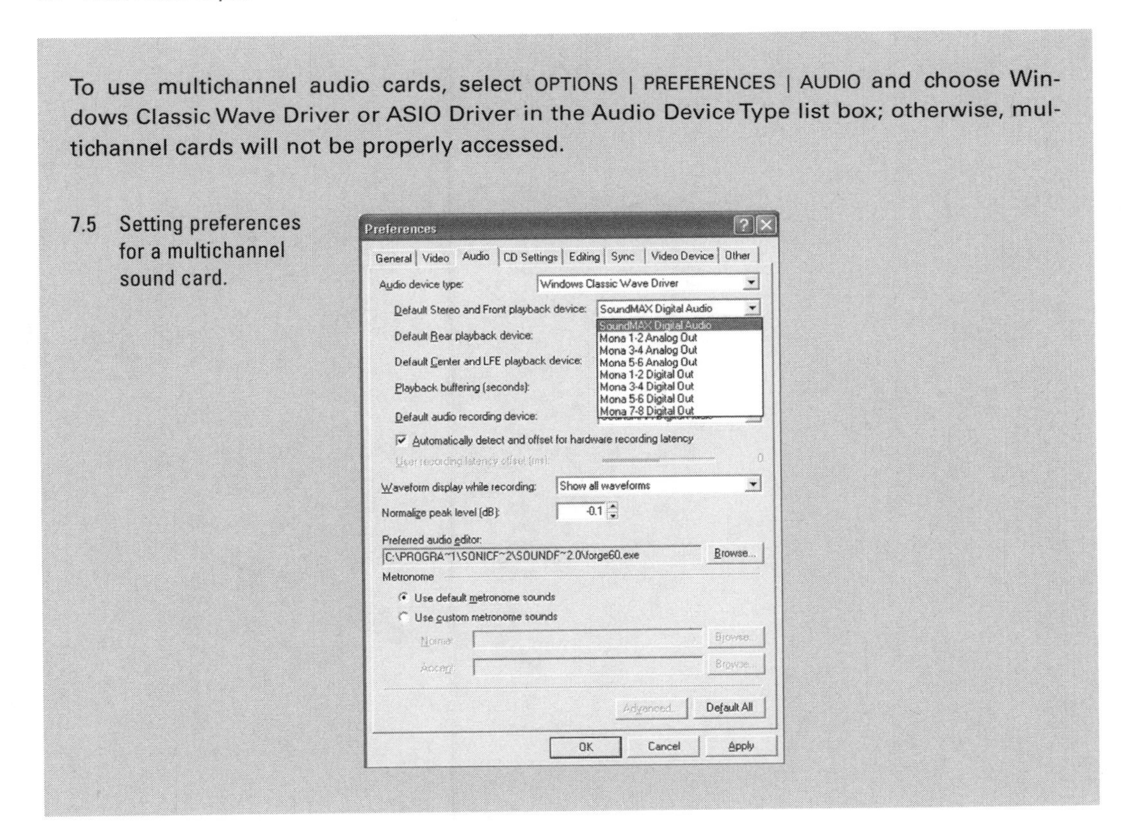

To use multichannel audio cards, select OPTIONS | PREFERENCES | AUDIO and choose Windows Classic Wave Driver or ASIO Driver in the Audio Device Type list box; otherwise, multichannel cards will not be properly accessed.

7.5 Setting preferences for a multichannel sound card.

If latency is encountered with Mark of the Unicorn 2408 interface, perform the following steps:

1. Select START | FIND | FILES OR FOLDERS.
2. Search for a file named `motu324.cfg`.
3. Open it in WordPad or Notepad.
4. Underneath the `DDHELP.EXE=17` line, insert the following line:

```
VEGAS.EXE=392
```

In general, multichannel sound cards do not have microphone preamplifiers (pre's) in them; however, a few models do come with them. In the event that your multichannel card does not have a pre-amp built in, you will need one in order to properly drive channel inputs. If you don't have one, audio will be very noisy and not very loud. Separate pre-amps are typically better; however, many of the built-in pre-amps, such as the M-Audio Quattro and PreSonus FIREstation, are excellent and for the investment, a terrific value.

Additionally, Vegas is capable of working with multiple sound cards. Consult your sound card manual for information on configuring more than one sound card.

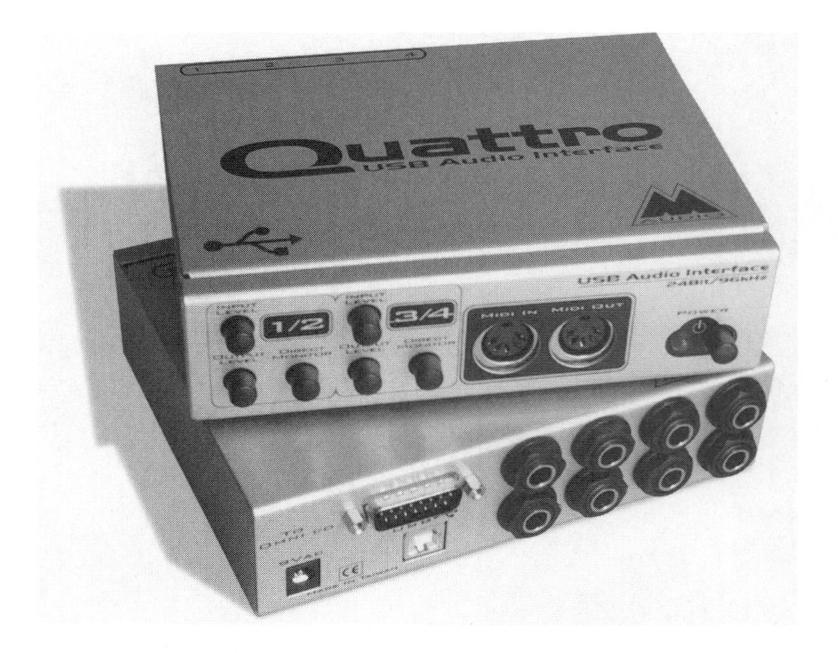

7.6 M-Audio Quattro.

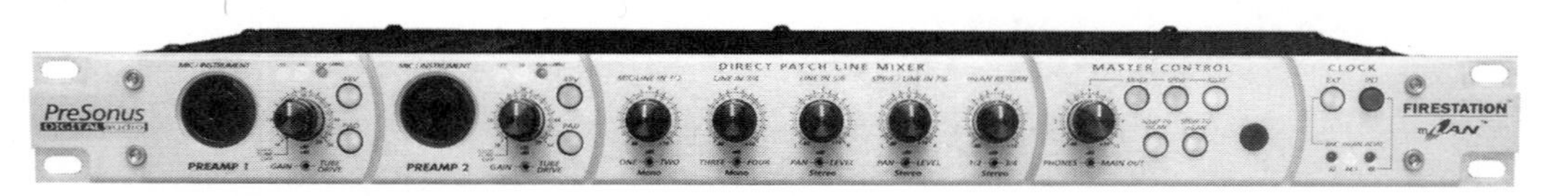

7.7 PreSonus FIREstation.

The volume control found in the Track Control pane is for playback volume only. Adjusting this during the recording process does not affect the recording process at all. Output from the sound card, sound card mixer, pre-amp, or other external input device determines the level of incoming audio.

To adjust the volume of incoming audio when not using a sound card mixer, select START | PROGRAMS | ACCESSORIES | MULTIMEDIA | VOLUME CONTROL. Select OPTIONS | PROPERTIES and click the button to adjust the volume for recording. In the Record Control window that is displayed, recording input volume can be increased/reduced.

Recording audio in the digital realm can be tricky, particularly if one is familiar with analog techniques and practices. In the analog realm, it's acceptable to hit 0dB and sometimes go slightly past 0dB. Tape has a saturation level that digital does not enjoy. Record audio in the digital realm with peaks hitting at around -.03dBFS and averages at around -15dBFS. Because of the low-noise/no-noise digital signal-to-noise ratio, it is also easier to expand audio levels later in the editing/recording process. Bear in mind that digital 0 and analog 0 are -20dB apart based on the Advanced Television Standards Committee (ATSC) standards (www.atsc.org). When using analog meters to view digital information, be sure that tones are used to match up analog meters to digital output. When working with audio for video broadcast, the analog broadcast standard is +4dB into a source impedance of 600 ohms. Traditionally house reference meters are set to 0VU based on the +4dB level. -.03dB is a good maximum level to target for maximum levels at input.

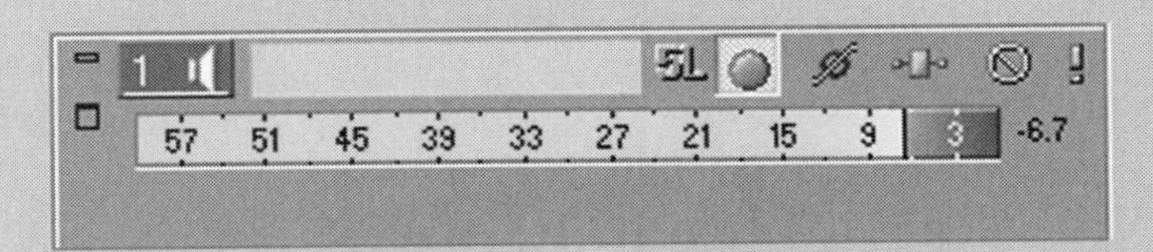

7.8 Levels peaking at approximately -6dB. The instrument is an acoustic guitar with no compression and is prone to unexpected peaks. Therefore, levels are marginally lower than they might otherwise be.

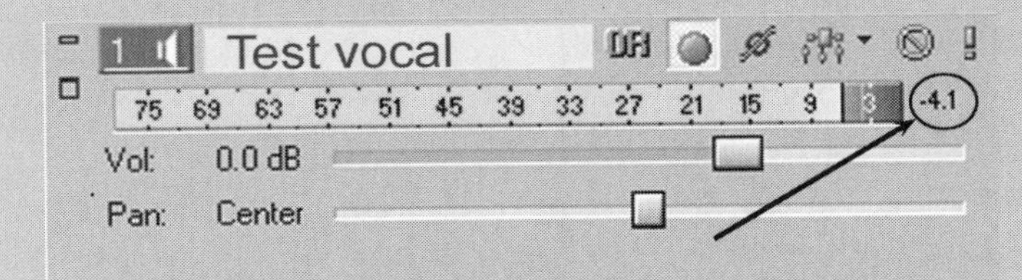

7.9 Level set correctly for the loudest point in recording.

Recording Multiple Tracks

Multitracking is what Vegas is all about. The ability to monitor tracks while recording is critical. Vegas provides this ability with most low-end sound cards and with nearly any multichannel card.

Recording and monitoring tracks is fairly fast and self-evident. Start by inserting a second audio track (SHIFT+Q) and assigning its inputs as you did for the original mono track. If the second audio track is to be mono, set it up this way now. Track 1 will play back through whatever sound card is assigned. Track 2 will play back through the default assigned sound card as well. Depending on the sound card, this process can demonstrate latency during the recording process. This process is also where a professional sound card is invaluable. Professional sound cards use ASIO drivers, which dramatically reduce latency in the hardware/software relationship.

In a computer audio system, latency means any delay or wait time that increases real or perceived response time beyond the response time desired. More specifically, latency is the time between when sound is input to Vegas, processed, and returned to headphones or speakers. A contributor to computer latency includes mismatches in data speed between the CPU and I/O devices and slow buffers. Drivers and buffers can be set in Vegas to reduce latency. Small buffers reduce latency but increase the risk of dropouts. Larger buffers lessen the likelihood of dropouts but increase latency.

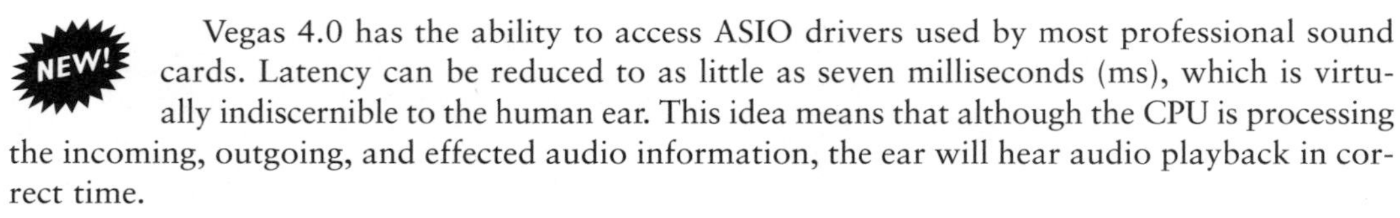

NEW! Vegas 4.0 has the ability to access ASIO drivers used by most professional sound cards. Latency can be reduced to as little as seven milliseconds (ms), which is virtually indiscernible to the human ear. This idea means that although the CPU is processing the incoming, outgoing, and effected audio information, the ear will hear audio playback in correct time.

Clicking the Record button (CTRL+R) on the Transport bar starts the recording process, while at the same time audio is playing back through the sound card.

Advanced Recording Techniques and Tools

Vegas has many advanced tools for recording audio. As mentioned in Chapter 1, Vegas has a *metronome* for monitoring a click track during recording sessions. (This feature requires that Vegas' properties be set for beats and tempos.) While this feature doesn't affect playback or record speed from the timeline, it does affect the tempo at which the metronome is heard. Metronomes are useful for video editors as well, as they can help set a cadence for video to be edited at.

Tip

Ask the scorist for the video for a tempo map, and Vegas can be mapped out accordingly, so that when the score is complete, the audio is dropped into place if tempo-based editing is the goal. The metronome will speed up and slow down according to the tempo map created in the production stages.

Tempos cannot be adjusted over time, as they are fixed in the setting specified in the AUDIO | OPTIONS dialog.

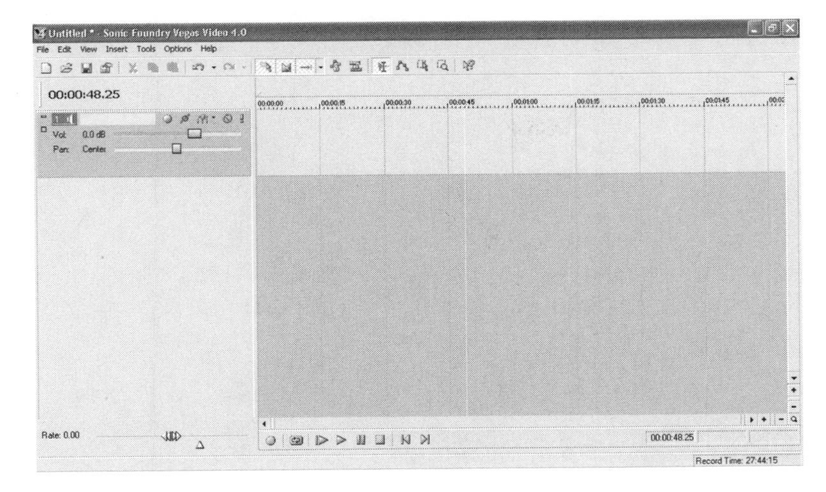

7.10 Setting a tempo grid in Vegas 4.0.

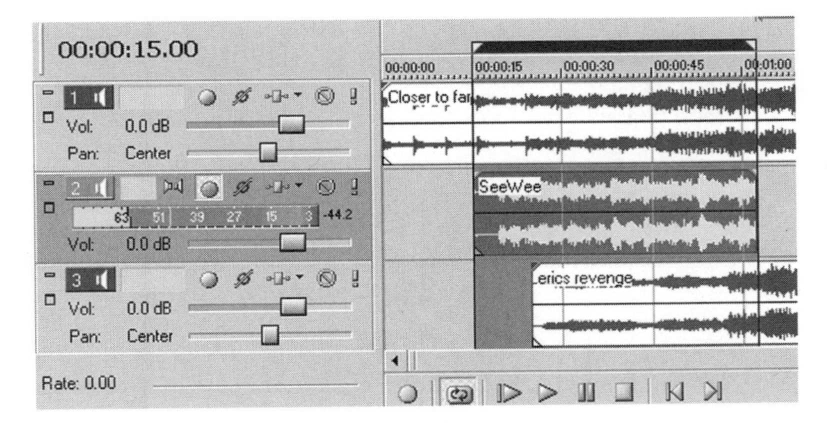

7.11 Completed tempo grid.

Recording with Takes

Rarely is a performance perfect every time; consequently, Vegas allows for takes to be recorded, just like in the movies when you hear "take 1, take 2." In Vegas, takes can be recorded in a number of different ways, as the number of takes is limited only by hard drive space.

Setting up a loop over the area where multiple performances are wanted instructs Vegas to continually loop over the area. If the area that requires multiple takes is a single event, double-click the event, which will set up the loop. Press Q or click the Loop button on the Transport bar.

Click the Record button on the Transport bar, and Vegas begins recording. To stop recording, press the SPACEBAR or click the Record button again. A new take will appear on the timeline, living over the old audio. The old audio is still there. Pressing T or right-clicking and selecting the appropriate menu item toggles back and forth between the various takes. Takes can also be directly monitored, by stopping playback, right-clicking, and selecting Choose Active, and then clicking Play in the dialog box.

7.12 Using Takes is an efficient method of recording a section over and over, keeping only the best performance.

Takes can be renamed to help identify them. Takes can also be opened in the specified audio editor for further editing and then saved as a new take or saved as the name of the take opened in the audio editor. Takes can be deleted from an event as well, leaving only the desired audio in the event.

Takes can also be added to the timeline in yet another fashion, albeit not necessarily in time. Right-click and select multiple audio clips in the Explorer or Media Pool, drop them on the timeline, and when the menu is displayed, select Add as Takes to put all files into one take. Takes can then be selected using either the right-click method or scrolling through takes by pressing T or SHIFT+T. This method is handy for just about anything, as it allows for toggling back and forth.

Takes are very handy for cutting together the best performance and for having the performer record the same piece several times and then cutting the unwanted parts of a take. Press S to split takes and then choose the best of each line. This process is known in the industry as comping. With Digital Audio Workstation/Non-Linear Editor (DAW/NLE) software, comping is easy and fast. In the past, engineers had to take several tracks, and using mute buttons and fader moves, comp parts of several similar performances together to a new track. With Vegas, it's all done on one track, with no difficulty like there is in the analog world where it either needs loads of hands on the mixer or good automation.

Pre- and post-roll settings are available in Vegas 4.0. Pre- and post-roll also act as punch in/punch out settings and can be combined with takes. To create a pre- or post-roll in a single event, split the event where the record in/out points are desired. Then create a loop or selection that begins before the split point and ends after the split point. Use the Loop tool if multiple takes are desired, and use a time selection only if a single punch in/out is desired.

Then click the Record button (CTRL+R), and Vegas begins playing at the beginning of the time selection, begins recording at the split point in the event, ends recording at the split point at the end of the event, and stops or loops at the end of the time selection or loop point.

Recording actually occurs through the entire selection. This process allows you to expand the event out in either direction if you needed to get a pickup note.

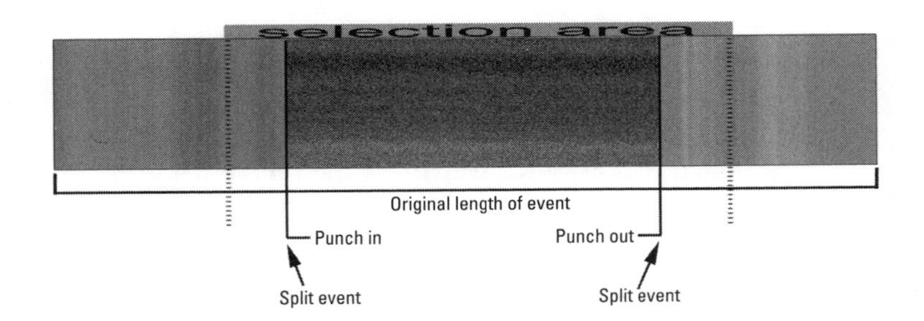

7.13 Defining pre- and post-roll on a single event.

Assigning Buses

Buses can be used to route audio to anywhere it must go. A bus can feed another bus, creating virtually limitless routing if necessary. Vegas is capable of handling up to 26 simultaneous buses, which is the same as the alphabet.

This fact means that up to 26 channels of audio can be streaming out of the computer at once, if enough hardware devices are owned. It takes four to eight channel converters to meet this output, and to leave six output channels empty. The norm is to have 24 channels of output.

This feature does not mean Vegas is limited to 24 channels of audio. Not at all. Vegas is limited to up to 26 channels of audio busing, but the number of audio tracks is limited only by the processing power of the computer.

If several tracks are on the timeline, insert buses into the mixer by clicking the Insert Bus button.

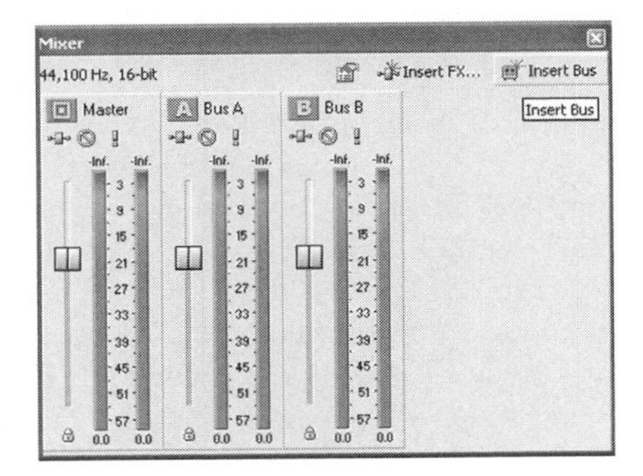

7.14 Adding a bus to the mixer.

On the Track Control pane, click the small square that appeared when the first bus was inserted. Choices for busing are listed here.

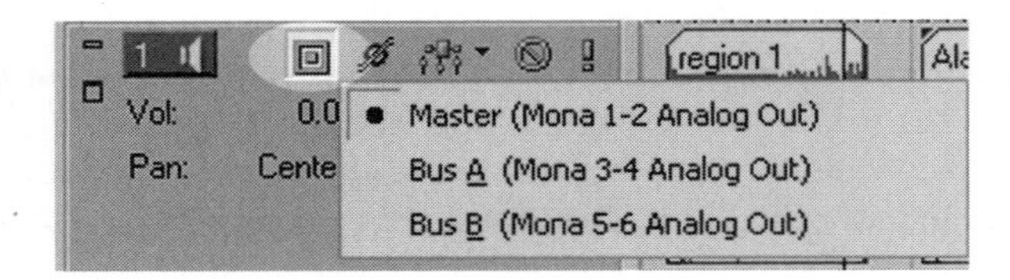

7.15 Select bus routing by clicking the square on the Track Control pane.

Select Bus A. Audio on that track is now sent to the Bus A mixer section. For example, Bus A could be sent to the normal right/left outputs of a mix, or it could feed a rear channel of a surround system. Alternatively, it could feed a signal processor, such as a flanger, reverb, or delay; perhaps it's a submix that eventually routes back to the master output.

Assign the bus outputs in the Mixer window in the same manner. Click the small square that is found on the bus control. On track 2, assign the track to the same Bus A output. This step routes Tracks 1 and 2 through the Bus A outputs.

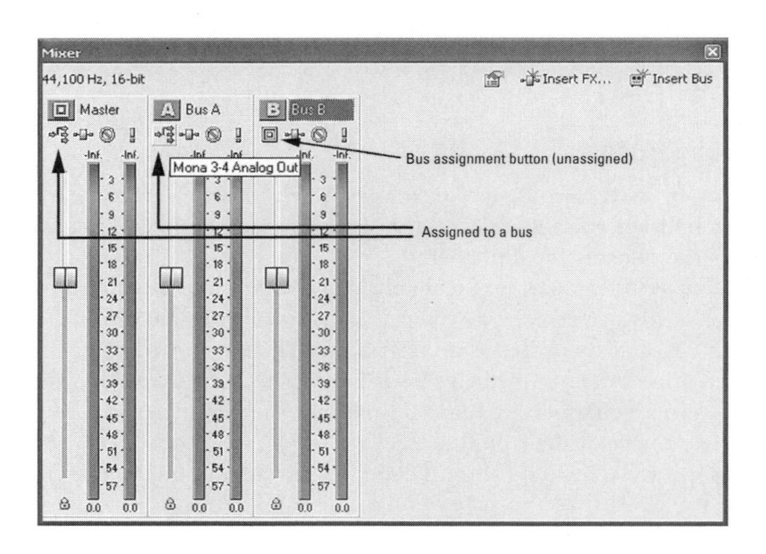

7.16 Assigned and unassigned buses.

Buses can be used for submixing, such as all drums going through one submix, all keyboards through another submix, and all rhythm instruments through yet another submix. This feature is of value as then all drums can be pulled down or up in volume without affecting the mix levels of each instrument/track or all keyboards can be raised or lowered. Signal processing is less taxing on the CPU if processing is done at the bus level as well. Putting six compressors on a drum kit is much more processor-intensive than putting one compressor on a drum submix. Generally, kick and snare drums are left out of submixes, as are lead vocals. These primary instruments/tracks are typically controlled individually.

Using Buses for Routing Effects

Buses can be used to route effects as well. Effects can be placed on individual tracks, but as the number of tracks grows, the load on the processor increases. It's a good idea, therefore, to place effects on buses to affect several tracks at once. This process can be done using direct track assignment or the multipurpose fader, or through assignable FX envelopes. Envelopes can be used to automate effect levels, sends, and returns.

In the Mixer window, click the Insert FX button, choose a delay effect, and set the effect to the parameters you'd like to hear. Now right-click in the Track Control pane and choose the bus to which you assigned the delay. An envelope is displayed in the track timeline. This is the send to for the specified bus. Raising the line will increase the amount of audio sent to the delay and lowering it will decrease the amount of it. Double-clicking the envelope places a handle/node on the envelope, allowing for automatic send changes. If a delay is needed only on a specific phrase or moment in an event contained in the track, the envelope can be pulled all the way down until that moment when the envelope will be raised to the desired level. Handles or nodes can be right-clicked, and the Set-to percentage or volume can be specified that way as well.

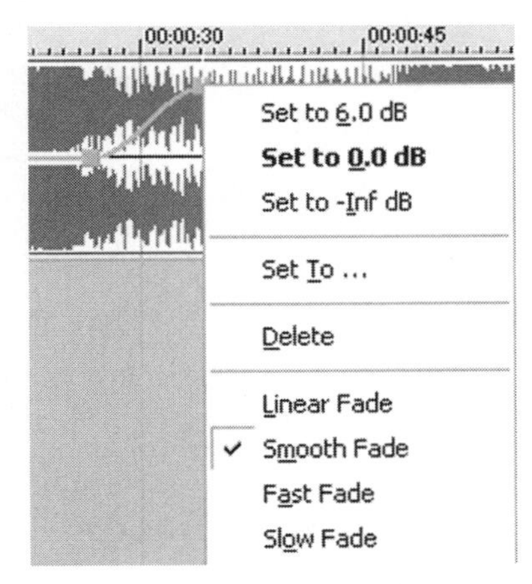

7.17 Inputting specific values from the right-click.

FX can be set as pre- or post-fader in Vegas. This option is found on nearly all mixing consoles and is important, particularly for sound designers. This option allows audio to be sent to the processor before the fader, so regardless of where the track fader is set, even if it's off, FX receives an audio signal. Post-fader is how most processing is done, in which reducing the output volume of a track also reduces the send to the FX. As an envelope fades on the track, the send to the FX fades as well.

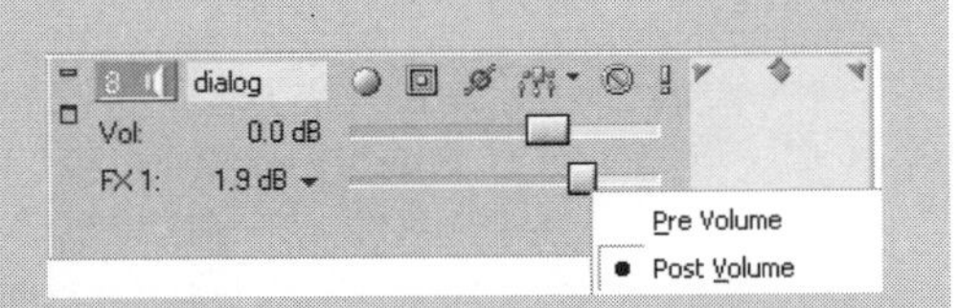

7.18 Right-click the FX send on a track to select pre- or post-fader send to the FX.

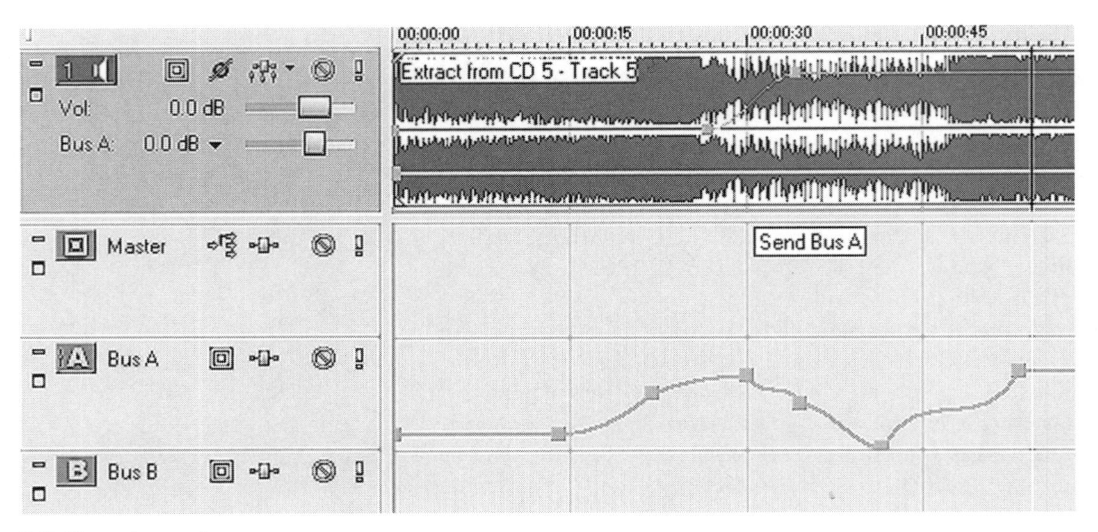

7.19 Vegas bus assignment.

NEW! Buses can now be viewed as tracks for automation. Bus outputs/returns can be automated for much greater mixing power. For example, if there were several tracks with background vocals, all vocals needed to be raised in volume at once, and all required a reverb to rise with the volume. The vocals could be sent to one bus, with the reverb sent to another bus. The reverb is controlled by an inserted bus send, and the vocal levels are controlled by an envelope assigned to another bus, just like a bus routing on a typical mixing board.

With busing and automation, mixes are capable of being very deep and exacting.

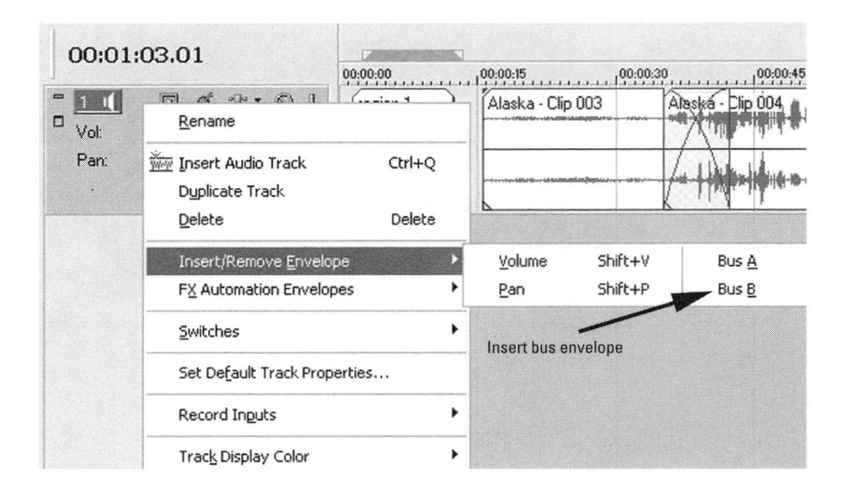

7.20 Analog mixer bus assignment.

Creating FX chains is an important aspect in Vegas. Often, presets in FX are simply not enough to recall. If multiple FX chained together with specific presets are needed, Vegas can save a chain of FX with presets. For instance, an EQ, a compressor, and a chorus can be used for vocal processing with each of the individual FX having their own presets. Rather than having to insert and choose presets for each effect manually, all three can be saved as a chain/package complete with presets.

To create a chain/package, perform the following steps:

1. Click the Insert FX button in the Bus, Mixer, or Track Control windows, which opens the Plug-In Chooser.
2. Choose the plug-ins you wish to use.
3. Arrange them in any manner by either dragging a plug-in before another or by using the Shift Plug-In Left/Right buttons found in the upper-right corner.
4. Click the Save As button and name the chain, which will now be stored as a package in the Plug-In Packages folder.
5. Click OK. Your preset is now stored and can be recalled at any time.

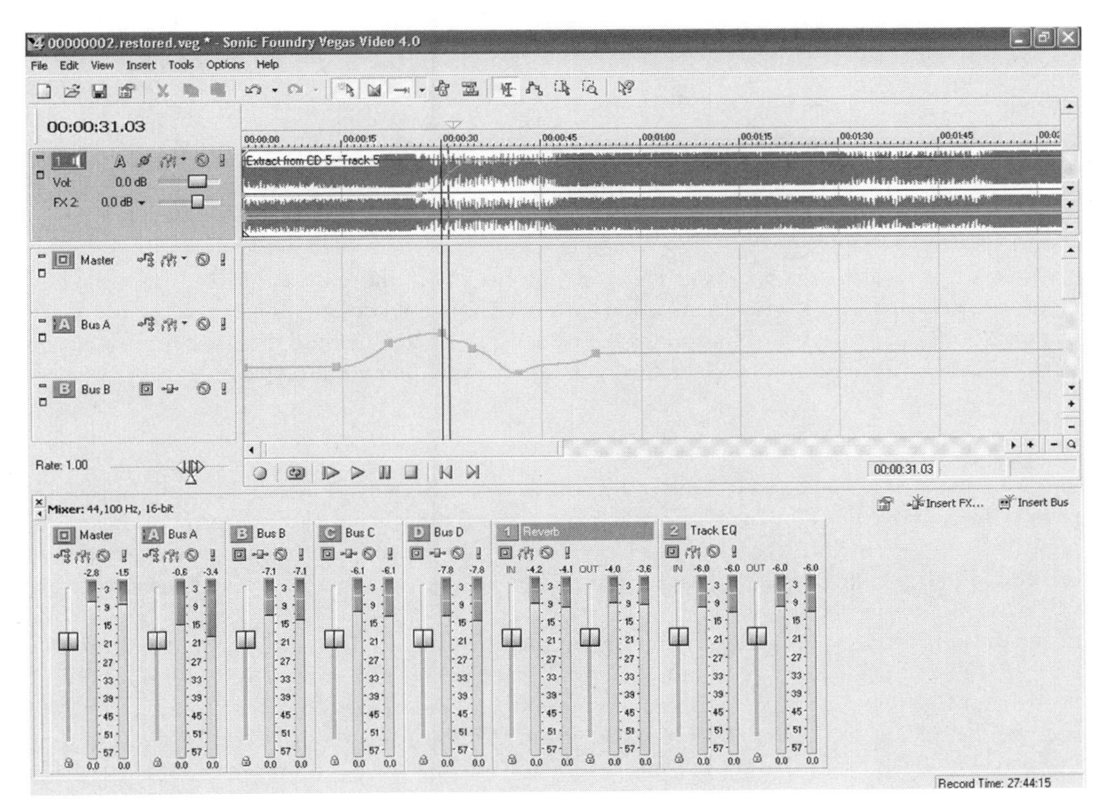

7.21 A single track feeding four buses and two auxes with automation.

Inserting Effects Without Using a Bus

Effects can be directly routed and controlled in Vegas without inserting a bus. This process is similar to an aux send that doesn't return to a bus on an analog mixer.

Click the Insert FX button in the Mixer window, which calls the Plug-In Chooser dialog. Plug-ins can be viewed by All, Chains, Automatable, Sonic Foundry, Third-party, or Optimized. All shows every Direct X plug-in installed in the system.

Choose the FX desired. If multiple FX are desired, press and hold down SHIFT while selecting FX. They will all be placed in the mixer as a series. FX selected this way cannot be individually controlled, as they all share the same send and are chained together.

FX can also be inserted individually. Every FX plug-in that is inserted in the Mixer window is also available to every track or bus. Keep in mind that this can be very processor-intensive during preview, although it will not affect the final render.

Use FX assigned to buses, as opposed to individual tracks when several tracks need the same effect. For example, a group of tom-toms, cymbals, and percussion might be able to share a reverb or delay. Rather than insert a reverb and delay on every track, send all relative tracks to a bus and insert the FX, controlled by the bus.

To do this, perform the following steps:

1. Insert FX.
2. Insert a bus.
3. Assign all desired tracks to the bus.
4. Select VIEW | SHOW BUS TRACKS.
5. Right-click the bus track and insert an FX envelope.
6. Set envelope as desired.

Now all tracks have the necessary FX, with automation to those FX controlled from the bus. Choruses, verses, fills, bridges, other sections of a song, or sound design for a movie can be unique in the way that FX are heard and managed as a result of this method, and yet processor and RAM resources have been spared the heavy load of having many plug-ins active.

Using Track FX

Every audio track in Vegas has a chain of FX installed on it already. A Noise Gate, EQ, and Compressor are found by clicking the FX button on each audio track. They can be heard during recording by using input monitoring, they can be removed from the track, or they can be appended to or replaced all together. These FX are optimized to keep processor resources at their peak but still place a load on the processor.

Set the default track properties by setting up the track the way you'd like it to default to. Set up an FX chain for the track FX, right-click the Track Control pane, and select Set Default Track Properties. Checking the boxes in the dialog that opens will cause the current track settings to become the default settings for future tracks inserted into the current project. Now every time Vegas opens, these track settings become the default, saving time and preventing the user from having to remember how the last mix was set up before beginning.

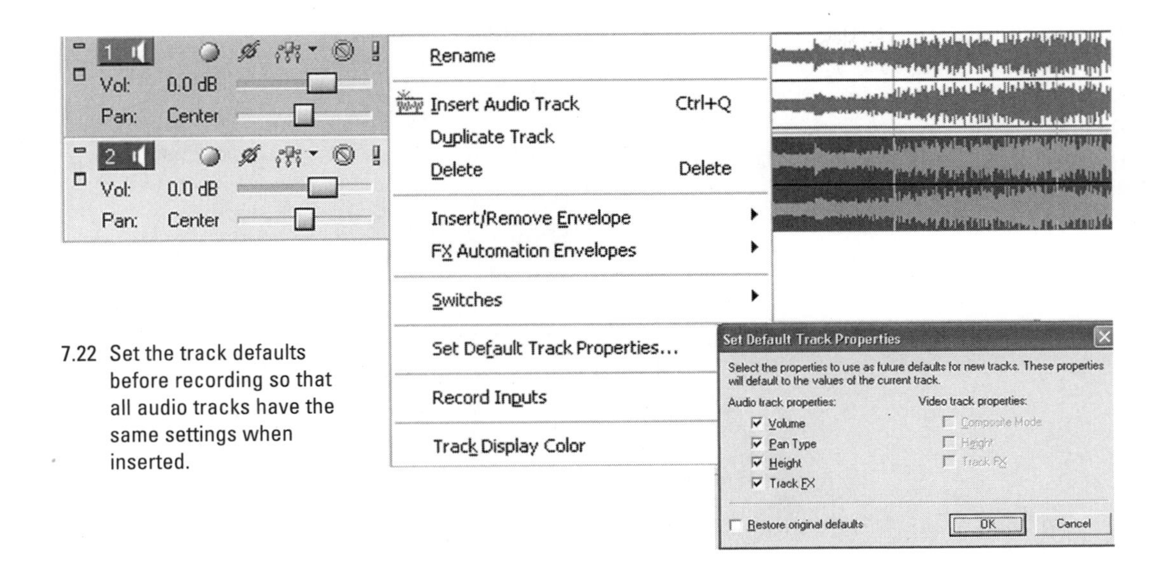

7.22 Set the track defaults before recording so that all audio tracks have the same settings when inserted.

Tip

Set up a blank project with all track settings in place, so that when a project is started, a template of several tracks is already built. Starting with 8 to 16 tracks, complete with commonly used compressor, EQ, and reverb settings, not only saves time but also impresses clients when they hear how quickly a tight sound can be dialed in. On the CD enclosed with this book, you will find a recording interface with tracks in place that can be used as a starting place for your preferences.

Recording Multiple Tracks

When working with a band or ensemble, several tracks might need to be recorded at one time. A multichannel sound card is best used in this situation. If a two-channel sound card is used, only stereo information will be recorded, and an analog mixer will be required to manage all of the microphones. With multichannel cards, an unlimited number of microphones can be input directly to Vegas, depending on the number of cards used. (Many manufacturers of hardware cards have limits as to how many cards can be stacked.) These can be individually controlled without using an external mixing device. To select multiple inputs with an individual microphone routed to each input, assign the input of each track as Mono and select either the right or left input of the sound card.

Remember, if your sound card does not have pre-amps built in, a preamplifier is needed for every input channel.

Select a channel input for each microphone, and with every channel that has an input assigned, be sure the channel is armed. Click the Record button (CTRL+R), and all tracks that are armed will record in sync. Enable input monitoring where necessary—the more tracks that are set for input monitoring, the heavier the processor load. Without using ASIO drivers, latency can be unreasonably high. Use ASIO drivers for input monitoring for best results.

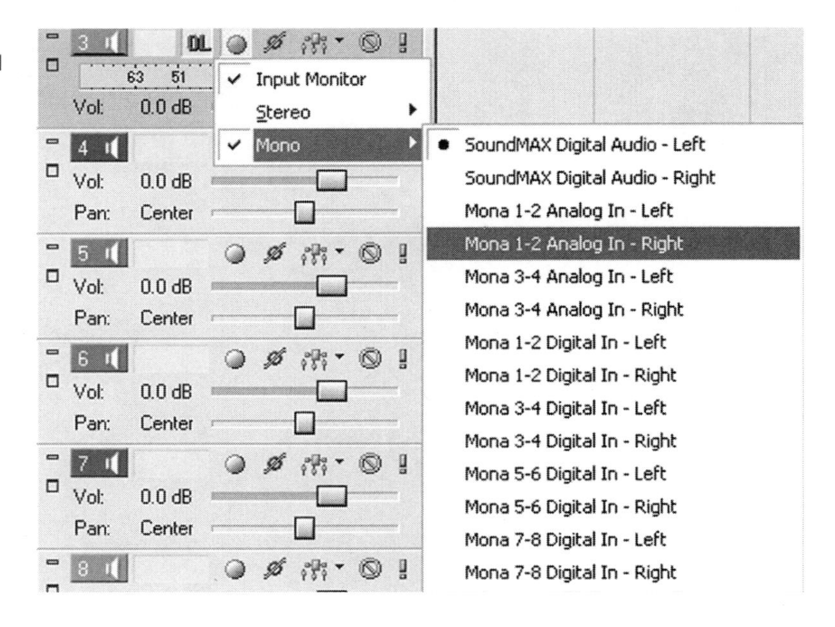

7.23 The Echo Mona has four analog inputs and eight digital inputs.

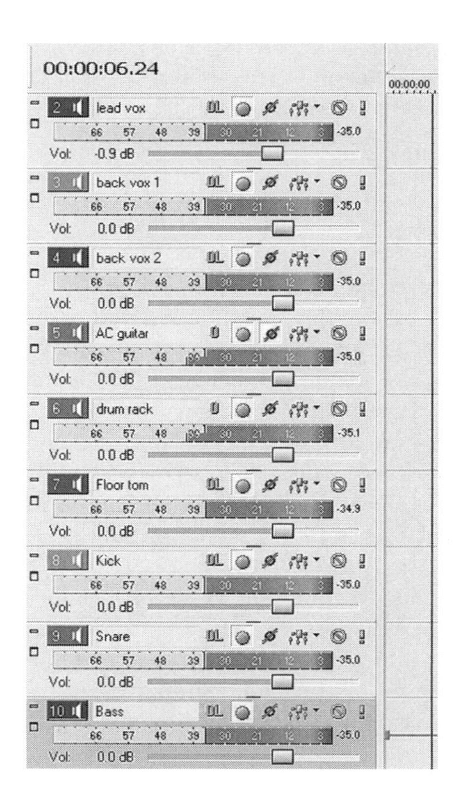

7.24 Typical session armed for recording.

Vegas now has the ability to support ASIO drivers, found in most professional sound cards. ASIO drivers are low-latency driver sets specified by Steinberg Media Technologies AG.

Tip

Pressing CTRL+A selects all tracks at one time, and clicking the Record button on one track simultaneously arms all tracks. Clicking the Record button again after the recording is finished disarms all tracks. Use this same technique to insert volume, pan, bus, or FX envelopes on every track.

Synchronizing Vegas to External Devices

Some studio setups can call for a multitrack tape machine, sequencer, or other external device to control Vegas, or conversely, Vegas can be the master device controlling external equipment if the proper hardware is in place. To do this, a device that can convert SMPTE to MIDI timecode (MTC) is required.

Most MIDI distribution devices are capable of converting incoming MIDI timecode to SMPTE or outgoing SMPTE timecode to MIDI timecode. Most tape machines will accept either MTC or SMPTE. Some accept both.

If Vegas is to be the master device, Vegas needs to generate timecode. Set this up in the **OPTIONS | PREFERENCES | SYNC** dialog.

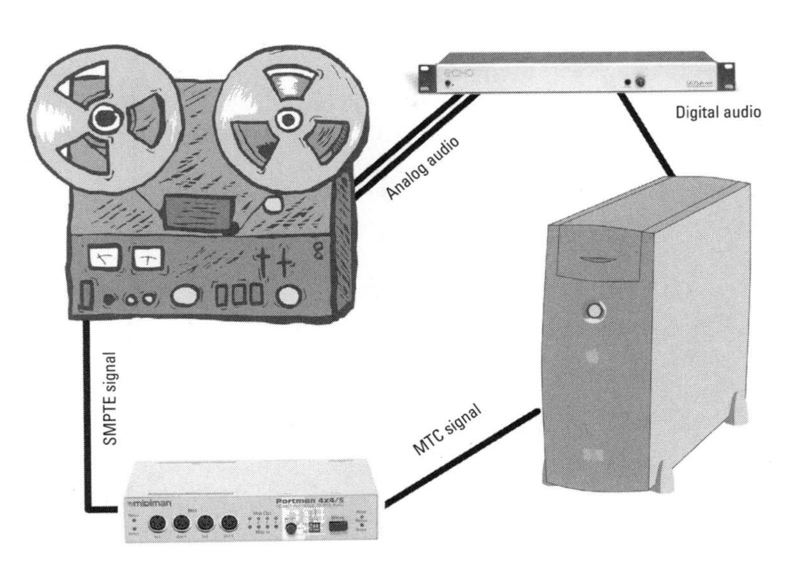

7.25 Setup of a timecode device to connect Vegas and external recording devices.

Tip

If you click Play in Vegas and it starts/stops an external device, Vegas is known as the *master*. If starting and stopping the hardware starts and stops Vegas, Vegas is the *slave*.

To generate a timecode that an external device can read, have the owners' manual of the external device handy so you can learn the incoming timecode specs that it requires. Most devices default to 30 frames per second, while other devices call for 29.97 non-drop. The standard for most professional audio equipment is 30 fps. DV has a different standard, based on American television standards.

Vegas generates timecode that is sent to the conversion device, where it is distributed to whatever external devices that Vegas needs to control. Timecode connection is necessary if Vegas will be used to edit external media that will be returned to the external recording device after edits are completed.

In the OPTIONS | PREFERENCES | SYNC | ADVANCED dialog, Vegas has an option to use either an internal clock from the CPU clock or the clock generated by the sound card. Most instances will call for the clock to come from the sound card. Vegas also has the ability to send full-frame messaging to external devices. Full-frame information can speed up location in some external tape machines. Not all devices can respond to full frames, and some devices can become confused. Check the documentation that came with the device that Vegas is controlling.

If Vegas is to be a slave to timecode generated by an external device, Vegas must be set to trigger from timecode. In the OPTIONS | PREFERENCES | SYNC dialog, set the trigger from the device sending timecode into the computer. Be certain that Vegas' frame rate setting matches the frame rate being sent from the external master device. Otherwise, audio will drift ahead or behind the master when Vegas is played in time with the external master. For Vegas to operate consistently with a word clock connection, *trigger* must be used. Trigger refers to Vegas seeing a point in timecode, at which point it will begin to play. It is not locked to the timecode specifically. *Chasing* timecode means that Vegas remains synchronized with incoming timecode and follows the timecode as best as possible. Vegas has a *Freewheel* mode that allows Vegas to continue to play for a period of time even if timecode input is occasionally lost, playing until it sees timecode again that it should be synchronized to. If timecode is consistently lost, check the hardware or signal flow from the master device.

Sync is set up in the Preferences dialog but must be enabled in the Options menu (CTRL+F7).

Tip

Use Vegas to edit audio from a tape machine and apply EQ, noise reduction, and processing. This process saves buses on an external machine, frees up reverbs and other processing hardware during a mix, and generally affords an opportunity to make a smaller studio have the capabilities and power of a much larger recording facility.

To view incoming or outgoing timecode, right-click the time display, which defaults to the upper-left corner of the timeline. Choose which timecode you wish to display: time at cursor, incoming, or outgoing timecode.

The timecode window is docked by default but can be moved to a convenient space on the timeline or wherever it is most visible. It cannot be completely removed from the workspace.

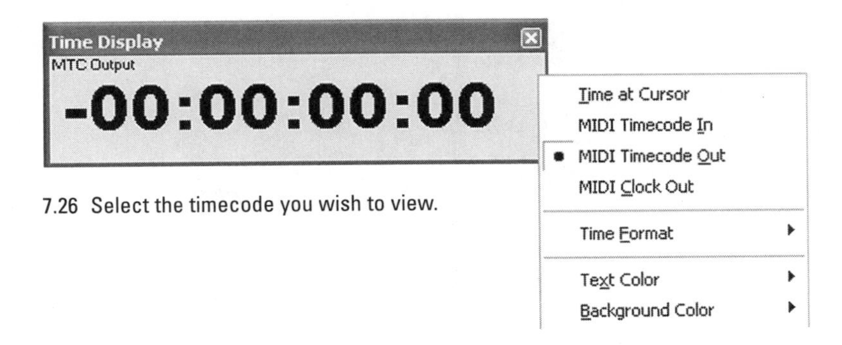

7.26 Select the timecode you wish to view.

Editing Audio in Vegas 4.0

The audio editing tools found in Vegas are among the best in the world. Moreover, the audio tools in Vegas are superior to any audio tools found in any NLE system available today. Playback and recording tracks are limited only by the amount of available CPU, hard disk space, and RAM.

Placing Audio on the Timeline

Audio recorded is immediately found on the timeline and is also found in the Media Pool. However, it's often necessary to place audio on the timeline that might be a take, from a stock music collection, an ACID loop, sound FX, or other stored audio file.

To do this, open the Explorer in Vegas and find an audio file to place on the timeline. Double-clicking the file inserts the file directly on the timeline. Right-clicking a file opens several options, such as opening the file directly in Trimmer, which allows audio sections to be trimmed before putting media on the timeline. After audio is placed on the timeline, it becomes an event.

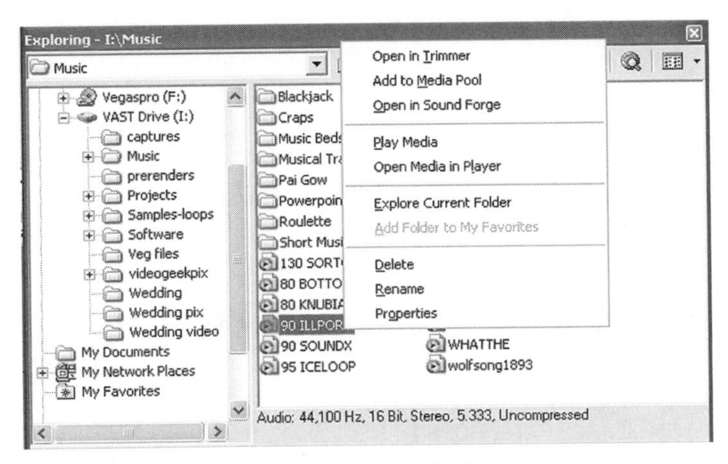

7.27 The Explorer in Vegas looks similar to Window's Explorer.

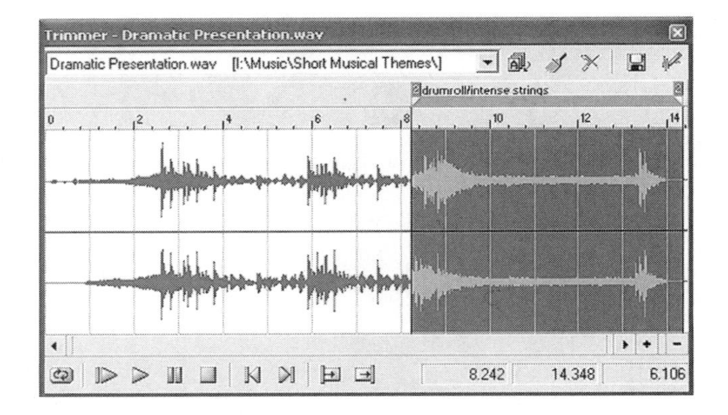

7.28 Audio opened in the Trimmer, with selections chosen to be added to the timeline.

Right-clicking also offers the option of placing the audio in the Media Pool or opening the file for editing in a third-party audio editor or in Sound Forge, depending on settings in the OPTIONS | PREFERENCES | AUDIO dialog. Opening the file in the Trimmer is a simple workflow that allows phrases or sections to be marked and added to the timeline.

After making a selection in the Trimmer, press the R key. This creates a namable region in the Trimmer that stays with the audio and will be seen anytime the audio is opened in the Trimmer. After creating the region, place the cursor on the timeline where the audio is desired and press A. This step adds the media to the timeline, where it now is an event on a track. Selected audio can also be dragged to the timeline from the Trimmer. Do whichever is comfortable for you.

Audio in the Trimmer can also be inserted in Ripple mode (see "Using Ripple Editing" later in this chapter for more details) by copying the audio in the Trimmer and enabling Ripple or by copying and pressing SHIFT+CTRL+V. The audio is pasted as an event at the cursor, and subsequent events ripple down the timeline for the length of time equal to the size of the pasted event.

Extracting Audio from CDs

Vegas has tools to allow audio to be extracted from CDs. While a single track will play back from a CD, multiple tracks or clips from a sample library that are scattered all over a CD might not play back so well, due to the limited speed of a CD or CD-R. There are two menu areas where Vegas can be instructed to extract audio.

The first is the Media Pool. Clicking the Extract Audio from CD button opens a dialog where users are asked to choose which tracks are to be extracted. Audio can be extracted three ways:

- By track (individual indexes can be selected)
- By time selection (a range of time for extraction can be specified)
- By the entire CD (the entire CD is extracted to the hard drive as one large file)

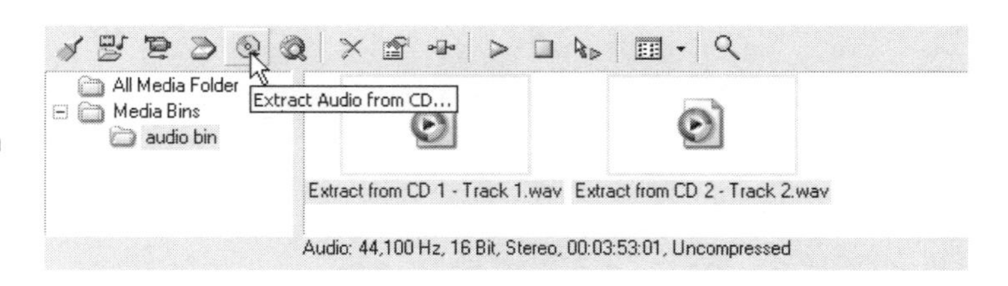

7.29 Clicking the Extract Audio from CD button calls up a dialog requesting information as to where files should be extracted.

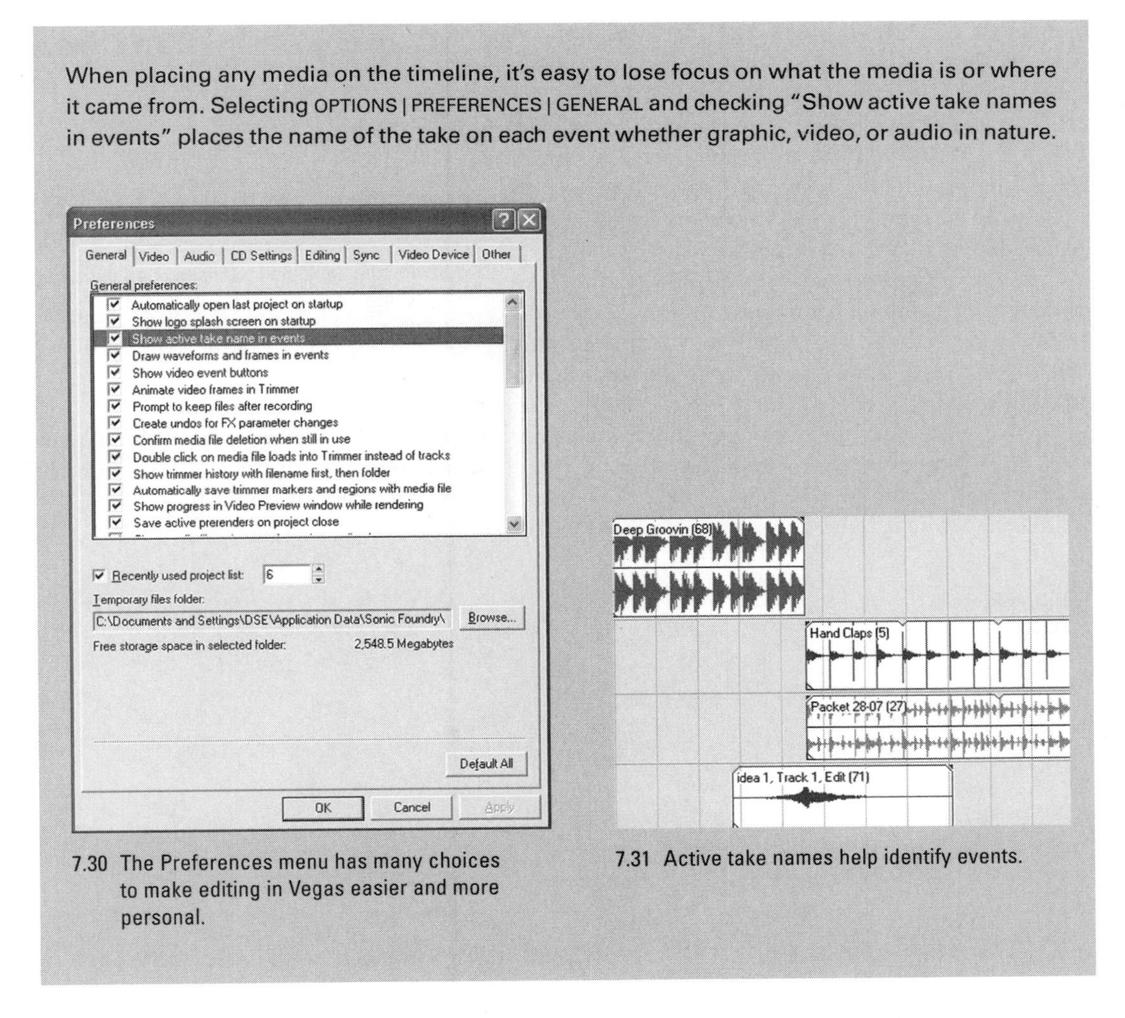

When placing any media on the timeline, it's easy to lose focus on what the media is or where it came from. Selecting OPTIONS | PREFERENCES | GENERAL and checking "Show active take names in events" places the name of the take on each event whether graphic, video, or audio in nature.

7.30 The Preferences menu has many choices to make editing in Vegas easier and more personal.

7.31 Active take names help identify events.

With audio on the timeline as events, it is ready to be edited. Basic edits would include cutting, copying, pasting, fading audio in and out, cross fading, splitting, and moving a file.

Cutting, copying, and pasting are handled in Vegas as in most applications. To cut an event, select the event and press CTRL+X or select EDIT | CUT after selecting the event. Events that are cut are available from the clipboard. To copy an event, select the event and press CTRL+C or select EDIT | COPY after selecting the event. To paste an event, first cut or copy the event and then place the cursor where the event is desired to go. Pressing CTRL+V pastes the event(s) to the timeline wherever the cursor is located. Cut, Copy, and Paste commands are also accessible by right-clicking any file and choosing one of the commands from the context menu.

Vegas also has a Paste Repeat command that allows events to be pasted multiple times. Events can be pasted end-to-end the number of times specified in the dialog or can be spaced at intervals defined in the menu.

Selecting events to the end of the timeline is an option when all events need to be moved up or down the timeline as a group. Rather than SHIFT+CLICKING each file, right-click only the first file in the series, and from the menu choose Select Events to End. This process will go all the way to the end of the timeline and select every event on that track after the cursor/selected event.

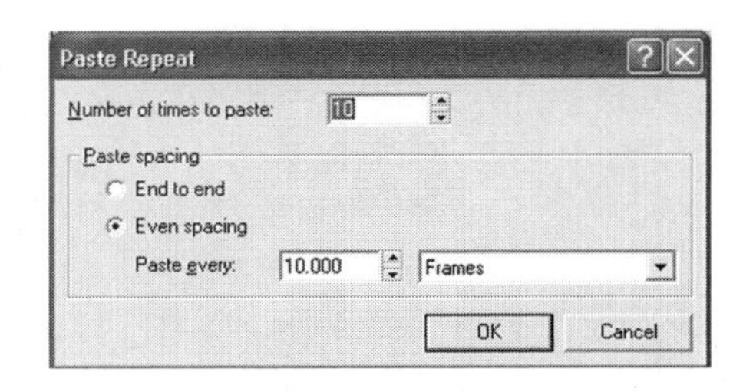

7.32 Pasting multiple instances of an audio event.

To fade an audio event in or out, it's as easy as clicking and grabbing the upper corner on either end of the event. A small quarter-circle will appear. This circle indicates that the tool is prepared to fade audio in or out.

Splitting an event or series of events that are on different tracks is accomplished by placing the cursor where the split should occur. If no single event is selected, all vertical events on the timeline will be split on the point where the cursor is. Press S to split the audio files.

If an individual event is selected, only that event will be split.

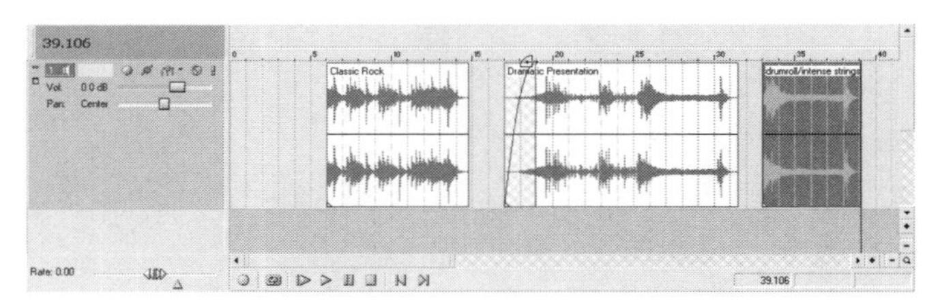

7.33 Creating a fade in.

Fades can be customized by right-clicking the fade and choosing from the menu of five different types of fades. If one of the five fade types still doesn't provide the desired sound, use the Audio Envelope tool to create the desired fade type.

7.34 Different fade types available in Vegas.

Fade Type
Open in Trimmer
Open in Sound Forge Ctrl+E
Open Copy in Sound Forge
Apply Non-Real-Time Event FX...

Creating a *crossfade*, in which two audio events overlap each other and one fades out as the other fades in, is achieved by sliding the two events over the top of each other for the desired length of fade. Automatic Crossfades must be enabled on the toolbar. Although defaulted to Enabled, check to be certain that the following button is pressed:

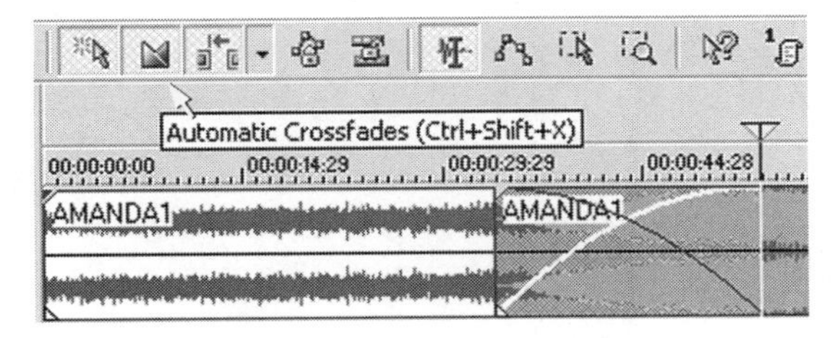

Double-clicking inside the crossfade selects the length of the fade. The exact timing of the fade can be read in the Time Display window at the lower-right corner of the timeline.

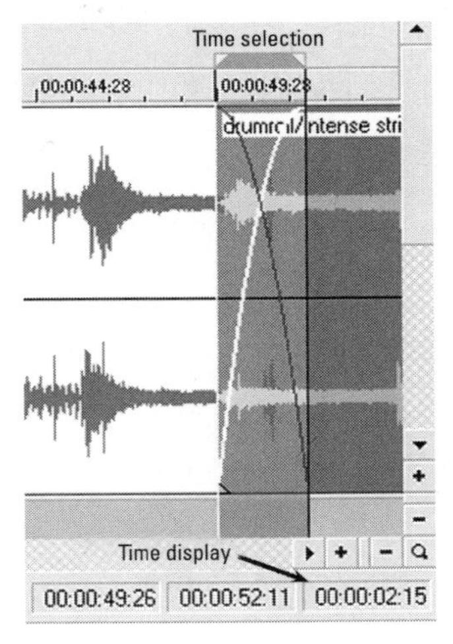

7.35 Double-click a crossfade to make a selection. The length of the crossfade is displayed in the Time Display window.

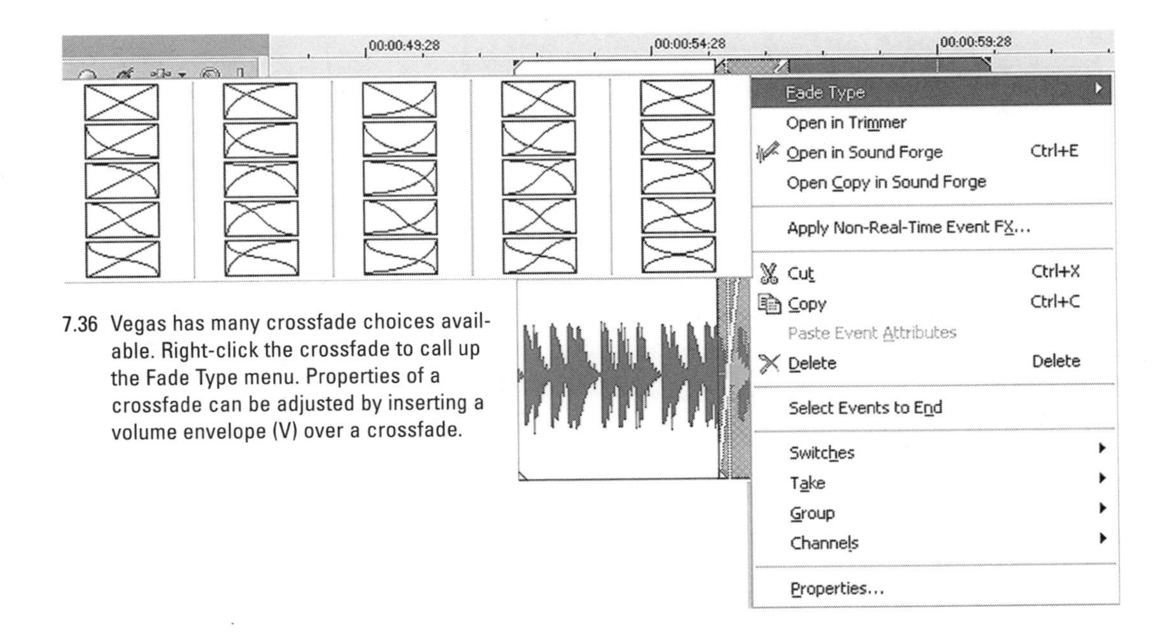

7.36 Vegas has many crossfade choices available. Right-click the crossfade to call up the Fade Type menu. Properties of a crossfade can be adjusted by inserting a volume envelope (V) over a crossfade.

Vegas has the ability to create auto-crossfades of a preset length. In the OPTIONS | PREFERENCES | EDITING dialog, crossfade times can be preset. Set the desired length for crossfades by time or by frames and select multiple files in the Explorer by pressing and holding CTRL and selecting media by clicking them. Drag them to the timeline, and the media will drop in as automatically crossfaded events. This same behavior applies to video files.

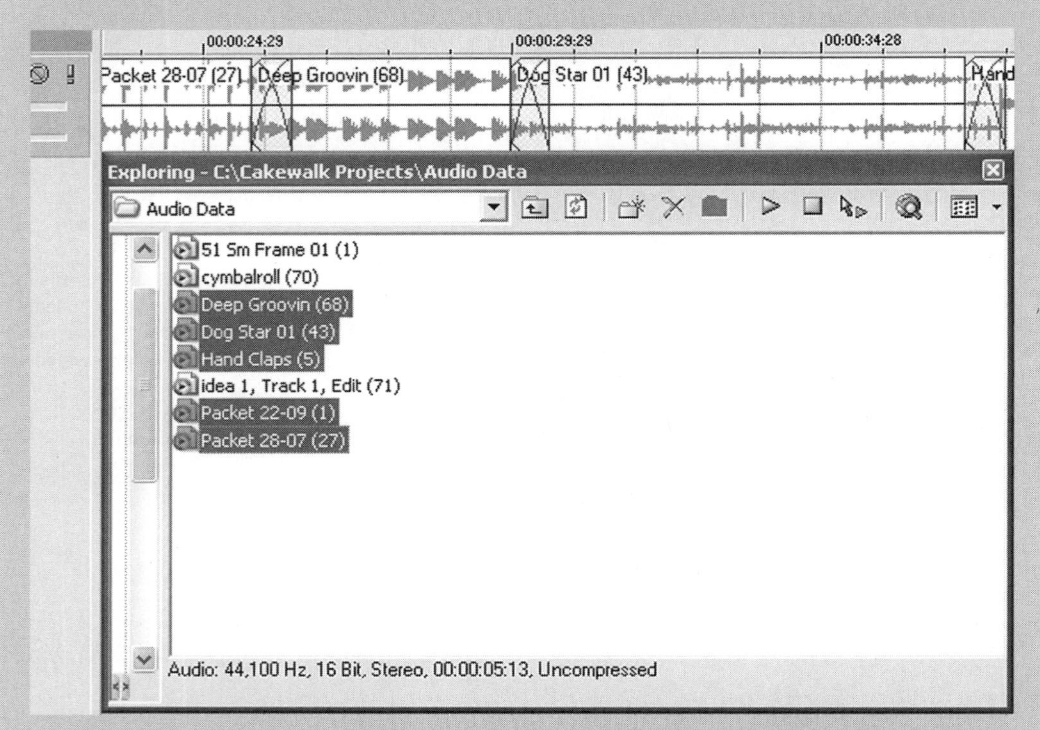

7.37 Several files can be dragged to the timeline and automatically crossfaded by selecting several files in the Explorer.

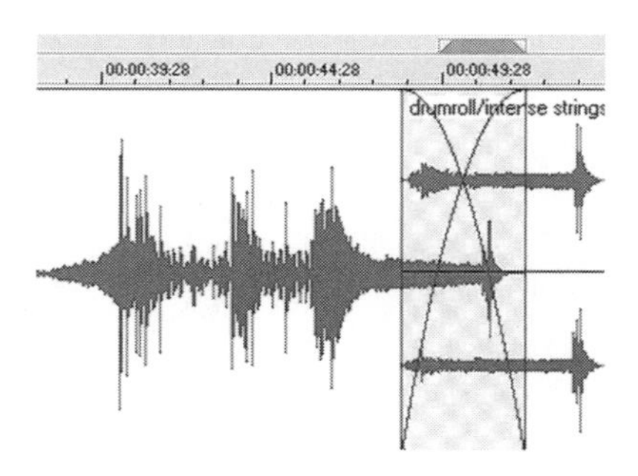

7.38 A mono event faded into a stereo event.

When interviewing with a DV cam and using one mic to the interviewee and one to the interviewer or another interviewee, the two channels are captured as one audio file with right and left information. With most NLE systems, separating channels requires editing in a separate sound editor. In Vegas, place the two-channel audio on a timeline. Duplicate the track by right-clicking the Track Control pane and selecting Duplicate Track. Click the upper of the two identical tracks. Right-click the audio in the upper track and select CHANNELS | LEFT ONLY. Right-click the lower track and select CHANNELS | RIGHT ONLY. There are now two tracks of mono audio on the timeline, capable of being edited and processed separately. This method is a fast means of accomplishing two tracks with separate control.

All tracks added to the timeline are stereo. Only events are either mono or stereo. A mono signal, however, will remain mono when dragged to the timeline. Mono and stereo signals can occupy the same track, and a mono event can be crossfaded over a stereo event on the timeline.

Audio that is stereo, but should be mono, is quickly converted by right-clicking a stereo file and choosing Combine from the pop-up menu. Files that have been combined can be restored by selecting Both. Audio can also be selected as Right Only or Left Only, and channels can be switched by selecting Swap.

Audio that is not at nominal level can be normalized non-destructively by right-clicking an audio event and selecting Properties. Check the Normalize box, which will bring audio to a level normalized as specified in the **OPTIONS | PREFERENCES | AUDIO EVENT** dialog. On the disc in this book, a JavaScript is found that can be run as a script in Vegas 4.0, normalizing all audio on the timeline as a script rather than normalizing all audio manually. The file is located in the `\scripts\` folder as `normalizeall.js`.

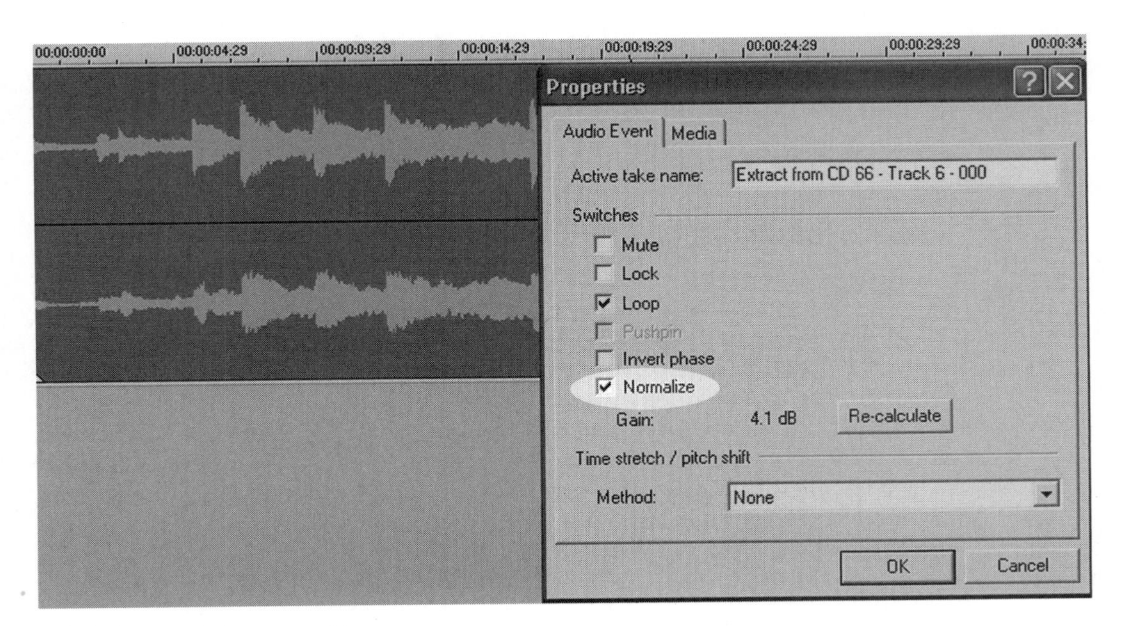

7.39 Audio event before normalizing.

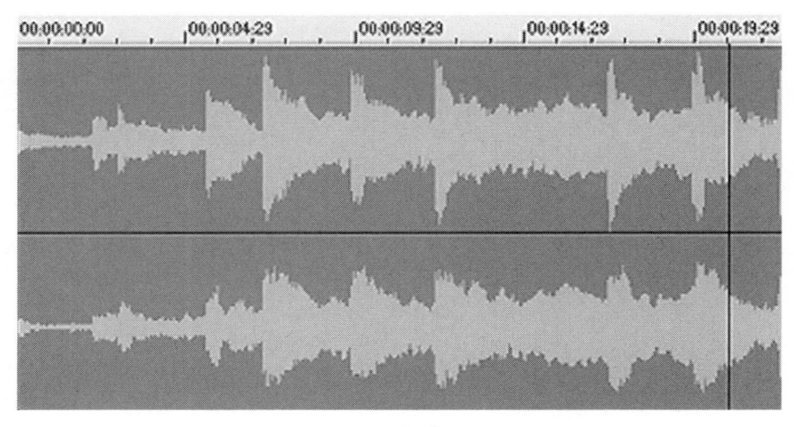

7.40 Audio event after normalizing, a 4.1dB increase in gain.

Audio events have an individual gain reduction on each event as well. Place the cursor at the top of the audio event and watch for the cursor to turn into a hand icon. Then click and hold, dragging the hand icon downward. Vegas indicates the gain level reduction as the gain line is pulled downward.

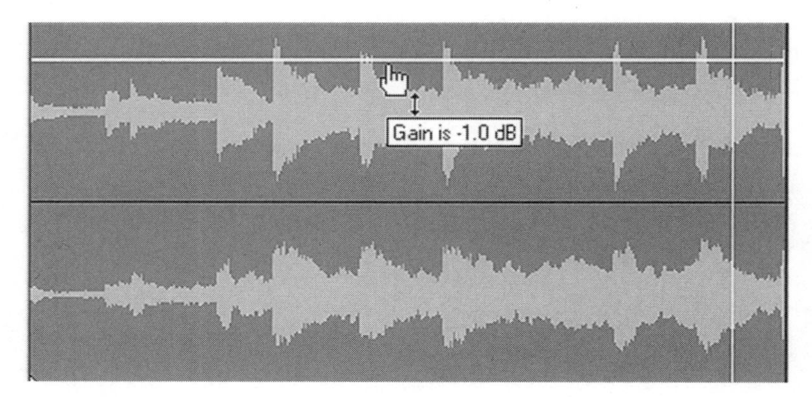

7.41 Reducing the gain of an audio event on the timeline.

Right-clicking an audio event calls up several menu options. One of them is the ability to loop an event, which is different from setting up a playback loop for previews.

Right-click an event, choose Switches, and check the Loop box. This step allows the audio event to be dragged out infinitely, and the audio will constantly loop. Vegas inserts an indicator that looks like a small divot or carrot in the event at the point of the end/beginning of a loop. Looping is terrific for repeating sounds that might fade together or for using an ACID loop that should repeat. Looping is set to On by default.

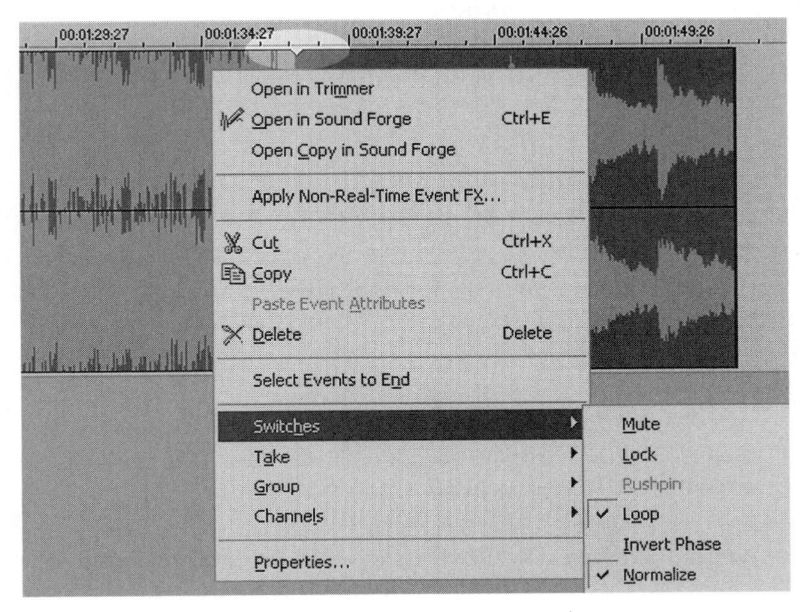

7.42 Setting an event for looped playback.

Looping is helpful in most situations; however, it can also work against a mix in some instances. If a reverb or delay is desired to continue reflection for a long time after an audio event ends, looping must be disabled, and the file dragged out for the length of time that the reverb/delay is desired. Otherwise, the plug-in sees the end of the audio event and thinks it needs to shut off if the end of the project is reached. Dragging out an empty portion of the event with looping turned off allows the reverb or delay to continue to be audible.

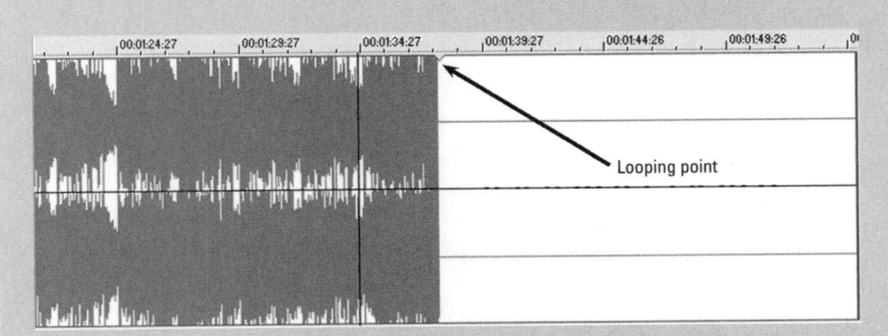

7.43 Event dragged out with looping off. Because looping is disabled and the audio has ended, Vegas treats the stretched event like empty media.

Inserting Markers

Markers are often needed to indicate a specific point on the timeline that requires an edit, to serve as a cue, to insert metadata, and for many other helpful uses. Vegas 4.0 allows for three different types of markers:

- **Markers**—reference points to be returned to later, which identify timing or act as a reminder for later edits. Markers can be inserted on the timeline or in the Trimmer. Markers can be named at any time.
- **Regions**—defined sections on the timeline. Regions can behave as permanent indicators of a selection either in the Trimmer or on the timeline. Regions can act as reminders of specific sections to be returned to later or can act as edit length indicators in the Trimmer. Regions can be named at any point in time. All training materials on the CD contained in this book are built with regions.
- **Command markers**—used to insert metadata in streaming media files. Text data, such as lyrics, promotional information, URL locations, and closed captions, can be embedded with command markers. (See Chapter 10 for more information.)

Markers and regions can remain with the audio file when inserted in the Trimmer tool. The OPTIONS | PREFERENCES | GENERAL dialog has a check box to automatically save markers and regions in the Trimmer tool.

To insert a marker, place the cursor where the marker location is desired. Press M, right-click the Marker bar, or select INSERT | MARKER. This step places a marker on the timeline. Markers can be deleted by right-clicking the individual marker or by right-clicking in the Marker bar and

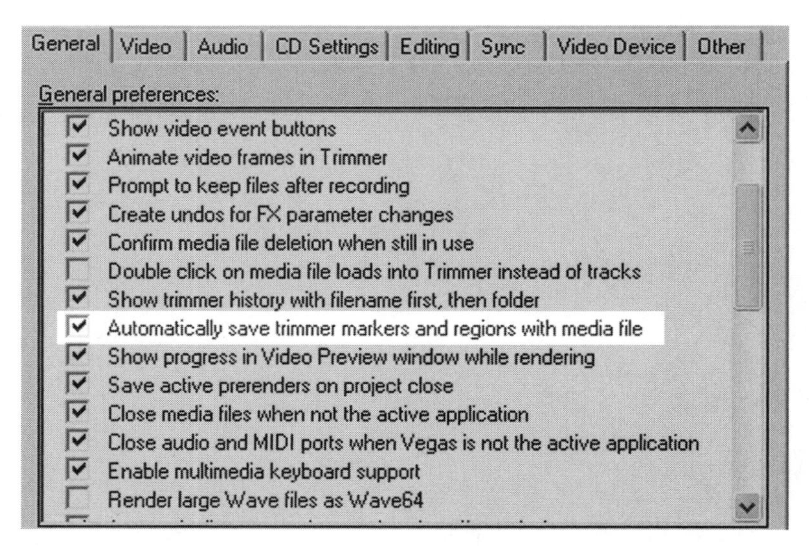

7.44 Be certain that the check box for "Automatically save trimmer markers and regions with media file" is checked to save the indicators in the Trimmer.

choosing Delete All Markers. A group of markers can be deleted by making a selection on the timeline that contains the markers to be deleted, right-clicking in the Marker bar, and selecting the menu option Delete All in Selection. Markers can be named when inserted or later by right-clicking a marker and selecting Rename. Pressing ENTER or clicking out of the box on the timeline sets the marker's name.

Markers can be navigated by pressing CTRL+LEFT ARROW/RIGHT ARROW. This step will jump the cursor from marker to marker.

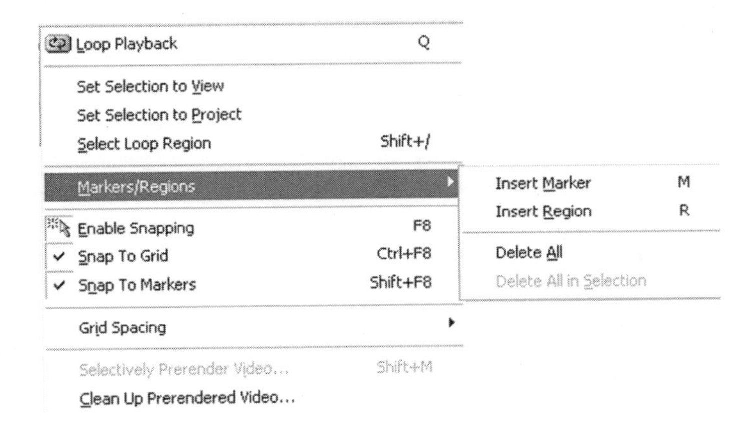

7.45 Inserting, deleting, or editing markers can be done by right-clicking the Marker bar.

Regions are defined by creating a selection on the timeline and pressing R or by selecting INSERT | REGION. Double-clicking an event creates an automatic selection for the length of the event. The region can be named in any way that you choose. Regions can be renamed at any point.

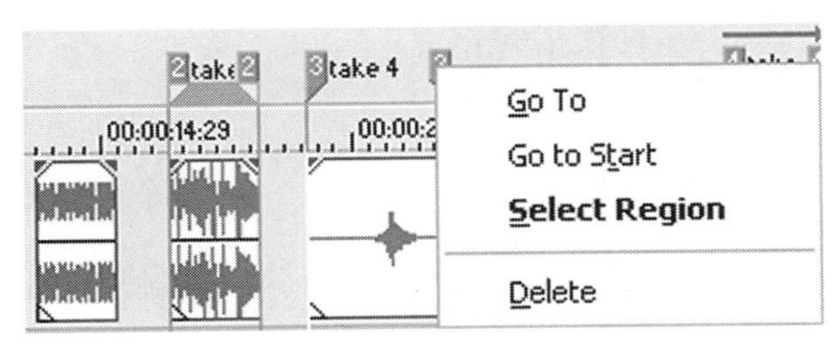

7.46 Cursor behaviors inside a region can be selected by right-clicking a region marker.

Tip

Use regions to indicate mini-projects or mini-sections, such as a verse, chorus, or bridge. This option makes it easier to find full sections or loop full sections during the recording session. It also makes it easy to fly in a repeat chorus or bridge later on, if necessary.

Regions can be navigated or toggled. Regions can be navigated by selecting a corresponding number on the keyboard (not the numeric keypad) or by pressing CTRL+LEFT ARROW/RIGHT ARROW. This step allows rapid navigation within Vegas of various regions on the timeline.

Command markers are created when the cursor is placed at the desired point for a metadata event to occur, and C is pressed or INSERT | COMMAND is selected. This step calls the Command Properties dialog, in which the desired type of metadata is selected.

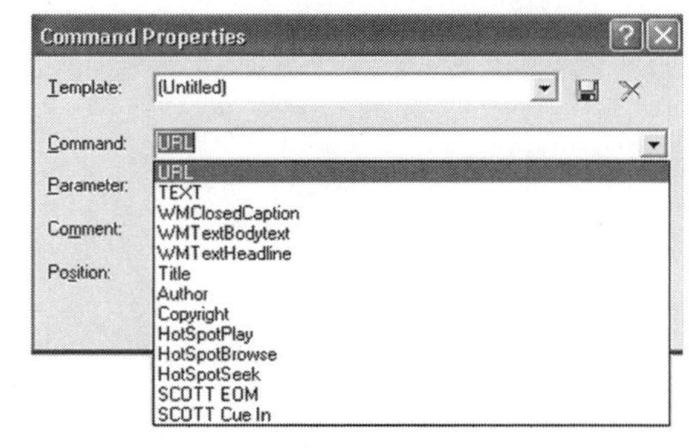

7.47 Different types of metadata that can be inserted.

Metadata allows for media to become rich media or media that is more than just audio. (See Chapter 10 for more information.)

The command marker can have a different name than the properties associated with the marker. Right-clicking the command marker provides a menu for renaming the file. Command markers can be deleted individually by right-clicking the marker and selecting Delete. Right-clicking in the Marker bar and selecting Delete All does not affect command markers.

Rubber Audio

Vegas has a feature known as *Rubber Audio*. Audio can be stretched by grabbing the right side of an audio event, pressing and holding down CTRL, and dragging the event to the right. This option is useful to correct a slightly off-sync voice or to expand or compress an audio event to make it fit within a specified period. Audio that is expanded/stretched much more than 10 percent, however, will sound very strange indeed. Perhaps long stretches are valuable for a specific effect; however, the audio is quickly recognized as affected after it has gone past the 10–15 percent stretch point. Use the Rubber/Time Stretch feature only as a means of achieving sync on an out of sync and time format event or as a special effect.

Tip

`Rubber Audio.veg` is a project on the CD that allows you to hear what rubber audio/stretched audio sounds like at various speeds and pitches.

Time stretch can also be accomplished by right-clicking the audio event, selecting Properties, and adjusting the length of the file.

Audio pitch/tuning can also be adjusted in this same dialog box.

Use Semitones to move pitch in musical half steps and use the Cents dialog to change by hundredths of semi-tones. For example, a pitch shift of -7 semitones is equal to a fifth interval down from the tonic, or original pitch. Shifting a pitch by +4 is a minor third up from the tonic or original pitch.

Tip

Sound designers will find the Pitch Shift/Time Stretch tool invaluable. Using household sounds, such as blenders, flushing commodes, gelatin being slowly sucked out of a can, and other creative sounds, coupled with the Pitch Shift/Time Stretch feature yields usable sounds for any kind of video or audio production. Adding reverbs, delays, flanges, and panning information to these sounds further enhances value and realism.

Variable Speed Audio in Vegas

Currently, Vegas does not have the ability to speed up, reverse, or slow down sound on the timeline. If audio speed must be gradually reversed or slowed down, the file must be opened in either Sound Forge or another audio editing application. If a multichannel sound card is available on the system, however, there is another option.

Create a new bus and route the audio to be sped, altered, or reversed to this bus. Create a new track and select the output of the new bus to feed the new track. Click Record and play away with the Playback Rate control in Vegas. This process is a lot more accurate and fun than using the J, K, or L keys or a Contour ShuttlePro/SpaceStation AV. The altered playback shows up on the new track where it can now be filtered, EQd, or otherwise processed.

This process can also be accomplished with the Windows Media Recorder by selecting the inputs from the Windows Media Mixer.

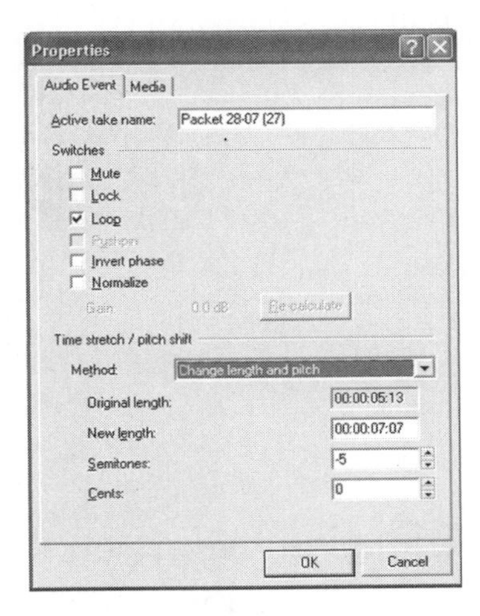

7.48 Changing pitch of an audio event can bring a slightly off-key singer into tune or create that "news television" sound for a slightly thin voiceover.

Snapping Audio Events

Vegas can snap audio events to a marker, region, grid indicator, or to an abutting event. Two events on the same track always snap together if snapping is enabled.

Two events can also be snapped together on different tracks, however. To do this, follow these steps:

1. Double-click the event to which you want the next event to snap. This step creates a time selection.

2. On the track that holds the event to which the selected event should be snapped, move the event toward the front or back of the region or selected event. Vegas auto-snaps the moved event to the first event that is selected, regardless of tracks. Several events can be snapped at once.

Events can also be snapped to a grid line. Pressing F8 enables snapping, and pressing it again disables snapping. If snapping is not enabled, timing grids are not visible. To snap an event to the grid, press CTRL+F8. To snap an event to a marker, press SHIFT+F8.

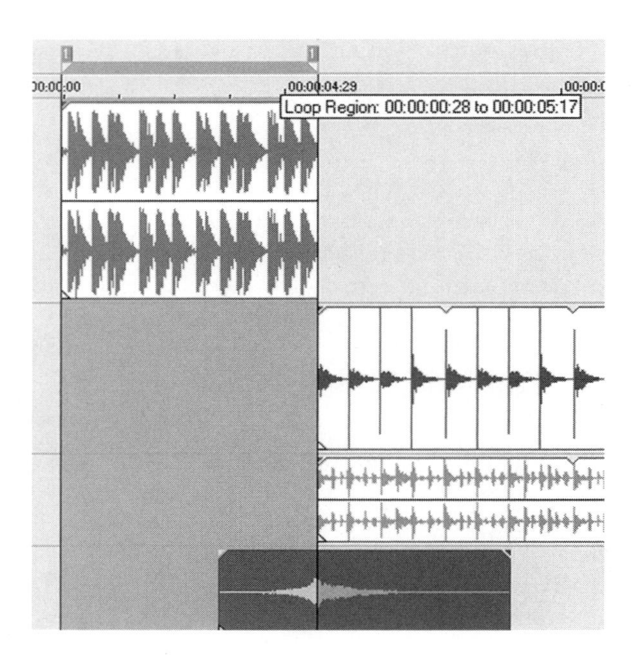

7.49 Snapping several events to a selected track event.

Trimming Events on the Timeline

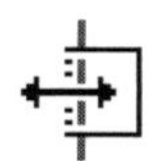

Events can be trimmed in the Trimmer window before being placed on the timeline. They can also be edited for length directly on the timeline as well. Position the cursor over the start or end of an event, and the cursor changes to the trim cursor.

Grab either the start or end of an event by clicking and holding, and drag the end of the event to the left or the start of an event to the right. The event shortens as it is dragged/trimmed.

If auto-ripple (CTRL+L) is enabled, any trimming on the timeline will result in events following the trim filling the space/hole left behind because of the trim.

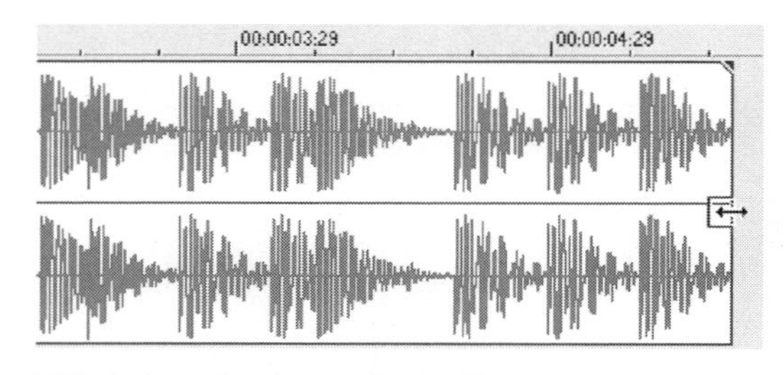

7.50 Before trimming on the timeline.

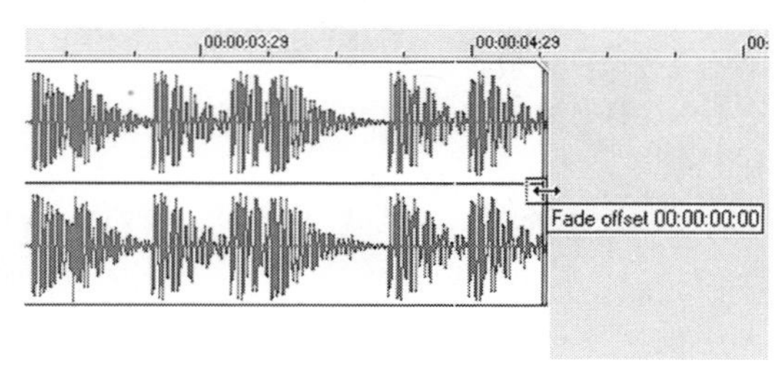

7.51 After trimming on the timeline.

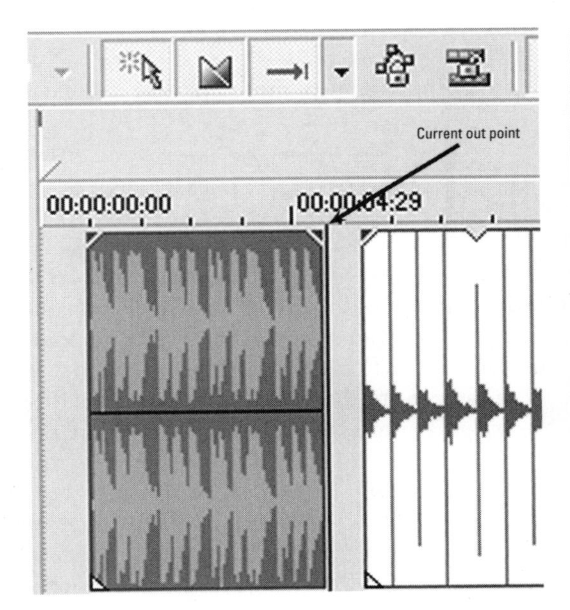

7.52 Trimmed on the timeline, media following trim filled in time vacated as a result of trimming on the timeline.

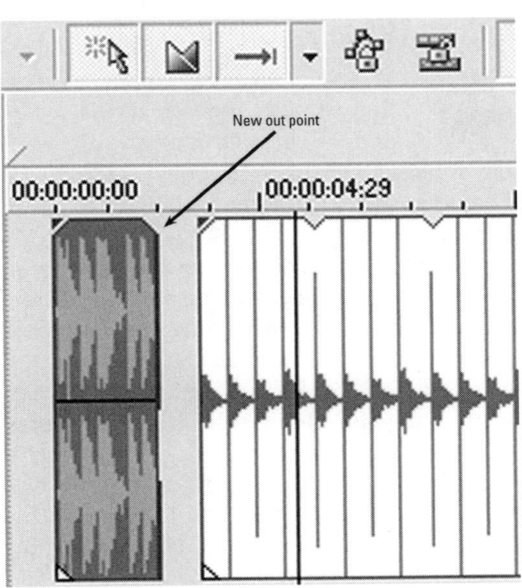

7.53 Rippled on the timeline, media following trim filled in time vacated because of trimming on the timeline.

Another option to trimming on the timeline is to slip or shift the contents of an event. By pressing and holding down ALT and dragging right or left inside an event, audio can be slipped forward or backward in time without moving the actual event. This feature is great for moving audio within a specified time space without having to be concerned about losing space in time. The event remains a placeholder while audio can be fine-tuned. This feature is also useful for fitting audio in a stipulated time space, where time is specified but the contents can be determined later.

Deleting Events from the Timeline

Selecting events by clicking them and then pressing DELETE removes them from the timeline. Multiple events can be selected by pressing and holding down SHIFT and clicking events to be deleted. Deleting events removes them from the timeline but does not eliminate the empty space that remains unless auto-ripple is enabled (CTRL+L). If an entire space in time across multiple tracks is to be deleted, make a selection on the timeline. Select Delete (CTRL+X). (Selecting Delete does not place deleted media on the clipboard.) If Ripple is enabled, all events remaining on the timeline shift to fill the empty space; otherwise, the time selection remains behind and is left empty.

Events or time can also be deleted by selecting EDIT | DELETE.

Tip

If in the process of deleting events where ripple is needed but not enabled, use a post-edit ripple to close deleted areas. Post-edit ripple behaviors are selected in the Edit menu or by pressing F, CTRL+F, or CTRL+SHIFT+F.

Using Ripple Editing

Vegas 4.0 has new Ripple features that make rippling much more intuitive and easy. Entire tracks, multiple tracks, and individual events can all be rippled collectively or individually. Enable rippling by selecting the Auto-Ripple button.

Select Affected Tracks from the three options displayed in the drop-down menu. Drag several events to the timeline, butting them together. Trim the outpoint of the first event. All events subsequent to the first event slide to the left, remaining butted up against the first event. At random, place markers on the timeline. Now drag the out point of the first event back to where it was when originally placed on the timeline. The events subsequent to the lengthened event slide forward in time for the exact length as the event being dragged moves. However, the inserted markers stay where they are.

Return to the Auto-Ripple drop-down menu shown above and select Affected Tracks + Project Markers, Keyframes, and Envelopes. Drag the out point of the first (or any other event) to the left, and this time when all events slide to the left, regions, markers, envelopes, and all other attributes of all tracks slide to the left as well, keeping the project timeline intact. This process is called a *ripple edit.*

Inserting from the Trimmer is easy and fast as well. Select or regionalize audio in the Trimmer and either press A with rippling enabled or, if rippling is not enabled, press SHIFT+CTRL+V. The audio inserts as an event with all subsequent audio sliding down to make room for the new event. Ripple editing is useful in quickly cutting together on-air commercials, or in a musical setting, it's excellent for cutting out entire sections or bars of music.

7.54 Vegas draws an arrow indicating the direction of the ripple that also demonstrates the length of the ripple.

Grouping

At times, it's helpful for entire sections or types of events to be grouped or locked together so that when one is moved, all of them move. Grouping events is quick and easy in Vegas. Select a number of events and press G. This step ties all of the events together so that when one is moved, all the others move with it. For instance, perhaps all events related to a chorus, bridge, or verse might be grouped so that when one section is moved, all related events to that group are moved with it.

To remove an event from a group, select it and press U to ungroup. Grouping can be temporarily ignored by clicking the Ignore Event Grouping button on the toolbar.

Groups are great for applying edits to multiple events at one time. To apply an edit to all members of a group, perform the following steps:

1. Press SHIFT+G or right-click one member of the group.
2. Select GROUP | SELECT ALL. You can cut, copy, or paste edits to the entire group. Grouped, but unselected events, are not edited together.

Events can only belong to one group at a time. Adding an event to a group essentially dissolves the existing group and creates a new one. As many groups as needed can be created, yet no two groups can contain the same event.

7.55 Grouping can be temporarily ignored.

Working with Audio Plug-ins

Vegas has a tremendous assortment of plug-ins that come with the application and that provide massive mixing and editing power. Particularly necessary in the digital audio world, there are multiple types of compressors and equalizers, plus reverbs, delays, flangers, noise gates, resonators, and more. As a result of the open Direct X platform, plug-ins from companies such as WAVES, Ozone, Sonitus, Cakewalk, and dozens of other manufacturers, can be used in Vegas.

Sonic Foundry was the first adopter of what was once called Active Movie, which later became Active X, which then once again changed names to Direct X. Sonic Foundry supported Active Movie in their 4.0a version of Sound Forge. Because of Sonic Foundry's pioneering work in the plug-in interface market, most every video editing application, as well as audio application, supports the Direct X audio platform.

Plug-ins, such as compressors, reverbs, delays, and noise reduction, all take the place of hardware counterparts in the analog world. In fact, many software plug-ins look just like their hardware counterparts, and, in some cases such as WAVES, the software tools were so popular that the company followed up the software tools with hardware tools for the non-digital environment!

Sonic Foundry offers nearly 30 plug-ins with Vegas 4.0. Rather than examine each of them, we'll look at what each family of plug-ins does.

Dynamic range plug-ins offer the ability to ensure that audio doesn't exceed the dynamic range capabilities of digital audio. This includes compressors, limiters, and combinations thereof, such as WaveHammer, Sonic Foundry's final-step plug-in.

While digital audio has an extreme dynamic range, it is also less forgiving than its analog counterpart, in that the zero point in the digital world is cold, harsh, and unforgiving. In the analog world, the dynamic range is fairly limited by any comparison, ranging in the upper 60dB area depending on the quality of equipment used. Twenty-four bit digital recordings have a dynamic range of 138dB, or more than double the dynamic of analog. Dynamic range plug-ins help reduce dynamic range, while maintaining most of the nuances of a fairly dynamic performance. They could be considered automatic volume controls. By limiting, or subtly compressing the exceptionally loud portions of a performance, the performance still maintains its soft to loud moments but at the same time doesn't exceed the preset point. More on dynamic range control is found in the "Mixing Techniques" section of this chapter.

Sonic Foundry Dynamic plug-ins include:

- Express FX Dynamics
- Graphic Dynamics
- Multi-band Dynamics
- Track Compressor
- Vinyl Restoration
- Wave Hammer

Frequency-based plug-ins/FX function by varying the amount of volume of a given frequency or group of frequencies to change the way a recording sounds. Equalizers, wah-wahs, phasers, and similar tools work on either moving a frequency's amplitude up or down to change the audio's sound or timbre. Wah-wah and phaser sounds sweep/shift the amplitude and frequency over time, giving the sound of motion, similar to the concept of a Doppler effect. In fact, some plug-ins are written to simulate the Doppler effect.

Frequency-based plugs include:

- Wah-wah
- Graphic EQ
- Paragraphic EQ
- Smooth/Enhance
- Track EQ

The Doppler effect, discovered in about 1845 by Christian Doppler, an Austrian mathematician, demonstrates that shifts in frequency occur as the article generating the sound moves. This effect is how our ears perceive sounds coming toward us and moving away from us. Listen to a train as it approaches. The frequencies contained in the sound that the train makes shifts as it moves toward you or away. The shifting frequencies result in shifting pitch as the object moves in either direction.

Time-based plug-ins/FX are based on just that: time. Delays, reverb, chorus, and flangers are all based on delaying, recombining, and/or reflecting a moment in time. When a sound originates, roughly 40 percent of its sound that we perceive is based on reflection of sound in a given space. In other words, singing in the shower is so much more fun than singing in a closet because the reflections of the voice in the shower hit the ear much more quickly than the reflections in a closet do. Some types of music are predominantly based on reverbs and delays, used to create emotional expression and to fill holes in a song. Without long delays and reverbs, many of the ballads and arena anthem rock songs would sound empty and powerless because the lead vocal and chorus vocals would lose their presence and power through the absence of filled time and massive reflection. Time-based effects are used more than any other effect to create emotional responses. Dripping water combined with huge reverb settings is part and parcel of nearly every horror movie made, and the simulated sounds of space (space has no air and therefore sound cannot pass) also include reverbs and delays.

Time-based plug-ins include:

- Flange/Wah
- Multi-tap Delay
- Reverb
- Simple Delay
- Acoustic Mirror*
- ExpressFX Chorus*
- ExpressFX Delay*
- ExpressFX Flange*
- ExpressFX Reverb*

* Optional FX from Sonic Foundry

On the CD enclosed with this book, you will find demonstration plug-ins from PSP Audio, WAVES, Ozone, and Sonitus. A wide variety of plugs are available for you to experiment with and enjoy.

Automating FX in Vegas

One of the most exciting new tools in Vegas 4.0 is the ability to automate FX. Many of the Direct X plug-in tools have the ability to be automated. EQ settings, delay times, reverb decays, and much more can now be controlled by Vegas' automation envelopes. Many of the high-tech sounds demanded by production music today require either several hands at the mix console or automated FX control.

For instance, if a vocal line reaches a break just before a chorus and a long delay is needed, automated FX can change the length of the delay time as opposed to using another FX send just for the vocal breaks. Perhaps an EQ sweep just as a car passes from right to left, creating the illusion of a Doppler effect, would make the sound more believable. Frequency sweeps added to drum mixes or to a main instrument sound become part of the instrument itself, creating an unmatchable identity. Remember "Axel F," the theme song from *Beverly Hills Cop*? The sweeps heard on the bass lines in that song were created by synthesis, but the sound itself was a huge part of the song. Techno music is nearly always automation-dependent because the repetitive nature of the music.

To hear an example of automation, open the FX Automation project on the CD. On track 1, you can observe an inserted automation envelope by right-clicking the Control pane and choosing FX Automation Envelopes. Notice in the Track EQ settings that Band3 Gain and Band3 Bandwidth boxes are checked. Click OK.

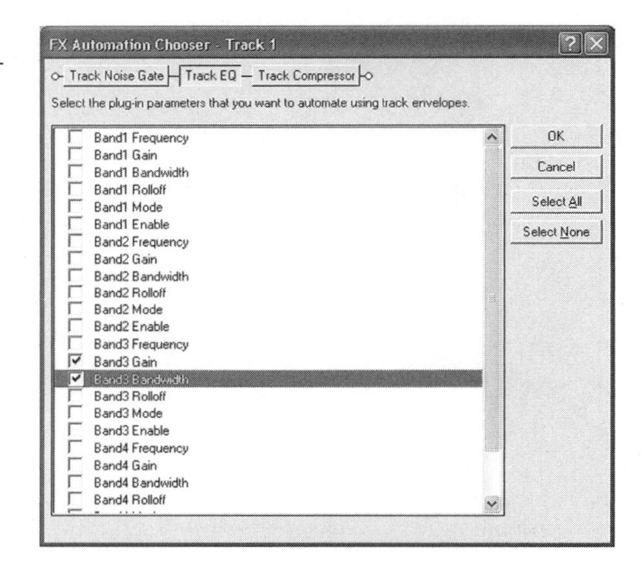

7.56 Select the boxes related to the automation parameters you wish to control.

A pair of dark-colored envelopes are on the track in the timeline. These are the bandwidth and gain controls for the EQ that you have set for automation.

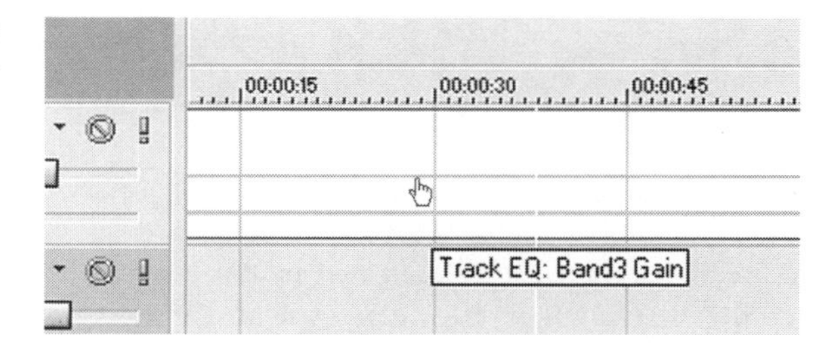

7.57 Automation envelopes appear when Automation parameters are selected.

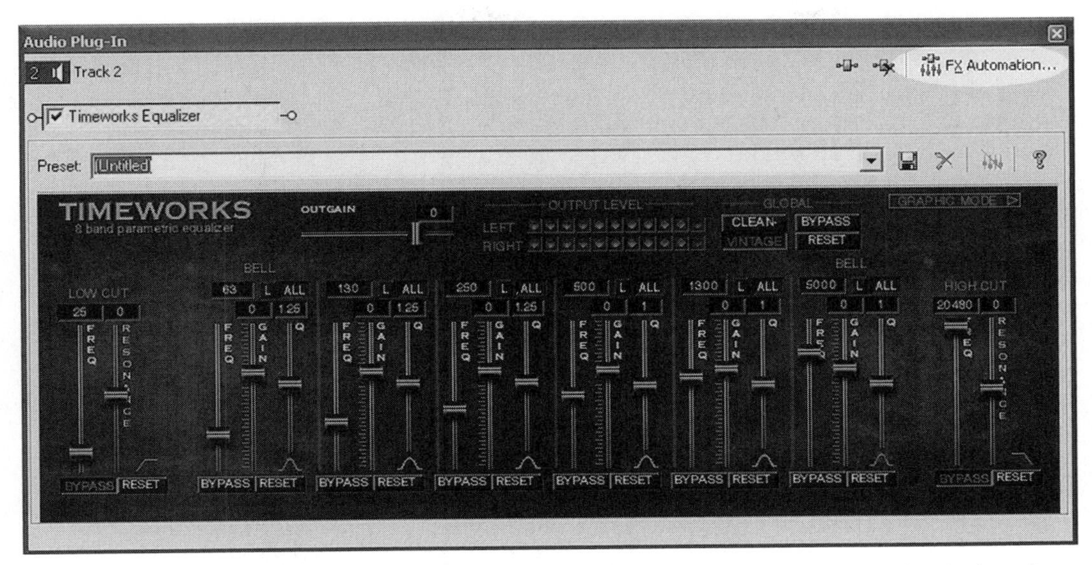

7.58 The FX Automation button opens up a dialog where tool automation features can be viewed and selected.

Every two seconds, insert a handle on the Band3 Gain envelope or the envelope that appears in the middle of the track. (The project grid is laid out in seconds.) In between the handles on the Band3 Gain envelope, handles are already inserted on the Band3 Bandwidth envelope. Now drag every other handle of the Band3 gain handles down to the bottom of the track, or -15dB worth of adjustment. Experiment with the type of fade by right-clicking the handle and selecting different fade types. You'll be astounded at the degree of control this offers the sound.

Not all Direct X plug-ins have the ability to be controlled by automation envelopes. iZotope, Sonic Timeworks, WAVES, Wave Arts, and the plug-ins found in the Automated plug-ins folder are all examples of automatable plug-ins. If a plug-in is capable of automation, a selection button will appear in the upper-right corner of the plug-in. When the button is clicked, the parameters available for automation will appear.

Some plug-ins have to be added to a chain at least once before they are recognized as automatable. Don't be alarmed if you don't see them in the automatable folder right away.

Working with Other Types of Plug-ins

Noise Reduction

In 1995, Sonic Foundry released a product called Noise Reduction 1.0. This new tool revolutionized the music industry. As in the past, only very expensive hardware tools or the Digidesign Intelligent Noise Reduction (DINR) tools found in ProTools or No Noise from Sonic Solutions were available. One was too expensive and only occasionally effective, the other not satisfactory in extremely noisy circumstances. Sonic Foundry's Noise Reduction tool gave working musicians the ability to clean up poor sounding rooms or environments.

Noise Reduction uses an algorithm known as Fast Fourier Transform (FFT) as part of the power behind how it functions. This algorithm is large, requiring millions of calculations based on information selected in the noisy portion of the file. A basic explanation is that when noise is present in a sound file, the noise can be isolated in sections between words, instrument attacks, or in heads or tails of a file. The noise is then sampled, and the algorithm applied to the rest of the audio file, removing the sampled sounds but leaving behind the original sound. Sometimes, if the sample is too large or contains too much information (sounds other than noise) or if the noise sample is frequency unstable (sound is moving), the process can remove some of the original, pre-noisy sound, resulting in artifacts. These artifacts often resemble the sound of rushing water or a waterfall. Other times, it has a metallic, robotic sound. In fact, some films have used this extreme noise reduction technique to create alien sounds. In this section, however, we're interested in removing noise to make a file sound cleaner. In short, Noise Reduction is similar to a frequency-depended noise gate that is divided into several hundred bands. The noiseprint determines the threshold or open level of each gate, and any time the volume of a frequency is below that threshold, it is suppressed/silenced.

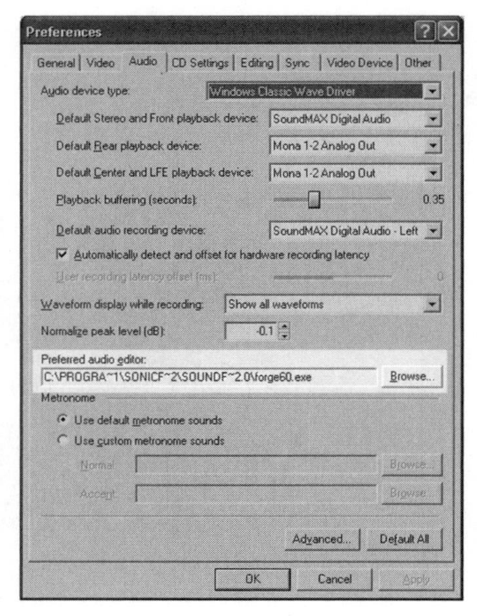

7.59 Select the preferred audio editor in the **PREFERENCES | AUDIO** menu.

Noise reduction is particularly effective when trying to remove a constant noise source, such as an air conditioner, refrigerator, or AC hum. It is least effective on moving sources, such as constantly passing traffic, running water, or varying wind noise.

A demonstration version of Noise Reduction is located on the CD in the back of this book. Install it now, if it has not already been installed.

To apply Noise Reduction to a file, first open the file in a Direct X–capable audio editor by right-clicking in Vegas and choosing "Open copy in (audio editing application)." This option should already be configured if you followed the steps in Chapter 3. If you have not done so, select **OPTIONS | PREFERENCES | AUDIO** and use the Browse button to locate the executable for your favorite audio editor. It is located in the `C:\Program Files` directory in most instances.

On the audio file, select a very small slice of audio that contains only the noise that needs to be removed. When selecting a noise sample, less is more. Fewer frequencies in the sample make for a more accurate reduction in those frequencies.

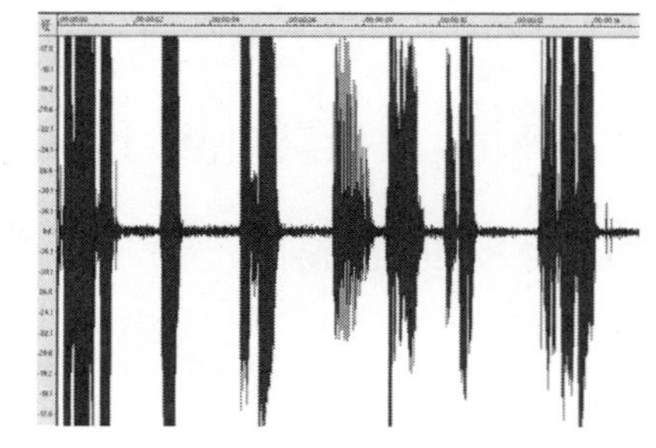

7.60 When the file opens up in the sound editor, it looks something like this. Notice the noise between words spoken.

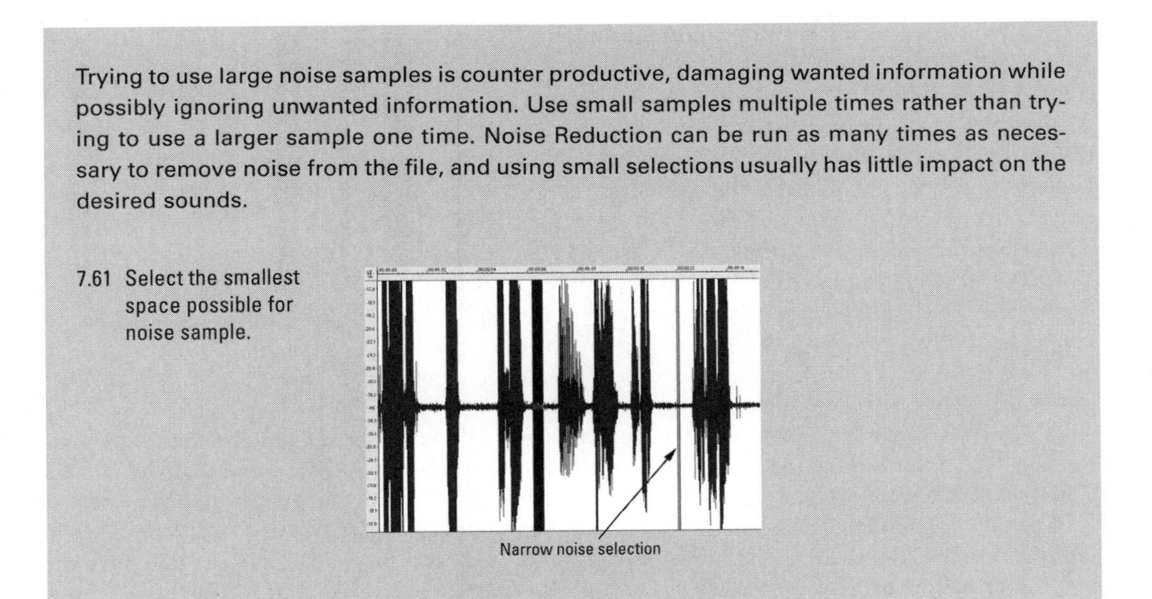

Trying to use large noise samples is counter productive, damaging wanted information while possibly ignoring unwanted information. Use small samples multiple times rather than trying to use a larger sample one time. Noise Reduction can be run as many times as necessary to remove noise from the file, and using small selections usually has little impact on the desired sounds.

7.61 Select the smallest space possible for noise sample.

Now open Noise Reduction from the DirectX plug-in host, and the screen shown in Figure 7.62 will be displayed.

Check the Capture Noiseprint box and click the Preview button. This step allows you to hear what section has been selected. At the same time, a noiseprint has been captured. Click the Noiseprint tab if you wish to see what the noiseprint looks like. Thousands of handles are on the noiseprint, so that custom or exceptionally accurate control of the noiseprint can be had. It is generally best to allow the plug-in to function automatically, however. Playing with these handles does offer the opportunity to create some fairly inventive and radical sounds.

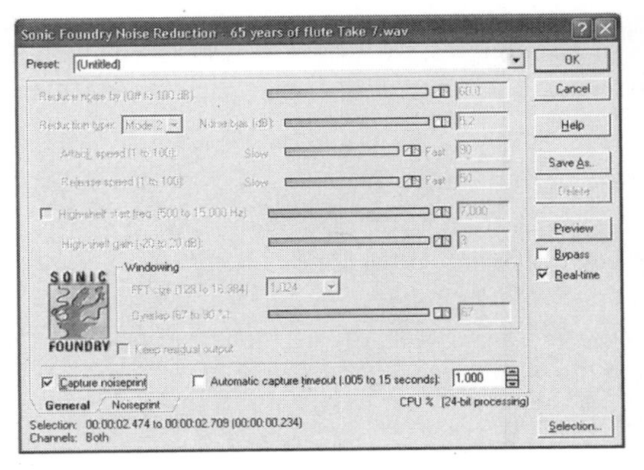

7.62 This is the main window for Noise Reduction, where all control of noise removal takes place.

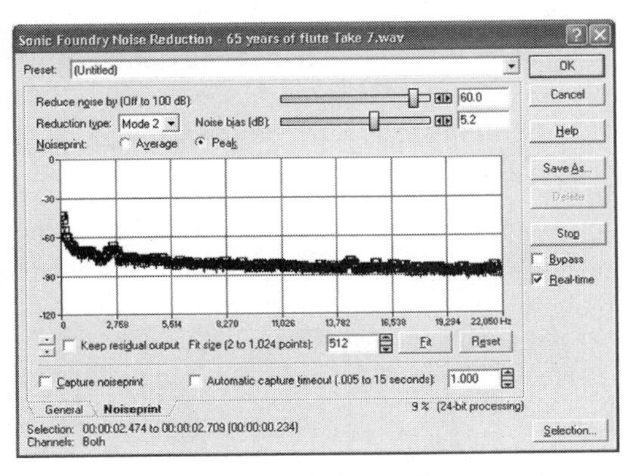

7.63 A view of how Noise Reduction interprets the noise information.

After the noiseprint has been captured, the area that the filter should be run on must be selected. Noise Reduction allows for selectable time spans, entire files, regions, or specific areas to be processed. Generally, files will be the entire area, but you can specify an area if so desired. Generally, choose All Sample Data because if noise is present in one area, it's typically present in the entire file. (Other host applications can operate differently.)

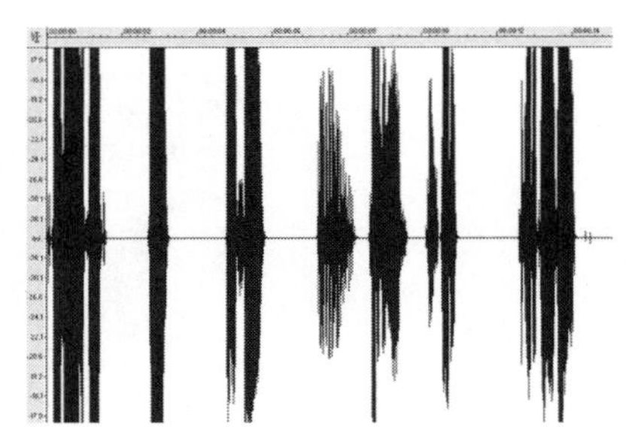

7.64 File after noise reduction applied.

Click OK when you have determined what time selection of the file should be affected. Now you'll be able to move the Reduce Noise By slider, listening in real time as Noise Reduction processes the noise out. Typically, it's rare to get more than -50dB of noise out of a file, but occasionally more aggressive settings will work.

A useful feature in Noise Reduction is the Keep Residual Output box. Checking this box inverts the signal, so that the information being reduced/eliminated is heard and the audio that will be kept is masked. This process allows you to hear if too much of the original signal is being affected or not.

After the desired level of noise is reduced/eliminated, click OK. The Noise Reduction plug-in will process the file. Closing the file in Sound Forge will prompt you to save the file. Save the file, and it will be deposited on the timeline in Vegas in the exact place that it came from, but it is now named the original filename with *take(number)* in the filename. It can be toggled back and forth from the original audio file for comparison.

Some files require multiple passes of Noise Reduction in order to obtain the best results without affecting the original audio/desired audio.

The CD in the back of this book includes a project called Noise Reduction Project. Open this file. You will see two video files; the first has had no noise reduction on it, the second has been noise reduced. If you monitor the project through a good set of speakers or through any headphones, the hum in the background is very clear. The noise source is a refrigerator, three rooms from where the interview was being conducted. In the case of the second file, -70dB of noise was taken out, without creating artifacts in the desired audio file. Even if you don't have speakers that can accurately reproduce the noise, pay particular attention to the master meters.

Practice on the first file. If you don't capture a small enough segment for your noise file, you'll most likely create artifacts in the speech. FFT processing doesn't require a large slice; use a 100ms slice of noise at most for your sample.

Noise Reduction can be used for other purposes, such as creating unique sounds over a drum loop, voice, or sound effect. Experiment with it and find what works creatively for you. Drastically removing the high end by sampling the decay of a high hat or snare can give unique personality to the sound of any drum loop. Try using Noise Reduction on a heavily compressed speech file and make any human sound become more robotic.

As a side effect of Noise Reduction's capability, it is not intended to be used as a real-time effect. It is exceptionally processor-intensive, and all but the fastest computers with loads of RAM will stutter and pop. This plug-in is best used as a destructive edit, as a take over original audio, or used in an audio editor as a destructive or take-based event.

Mixing Audio in Vegas 4.0

Mixing audio is a practice based entirely on personal preference, but some standards and techniques are used by most engineers and producers to achieve mixes that meet with industry expectations. Each person's ears are different; this is what makes one producer more popular for various kinds of music than another. Mutt Lange, a very well-known producer of heavy metal music and pop-rock, is one of the few to jump from one musical style to another. (Marrying country legend Shania Twain didn't hurt.) Peter Gabriel is renowned for his production of world-pop music, just as William Aura is known for his adult contemporary work. Alan Parsons, Tom Lord-Alge, Ross Collum, and Danial Langios are all unique and well-known producers/mix engineers. All of them credit their years working with the equipment and producers in the studios they cut their teeth in as third engineers, janitors, or receptionists. All of these producers got to where they are not only because of their inspired sense of what people want to hear and how to bring an artist's talents to the fore but also because they knew the basic principles and practices of mixing audio. Knowledge and a practiced ear are the two most important things to creating a great mix. Knowledge is easily found. Having a practiced ear can take years. The basic principles of understanding a mix are fairly simple; from there, it simply becomes a matter of expanding on experience.

Mixing Techniques

Setting up a mix is entirely dependent on the contents of the track, whether it's instrumentation, bed music with a voice over, Foley, sound design, or full-blown orchestra. Bass frequencies tend to be more centered in a mix than are high frequencies, where a right-to-left spread is more evidently heard. Bass frequencies below 250 Hz aren't really even directional but rather omni-directional. Higher frequencies are very directional, which is why bass drums and bass guitar/synthesizers are generally mixed to the center of a mix and higher frequency sounds are spaced throughout the right-to-left areas of a mix. An easier way to think of it is to look at a photo of a beautiful mountain vista. The mountains form the base of the image, and the eye is drawn to the bugling elk in the forefront, slightly off center. You notice the green trees and the stream or brook flowing through the picture. The primary subject of the picture is the elk in the foreground, but without the surrounding beauty, the elk looks rather stark. Creating a good mix is just like that: lots of individual elements that draw the ear to a primary element. Try to picture a mix visually with the score or performance in your mind's eye, which will help achieve the end goal.

Fine-tune mixes by holding down the left mouse button and use the 2 or 8 key on the numeric keypad to fine-tune any envelope up or down. The 4 or 6 key moves a node forward or backward on the timeline. The amount of control is determined by the depth of track focus. With maximum zoom, superfine envelope increments are possible. The Quantize to Frames (F8) option might need to be disabled to have accurate access at maximum zoom.

The same keyboard trim features affecting the video timeline are also found in the audio toolset. Use the 7 or 9 key on the numeric keypad to select an event edge; use the 1 or 3 key to edit longer or shorter by samples. The depth of track focus determines how fine the keyboard edits are made. Zooming in deep on a track causes up to sample-by-sample edits; zooming out on the track edits larger sections of audio. (See Chapter 5 for more information on keyboard editing.)

Some mixes are built from the bottom-end elements first, finishing with the primary element, and other mixes are built starting with the primary element and the other sounds are built around it.

For working with a dialog-based mix for film or video, this is a good way to start, getting the voice to a level that is comfortable. It should be loud, but never crossing the -3dB mark so that room exists for other elements to maintain their dynamic expression. Adding a compressor to the primary element is fairly standard practice. If the primary is a voice, start with light compression settings of 1.5:1, or 2:1, working with the threshold to suit the mix. Be cautious of squashing the sound too much, as this element is the most out front one of the mix.

Next, begin to place the foundation elements, such as static walla, traffic, or background noise. In a musical context, these elements would be the kick drum, bass guitar, or other bottom-end elements. If the bottom sound is muddy or tubby, remove some of the frequencies in the 300 Hz region. In a musical mix, add a little 1.5 KHz for some snap in the kick drum and bass guitar. In a mix for video with dialog, leave some of the upper frequencies out, perhaps even dampen them a bit, so as not to fight with the frequencies in the primary element/front voice. Use a compressor, starting with a basic setting of 3:1 or 4:1, and work with the settings from there to keep the mix from becoming too dynamic in the low frequencies. Send all the bottom/foundation elements to one submix/bus, with an additional compressor if necessary on the bus.

The term *walla* stems from the early days of radio, when extras were brought in for a radio or film show, and actors would say "walla-walla-walla" over and over, with just a few people saying the words out of sync with each other to create the sound of a murmuring crowd. No individual words were spoken at all. To this day, background dialog is called walla, but actors no longer use the fixed word format.

Sound design comes into the mix next, with motion and depth filling the speakers with moving elements that can be musical or not. Either way, the sound should be a filling sound that underlies the other pieces of the mix. In a music-oriented mix, this is the rhythmic element. Try to avoid panning this sound hard right or left if possible. More of a 9 o'clock and 3 o'clock is the ticket, washed wide rather than being too loud, so as to take away focus from the aural subject. Guitars, synths, and background vocals all fit into this space too. Be cautious of wanting too much sparkle or bottom end in these elements and leave room for the primary sounds and foundation sounds. Use reverbs or delays to wash these sounds across the sonic canvas, rather than increasing the volume of an individual instrument. These sounds contribute a sense of color or timbre of the musical element. If everything were taken away except this element and the lead element, there would still be something worth listening to. If inconsistent sounds are in this element, try to bring them in and out gently rather than with surprise. Otherwise, they'll detract from the front elements.

Next placed are the special FX or signature sounds of a piece. In the musical context, this is the moving sounds of a synth that has a signature to mark the song, rather than a synth that is emulating a traditional keyboard. In a video context, these sounds are the cannon fire, bullets flying, spaceships, aircraft swoops, or other action sounds. These should be placed to move right to left or front to back in a surround mix. These are the exciting elements of a stereo or surround mix.

Remember, these tips are mere rudiments. From here, the mix can be tweaked according to the individual ear.

Compression

One might say that it is simply impossible to do a good digital mix without compression, yet compression is one of the most misunderstood parts of the mix and audio process. First, it's important to understand the difference between a compressor and a limiter.

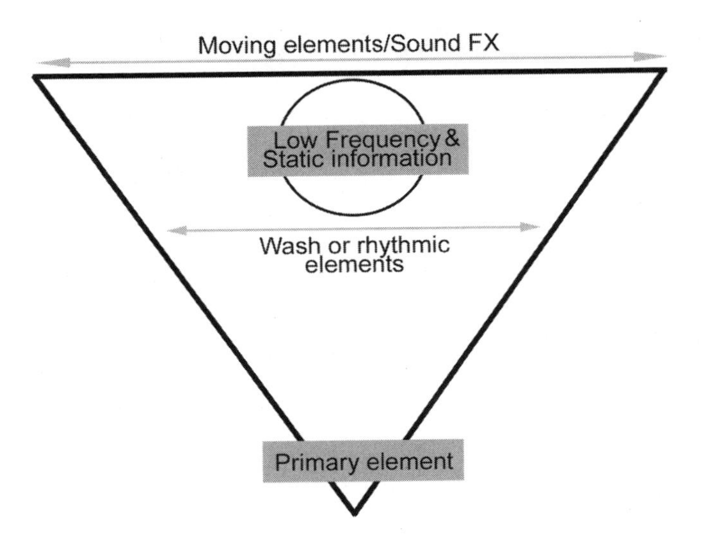

7.65 An overhead view of mix elements.

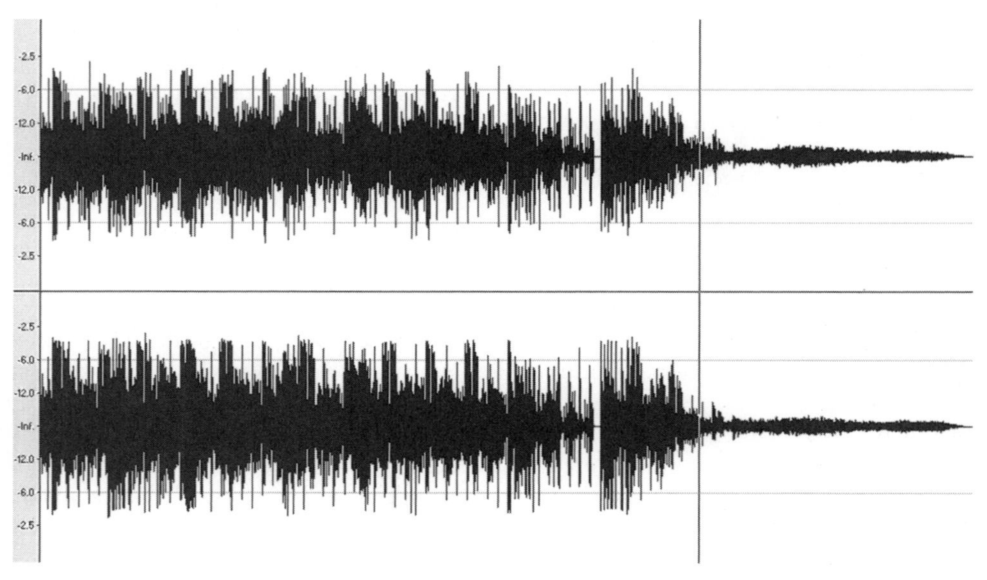

7.66 Audio before being limited.

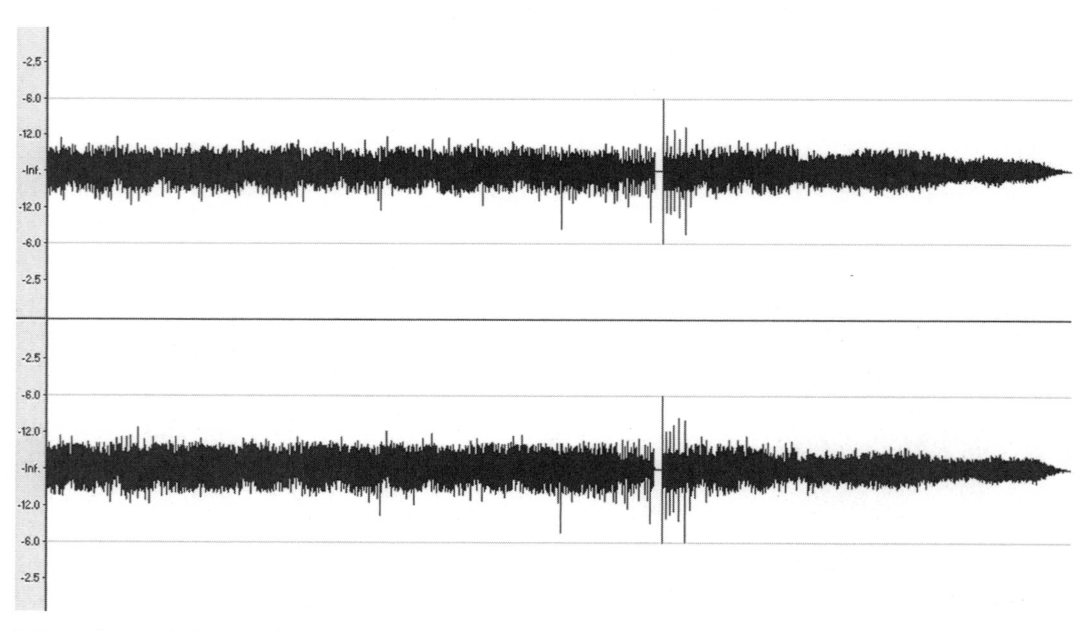

7.67 Audio after being hard limited.

A *limiter* prevents audio from passing at a predetermined volume. Any audio that attempts to pass the preset level is squashed to fit the preset level. Limiters are generally not good for music or audio for video mixes but are more suited for situations in which not all elements of the media can be controlled.

There are hard limiters and soft limiters. The difference is how much of a dynamic the limiter might be set at before the hard limit point. In other words, a limiter might be preset to ensure that nothing greater than -3dB is allowed to pass, but the majority of the audio information is at -5dB with lots of dynamic that wants to go much louder than the -3dB allowable maximum. Using a setting that instructs the limiter to start reducing levels at -9dB allows some dynamic feel to remain in the audio track. Limiters often squeeze the high end and life out of a mix, making it dull and boring. Television commercials are often hard limited and compressed to give maximum volume (one of the annoying things about the overall volume of a television commercial).

The term *soft limiter* is really just drawing a comparison to a *hard compressor*. Most compressors also have limiting ability built in.

Compressors operate a little differently from a compressor limiter, being more elastic or malleable only in the way they control dynamics on release. A limiter is a compressor with a ratio of 10:1 or greater but usually with fewer adjustable attributes than a compressor. Compressors can make audio louder, while at the same time making sure that the dynamic range is reduced. Generally, a compressor is used to keep dynamics as accurate as possible, while at the same time preventing audio from crossing a chosen threshold. This feature makes it easier to get an event/signal louder in a mix, while still maintaining control over the apparent signal.

For example, compression might be required for a singer or voice-over artist in the studio who moves toward and back from the microphone. Recorded levels are not consistent. Perhaps the same vocal artist has very quiet to very loud passages in their speech. A compressor evens out the differences between the two sections.

A compressor is used by guitar players and bass players to maintain the sustain of a plucked or strummed note/chord. A compressor smoothes down the attack on the strings and then opens up/releases gradually by allowing the note/chord to appear as though it is going longer than it would have otherwise been heard.

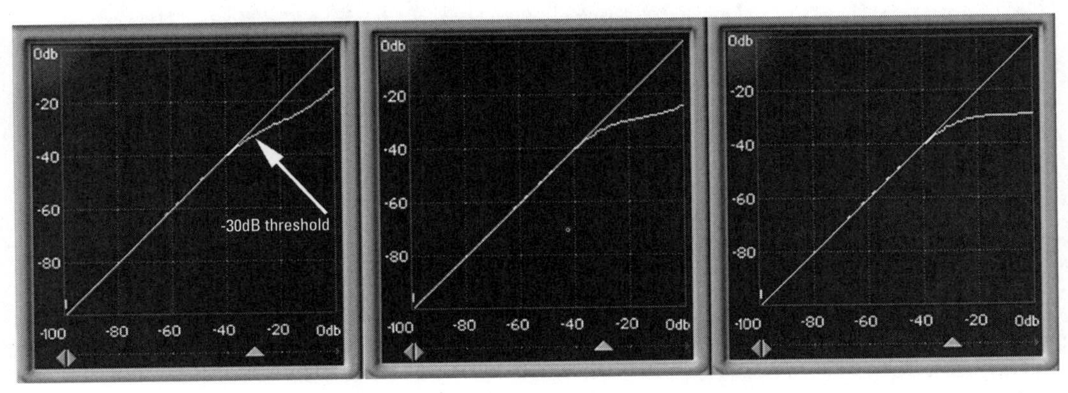

7.68 Threshold set at -30dB, with variable compression ratios: 2:1, 5:1, 10:1. Notice the compression curve at the different settings.

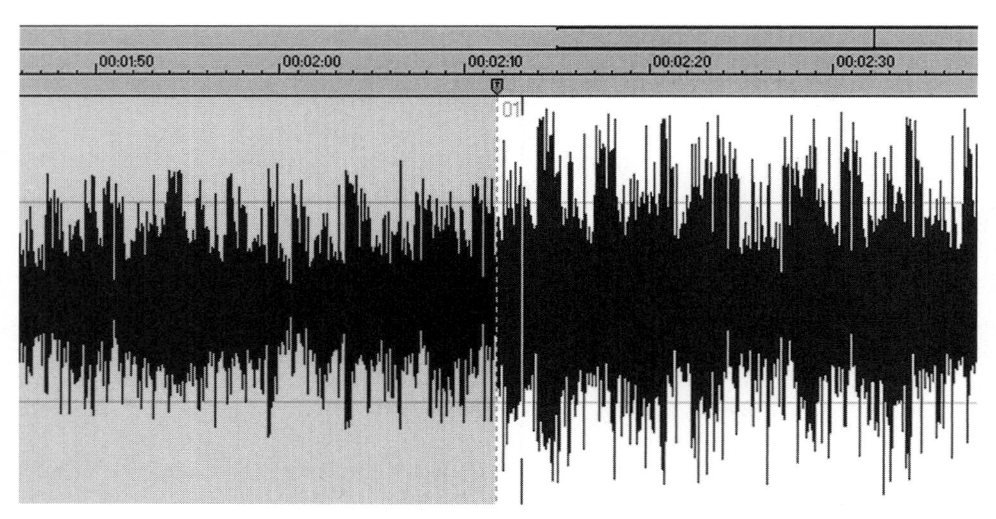

7.69 Mastering level compression at 2:1, with threshold at -8dB allows the signal to be louder overall because of the reduced dynamic range.

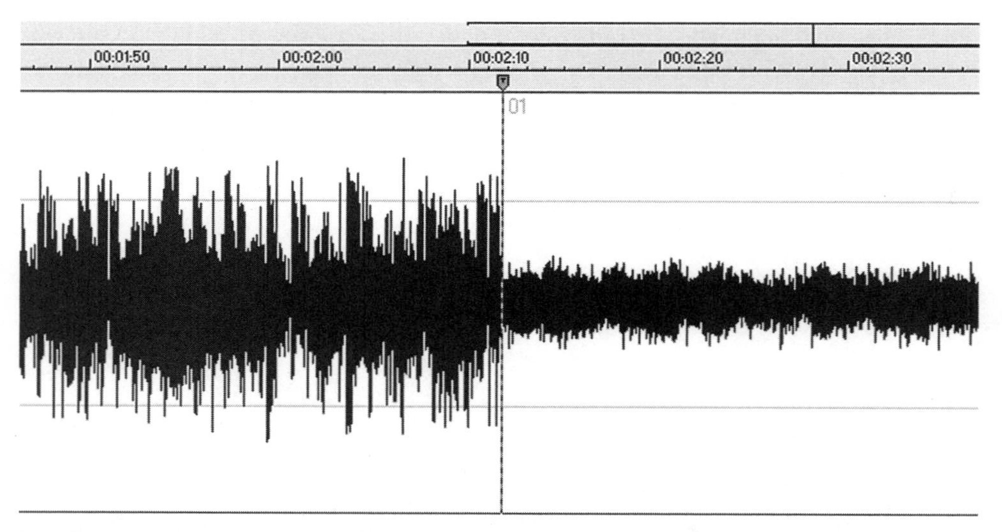

7.70 Compression applied at 2:1, with threshold at -24dB. Beginning the attack earlier in the signal with a lower threshold makes a huge difference in the behavior of the compressor.

Here's how a compressor works. Comprehending the ratio, threshold, attack, and release are the most important pieces to understanding compression.

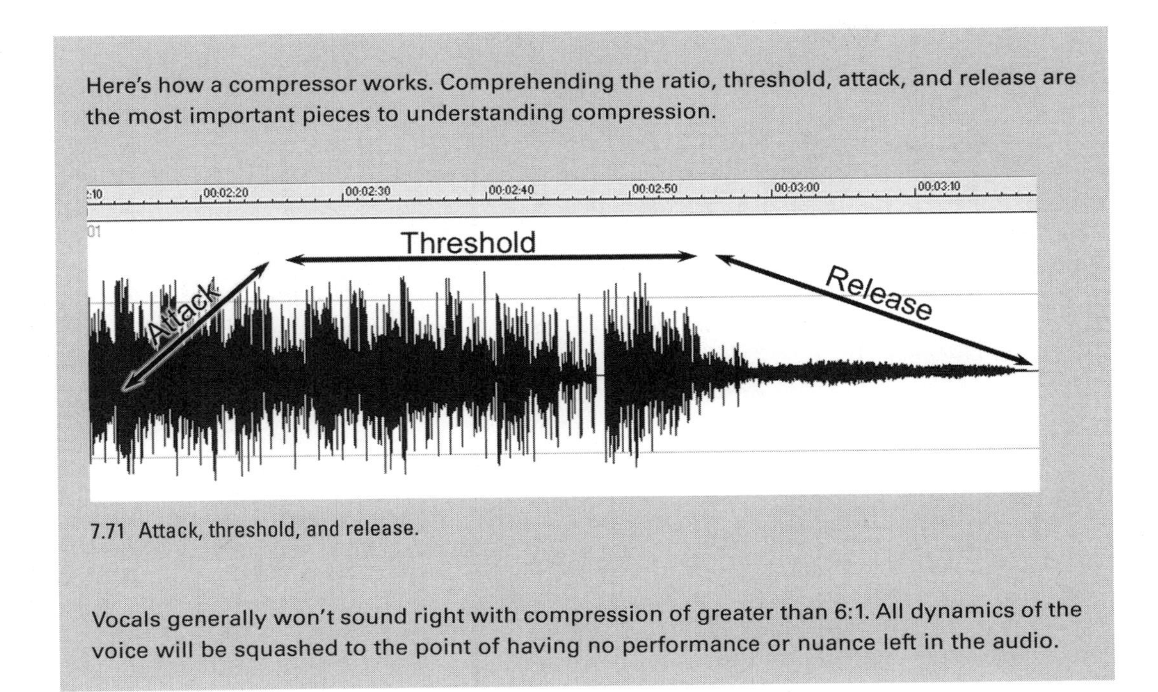

7.71 Attack, threshold, and release.

Vocals generally won't sound right with compression of greater than 6:1. All dynamics of the voice will be squashed to the point of having no performance or nuance left in the audio.

On the CD included in this book are five demonstrations of audio and audio for video. Listen to the differences in how the audio cuts through on the repaired sections.

Sonic Foundry provides a multiband dynamics processor in their plug-in arsenal. This multiband tool is a combination of four compressors with frequency-dependent settings. Rather than compressing an overall instrument or mix, individual facets of the audio can be affected while leaving other portions alone. For example, a snare drum might be exactly the sound you want but is too loud at one particular frequency or range of frequencies. The multiband dynamics plug-in, or EQ-based compressor, gives the ability to compress only the offending frequency or range of frequencies without affecting the rest of the audio information.

De-Essing

A component of both spoken and sung word is sibilance. *Sibilance* is the hissing sound made when speaking or singing *s* or *sh*. Difficult to repair, this issue, known as *essing*, creates distortion or apparent distortion on cheap speakers and high-end systems alike, regardless of how much compression or EQ are applied. In fact, EQ is the last resort to repair essing, and compressors are only one step ahead of EQ when it comes to de-essing sound. A de-esser is just a band-specific compressor.

When sibilance from *s*'s get in the way of the mix, open up the Sonic Foundry Multiband Dynamic plug-in and start with the De-Ess preset. Reset the attack time to .5 milliseconds. Reset the decay time to 40 or 50 ms. For a male voice, set the frequency center to 5 KHz, and for female vocals, set the frequency center to 6.5. True sibilance is usually centered between 8 and 10 KHz. Both of these settings are starting points. It might be that further adjustment is required.

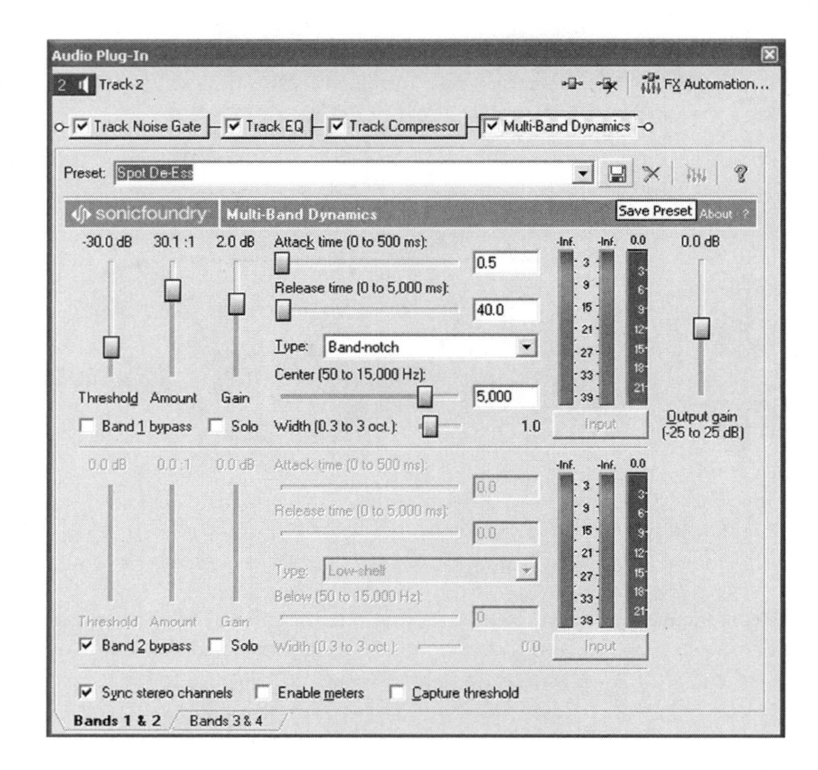

7.72 Saving the De-Ess preset setting preserves it for another session, or other de-essing in the same project.

The compression ratio should be set at 30:1 to start with, and the threshold set at -30dB. Now start a loop over a section where sibilance is present and adjust the multiband settings as the loop is heard. Be cautious of applying too much of a ratio; a singer or voice-over can quickly gain a lisp if it's set too high. Sometimes as little as 6:1 is enough. Experiment with raising the threshold and reducing the ratio until the desired fix is found. Save this setting as a preset by typing a preset name in the preset box and pressing the icon that looks like a floppy disk. This preset is now part of your toolbox.

Using Equalization

Equalizers, also known as *EQ*, are one of the most oft turned-to tools in the plug-in arsenal. EQs are used to cut frequencies of fat sounds that occupy space in the audio spectrum that other sounds can also occupy, creating confusion about what the ear is hearing. EQ is also used to correct problems with a tracking room, poor frequency balance in a microphone, or sometimes simply to enhance a performance or characteristic of an audio event. While invented initially to flatten sound over the telephone, today's EQs are predominantly used to unflatten sound, and so in jest, some engineers call these tools *un-equalizers*.

The use of an EQ must be judicious. Too much boost of any frequency can cause overload of the outputs, can cause a compressor to work harder than it needs to because of excess volume boost, or can even create distortion in the harmonics related to a sound.

EQs typically come in two formats: graphic and parametric.

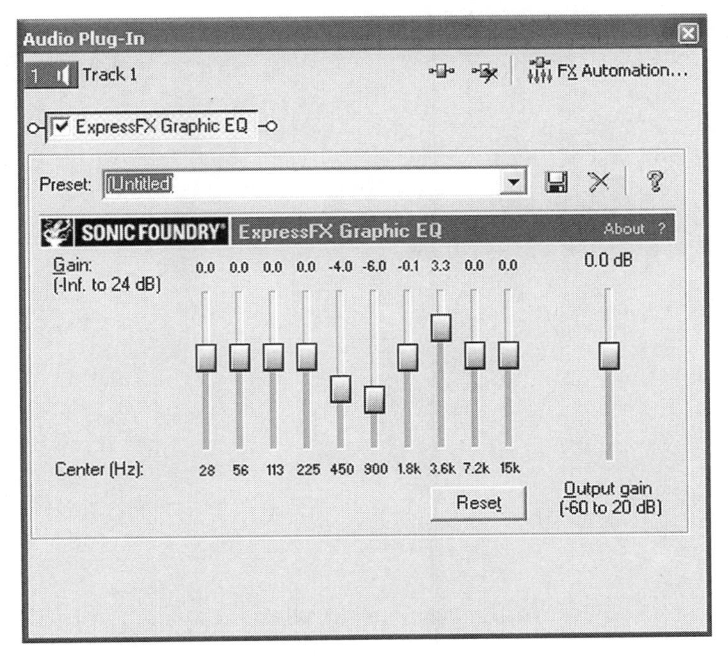

7.73 Graphic equalizers set at octaves.

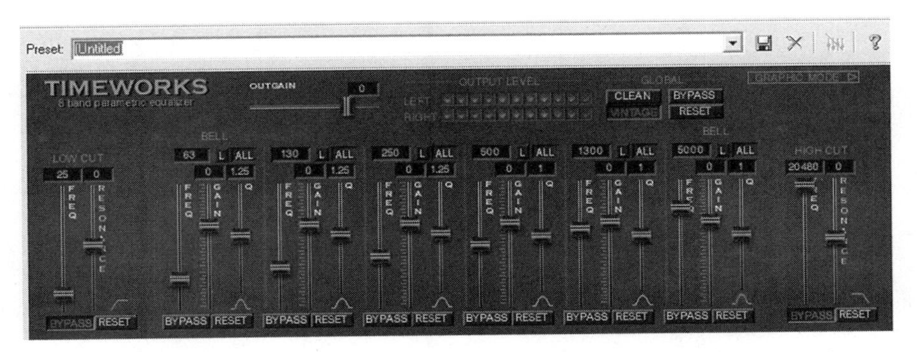

7.74 The Timeworks Parametric EQ resembles a Graphic EQ. However, frequencies can be selected, and the bandwidth or Q of the frequencies can be defined.

With *graphic EQs*, frequencies are preset, and only the gain of specific frequencies can be increased or reduced. It is found in octave, half octave, and one-third octave in most instances. An octave is determined by a frequency multiplied by two. The octave of 125 Hz is 250 Hz, the next octave being 500 Hz, and so forth. The half-octave of 125 Hz is 187.5 Hz, and the one-third octave of 125 Hz is approximately 160 Hz. Frequency centers are International Standards Organization (ISO).

Vegas has four different graphic EQs available as plug-ins.

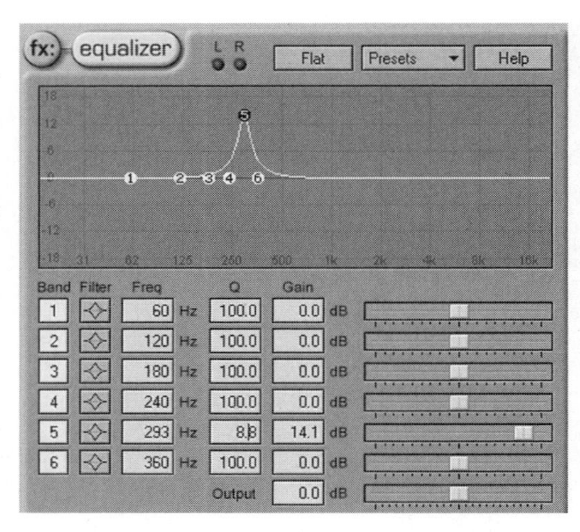

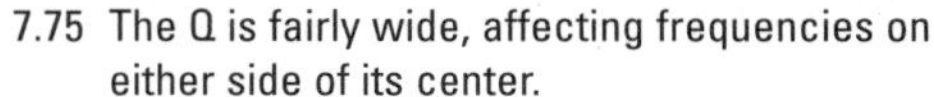

7.75 The Q is fairly wide, affecting frequencies on either side of its center.

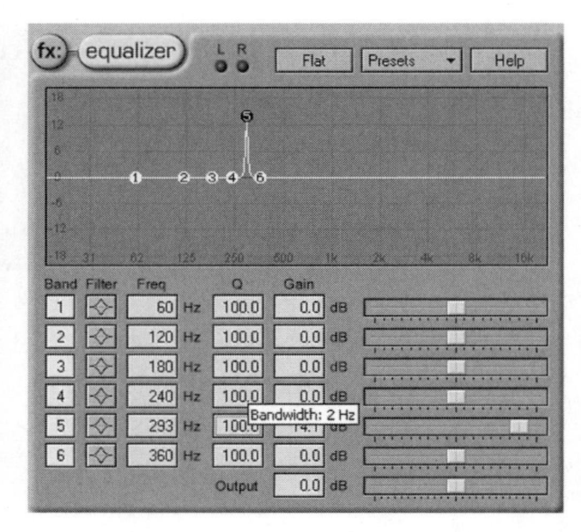

7.76 The Q is very narrow, affecting only the immediate frequency selected.

With *parametric EQs*, users can select a specific frequency center, select the amount of bandwidth that the frequency has control over, and control the gain or reduction of that same frequency band. Usually parametric EQs allow several frequencies to be specified and controlled.

A derivative of both graphic and parametric EQs, *paragraphic EQs* are mixed mode equalizers that generally have fixed frequencies for the bottom and top frequency spectrums, and midrange frequencies that can be shifted. This type is rarely seen in software-only forms.

Tight Q is useful for cutting out a specific frequency that can be a hum, distortion, or simply a problem frequency. Tight Q is also valuable for enhancing specific frequencies, such as boosting presence in a vocal to give an airy quality in order to obtain the pop vocal sound that is popular today.

Tip

On the CD, open the Equalization Demonstration project. The file contains the same event in duplicate. The first version is the broadcast version; the second has presence and compression added to it using a parametric EQ with a boost at 4.1 KHz. Notice the difference between the two. Either of these two mixes is acceptable for final output, but a difference exists in how the two are perceived.

Using Reverb

Reverb can be used for any kind of audio mix, whether for video or audio, spoken word or music. It is also one of the most overused effects in the palette of tools available. Reverb is used to artificially reproduce various room environments. In the past, music had to be recorded in large chambers or in rooms with large reflective areas. The reverb created in such large rooms helped blend sounds together, creating a cohesive yet individual expression of instruments or vocal authority. Ever notice how the voice of God in film or recording always is accompanied by a large reverb? As technology and creativity grew in the music industry, reverberation was emulated in tubes, springs, complex room systems, eventually electronics, and now in software. Reverb can express a vastness and spatial distance or be part of an intimate communication, made unique by how certain consonants, syllables, or instrumental inflections sound within a reverberant sound. Every environment common to the world has a sound. What makes up this sound is the reflections within that environment. Singing in the shower makes even a poor voice sound rich, while that same voice, singing in a closet sounds dull and lifeless. The shower is full of hard surfaces, while the closet is full of soft surfaces. The hard surfaces effectively create a doubled voice sound that the ear hears immediately, where the soft surfaces absorb nearly all sounds, preventing almost all reflections and making the sound harder to hear. Large rooms with reflective surfaces, such as cathedrals, auditoriums, and long hallways, not only reflect the sound, but reflect it repeatedly, expanding and elongating sound, while all the while softening the sound, giving an illusion of infinity.

Somewhere between the sound of a closet and cathedral, there are reverb settings useful to nearly every situation. Vegas has lots of great reverbs built in, and incredible software reverbs are available from several third-party manufacturers, such as PSP Audio software, WAVES, iZotope, Anwida, and more.

Several reverbs are included on the CD with this book, including freeware from PSP Audio software.

Overusing or overly loud reverb can wash over too much of the audio spectrum, making audio sound muddy and without clarity. Use carefully.

Consider using reverb in very short, quiet settings for interviews that might be too dry from being in a very tight, quiet room or that might have some ambient noise that can't be washed with other sounds or beds. Reverb is also useful for making a mono audio signal appear to be stereo by creating artificial reflections that reach the ear at different times. Feed a mono voice or instrument into a bus or FX send that is only the reverb and then route back to the master output. Make sure the reverb is fairly short and has very little pre-delay. Panning the reverb/ sending the reverb to the same location as the dry sound also is a great technique for keeping the sound large, but located. Pan the original signal to where it sounds right in the mix before adding reverb. Use an FX send to route the dry sound to the reverb. Pan the reverb to the same space. This technique works well with vocals mixed with a music bed, as well as with single instrument sounds, such as guitars, snare drums, keyboards, and other single-point source instruments.

Be cautious about reverb sounds that are too bright. While they usually warm up a voice, bright reverbs also tend to elaborate sibilance and can easily create distortion in the high end. Use a de-esser or EQ before the reverb if this problem occurs.

When mixing audio, sometimes a need arises to lower the audio volume quickly, whether for a phone call or visitor or because it's too loud for a conversation. Vegas 4.0 introduces a Dim button that dims audio by 20dB. This option ensures immediate volume reduction without changing the audio output level sliders and without affecting render output. The Dim button is found just above the Master Volume slider or can be enabled by pressing CTRL+SHIFT+F12.

Mixing for 5.1 Surround Sound

Surround sound mixing brings an entirely different perspective in prepping and setting up a mix. Rather than being concerned with just two speakers in a stereo field, there are now six speakers, all individually controlled. The term 5.1 is derived from there being a right/left speaker for the front and rear, a center channel, and a subwoofer (the .1 of 5.1). Each speaker has its own input, and setting the mix for each speaker can quickly become a challenge, as the stereo field all of a sudden becomes small. Moreover, having extra sound points excites most any engineer who hasn't had experience in the surround realm, and it's very easy to create an over-the-top mix, which while sounding exciting in the control room, loses its excitement in the listening environment. Sometimes mixes don't even make sense to the ear when taken in the context of a non-visual setting. In short, it's easy to over-emphasize the rear channel speakers simply out of the novelty of having them.

Vegas 4.0 has the ability to do 5.1 channel surround mixing. This type is the standard for most DVDs, and many music recordings are moving into the realm of 5.1 surround. The Eagle's *Hell Freezes Over* CD was one of the first immersed-experience CDs on the market and is an excellent example of a 5.1 surround sound audio recording.

The room must be set up correctly for mixing in 5.1 surround. Having a 5.1 sound system from the local office supply and a 5.1 surround card is barely going to suffice and will not suffice at all if the room is not set up correctly.

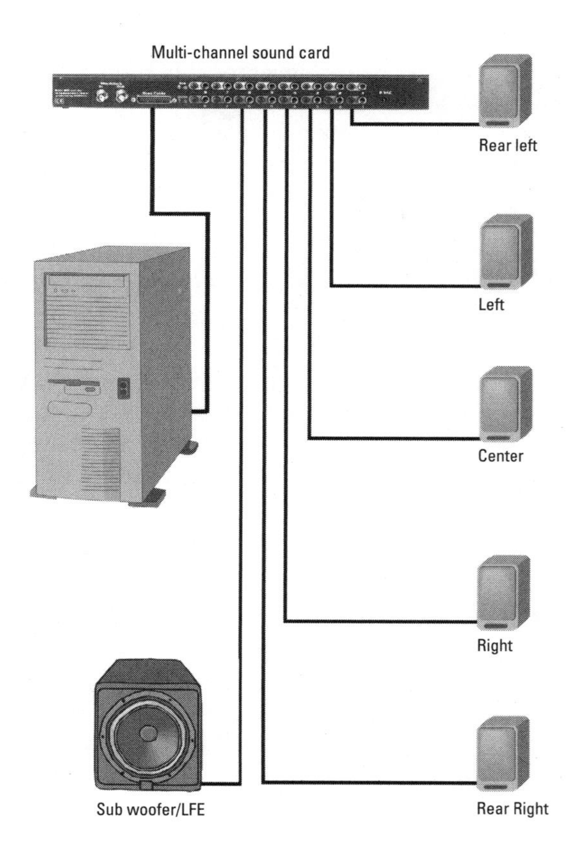

7.77 Wiring the surround system properly is important. Use high-quality cables to connect speakers to a sound card.

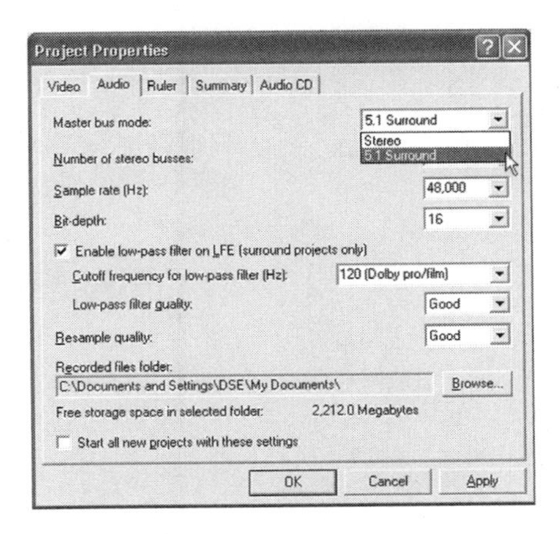

7.78 Select the surround setting to create buses for the surround mix.

On the CD contained in this book is a test clip, which allows audio to be tested on the output of all speakers in the room. It's critical that they all be equal in volume to achieve a balanced mix. Use these tones to test the output.

Do this process by placing the tone as an event on the timeline of Vegas 4.0 and then open the FILE | PROPERTIES dialog. Select the Audio tab, and, in the Master Bus Mode drop-down list, select 5.1 surround.

This process automatically inserts buses in the Mixer docking window to allow for mixing levels to front, rear, and center speakers, plus a low frequency enclosure (LFE) control. It also adds a surround panner to each audio track on the timeline.

Set levels on the soundcard's mixer at 0dB to ensure that Vegas is the only mixing device being used to control levels. Otherwise, it's possible that rendered files will not be accurate to what is being heard. Having unequal sound card mixer settings guarantees an unbalanced mix.

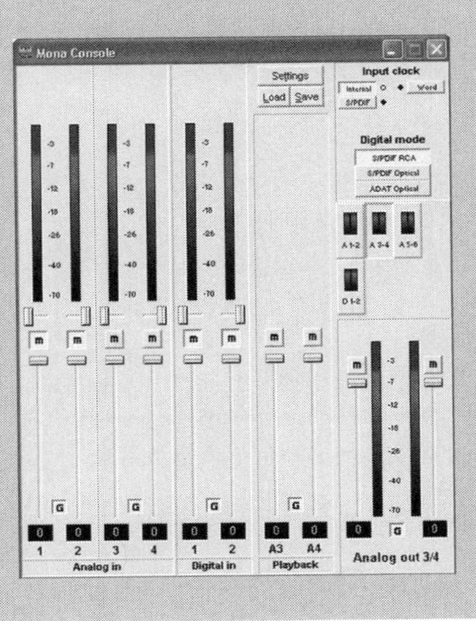

7.79 Some soundcard mixer control levels outside of Vegas. Check this carefully for accuracy in the mix.

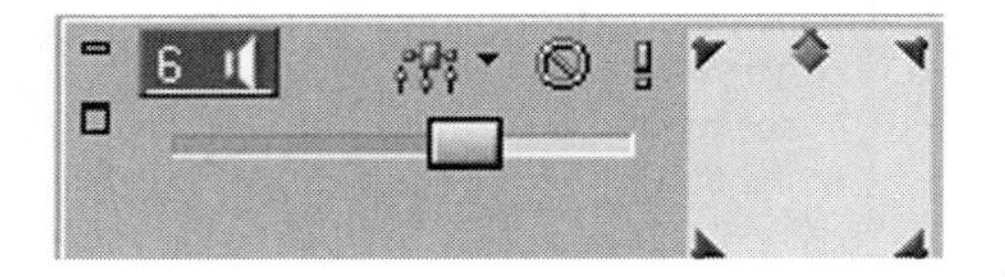

This panner control is how Vegas maneuvers audio around the audio space. Notice that five speakers are in the panner. Clicking the LFE button in the upper-right corner of the Panner control sends the channel only to the LFE output.

Audio is panned from front to rear and right to left using keyframes. Vegas uses keyframes in the audio tool slightly differently from how it uses keyframes in the video tools. Automatic mixing makes this very effective and powerful; rather than using hardware to set up the mix, audio positioning can be exactly keyframed and edited at any point in the mix, with Vegas graphically showing the placement of audio at any point.

After opening the surround panner, right-click the Track Control pane associated with the open surround panner. Select Insert/Remove Envelope (SHIFT+P) from the menu and then Surround Pan Keyframes from the submenu. This process allows for keyframes to be placed on the timeline, instructing audio to pan to wherever you wish it to be, at any point on the timeline.

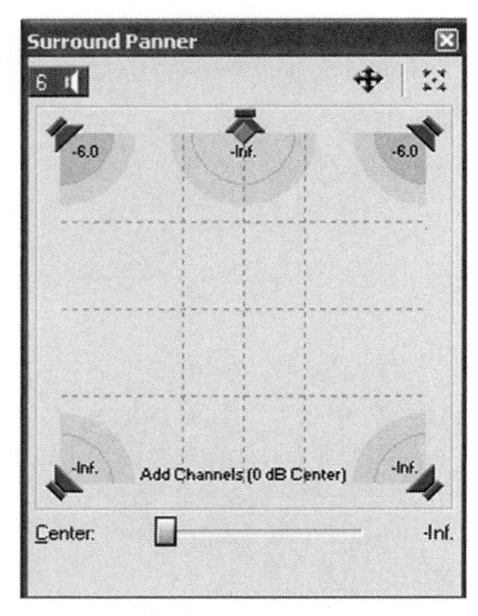

7.80 Double-clicking the surround panner on the Track Control pane splits it off as a movable window, making panning easier to define.

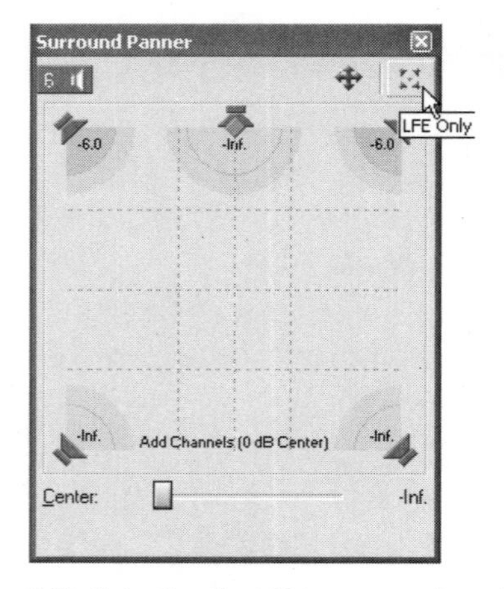

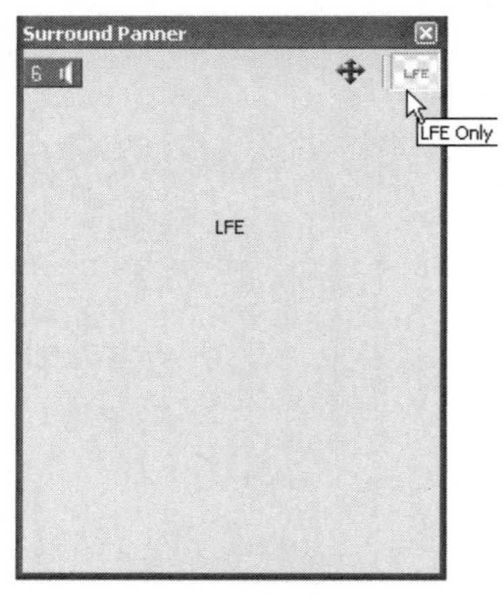

7.81 Selecting the LFE button sends audio only to the LFE channel output.

With the surround panner open, follow these steps to move audio in the listening space:

1. Place the cursor where movement should occur or should finish, if the location is different from the default front-center position.
2. Move the orange dot found in the Surround Panning box to where the audio should be at that moment. Audio will slowly pan to this point from where it originated at the first keyframe. Notice that the reference position is followed by small dots, indicating how the audio will move.
3. Place the cursor on the next time point to which audio should be moved.
4. Move the reference position in the surround panner to the desired point.
5. Repeat until all panning is complete for the project on that track. Keyframes/audio position can be edited later.

When panning stereo signals, moving right to left can create problems when audio becomes louder on one side than on the other side when audio is off balance during a pan. Add Channels, the default setting, can easily create distortion. The Constant Power setting moves audio around the audio spectrum at equal volume with no change in volume at any point. Notice that Vegas displays the relative volume for each channel in the Surround Panner tool. Experiment with various settings during panning operation if audio seems to jump in volume during a pan.

Vegas has separate control of the center channel per track, rather than assigning it to a bus as some 5.1 systems do.

Right-clicking keyframes in the Surround Pan timeline opens a menu, which shows choices for the velocity attributes of how the keyframe will move sound. Hold freezes audio until the next keyframe. Smooth gently speeds up as it leaves the keyframe and slows down as it approaches the next keyframe. Fast ramps the audio quickly to its next keyframe. Slow ramps the audio to the next keyframe fairly slowly. Linear provides a straight movement. This is the default keyframe setting in Vegas 4.0. While not always easy to see, the various keyframe settings draw the audio path differently in the Surround Panner window.

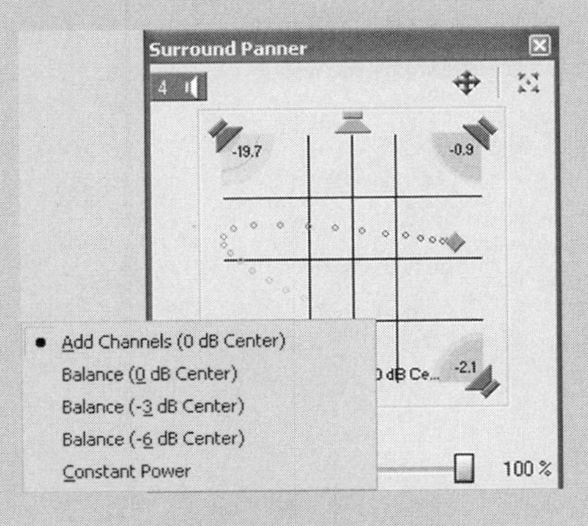

7.82 Right-clicking the surround panner offers choices regarding the audio power at a specific point in the pan.

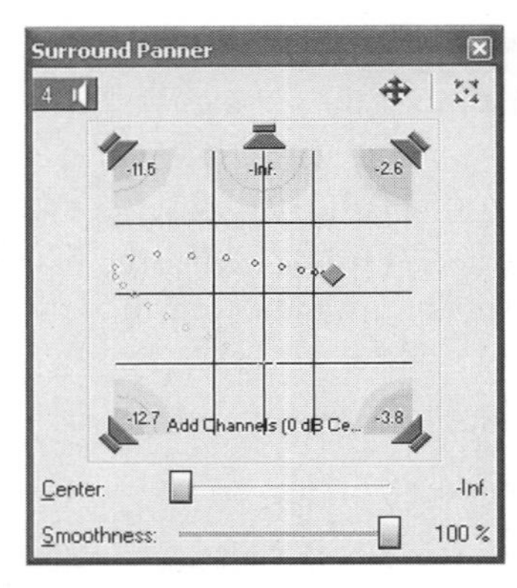

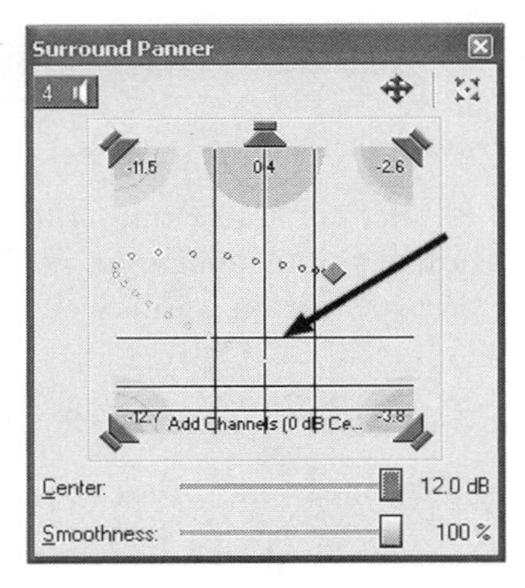

7.83 Center channel is at inf by default. Notice how the sound field draws to the rear as volume from the center speaker increases.

The center channel is at zero volume by default. As the volume is increased in the Surround Panner window, however, notice that the sound grid draws backward, as the power/volume increases in the center channel.

When using motion in the center channel, experiment with the Smoothness setting. If dialog or center audio is fairly transient in nature, such as intense dialog, fast-attack guitar playing, or other sharp sounds, and the smoothness is set too high, the audio will sound very slow moving from center channel to the left or right. If set too low, audio will appear to jerk into other speaker fields. This issue varies with the listening environment, which again demonstrates the necessity of having a good monitoring system.

The room should be set up with equal distances between audio speakers (audio monitors), and it should also be damped and have reflective surfaces minimized. (See Chapter 1 for details on setting up the room.) Monitors should be equidistant to the listening space, measured from where you'll be sitting.

Try to set the room up similarly to a home listening environment, as this environment is how most recordings will be heard.

> It is possible to mix through a typical home stereo system, if the receiver allows for separate audio channel input. Many receivers/amplifiers today do. Run cable from each output on the sound card to the inputs on the receiver back. Using high-quality cable is critical if radio frequencies (RFs) and other noise in the room is to be avoided. Mix in the room, perhaps referencing a favorite or similar DVD.
>
> Additionally, many professional sound cards have TOSLINK outputs/optical outputs that can feed consumer receiver/amplifiers for monitoring over home theater systems. This process is not advised, as home theater systems are anything but flat, and it is very difficult to create a quality mix on anything but reasonably flat speakers.

The primary elements of a surround mix are:

- Dialog
- Music
- Sound design
- Effects

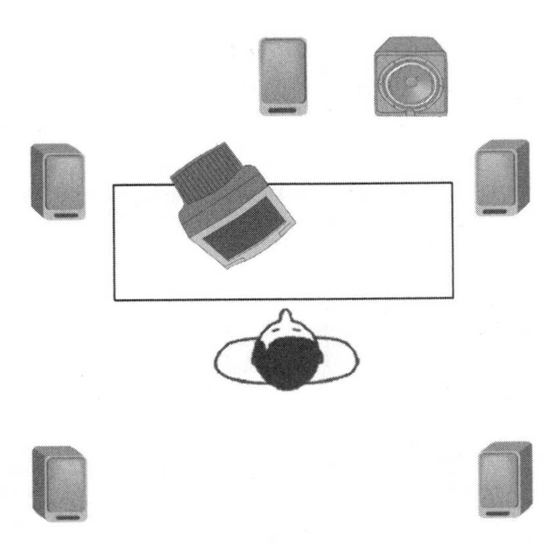

7.84 Monitors should be equidistant to the listening space, measured from the listening position.

Dialog is obviously the most important consideration in most situations. Typically, it is in the center channel and should receive attention first. When setting up the mix, find the peak point of the dialog and then lower the dialog by -6dB relative to the peak. This process allows for headroom/increase in the dialog channel, as it usually is needed after the final mix is prepared. If the peak is lower than -3dB, consider normalizing that event or perhaps raising the level at the track.

Conversational dialog or direct dialog belongs in the center channel in most instances. Working with the center channel presents some challenges, particularly in music mixes. If sound isn't properly handled in the center channel, listeners on a stereo system or a four-channel system missing the center speaker will hear strange artifacts in sounds mixed for the center channel. Adding a slight reverb to the center channel is always good practice, even if it's merely a small room bounce with no pre-delay. Walla or background conversation goes in the left/right or rear left/right, so it won't be brought into the mix at this point.

The next element is music in most instances. Music typically is placed in the front right/left monitors, with some elements moving to the rear speakers, depending on the point of view or movement of the camera, the sound design, or moving dialog.

This point is where the actual mixing starts. Bring the music up to the level that it should be, monitor at a comfortable level. Listening too quietly makes the mix out of balance in both levels and equalization. Listening too loudly means your ears fatigue fairly quickly. Low-quality speakers also tire the ear, so use the best monitors that you can afford.

Active monitors are typically cleaner, flatter, and easier to work with in a surround environment.

Begin bringing in any sound design underneath the music, leaving the dialog channels alone for the time being. Mix the music and the sound design, working the sound design to the rear channels. Be cautious of too much movement in the rear channels as it quickly becomes distracting to the ear.

As an example of a mix, we'll take a city-scape shot with a dialog in the foreground and examine some of the aural potential, as shown:

- **Center speaker**—contains dialog between two people. Low frequencies rolled off/high passed from 120 Hz and above.
- **Left/right front speakers**—contain traffic, walla, moving audio, bed music, and sound design. Low frequencies rolled off/low passed from 120 Hz and above.

- **Left/right rear speakers**—20/30 percent sound design, 10/20 percent bed music, traffic noise panning from front at camera movement or when dialog permits. Motion front to back related to transitions in video. Music moves to rear speakers during dialog, as scene elements require.
- **Subwoofer/LFE**—contains dialog elements, bed music elements, and sound design. High passed at 120 Hz by default in Vegas' surround setup and as ISO. Although this is a standard, experience suggests that a high-pass filter, rolled off at 24dB per octave, starting at 80 Hz, be inserted in the LFE channel.

Be cautious of moving audio into rear speakers during dialog.

In this scenario, when dialog is the focus but music is required along with traffic and walla, move the music to the rear before moving traffic and walla to the rear. Cut midrange in walla and traffic to make room for dialog. With dialog taking place, the cut frequencies are not noticeable unless the mid-cuts are too drastic.

Use smooth keyframes for sound design and music that is moving to the rear. Don't be afraid to use hard keyframing/abrupt shifts in audio placement, based on hard cut editing. Smooth visual transitions call for smooth aural transitioning. Hard visual transitions can use either hard aural transitions or smooth transitions, based on what is happening on-screen.

Overall, do not tread on the dialog mix. As keyframes start to fly bed/design/FX audio right to left or front to rear, it's easy for the dialog to get lost. This point is when the extra 6dB of headroom left in the early stages of the mix are invaluable. If you find that 6dB isn't enough, something is not correct in the other audio mixes.

View buses as tracks in Vegas. Doing so provides automated mixing of the buses, offers easier surround control, and is visually more comprehensive to follow. Assign dialog, bed music, FX, and sound design to separate buses. If a single track doesn't provide enough control of a sound/series of events, duplicate a track and keep the volume out of the track until it's called for in two aural locations. At that point, raise the duplicate track volume assigned to the correct bus.

When planning a project that will include a surround mix, camera shots can be a consideration as well. Panning the lens toward and away from a scene makes an easy transition both visually and aurally. Dollying into a scene makes a wonderful rear-to-front transition. Panning from the bottom of a scene with upward motion or panning down from above a scene into the subject makes wonderful visual transitions that sound readily enhances.

> Do not remove all low end from the right/left/center/rear mixes. If there is no bass in these mixes, listeners in a collapsed environment (stereo only) will hear no low end. Most everything in the mix goes into the LFE mix anyway, but if there are exceptionally important low frequency elements, such as gunfire, cannons, and extremely low end in sound design, it's advisable to duplicate those tracks, roll the low end out of one copy of the tracks, and roll the high end off of the other copy. Send the copy that has high end removed to the LFE channel to supplement the bottom, while not distorting the other surround speakers and ensuring that audio collapses properly in a stereo listening situation.

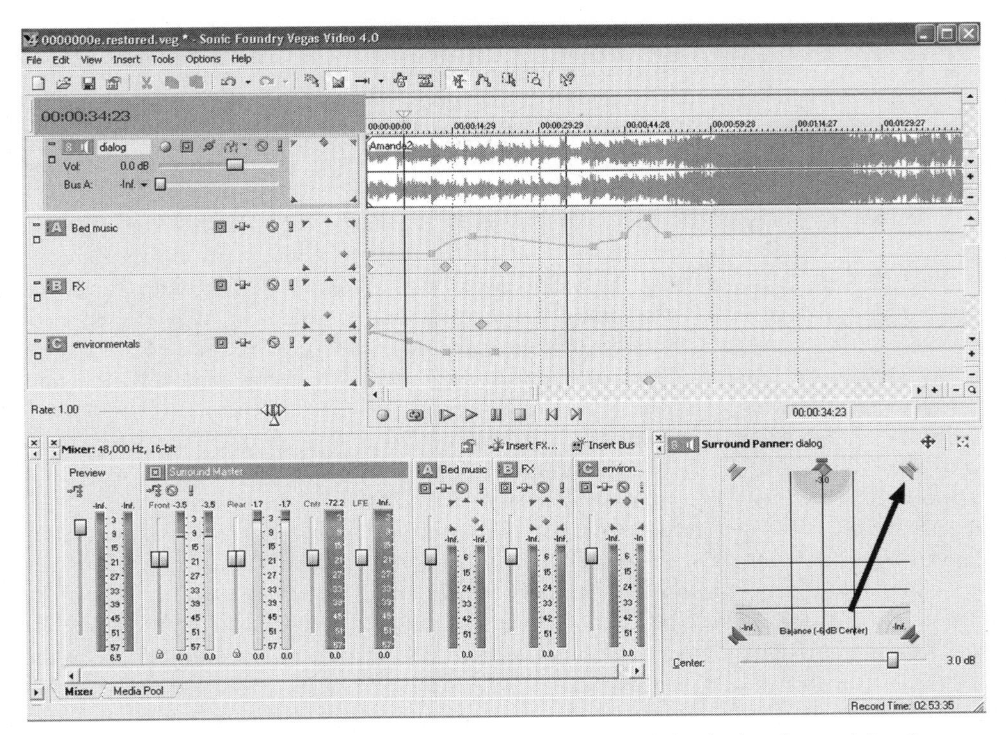

7.85 Viewing buses as tracks offers tremendous control. Notice each bus in the mixer and that the left/right speakers are disabled in the dialog bus.

Mixing Surround DVDA

Musical DVD is an entirely new soundstage with surround sound, offering musicians, producers, and engineers entirely new methods of presenting their art. The quality of digital versatile disc-audio (DVDA) finally goes past the consumer-hype words of "CD quality." CDs are 16 bit/44.1 KHz, while DVDA is 24 bit/48 KHz.

Mixing audio-only for surround presents new opportunities, while also presenting new challenges. Caution must be taken with placement, EQ, and sonic spacing in order to keep the mix cohesive and credible. One major advantage of surround is that it requires far less compression as currently called for in a stereo-only environment.

In a recording of a vocal-based group, the lead vocal takes center space. It's very important, however, that the lead vocal not be assigned exclusively to the center channel, or collapsed listening environments (stereo-only) will hear artifacts and ghosting of the vocal. Using a slight reverb on the vocal helps dilute this phenomenon, if the reverb is added to the right/left mixes.

Percussion, acoustic guitars, and ethnic instruments all share the listening space of left/center/right/center quite well, which opens up more frequency space for vocals or lead instruments.

Picturing the soundstage as it would appear to a concert-goer helps build the mix. Drawing the mix in advance can't hurt and often provides a great visual reference for an engineer.

Consider pairing left/center, and right center, rather than thinking of left/right as separate elements. Treat the rear speakers predominantly as assignments of the reverbs, delays, and audience responses. Audience members don't often sit in the middle of a stage, so placing much musical information in the rear speakers is not often a good practice. Placing the audience, however, where a lead vocalist might stand is a wonderful place to be. Save the rear surrounds for effects and aural surprises.

Mix the bass guitar/bass synth to all speakers, without regard for the LFE enclosure. If more low-end punch is needed, bus the bass and any other low-frequency information to a separate output, feeding the LFE. While the bass channel in a surround system has 12dB more headroom than the remainder of a surround system, few consumer-grade subwoofers are able to handle the extreme volumes of low end, so be mindful of your listening audience. Save the LFE for exceptional low end, such as the five-string bass or synthesizer that really needs the punch. Kick drums should be sent to the LFE with caution.

Feed reverb returns to the rear speakers, along with any vocal or instrumental effects. Sending the vocal to two reverbs or delays with similar settings allows for separate control to front and rear speakers, giving listeners a more intimate seat in the listening experience.

Use the rear channels for any audience information, such as applause and walla. Don't forget to feed a little to the front speakers as well, or stereo-only listeners will hear odd anomalies at the end of songs in a live environment. Panning reverbs and delays to the rear as part of an effect makes the rear speakers become part of the art of the mix. Keeping the rear speakers at full volume, full of a surround element can quickly dilute the value of a surround mix. Use the rear speakers as a paintbrush, not as a canvas.

Always check a surround mix in a stereo-only environment as well to ensure that it is collapsing correctly. Headphones are great for checking this out. (Remember never to mix in headphones.)

Blue Man Group's self-titled CD, the sound track to the movie *A.I. Artificial Intelligence*, the Eagle's *Hell Freezes Over Tour*, or Jon Anderson's *Deseo* CD are all excellent examples of surround sound in non-visual settings.

Surround sound is part of the future, and eventually nearly every facet of video and audio production will call for it. Artists like surround productions because they are harder to pirate. Consumers enjoy 5.1 because it's a sweeter entertainment experience. Radio stations are starting to broadcast in surround, and surround systems are becoming fairly common in the car as well. All in all, the industry has many motives for moving in this direction. Engineers with surround chops will find their talents more and more in demand.

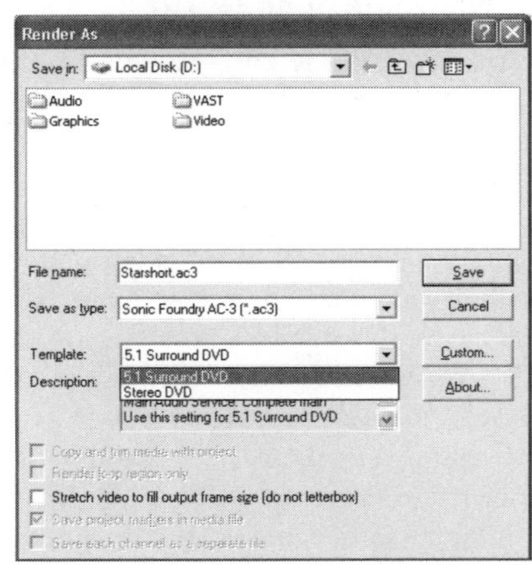

7.86 Select the AC-3 format in the **FILE | RENDER AS** dialog, which has a stereo or 5.1 option.

Encoding Audio to AC-3

Encoding audio to AC-3 is included in Vegas+DVD. AC-3 is a compressed audio format that allows audio for DVD or DVDA. DVDA typically doesn't use AC-3; it uses meridian lossless packing (MLP). The whole point of the separate format is to allow ultra-high quality audio to be compressed to fit six channels of information onto a DVD for video or audio. This process allows for greater bit rates of video on a DVD and allows for higher sampling rates for audio on a DVDA. AC-3 stands for Audio Coding—Third Generation. AC-3 is licensed from Dolby and follows their specifications. Sonic Foundry became one of the first licensees in the PC environment with their Soft Encode product in 1996.

After a mix is finished in Vegas, it's a simple matter to select the AC-3 encoding option in the **FILE | RENDER AS** dialog. Audio is output that is then imported to a DVD authoring application. If you are working with DVD Architect (available only in the Vegas Professional package), AC-3 becomes a selection in the final creation of the DVD or can be imported as a separate stream. Vegas does not allow the monitoring of AC-3 files, so be certain that the mix is correct before encoding.

7.87 Dolby and the double-D symbol are registered trademarks of Dolby Laboratories.

Creating and distributing an AC-3 file technically requires any packaging to contain the Dolby Digital trademark. For more information, visit `www.dolby.com`.

Compositing in Vegas 4.0

What Is Compositing?

Compositing is the combination or interaction of two or more graphic or video images to create a single finished image. Compositing can be as simple as laying a title onto a visual image or as complex as combining hundreds of photographs, video, and animated elements to create complex images.

Compositing in its most complex forms is absolutely an art; there are compositing artists all over Hollywood, Austin, New York, Sydney, and other major cities who do nothing but create complex imagery for major networks, film, and corporate media departments. The fantastic imagery seen on ESPN, CNN, and in most major motion pictures is the result of these artists' work. It is most likely that not one major motion picture has been released in the past few years that didn't contain significant compositing, whether it's titling, green/bluescreen work, or overlaid imagery. *The Matrix* is an amazing example of how greenscreen work is brought to the big screen through the creative use of composited images, as is the *Lord of the Rings* and even the heralded *Titanic*. Many of the full ship images in Titanic were composited together, including animations of people walking, driving, and loading the ship. Most of the scenes including fish are composited and animated. So in short, compositing is very much about eye candy and illusion.

Basic Compositing

Compositing in its most basic form is titles laid over images. Vegas has a fairly powerful titling toolset, capable of creating all but the most complex of titles.

To insert a title over an image, first place a visual event on the timeline. Right-click the Track Control pane and select Insert Video Track or select INSERT | VIDEO TRACK.

In the new video track, right-click the timeline and choose Insert Text Media from the menu.

The Title tool opens with the Edit tab displayed, showing the words *Sample Text* inside. This area is where the words found in the title are inserted. Text size, font, and attributes, such as bold, italic, and justification are also selected here. These attributes and the Edit window work just like a word processor.

8.1 Right-click brings up this menu, then choose Insert Text Media to insert a title.

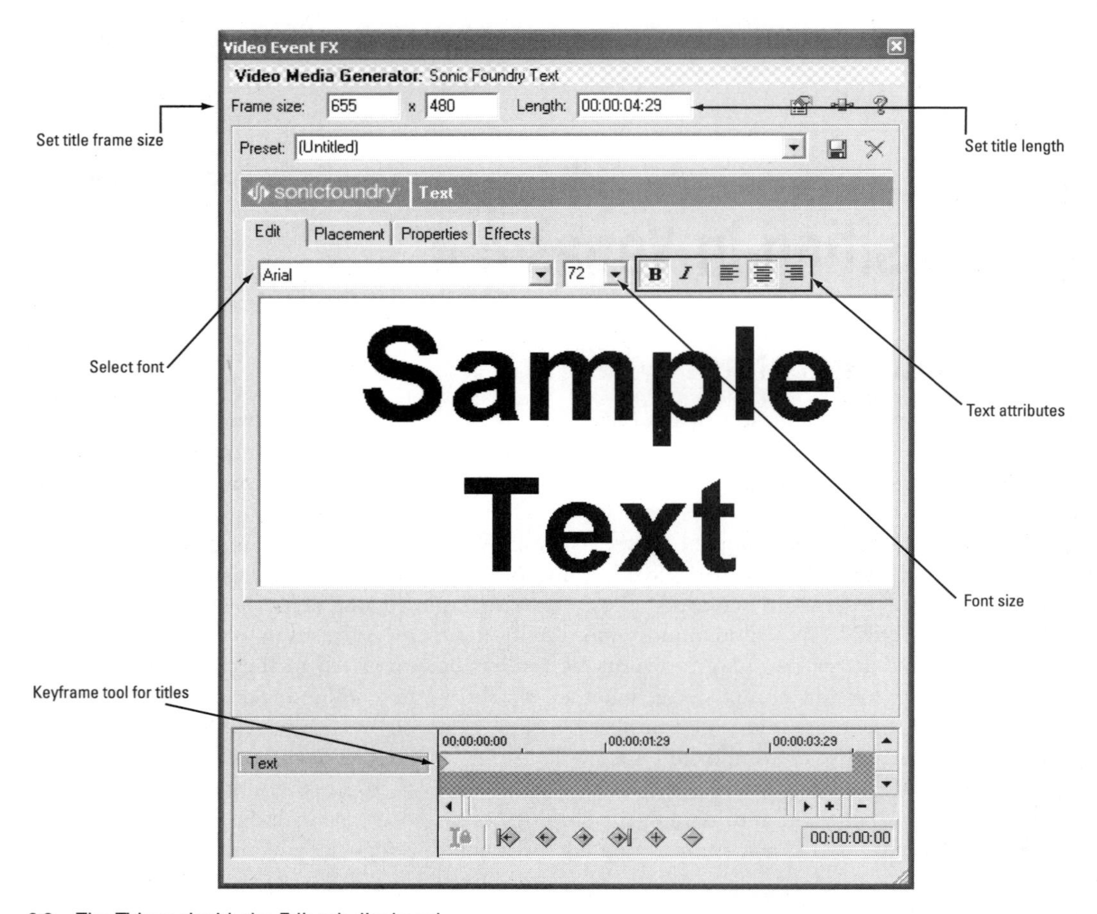

8.2 The Title tool with the Edit tab displayed.

The frame size can be selected at the top of the titling tool. In Vegas 4.0, generated media always defaults to 720 × 480, (non-square pixels) but appears correct as to frame size. (If you are creating titles outside of Vegas for NTSC DV, the frame size should always be 655 × 480, and for PAL the frame size should always be 704 × 576.) This factor compensates for the non-square pixel aspect ratio of DV. (See Chapter 2 for more information on non-square pixels.)

The media length is set by default to whatever the length of stills is set at in the OPTIONS | PREFERENCES | EDITING dialog; however, the length of the media can also be selected in the top of the Title tool, overriding this value.

Often, the setting in the Options menu is for stills, yet titles need to run longer. It's speedy to extend a title by specifying a longer title in this dialog. Titles, of course, can be looped to any length.

Presets are available in the titling menu. These presets are there to assist you in creating great titles out of the box, but for truly powerful and professional-looking titles, you need to learn the features of the rest of the titling tool.

Titles can be positioned anywhere on the Text Placement dialog, which is under the Placement tab.

To manually place a title, click the text in the window of the Edit tab and then drag the title to the desired location. By default, a safe area is defined at 10 percent. This safe area indicates the area in which a title should live if it's being shown on television. Televisions have an area that is partially covered by a bezel in addition to being unscanned. This area hides approximately 10 percent of the video signal. Some televisions cover more and others less. The safe zone can be defined from a zero size area to a 30 percent safe area, by clicking in the Safe Zone text box and entering the desired safe area size. Ten percent is a typical safe area for NTSC and PAL video. Streamed media requires no safe area.

Title placement can be keyframed if the title should move onscreen. (See Chapter 2 for more information on keyframing.)

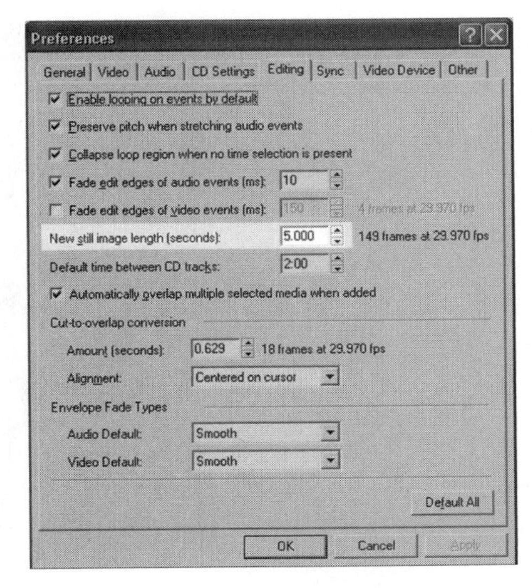

8.3 The Editing tab found in the **OPTIONS | PREFERENCES** dialog allows users to preset title and still default lengths.

Title Properties

Under the Properties tab, the color, tracking, scaling, and leading of the title can be controlled. Selecting the color is done by placing the cursor in the color palette and clicking. The cursor changes to a small target icon that can be dragged inside the color selector.

If a specific color is desired, the RGB values of the specific color can be entered into the color value boxes. Vegas also offers a color picker, which allows you to use the Eyedropper tool to select a color or range of averaged color.

The Eyedropper tool can be applied to any image on the screen. If a commercial color set is available, exacting color selection is possible.

Color can also be selected based on hue, saturation, and luminance (HSL) by selecting the HSL button just above the Eyedropper tool.

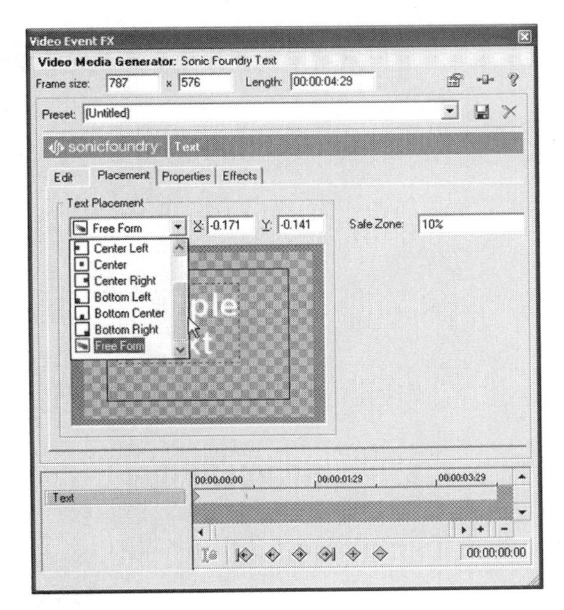

8.4 Titles can be manually placed anywhere, or presets can be used to place a title.

A background color can be specified if necessary. If the title is overlaid on a video image, a background color probably shouldn't be specified and should be left transparent instead. This feature can be used as an effect over video, and if the opacity of the background color is reduced, the visual image will show through the color, causing the background color to act as a color filter. All color aspects are keyframe-controllable.

The Text Properties area of the Properties tab in the Title tool allows for adjustment of the kerning, leading, and spacing of text. By default, auto-kerning is disabled. Check the button to enable it.

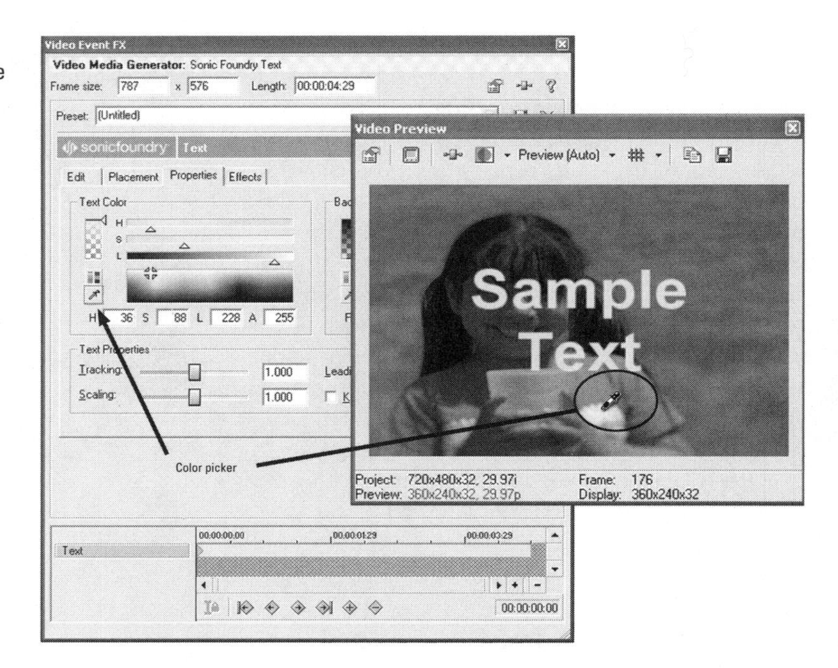

8.5 Using the Eyedropper to match the color title to the flower on the shirt.

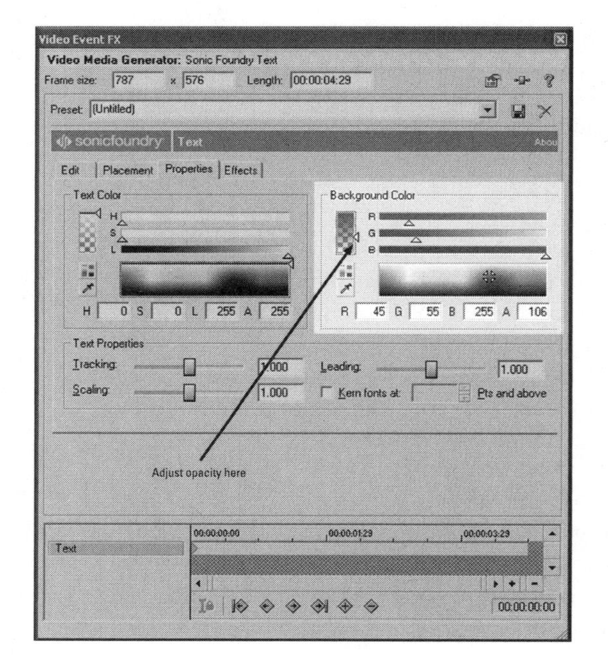

8.6 Selecting a background color's transparency/opacity.

Tracking controls the distance between letters in the text/title field. *Leading* controls the space between two lines. Leading fine-tunes a single line of text up and down. *Scaling* controls the sizing of the letters. All three text controls are keyframable. Scaling tends to be jerky in the scaling of text, so use with caution.

A check box is available for automatic kerning of letters above a preset, user-controlled size. *Kerning* is the specific spacing between a pair of characters. Some letters take up more space than others, and without kerning, sometimes letters appear to be too far apart in a word. Kerning generally alleviates this problem, although some fonts use more space than others.

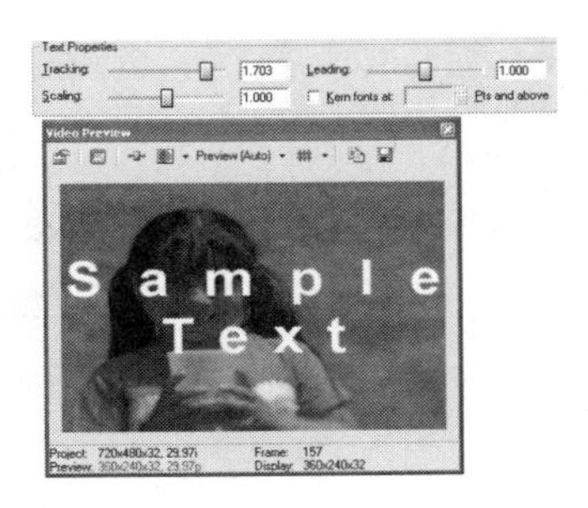

8.7 The tracking control.

Adding Effects to Text

The Effects tab offers a number of options to edit and enhance text. Effects often are used to make text stand out on otherwise busy visual screens or to create a specific mood.

Adding a glow to text can make letters show up when the underlying media is of a similar color. Glow is also often used to give text an image of depth, such as coming out of a vacuum.

Clicking the Draw Outline check box adds a glow to the text. This effect is great for outlining light-colored text with a slightly darker color, creating a slight illusion of 3-D.

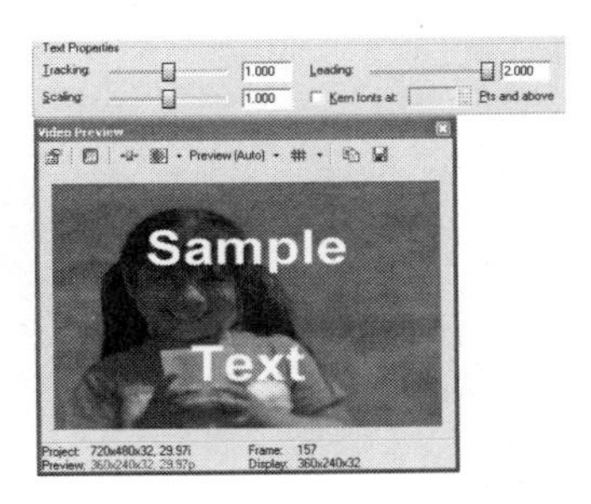

8.8 The leading control.

8.9 The scaling controls.

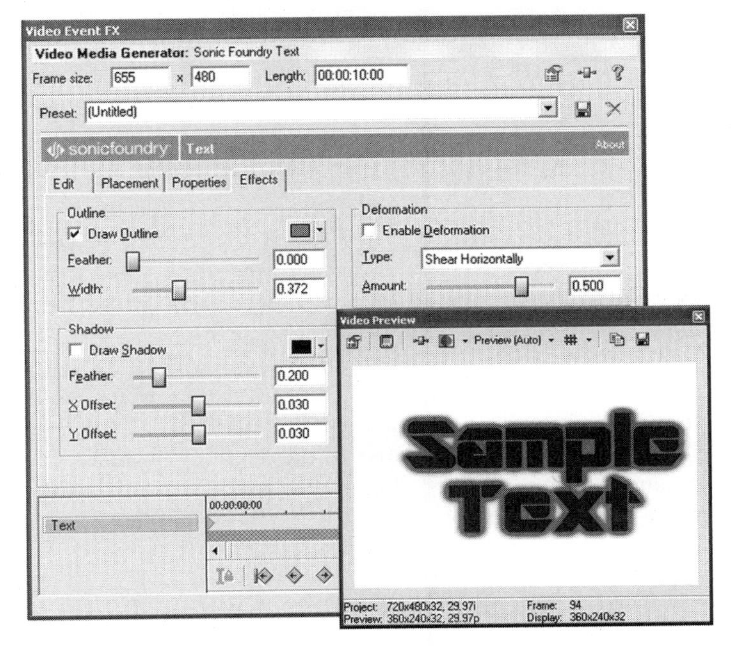

8.10 Using Outline to create a glow around text causes the text to stand out.

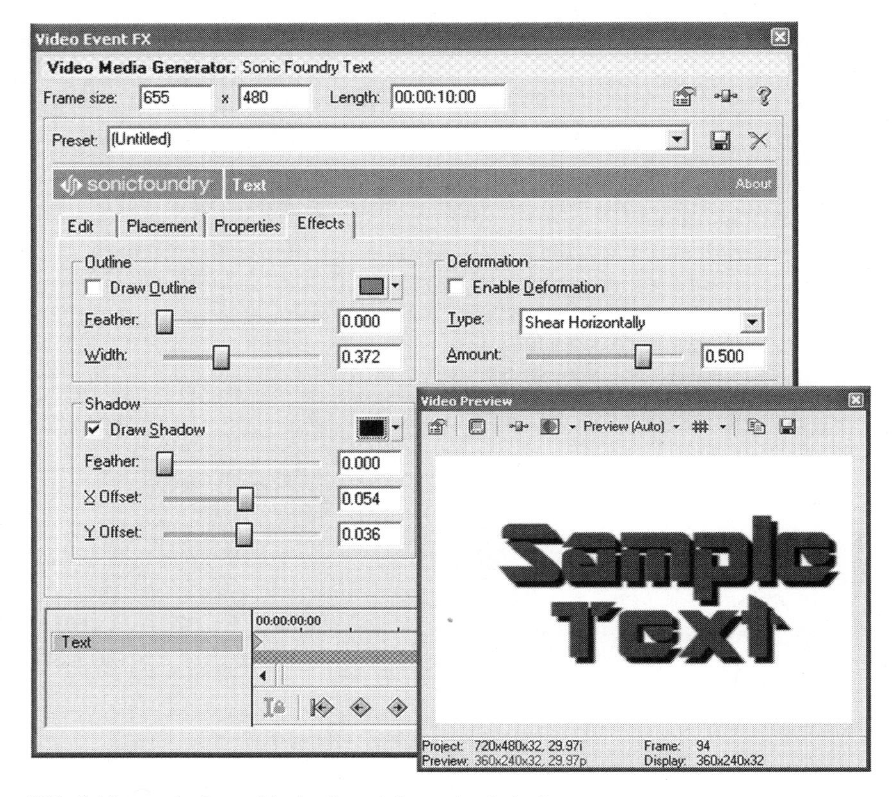

8.11 Adding a shadow adds depth and dimension to text.

The outline can be feathered, which smoothes out the transition between the text and underlying media and softens the edge of the glow. Use this effect to add a bit of flair to text that might otherwise be dull.

Drop shadow, or shadowing, is one of the most useful tools in the Effects tab. Click the Draw Shadow check box, and a shadow will appear beneath text. This shadow can be colored in any color found in the palette by clicking the Color button and then selecting either a preset color or choosing a custom color by using the color picker/eyedropper to select a color.

Shadows can be keyframed to add depth to a title, moving in a direction opposite the original text, which creates an image popular in Hollywood titles today. Colors and opacity can be keyframed and combined with glow to create very rich and deep title attributes. Feathering smoothes and softens the edges of shadows and adds perspective to the depth of the text.

The Deform tool can shear, bend, squish, or compress text. Text can be wrapped around images by curving, or text can be used to give the impression of urgency by keyframing a fast bend as the text comes on or off the screen.

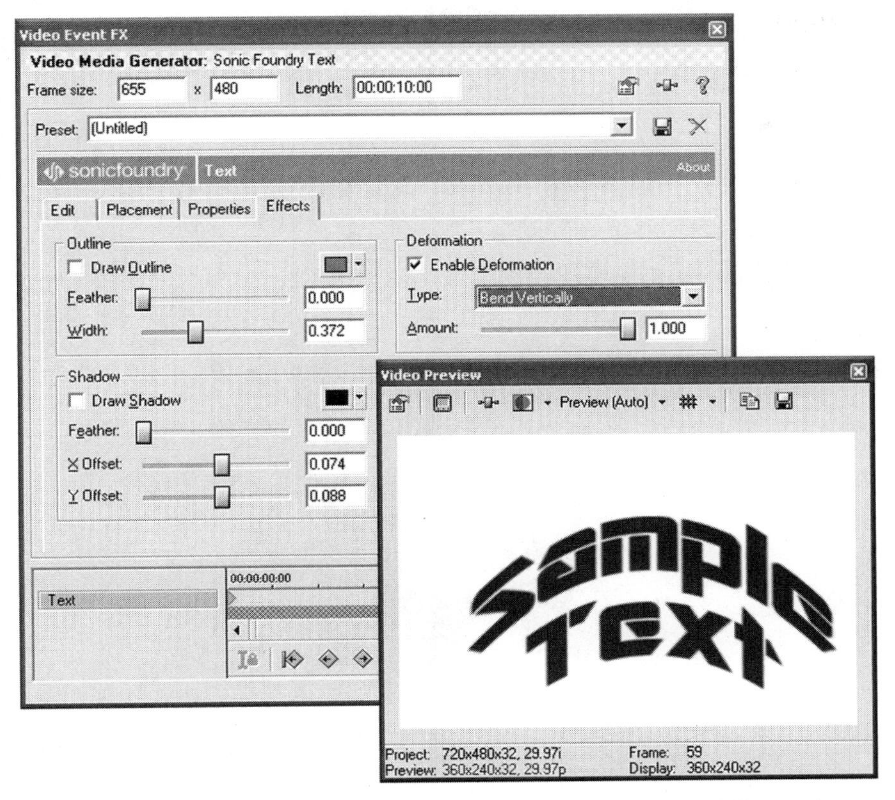

8.12 Bending text around an image can create an illusion of text and image being one image.

You can keyframe text with any one of the many effects in the tab and create flowing text that appears to be very deep, particularly when combined with other text effects.

Each of the three effect sections has slider controls for different text features. Double-clicking a slider returns the slider to its default "null" position. Clicking on the slider activates the effect, and the control can be controlled either with the mouse, or if active, can be moved with the RIGHT ARROW and LEFT ARROW keys for more precise control.

Learning to manipulate titles is an introduction to compositing, and titles are one of the most creative and yet difficult aspects of film and video production. In the past, complex applications were required to do in-depth titling, while today Vegas contains many of the same tools as the more complex tools offered in the past. With some creativity and inspiration, Vegas can be used to create Hollywood-level titles. In fact, many Hollywood-level films and television productions contain title work created in Vegas.

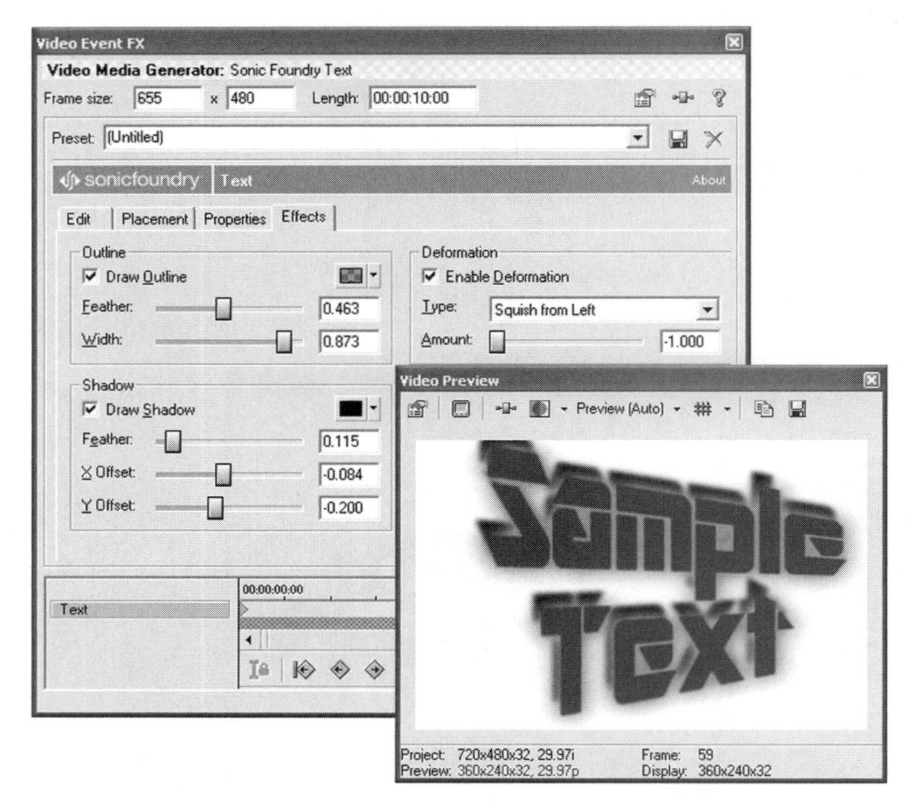

8.13 Inserting a mix of effects can create complex titles.

Using the Pan/Crop Tool

Vegas has a Pan/Crop tool that can be used on any video event. This powerful tool can be used to correct camera depth, create motion on a still image, match aspect ratio, or crop out unwanted material from an image. While simple in use, this tool is extremely powerful.

Open the `Pan-Crop.veg` project so you can follow along visually with this section.

After capturing video and placing it on the timeline, each clip will contain a small Crop symbol. Notice that in Figure 8.14 the image in the monitor screen does not entirely fill the video preview area. Clicking the Crop tool opens a dialog that allows the user to entirely fill the Preview screen and match the image view to the aspect ratio of the Preview screen.

This feature is particularly useful when creating videos of scanned-in still photos or cropping unwanted content from a video image. See Chapter 5 for details on photo sizes and scanning.

With the Crop dialog open, right-click the image inside the Crop window and select Match Output Aspect.

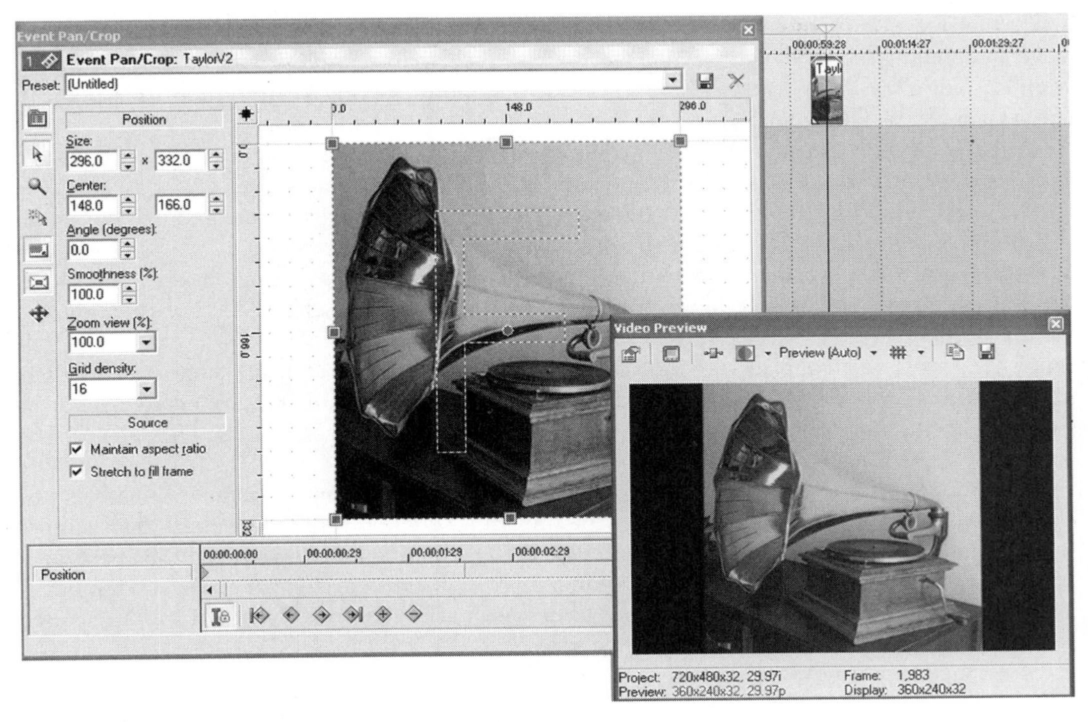

8.14 The photo does not fill the entire Preview window.

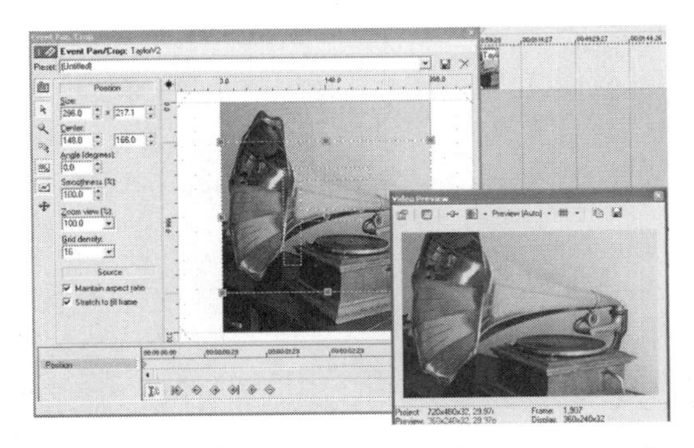

8.15 Selecting the project aspect ratio crops too much of the image out in some instances.

This action causes the image to be cropped to a size that will completely fill the preview area. Now, the Preview window is completely filled with no black background showing. The aspect ratio of the photo, however, appears to be correct even if the photo itself really isn't. This aspect ensures that all imagery mixed with DV and other graphics can be made to fill the viewing screen completely at all times, matching the aspect ratio of the project. This accomplishment, however, isn't without its specific issues. For example, we no longer can see the entire image that comprises the photograph shown in Figure 8.14. Notice that the gramophone has lost the overall view in Figure 8.15. The viewer can now see only the part of the gramophone that the editor allows them to see—useful in some situations, but not in others.

Wouldn't it be great if the still image could move, almost as though a camera had been used to capture the image rather than a scanner?

We can create that movement by the use of keyframes. Right-click again and select the Pan/Crop tool or click the Crop tool icon found on every clip or image on the timeline. Now, move the Selection/Crop window to a new position. The Keyframe tool automatically inserts a keyframe, telling the timeline that at a specified time, the photograph should be cropped differently. Zoom in on the photo and zoom out of the photo, placing keyframes at each point of movement. Notice the added keyframes in Figure 8.16.

Producers of PBS specials, BBC documentaries, wedding videos, and broadcasts apply this same technique to create the illusion of movement. Even if the image is exactly the correct aspect ratio, no one really likes looking at a static photo on a computer or television screen. Even the smallest bits of movement are far more interesting to watch than a static photo.

In the past, tracking cameras, or remote control–operated cameras that mount on a rail, shooting motion images over a still photo, have been used to create this effect. Ric Burns, celebrated filmmaker/documentary genius of *The Civil War*, *The Way West*, *New York*, and others, has employed still images coupled with motion to change the entire face of documentaries. This effect can now be easily accomplished with a scanner and the Pan/Crop tool. This tool is tremendously powerful and useful for putting together slide shows, wedding montages, or other photo-heavy presentations.

Another application uses the Pan/Crop tool to correct bad camera movement. It can also be used to generate camera movement where the camera is static or not moving at all. If the clip of a soccer game, for example, is tripod-shot and static the camera movement could be generated to tell the viewer where they should be looking. A word of caution: don't zoom in too far, or the pixels will become too large and the detail of the event will wash out. This tool should be used in a subtle way for video. It's a terrific precursor to a transition.

Still images can usually handle deeper zooms if the resolution is set high enough on the original photo. Typical sizes of photos generally are considered best at 72 dpi for video; in most cases, saving photos at double the standard resolution provides a clean image for zooming and panning. However, preview stills in your scanning software to ensure the best resolution for your needs.

In addition to the Match Output Aspect menu in the right-click dialog, additional choices include the following:

- **Restore**—restores pan/crop boundaries to the edges of the image. No pan/crop takes place.
- **Center**—centers pan/crop boundaries around the centering dot found in the middle of the crop boundaries.
- **Flip Horizontal**—flips images horizontally, making the right side become the left side. (This feature is great for changing a view. Open `multicamera.veg` for more information.)
- **Flip Vertical**—flips images vertically, causing most images to be upside down.
- **Match Output Aspect**—matches the image aspect ratio to the settings in the FILE | PROPERTIES dialog, which is usually 720 × 480 for DV editing.
- **Match Source Aspect**—matches the image aspect to the original aspect ratio. Use this setting when applying full photos over other montages and when it is acceptable for the photo's borders to be seen. This feature will still allow an image to be resized smaller; however, the aspect ratio is that of the original image, not of the project.

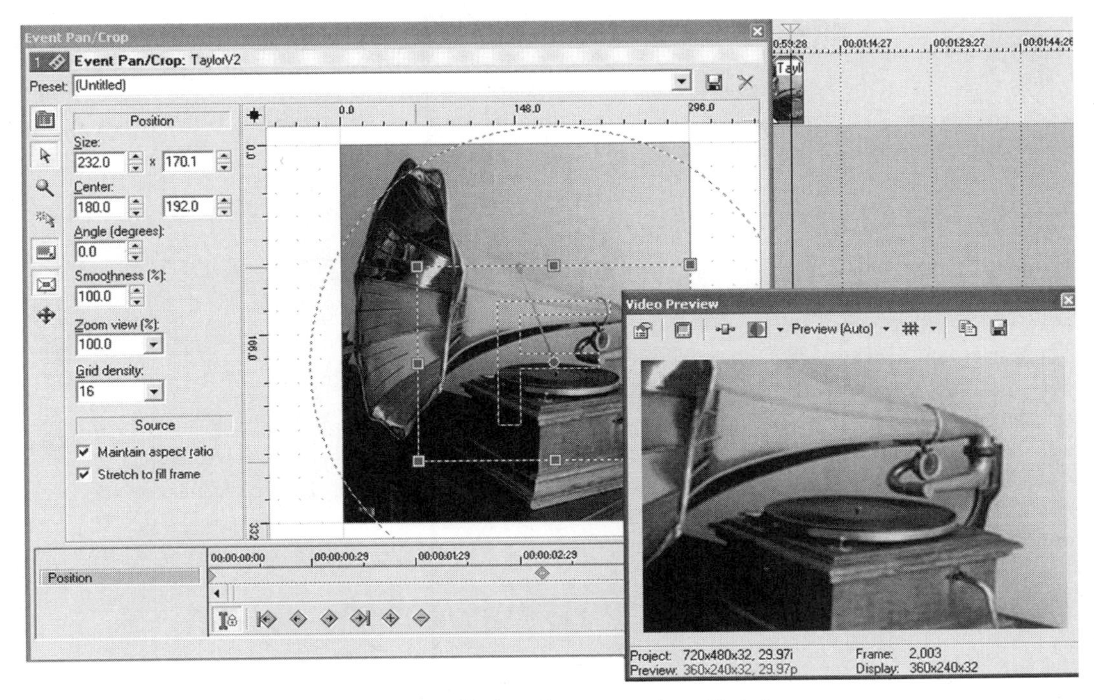

8.16 Notice the keyframes created to build the illusion of a camera-shot still photo.

Tip

Having photos at too high a resolution creates artifacts on the video image, containing too much information for a video image.

Standard-size graphic images, such as titles, backgrounds, and generated graphics, should be created at no higher than 300 dpi, or double the maximum zoom, and should be sized at 655 × 480 for insertion into a DV project. This configuration compensates for the non-square pixel issues found in DV. Always reduce interlace flicker for photos and consider using the Broadcast colors filter on stills if they are beyond legal colors. The Broadcast filter can be dropped directly on a still image event, or if the timeline contains a large number of stills, the filter can be dropped on the Project window instead. Instead of a Broadcast filter, you might want to try a Levels filter, set to convert Computer RGB to Studio RGB. Set this by reversing the Studio RGB to Computer RGB settings, which will ensure that the entire dynamic range is passed. (See Chapter 9 for more information on legal colors.)

The Pan/Crop tool can also be used to create 3-D–like imagery. By spinning an image, whether video, photo, or graphic, the illusion of 3-D can be created by adding a glow or shadow to the image.

In the second section of the Pan-crop.veg project, a flip or rotation is found on the project. Glow, shadow, and keyframed motion of the pan/crop is applied to the image.

Use the Pan/Crop tool to flip images horizontally, vertically, or to squeeze an image onto the screen. Partnered with the Spherize plug-in found in Vegas, this process is one means of creating unique titles or animations of transitioning video.

Using Track Motion in Vegas

On each video track, Vegas assigns a Track Motion tool. This tool is exceptionally powerful and goes far beyond its initial impressions.

The Track Motion tool in its most basic form can be used to create a picture-in-picture presentation, allowing one motion video event to be placed over another motion video event while reducing the size of either event and allowing both images to be visible simultaneously.

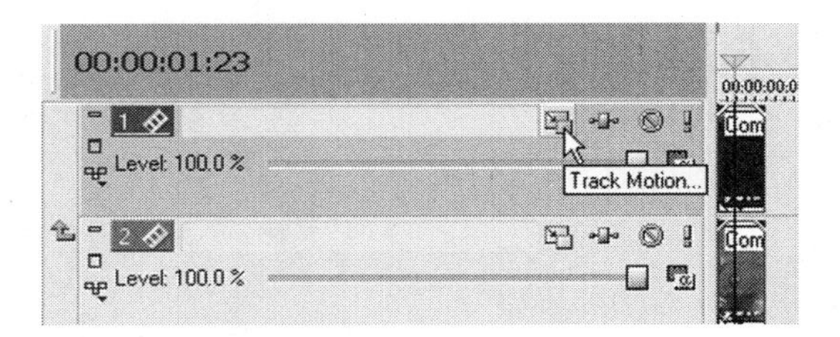

8.17 The Track Motion button is found on every video track in Vegas.

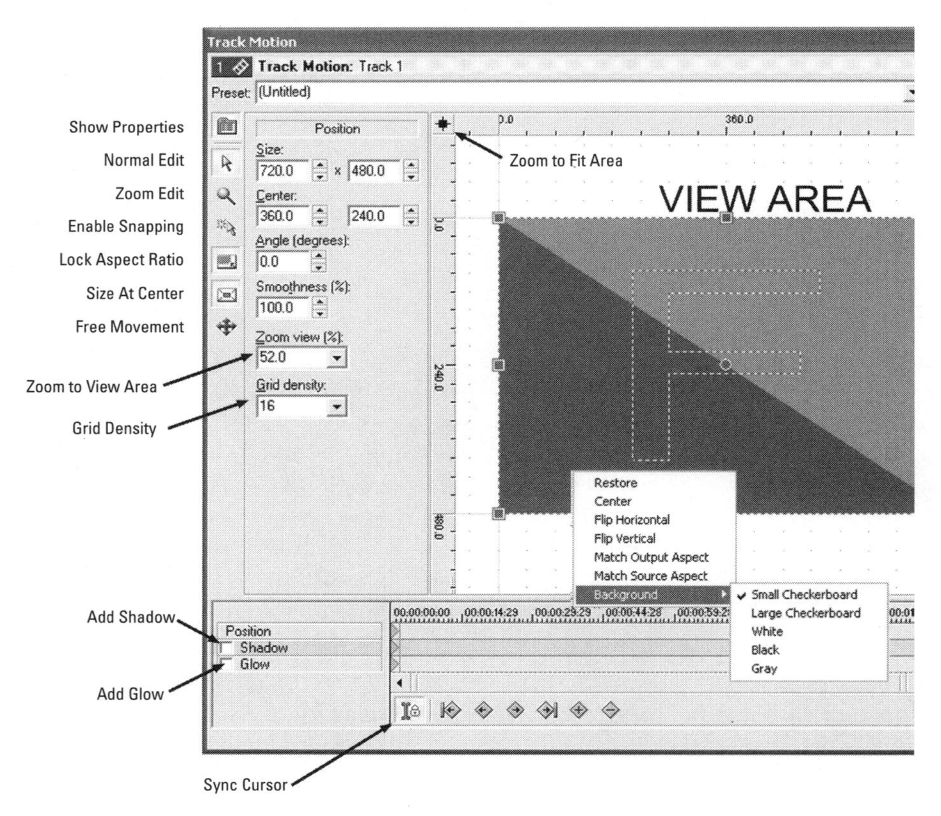

8.18 The Track Motion dialog has many choices with which to manipulate an image.

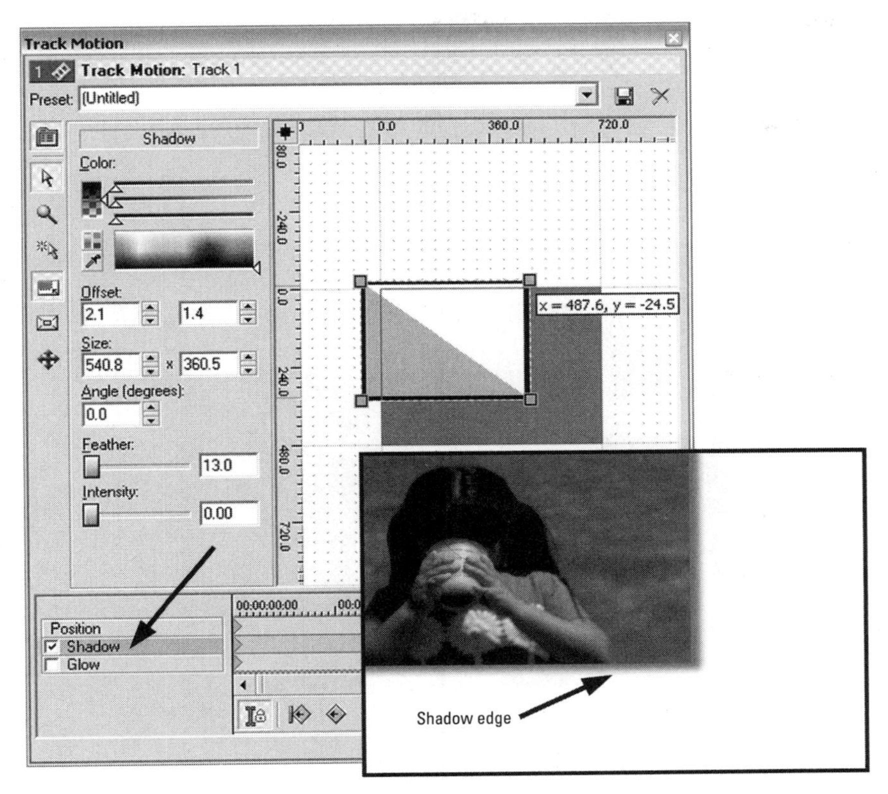

8.19 A shadow applied with the Track Motion tool.

Clicking the Track Motion button opens a dialog that is similar to the Pan/Crop dialog. In fact, these two tools share many of the same attributes but accomplish completely different tasks.

As an example, a shadow can be applied with the Track Motion tool, as shown in Figure 8.19. Shadow can be applied to any image with the Track Motion tool, even if the resize or motion features of the tool are not used. This fact will be useful when using text as masks as shown later in this section. It's also a simple task to apply shadow to a still image or graphics with a transparency by using this tool.

Starting with the top-left corner, a number of tools are found within the Track Motion dialog. The Show Properties button hides and shows the position, angle, zoom, and other views found on the left side of the Track Motion tool. The Normal Edit button, just below the Show Properties button, is the tool predominantly used to size, rotate, and manipulate the images in the track motion frame window.

The Zoom tool, which looks like a magnifying glass, allows for tight zooming on the frame window area and is very helpful when trying to align images in the frame to a small grid space or when creating exacting keyframes.

Enabling Snapping causes the frame in the Track Motion tool to be snapped to the nearest grid and is also useful for accurately locating the frame during editing.

The Lock Aspect Ratio in the Track Motion tool functions in the same way as the Pan/Crop tool. Unlike the Pan/Crop tool, however, the Track Motion tool causes the frame in the Preview window to remain in the correct aspect ratio, but size-reduced.

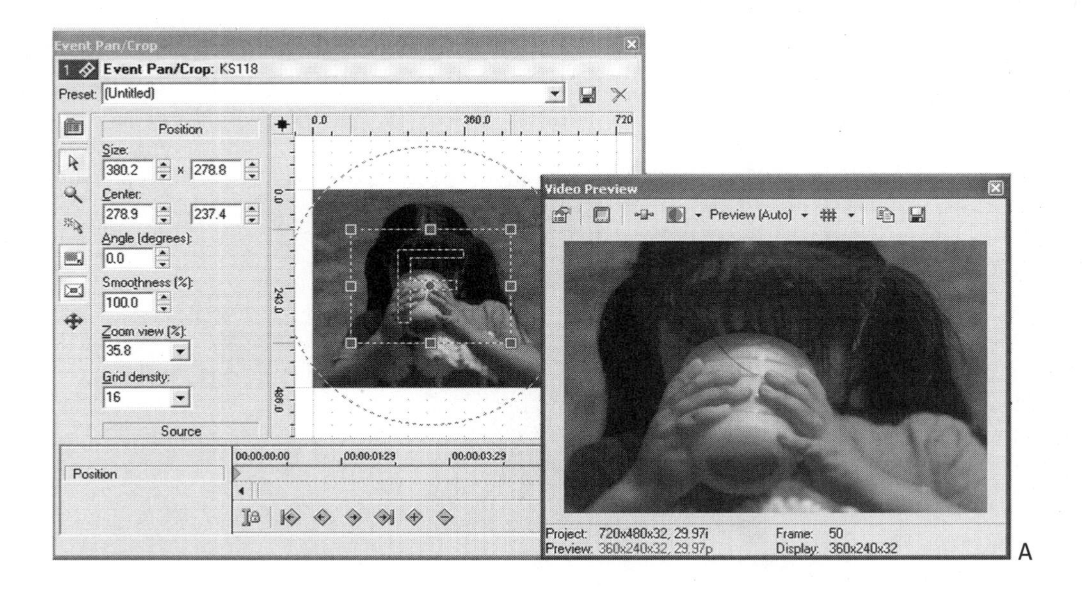

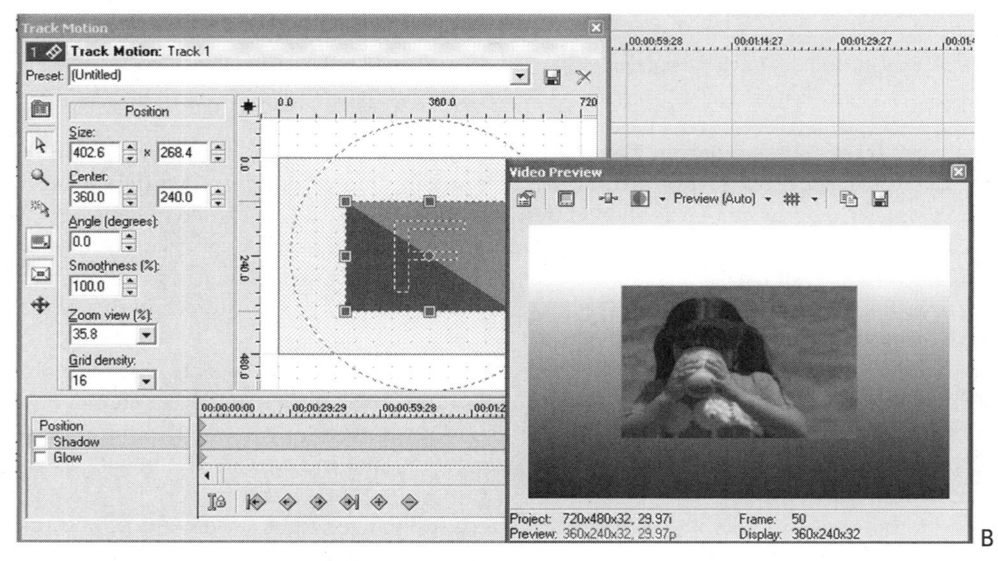

8.20 Similar settings in the Pan/Crop tool (a) and Track Motion tool (b) frame windows, but very different results in the Preview window.

The Lock Aspect Ratio function should be used in most operations of Vegas so that the vertical and horizontal aspects of the image remain equal. The aspect ratio, however, can often be used as part of a creative edit. Turning off aspect ratio and adjusting generated media, for instance, is great for creating lines or other shapes. (Generated media is discussed later in this chapter.) By default, this button should be enabled. Pressing and holding down CTRL while sizing a frame temporarily disables the Lock Aspect Ratio function.

The Size At Center button found on the left side of the Track Motion tool just under the Lock Aspect Ratio causes the frame and subsequent view in the Preview window to center around the dot found in the middle of the track motion frame. When you move the dot to another location, right-click, and then choose Center while this button is enabled, the dot will return to the center of the track motion frame. In addition, the frame image will move with it while keeping the dot where it was moved in relation to the original center of the frame. The center dot is actually designed to move a frame off center during movement, providing a pivot point for rotation of the frame.

Sizing a frame down to one-fourth screen and snapping at the centerline causes the frame to be part of a quad image on a screen. Figure 8.21 shows three images reduced in size and snapped to centerlines of grid. The gradient is generated media with a title overlaid.

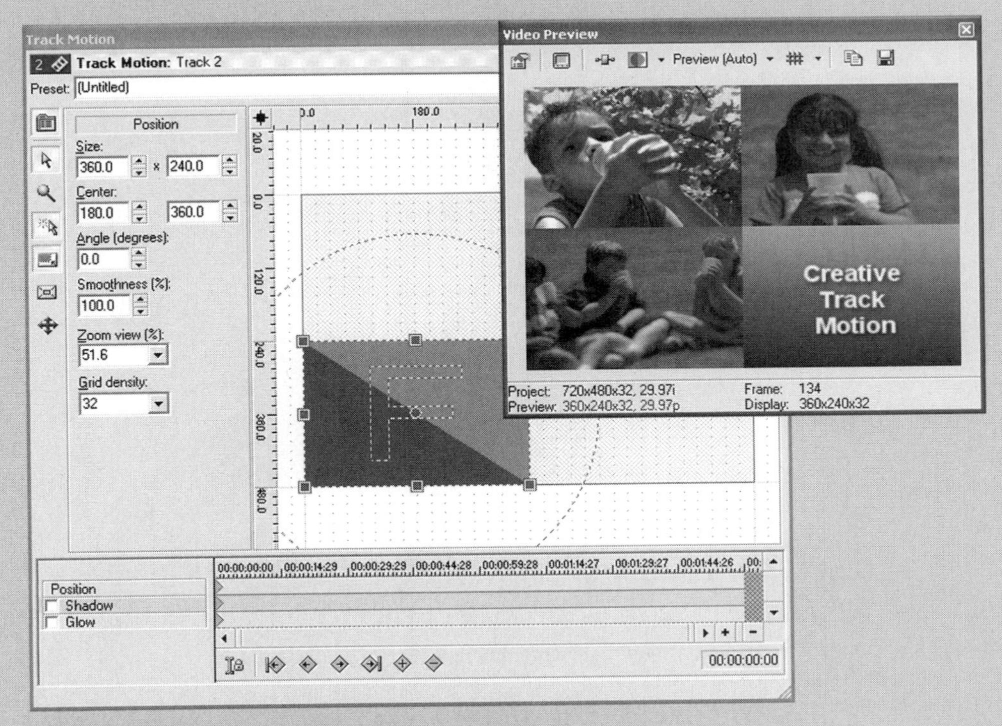

8.21 Three images reduced in size and snapped to grid.

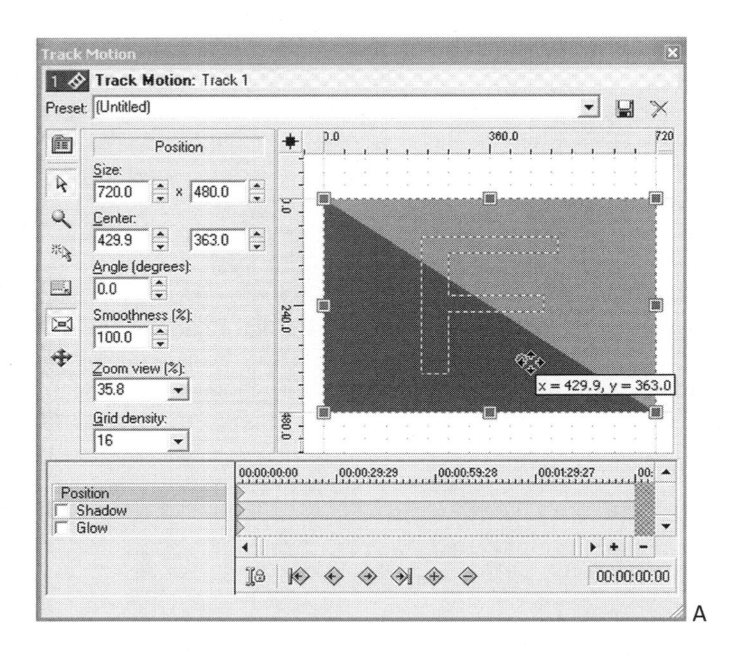

A

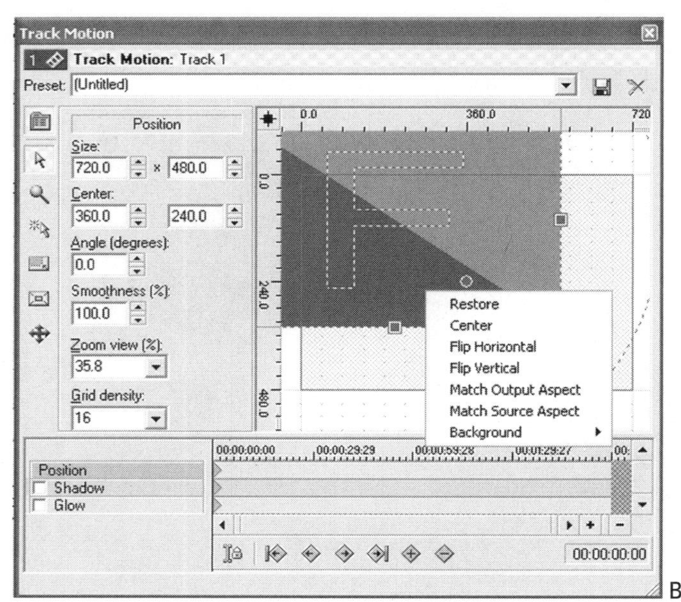

B

8.22 Moving the target dot to a location other than center (a) and then selecting Center with a right-click (b) causes the media to re-center around the target dot's new location.

Track motion is where most users become acquainted with the concept of deeper compositing, as opposed to just creating titles overlaid on images. Another common use of track motion is to create a picture-in-picture image. Creating a picture-in-picture is fairly simple and is a good exercise for getting started with manipulating images.

To create a picture-in-picture image, create two tracks of video, with events on both tracks. Place the cursor in the middle of the two events.

Track 1 is the overlaid event that will become the picture-in-the-picture. Select the Track Motion button on track 1.

Make sure Lock Aspect Ratio is enabled, grab the lower-right edge of the frame in the track motion frame view, and reduce in size, watching the Preview window as the frame and preview image are reduced. Reduce the track 1 event to the size of x + 315.0, y = 210.0. (Size is displayed near the cursor during resize or can be directly input in the Track Motion Properties dialog.)

In the Properties Center dialog, input a value of 180 for the left box and a value of 120 for the right box. These settings will place the video slightly to the right of extreme left and slightly below the top of the screen.

Preview the image by pressing W (rewind) and the SPACEBAR (play).

The track 1 event will be previewed in the upper-left corner, while playing over the top of track 2's event. If both tracks are video tracks, full motion will be seen in both events.

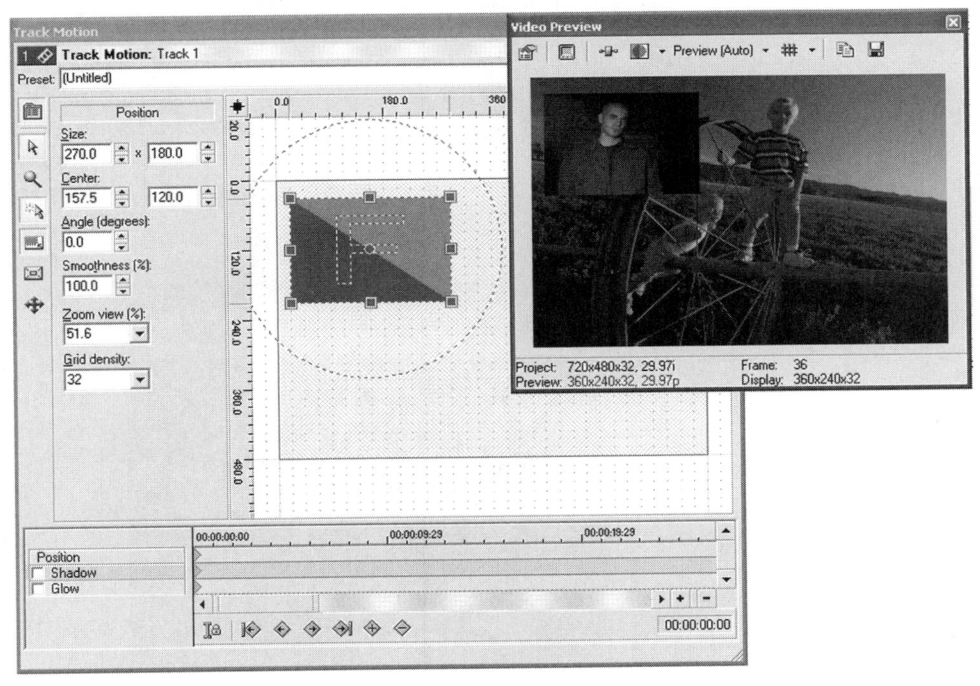

8.23 Picture-in-picture images are very fast to create with Track Motion.

> Be sure that the Sync Cursor button is disabled when working with this exercise unless the desired effect is to shrink the track 1 event to the proscribed size. Otherwise, the first keyframe, which in most cases will be a full frame view of track 1, will gradually shrink the track 1 event until it reaches the second/new keyframe generated by shrinking. It is easy to accidentally create unwanted keyframes during the track motion or pan/crop actions. In the event of a second keyframe being created in the Track Motion dialog, right-click the first keyframe found in the keyframe dialog and select Delete from the menu. This step will then cause the second keyframe to be the initial size of the event. You will quickly create a workflow based around your personal preference for managing these two tools.

The picture-in-picture image might seem flat or without dimension, as the two images are at exactly the same depth of field. To create a depth of field, insert a shadow.

Shadows can be inserted in any color, opacity, and size within Vegas. Open the Track Motion dialog again by clicking the Track Motion button on track 1. A check box for a shadow is located at the bottom of the dialog on the left side of the keyframe timeline. Checking this box adds a shadow to the frame in the Preview window. One of the more creative uses of the Shadow tool is to turn off the aspect ratio of the Track Motion tool when applying shadowing and to create shadows that stretch, move, shrink, or rotate while the image that the shadow falls from remains in aspect ratio. To disable aspect ratio on the shadow, which is enabled by default, press CTRL while moving the shadow about the screen. To disable it for the duration of the editing process, click the Lock Aspect Ratio button, turning this tool off. If the image on the timeline has an alpha channel, the shadow will work with the opaque image. This method is one way to create layered shadows under PNG, TGA, GIF, AVI, or other alpha-capable image formats. Shadows are keyframable for opacity, color, size, and position.

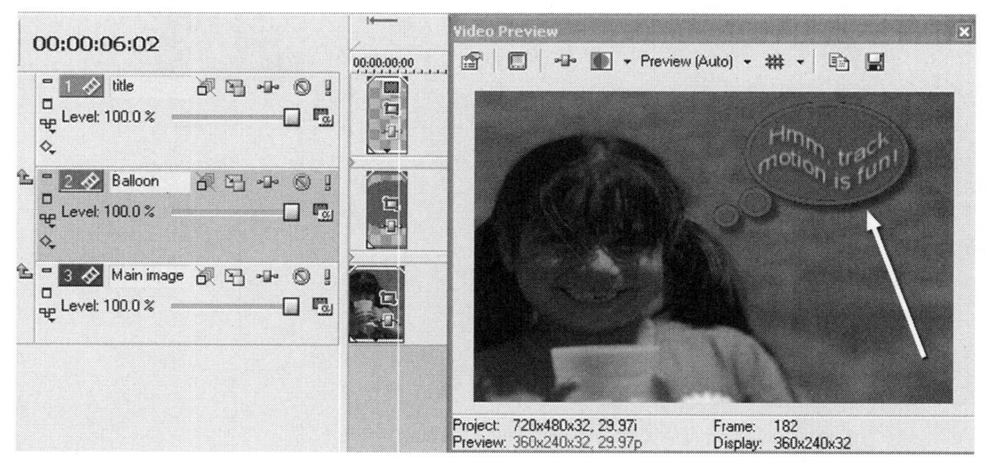

8.24 Shadow applied to PNG file containing alpha channel.

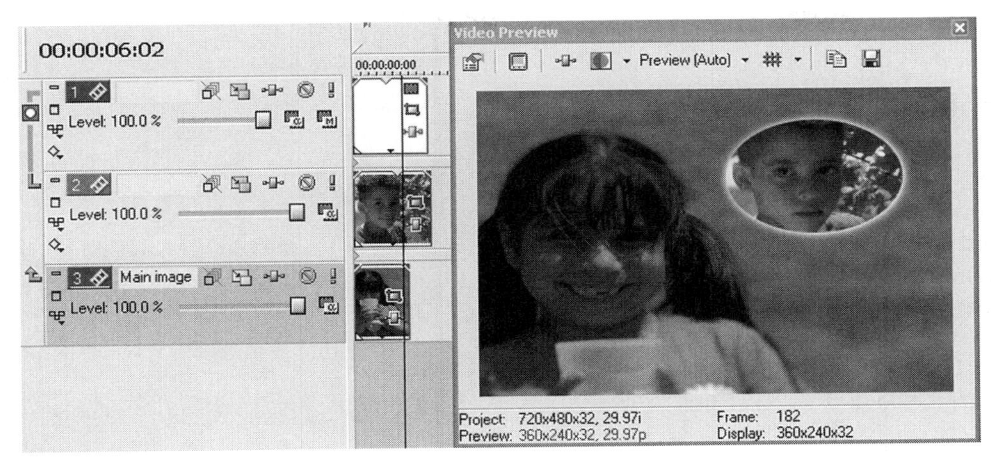

8.25 Glow can be subtle.

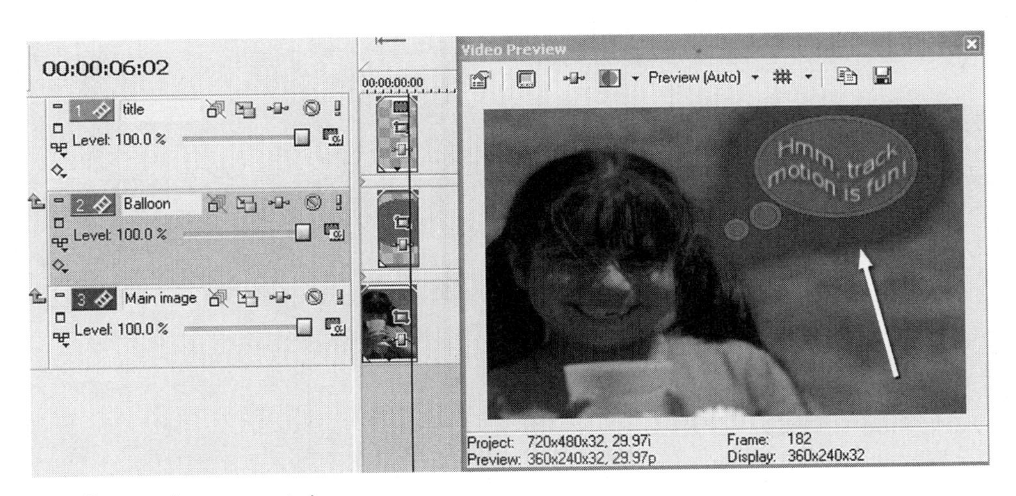

8.26 Glow can be exaggerated.

Glow can also be applied to images in the frame. Glow on its own, or combined with shadowing, allows for very complex image manipulation. Glow is adjustable in opacity, color, and size. Glow can also be feathered to blend into background or other composited images.

Using glow in a subtle manner around letters whose color properties closely match a background color can help offset the letters from the image underneath. Glow is keyframable for opacity, color, size, and position and can be used in a similar way to a shadow in that it can have a separate path from the image to which it is assigned. Using a combination of glow and shadow allows for two separate moving images beneath a single event.

The project seen in Figure 8.27 is on the CD in this book, labeled `composite1.veg`. This project can be used as a template.

The Track Motion tool can be applied to just about any project in some form or fashion, splitting screens, creating multiple screens such as seen in the opening of the *Brady Bunch* television series, used to fly titles in or out and any number of other creative uses.

The project displayed in Figure 8.29 can be found on the CD in this book and is labeled `composite2.veg`.

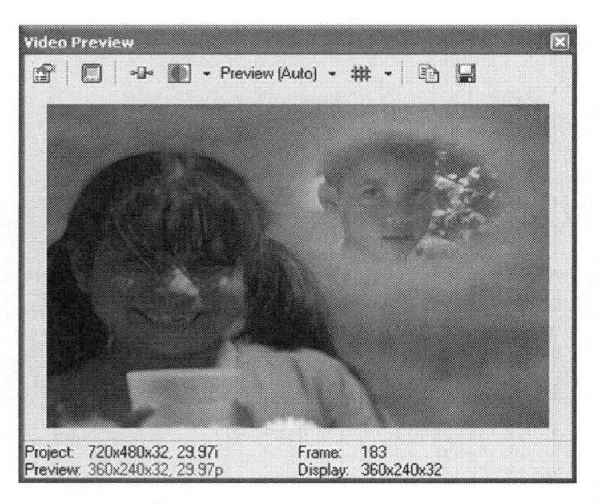

8.27 Glow combined with a mask and shadow makes for a seamless transition between the foreground and background events.

Generated Media

Vegas has the ability to create or generate media using nothing but plug-ins (no physical file is required) that can generate titles, solid colors, noise, gradients, checkerboards, and other graphic elements. You can quickly and easily generate colors for inserted media, masks, filling backgrounds under titles that don't have associated video or graphic events, and many other uses.

To learn how to create generated media, insert a new video track on a new project by right-clicking the Track Control pane and choosing Insert New Video Track or by pressing **CTRL+SHIFT+Q**.

Right-click in the timeline area of the new track and choose Insert Generated Media from the menu. The Generated Media dialog opens.

Select Sonic Foundry Solid Color. The Solid Color dialog box opens, and you can select the color of choice. For this exercise, choose Black.

After setting the color, set the length of time that the event should last and close the dialog by clicking the × in the upper-right corner of the dialog. A black event now appears on the screen.

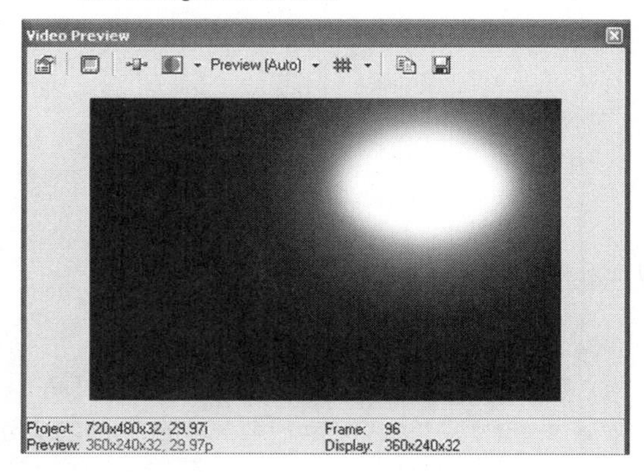

8.28 Mask soloed to demonstrate the cookie cutter, blur, glow, and shadow combined.

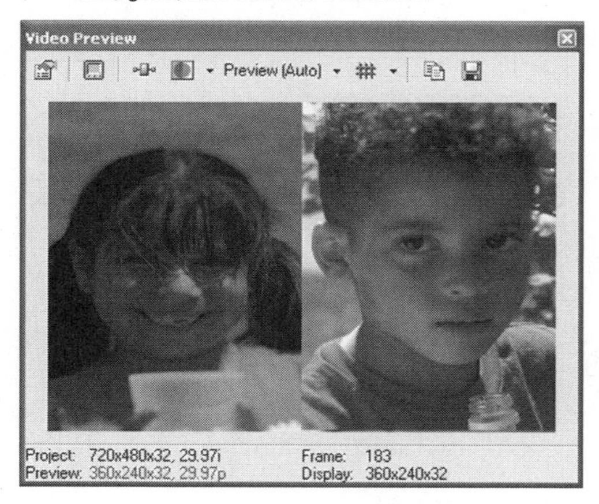

8.29 The split screen is achieved quickly with the Track Motion tool.

Click the Event FX button found on the event, or right-click and choose Event FX.

From the FX dialog, select two plug-ins: Add Noise and Black Restore. Both can be selected at the same time by pressing and holding CTRL down while selecting the second FX plug-in. On the top/header of the plug-ins dialog, you should see both FX names displayed.

Click the Add Noise box. In the Add Noise plug-in dialog box, uncheck the Animate check box, and slide the Noise Level slider to 1.0. In the Black Restore dialog, set the threshold for 0.600.

You have just created a star field from nothing more than computer-generated media.

To add more media to the star field to make it more realistic or interesting, insert a new video track above the existing track. Right-click in the timeline area and choose Insert Generated Media just as you did before.

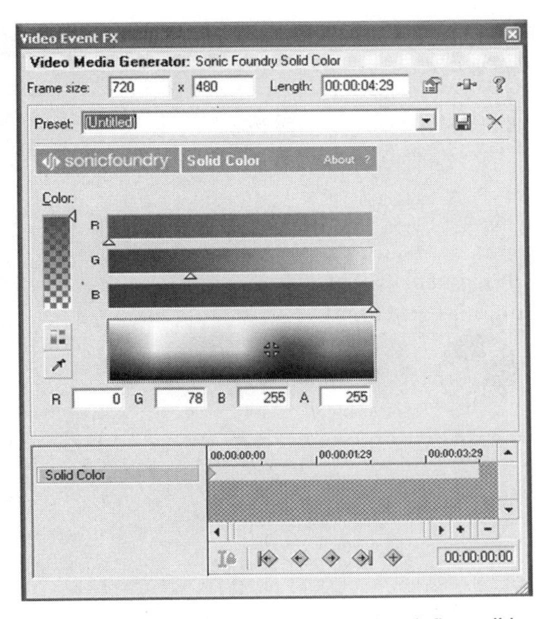

8.30 The Solid Color dialog box is used to define solid colors on the timeline.

Now follow these steps:

1. Choose Sonic Foundry Color Gradient.
2. Select Sunburst from the menu. Notice that by moving the target marker around in the Gradient Preview screen, the curve of the gradient can be adjusted.
3. Define the length of time that the event should occur. This length of time should be the same length of time as the previous event created. Now close the dialog.

The new gradient now covers the entire Preview screen. To show only the portions that are to be actually seen, select the Video FX tab on the bottom left of the screen or click the Event FX button found on the new gradient. Choose the Cookie Cutter FX plug-in.

In the Cookie Cutter dialog, select the Arrowhead in the Shapes menu. The screen can still be covered with the Sunburst image. Using the Size slider, size the arrowhead so it is not filling the entire Preview window. A size of around 0.050 is a good choice. Close the Cookie Cutter dialog.

Now open the Track Motion dialog on the track with the arrowhead. Place the cursor on the upper-left corner handle of the frame. The cursor will change into a circular shape, indicating that the Track Motion tool is prepared to rotate the frame.

Rotate the frame so that the arrowhead in the Preview window is pointing at the lower-right corner of the Preview window.

Click and hold in the middle of the track motion frame and slide the frame to the upper-left corner of the frame window. The arrowhead in the Preview window will move to the same location and should still remain pointed at the lower-right corner.

Enable the sync cursor in the Track Motion Keyframe section. In the timeline, place the cursor on the right edge of the generated media event. In the Track Motion tool, move the frame to the lower-right corner. A new keyframe will be inserted in the timeline as a result of the sync cursor being enabled.

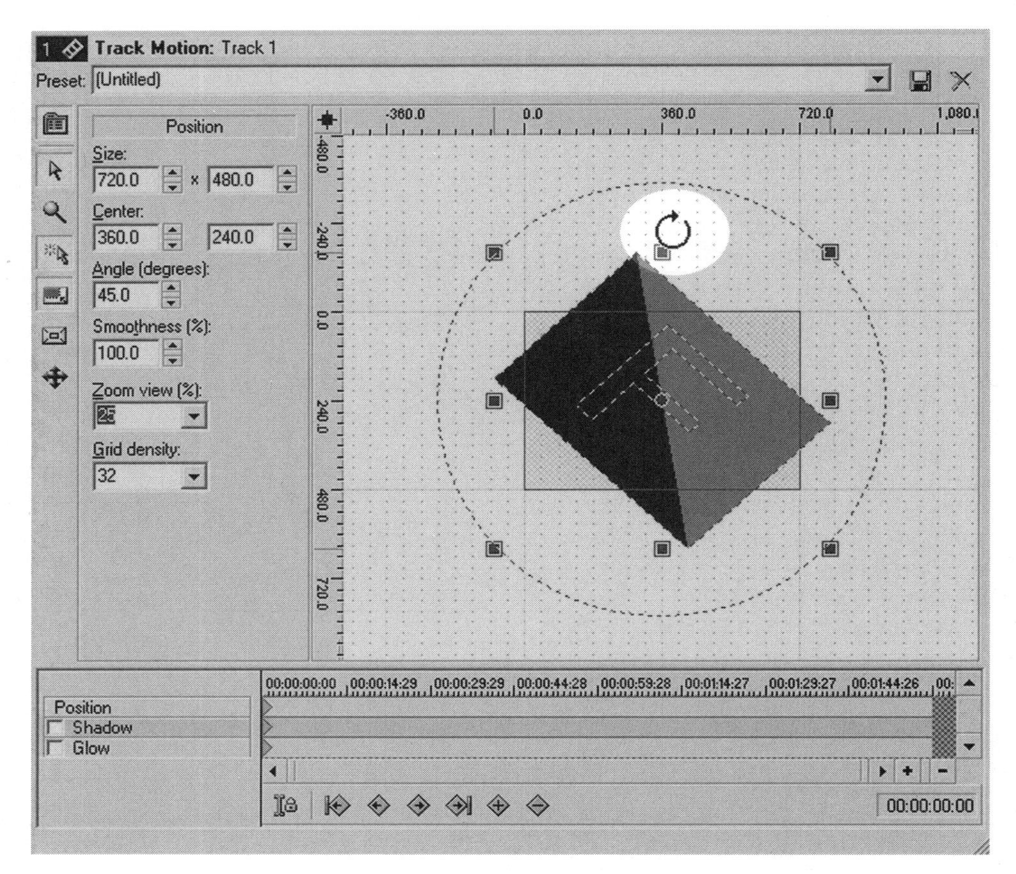

8.31 The cursor changed to a rotating cursor, used for rotating the frame inside the Track Motion tool.

Now press the SPACEBAR or click Play on the Transport control. You should see the arrowhead flying across the star field, and if the loop is enabled (press Q), the arrowhead/starship will fly across the screen from the upper left to the lower right.

Adding an orange glow and sizing it to cover only the back of the arrowhead will create a flame coming out of the back of the arrowhead/starship. By animating the feather, opacity, and intensity of the glow, the flames can come to life, moving at different depths and opacity, giving a realistic sense of video to an otherwise static image.

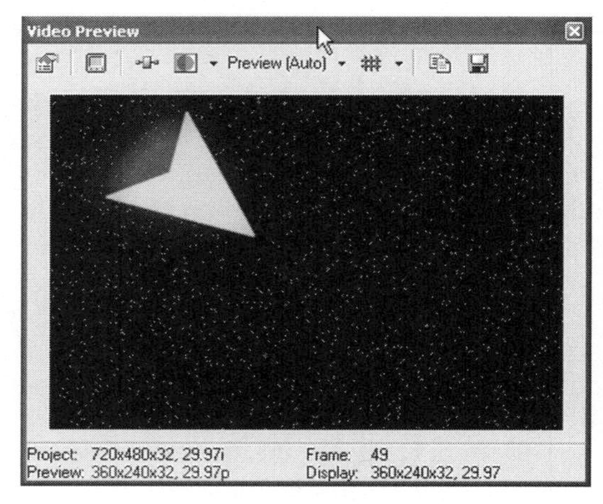

8.32 Flames created with glow give a realistic edge to the starship.

This project can be found on the CD included in this book and is labeled `compositerocket.veg`. The keyframes animating the flame are also found on the VEG file.

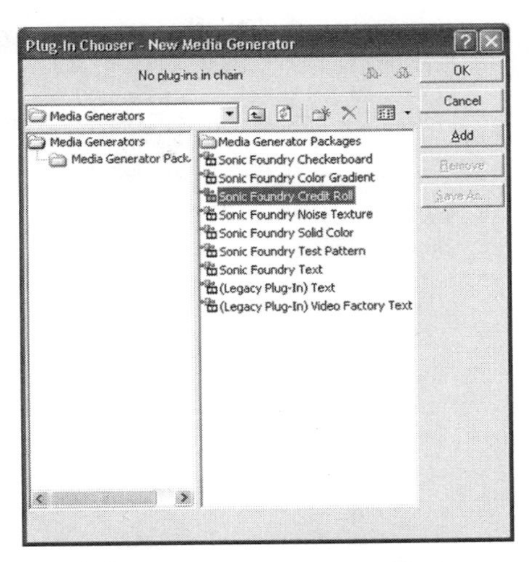

8.33 The Credit Roll plug-in is found in the Generated Media chooser.

Credit Rolls

Inserting a credit roll is one of the generated media options in Vegas. Credit rolls are used for more than only titling at the end of a video project; they are great for showing specific product information, creating introductions to a video, or creating fast-moving numbers or letters for a specific effect, particularly when used at reduced opacities.

Insert a credit roll by right-clicking a track and selecting Insert Generated Media from the submenu. In the Generated Media chooser, select Credit Roll from the listing of available generated media types.

The Credit Roll plug-in offers many choices to manage text in the credit roll. Three text settings are allowed in the roll: a Title Text (headline), Subitem Text (main topic), and Item Left/Item Right (sub-topics). Each of the three settings can have its own font size, color, style, and position.

The type of line can be adjusted by clicking next to the field in which the text will be entered. For example, a single line of text can be either a Title Text field, or it can be a Subitem Text field. The type of text is selected by clicking the small box next to the input field. Clicking in the field allows text to be input to the field.

In the Style dialog, kerning, spacing, and connecting styles, such as dashes or dots, can all be specified. All aspects of the credit roll can be keyframed. Because of the nature of the credit roll and generated media, there is no sync cursor ability.

The length of the event determines the speed of the credit roll, and the number of lines determines how fast the roll moves through the length of time. Longer event lengths with fewer lines create a very slow crawl, while short event lengths with several lines scroll more quickly.

From the Properties dialog, a timed sequence can be created. In a timed sequence, titles can be zoomed in and out, faded in and out, and slid from the bottom, top, and side depending on selection in the menu. The number of lines and length of credit roll event determines the length of time of each title image on screen.

By default, credit rolls are of transparent background. Background colors, transparency, and letter colors, however, can all be adjusted. Shadows and glows are not options in the credit roll.

Using the DebugMode 3-D plug-in, a bump map, or height map (discussed later in this chapter) can add a number of options and creative ability to the credit roll.

Tip

To add a shadow or glow to a credit roll, open track motion. Leave the track position at default but check the shadow or glow (or both) check boxes. These settings add a shadow or glow to the characters on the credit roll. These can be keyframed on or off or for position if the shadow or glow is desired to flow differently from the direction of the credit roll.

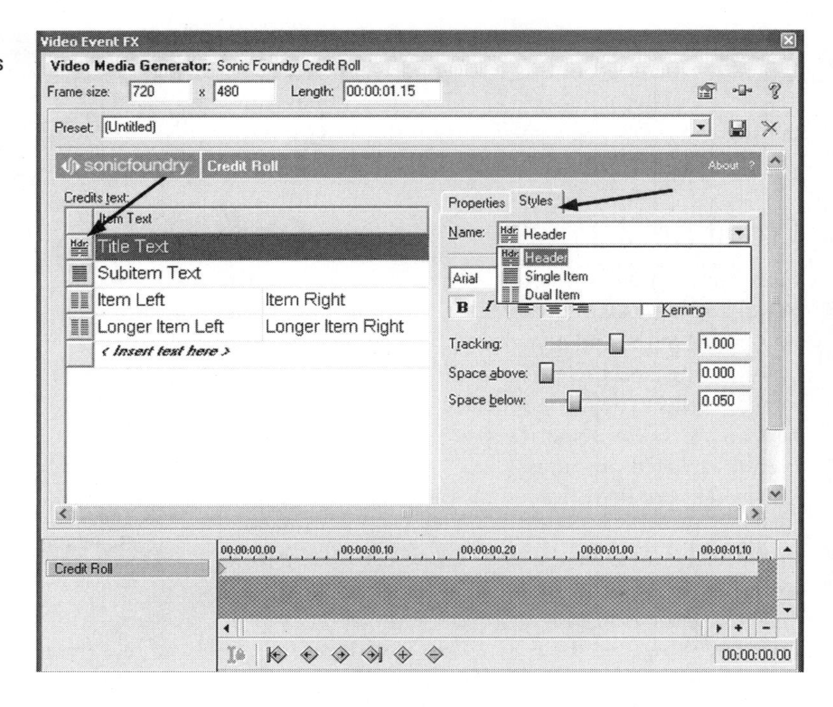

8.34 Selecting the Styles tab allows parameters of each of the three styles to be adjusted. Clicking the Style button on the left side of the Credit Roll plug-in chooses which text style the line will present.

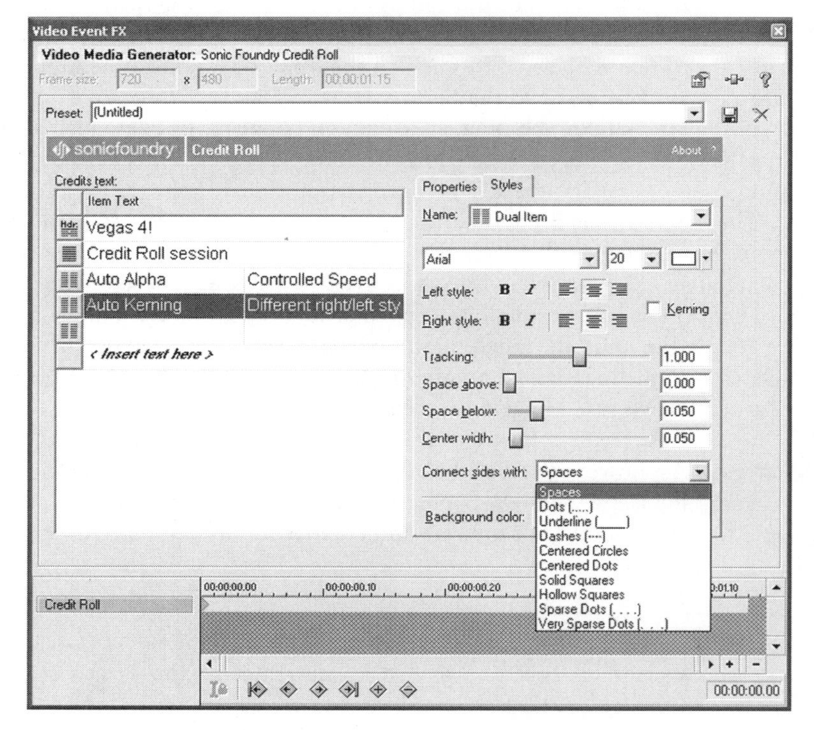

8.35 Connect the left and right credits on the scroll with dashes, dots, or lines.

8.36 Using the DebugMode 3D LE plug-in, text can be angled on the screen, resembling a wall or doorway. Height mapping can be used to wrap the text on a textured surface.

Creating/Using Masks in Vegas

Masks are exceptionally powerful tools for creating composited media. The generated media project created earlier contained a mask, inserted with the Cookie Cutter plug-in.

Masks can be made from just about any photo editing tool, font, or shape that exists. Vegas auto-senses an alpha channel from most images that contain them and has options to force recognition of an alpha channel if it is not automatically sensed.

To force Vegas to recognize an alpha channel if it is not seeing the channel automatically, right-click the event containing the alpha channel. Select Properties from the submenu that appears. Select the Media tab in the Properties window and at the bottom is a menu labeled Alpha Channel. Next to this, locate a drop-down list that offers several choices. Select Premultiplied from the menu.

> An alpha channel defines the areas you want to be transparent. Each image on the video screen is comprised of four channels: red, green, blue, and alpha. The alpha channel acts as a mask, instructing the other channels on how the pixels should blend or merge. Alpha channels can have gradations, allowing blending of multiple images on top of each other at various levels of transparency. Another way to understand how an alpha channel functions is to imagine a stencil laid over an object and a can of spray paint. The alpha channel reveals the events beneath much as a stencil reveals the object beneath. Alpha channels usually display transparent areas as white space, while hidden areas are displayed in black.

Masks can either hide or reveal information/images in a video project, allowing image areas to be defined within an event. Another name for a mask is a *matte* or a *key*. All terms mean the same thing: hiding or revealing an area so another image can show through, or hiding an area so that another image can be composed over the top of it. In any definition, a mask, matte, or key defines the transparent pixels of an image for superimposing or revealing another image.

Just as titles are one of the simplest forms of compositing, they are also one of the most viable uses of masking. Titling is an excellent method of learning exactly how a mask works.

Open a new project in Vegas. Insert three new video tracks. Choose a video or still file to serve as a background. To create the mask, place the video or still image on track 3.

Place a title on track 1. Do not use fonts that contain serifs or small lines. A font such as Impact or Arial Black will serve best as you learn this technique.

On track 2, place a video file that has high motion with bright colors for the best effect in the learning stage.

The text is a mask but is not seen as such just yet. On the extreme left side of the Track Control pane, there is a small arrow pointing upward, which is the Make Compositing Child (Parent/Child) button. Clicking this button will cause all areas of track 2 to become transparent. The only portions of track 2 that are seen are seen inside the letters of track 1. Click this button.

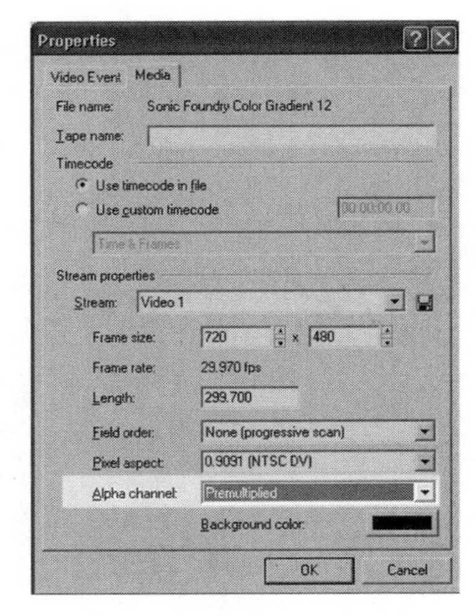

8.37 Right-clicking and selecting **PROPERTIES | MEDIA** allows alpha channel recognition to be forced if alpha channel is present.

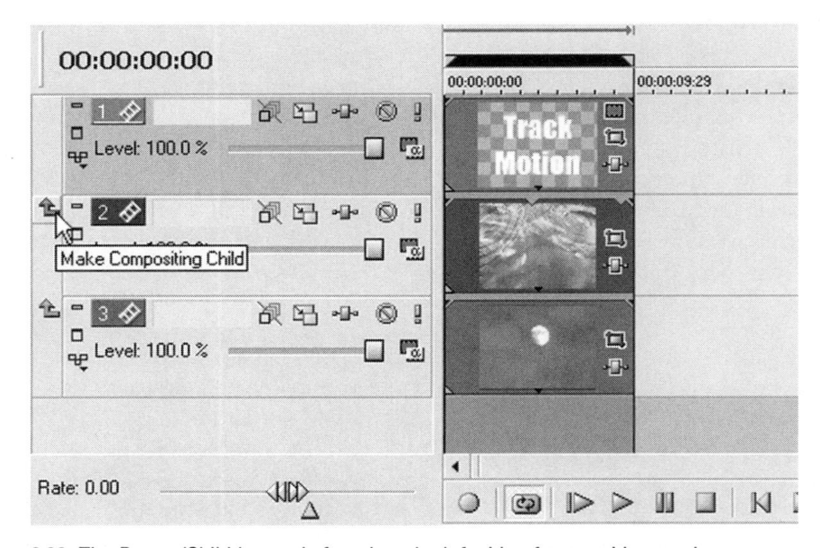

8.38 The Parent/Child button is found on the left side of every video track.

The event on track 2 disappears and becomes visible only inside the title letters. The letters are a mask; the letters become the *parent,* and the event beneath becomes the *child.* This process causes the parent to instruct the child how to be shown. Multiple tracks can be the child, but only one track can be the parent. As many parent and child relationships as necessary can exist across multiple tracks; however, but no child track can have more than one parent. If multiple parents are required of a child track, the child track must be duplicated.

8.39 The video of track 2 (the child) shows through the letters of track 1 (parent track).

To gain a better understanding of the masking process, change the color of the text on track 1 to black. It will cause the letters to become invisible. Reset the color of the title to white in order to continue this section.

This process is masking in its most basic form. Any image that contains transparent/alpha information can be used in place of a title. To continue the concept of manipulating the title appearance, however, keep this project on the timeline.

8.40 You can have fun while learning to make masks.

In track 1, right-click the title to edit it or click the Edit Generated Media button found on the title event. The Title tool will open. Navigate to the Effects tab and select a shadow on the text. The shadow can't be seen. Changing the color of the shadow will result in the shadow being seen.

To place a shadow on the mask, or placing letters in this instance, select the Track Motion tool on track 1. In the Track Motion tool, check the Shadow check box. A shadow will appear beneath the mask/letters on the Preview screen. This shadow can be keyframed in exactly the same way that a shadow in a normal track motion instance would be keyed. Color, size, feathering, transparency, and position can all be user-defined using keyframes.

Open your favorite photo editing application. In the application, create a new project/photo image that is 655 × 480 pixels for NTSC or 704 × 576 for PAL. Flood the entire work area with black.

Paint a design or shape in the work area, using white as the color. Save the design as a PNG file for best results. A GIF, JPEG, TIF, or Targa will also work.

Open a new project in Vegas. Create three new video tracks. Place the new image on the timeline on track 1. The explorer might need to be refreshed in order to see it in its newly saved location. Refresh the explorer by clicking the Refresh button on the Explorer toolbar.

Insert the media that should be showing through the mask on track 2. The length of the event doesn't matter, as the length of the parent event determines how long the underlying event is visible. (Of course, if the child event is shorter than the parent event, the child will not be seen for the full length either.)

On track 3, insert an event that contains video or a still image that functions as the background image.

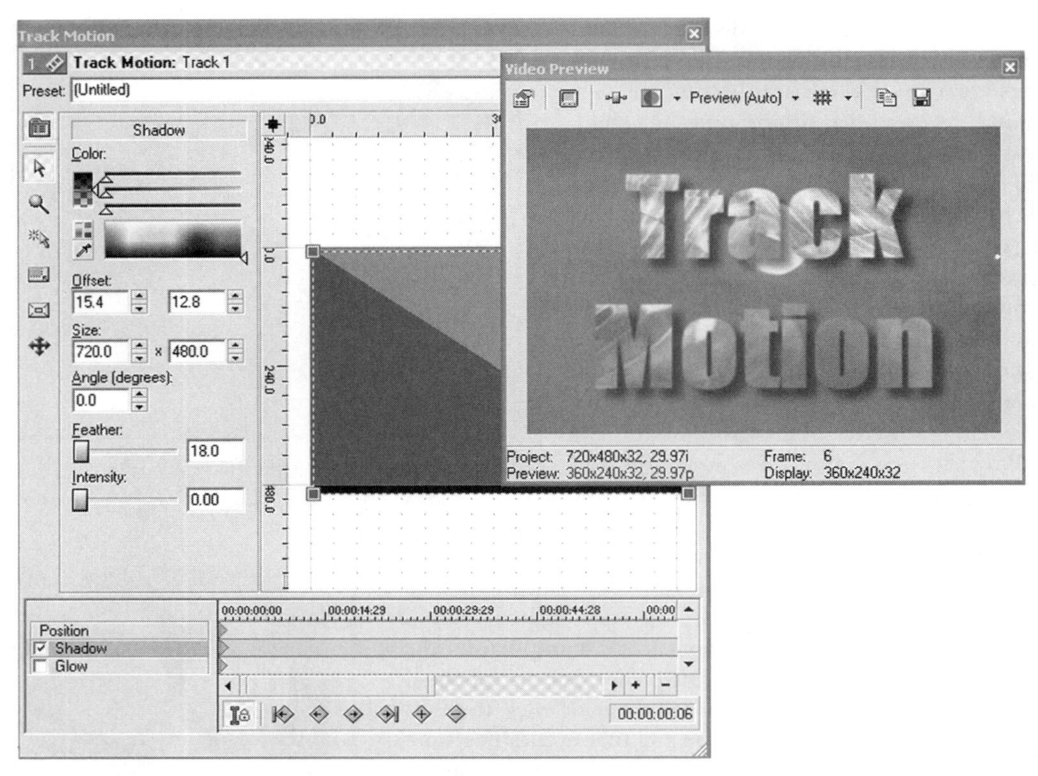

8.41 Adding a shadow to the mask with the Track Motion Shadow tool. Glow can also be added to a mask.

8.42 Any shape can be used to create a mask, so long as it contains a mask and an alpha image.

In a gradient mask, white areas are revealed, while black areas are hidden. When using other colors as the mask, the closer a color is to white, the more revealed it becomes, while just the opposite is true for gradient colors approaching black in value. However, when using other colors in a mask, those colors reduce the opacity of the underlying image.

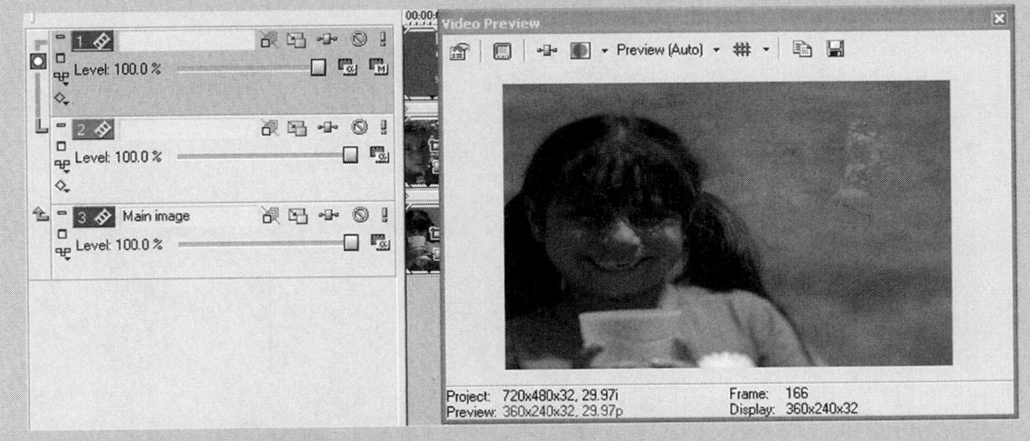

8.43 Black to white gradient changed to solid blue gradient, reducing the opacity of the image shown through. (See Figure 8.27 for the original mask appearance.)

Now click the Parent/Child button on track 2, and the image you created becomes transparent, except the visible mask areas, which now show the event on track 2.

Filters can be applied to the events on the child track, which will affect how the events on track 2 are shown. Applying blurs will blend the mask into the underlying events.

Another means of creating a mask in Vegas is to use generated media to define what will be masked out. One useful way to do this process is to create a mask that splits the screen in two or three parts, using blur and mask to create a completely smooth transition from one side to the other.

To create a split screen with a seamless transition:

1. Insert three video tracks. Track 1 is for the mask, track 2 for one half of the screen, and track 3 for the remaining half of the screen.

2. Insert the desired events on tracks 2 and 3.

3. On track 1, insert generated media by right-clicking in the track 1 area of the timeline and selecting Insert Generated Media. In the Media Generator plug-in dialog, select Sonic Foundry Solid Color and choose white as the color.

4. Select the Pan/Crop tool on the generated media by right-clicking or by clicking the Pan/Crop button found on the media. In the Pan/Crop Preview window, slide the frame to the right until it divides the frame in half. (The Center should be at 720 and 240.) Close the Pan/Crop dialog box.

5. On track 2, click the Parent/Child button on the left side of the screen. This step will cause the screen to split in half, showing track 2 on the left side of the screen, as it is being masked by track 1 and the pan/cropped image. If the images on tracks two and three are not centered, the Pan/Crop tool should be applied to the images to center them.

6. With the screen split, apply a Gaussian blur to the event on track 1 either by dragging the blur to the event from the Video FX docking window or by clicking the Event FX button on the solid color event on track 1. Apply the blur to the desired setting.

7. This process can also be applied in a series of two soft transitions as well, by adding two more tracks to the project. This step will display three split sections. Pan/Crop will need to be applied in opposite directions and in thirds of the screen.

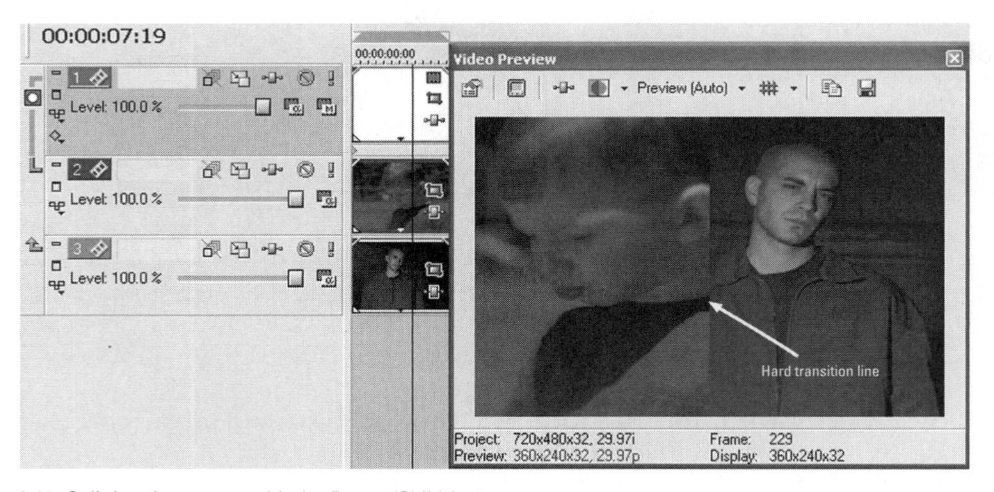

8.44 Splitting the screen with the Parent/Child feature.

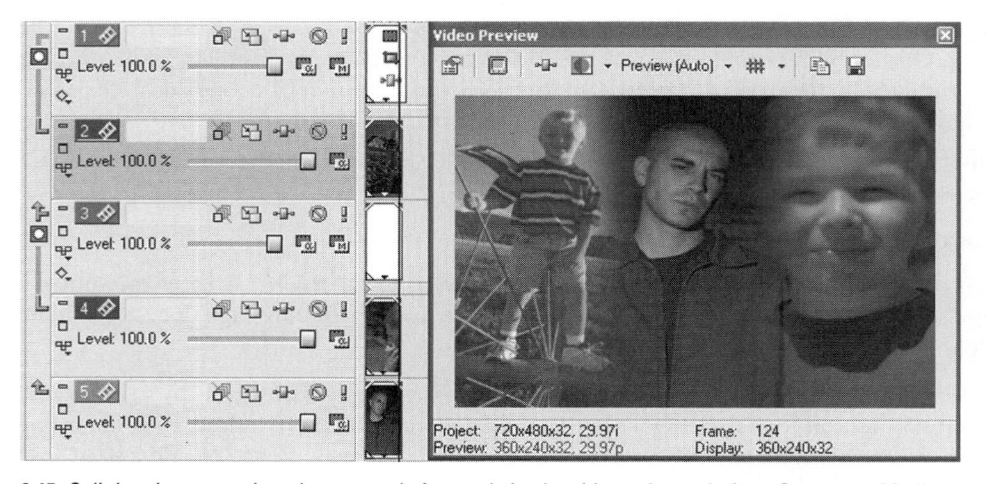

8.45 Splitting the screen into three parts is fast and simple with masks made from Generated Media.

Tip

The project shown in Figure 8.45 can be found on the CD in this book. The filename is `3screen.veg`.

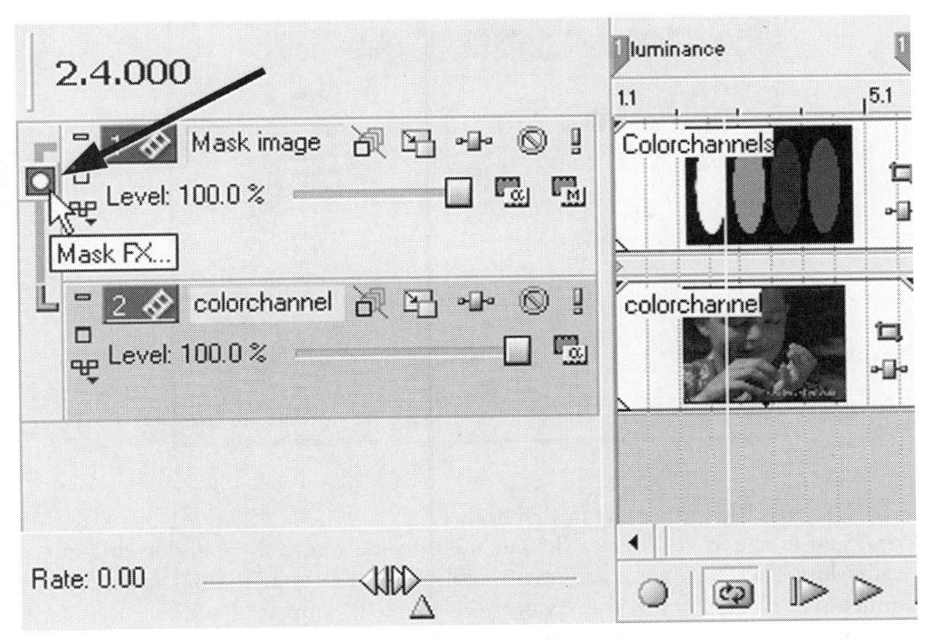

8.46 Access the Mask Generator by clicking the Mask FX button found on any Parent/Child combination.

Vegas also has a Mask Generator for every video track parented to another.

Five different mask modes are available in the Mask Generator dialog. Each mode blends or masks media differently, based on color or luminance found within the parent image, and are described here:

- **Luminance**—uses luminance in the image to determine transparency.
- **Alpha**—uses an alpha channel to determine transparency.
- **Red Channel**—uses the color red to determine transparency.
- **Blue Channel**—uses the color blue to determine transparency.
- **Green Channel**—uses the color green to determine transparency.

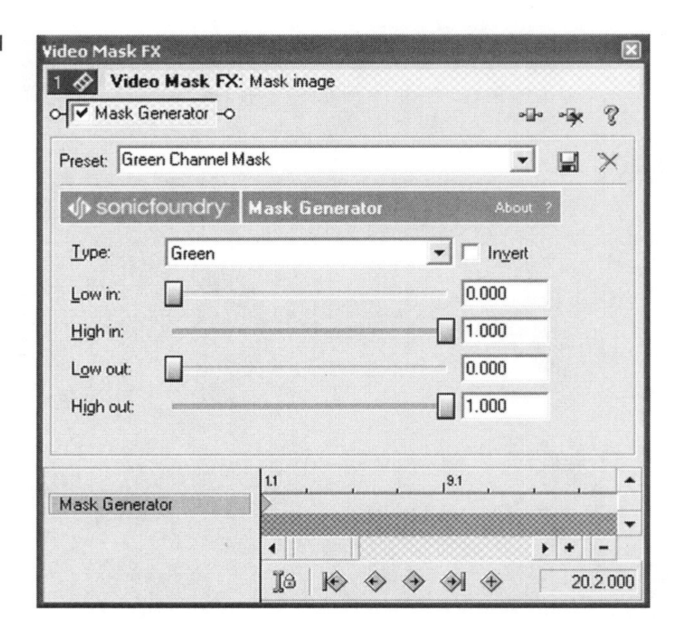

8.47 Specific color channel controls determine the transparency of that color's channel.

Open the `colorchannel.veg` file on the CD in this book.

Use the Low in/High in sliders to limit or delimit the information in the Child events in the Parent/Child relationship. These attributes are keyframable and can be used to limit/delimit transparency of a channel over time. Use this tool as one alternative to animate lines on a map, create handwriting on an image, or mask out a color channel over time.

There are many creative ways to use masks in Vegas. Using masks, a reflection of text can be quickly composited.

Open a new project. Insert six video tracks. Do the following steps in this order:

1. On track 6, insert a background image. For purposes of this exercise, use generated media, preferably a color gradient.
2. On track 5, double-click the media on track six to create a selection. On track 5, right-click to insert text media. For this exercise, keep the text to one single line. Use a creative font for best results.
3. On track 4, with the selection still active from double-clicking track 6, right-click and select INSERT GENERATED MEDIA | SONIC FOUNDRY COLOR GRADIENT. (If the selection is not active, double-click track 6 again.) In the Gradient dialog, select the Linear White to Black from the menu. In the Aspect Ratio Angle dialog box, enter a value of 34.0. In the keyframe timeline of the Color Gradient dialog, click the Last Keyframe Button to move the cursor in the dialog to the end of the selection/event. Rotate the Aspect Ratio Angle to -180.0, which will cause the gradient to rotate in the project. Now select the Parent/Child button on the right of track 5.
4. Right-click the text on track 5 and select Copy. Right-click in track 3 and select Paste. Vegas will ask if you wish to create a reference to existing media or if you wish to create a new copy. Select New Copy.

5. On track 5, select the Track Motion button. In the Track Motion dialog box, right-click in the Track Motion Preview window and select Flip Vertical. The "F" in the Preview window will now be upside down in the Preview window. Now resize the track preview to 945.1 × 568.3. Center the image at 360.0 × 360.0.

6. Still working on the event on track 5, insert the following event FX: Add Noise, Gaussian Blur, Deform, and Light Rays. These plug-ins create the illusion of a reflected surface. Set the Add Noise plug-in to 0.168, with MonoChromatic and Gaussian Noise boxes checked. In the Gaussian Blur dialog, set the Horizontal value at 0.023 and Vertical at 0. In the Deform dialog, set the Amount to 1.00. Check the Center Image box. Leaving the Left and Right sliders set at 0, set the Top slider value to -0.209, and the Bottom slider value to 0.510. Finally, set the Light Rays to the values shown in Figure 8.48.

7. Place the cursor in the center of the Light Rays keyframe timeline. Set the Strength value to .550 and move the Light Source indicator to the center of the preview box. Move the cursor to the end of the keyframe timeline in the Light Rays keyframe timeline. Set the Strength value to .250 and move the Light Source indicator to the far right of the preview box.

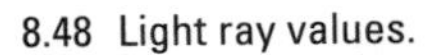
8.48 Light ray values.

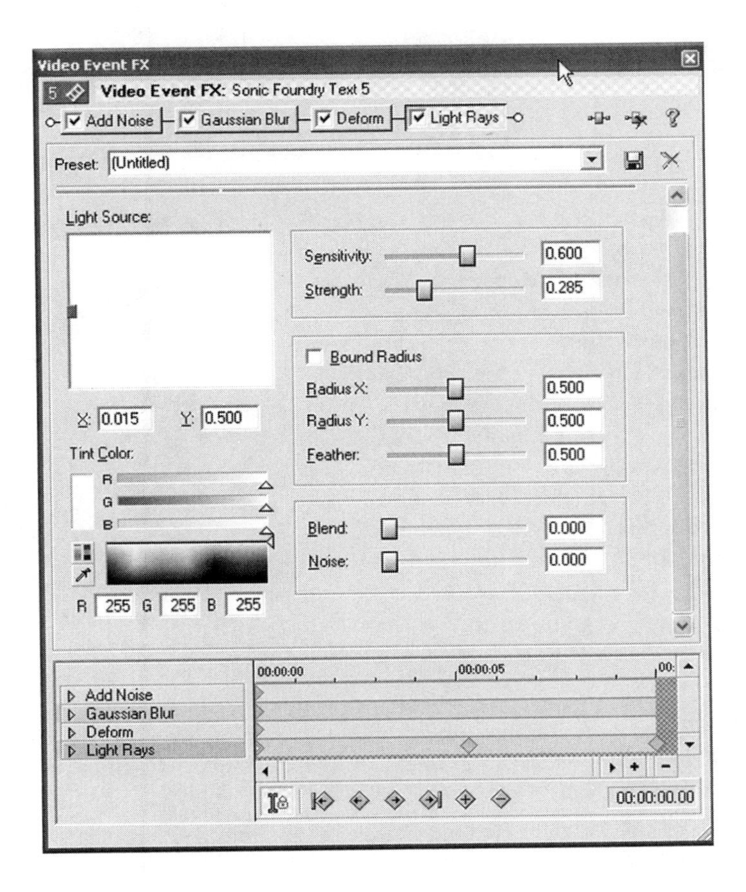

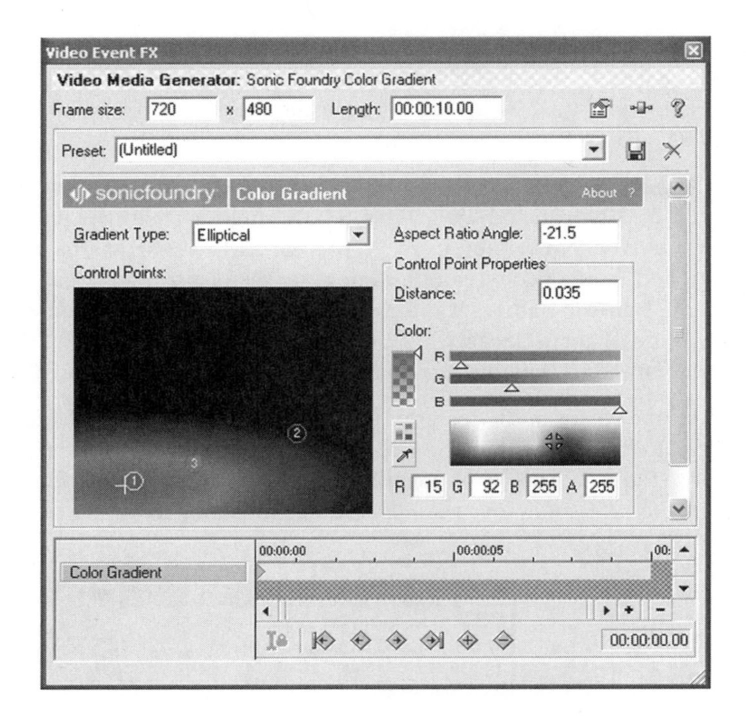

8.49 Setting values.

8. Right-click the mask on track 4 and select Copy. Paste this copy on track 2. Once again, Vegas will ask whether a new copy should be created or a reference to the original file should be made. Create a new copy. Click the Edit Generated Media button found on the generated media on track 2. In the Aspect Ratio Angle dialog, on the first keyframe, enter a value of 0.0. Now click the last Keyframe button in the keyframe timeline and enter a value of -180.0. On track 3, click the Parent/Child button to cause track 2 to be a parent to the text event found on track 3.

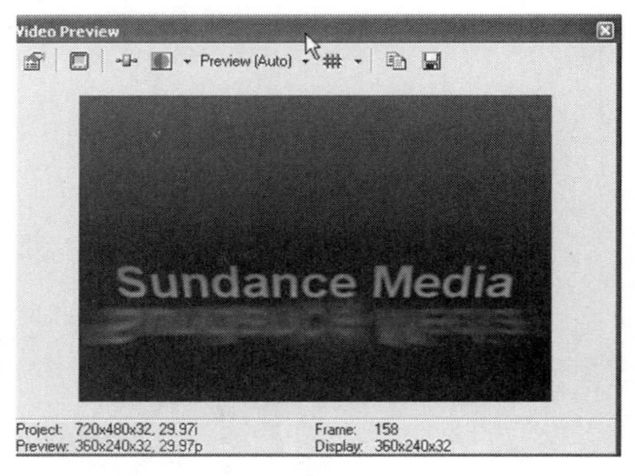

8.50 Creating a reflective surface is very fast, and requires no additional plug-ins.

9. Create a selection again by double-clicking any event on the timeline. On track 1, place a Sonic Foundry color gradient. In the Gradient dialog box, set the values to the settings shown in Figure 8.49.

Use the RAM Preview Render setting by creating a selection of all the media and pressing SHIFT+B. Vegas will take a moment to render the project to RAM. Be sure that RAM preview is set to the maximum available amount in the OPTIONS | PREFERENCES | VIDEO dialog. The default setting is 16MB of RAM. To render this project, you'll need at least 300MB of RAM available.

Compositing Modes in Vegas

Vegas has 14 different compositing modes. Source Alpha is the default mode, while additional modes can be used for a number of image manipulation and creative settings.

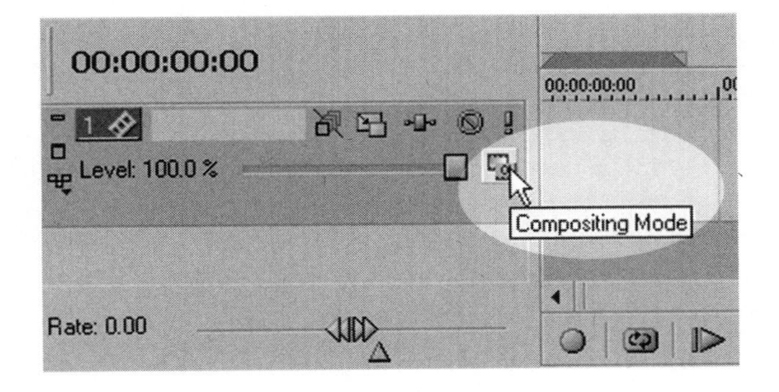

8.51 Compositing Mode button on each video track.

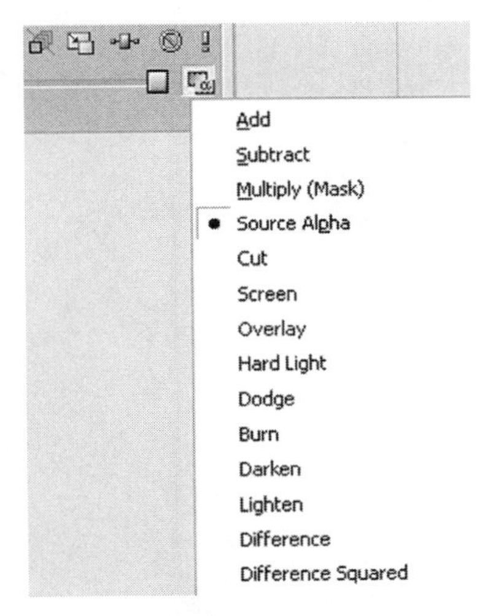

8.52 Various compositing modes in Vegas 4.0.

8.53 Original foreground and background events.

Compositing modes define the manner in which a higher track combines with a lower track. As higher-level tracks dominate, for purposes of reference, the higher-level tracks are referred to as *foreground*, and lower-level tracks are referred to as *background*.

Compositing Mode tools allow for rapid access to creating multilayered images quickly. Compositing mode parameters are not adjustable. Using compositing envelopes allow you to adjust compositing level, or color management plug-ins can be used to provide opacity, chroma (color), and transparency control over all events. For instance, in the Add mode, using the Color Correction Secondary plug-in, combined with blur, creates a moving sonic wave. Using the Screen mode creates a more organic property in the sonic wave, almost as if it were a moving cloud.

NEW! Vegas 4.0 has a new compositing mode feature. When tracks become parented to one or more lower tracks, a new icon will appear in the Track Control pane of the uppermost track of the parent/child set. This is the Parent Track Overlay Mode button.

The Parent Track Overlay Mode button provides options to manage masked events above other events.

Out-of-focus video is sometimes an issue with even the most experienced of videographers. Using the Hard Light composite mode and a plug-in or two, marginally out-of-focus video can be brought to a faked focus or at the least, a more sharpened image.

Place the problem footage on a track and duplicate the track. On the upper track, apply the Convolution filter and the Sharpen preset in the Convolution filter. Reduce the opacity of the upper track to approximately 50 percent. Various settings will apply to individual events. On the lower track, apply a Hard Light compositing mode.

Some events with good color balance can also benefit by applying a Hard Light overlay mode in the compositing modes selection.

8.54 Before and after applying Hard Light and Convolution Kernel filter.

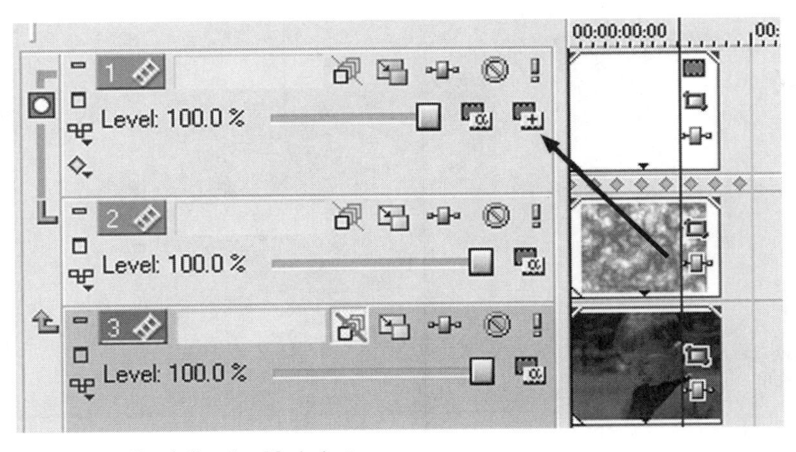

8.55 Parent Track Overlay Mode button.

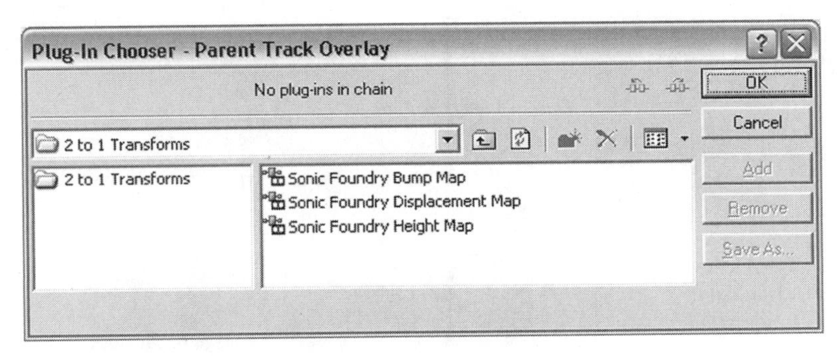

8.56 Parent Track Overlay mode.

When you select this button, there are two menu choices. The first is the default Multiply mode, and the second is Custom. Selecting Custom opens a dialog offering three compositing overlay modes: Displacement Map, Bump Map, and Height Map.

Displacement Map

This mode uses the parent image as a controller to offset the pixels in the composited child tracks along the X and Y axes. This process uses a two-channel displacement. In other words, pixels are displaced to the left, right, up, or down depending on their relationship to colors in the mask. The displacement map is also useful for following contours of shapes.

This plug-in can be challenging to control. Working with solid colors or gradients as masks will yield the best results, particularly when learning how to use this tool.

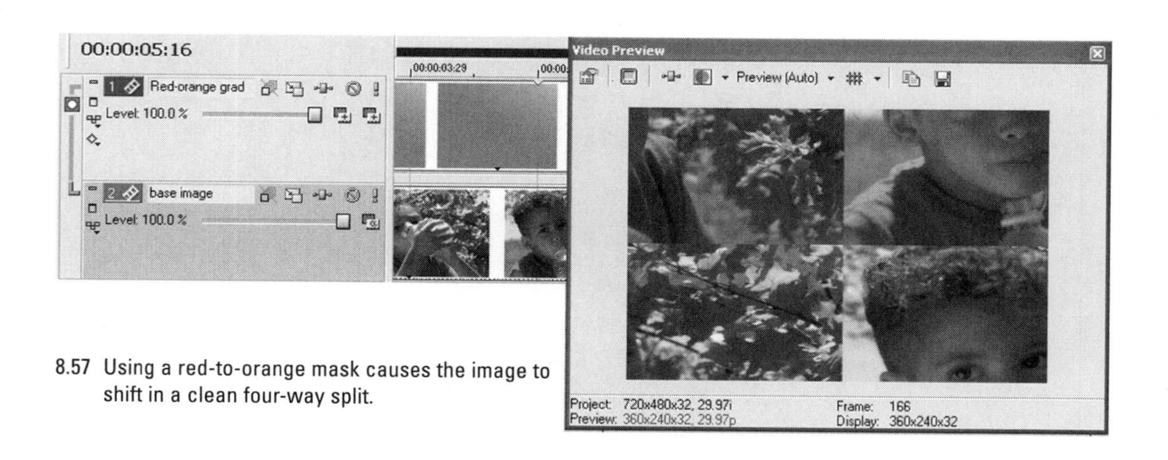

8.57 Using a red-to-orange mask causes the image to shift in a clean four-way split.

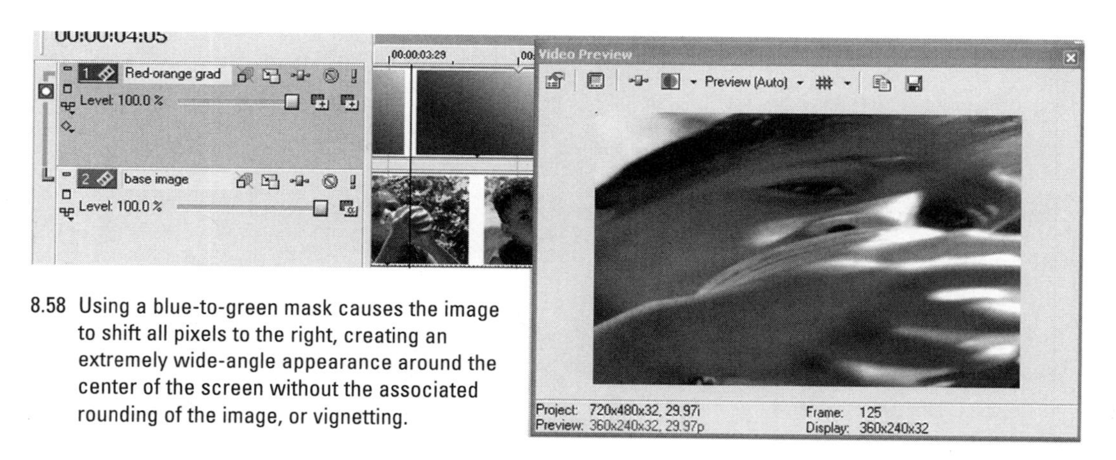

8.58 Using a blue-to-green mask causes the image to shift all pixels to the right, creating an extremely wide-angle appearance around the center of the screen without the associated rounding of the image, or vignetting.

8.59 Ordinary clouds shift to streaks of movement using the height map to shift pixels farther from their original position.

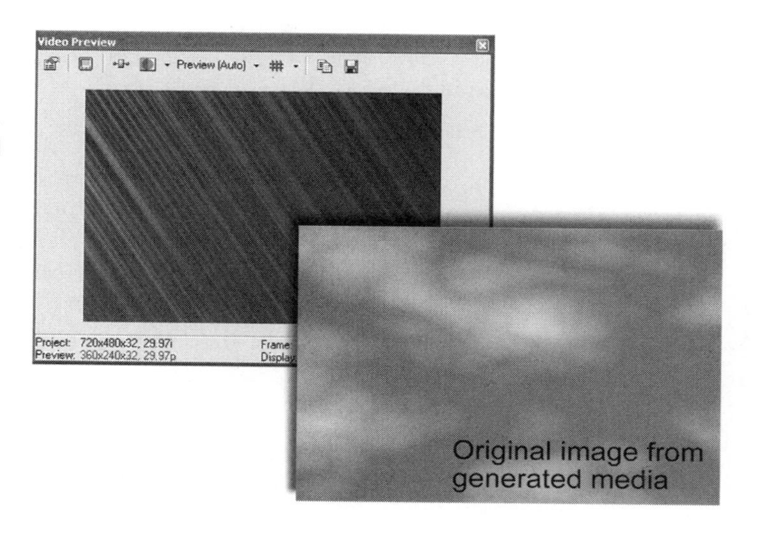

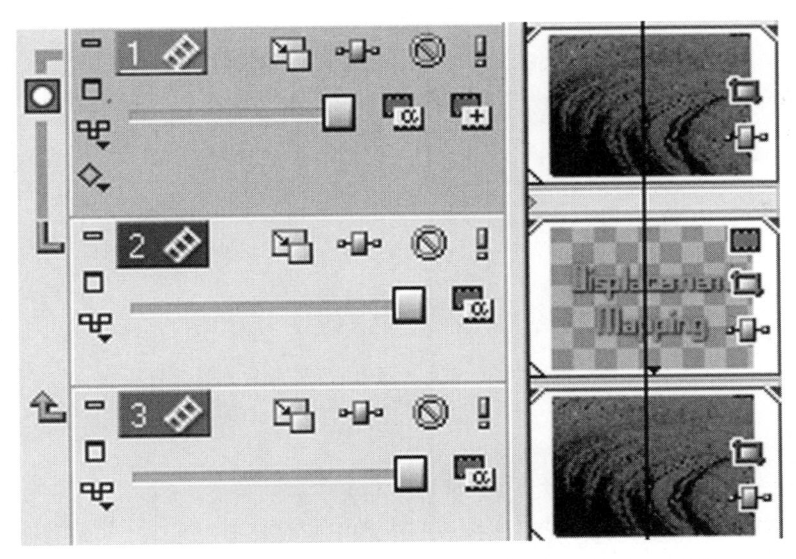

8.60 Three tracks are required to create the project illustrated in Figure 8.61.

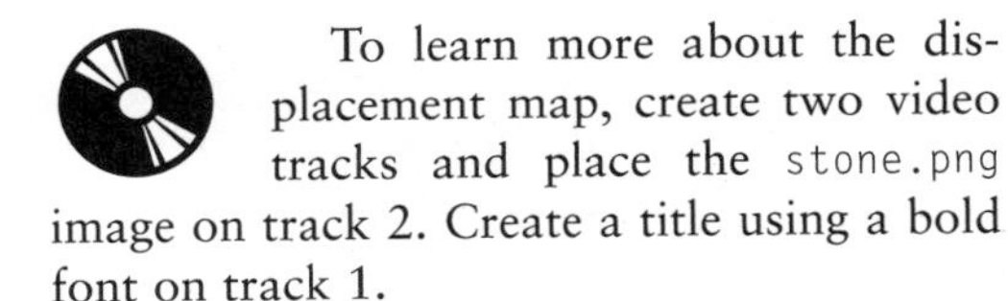

To learn more about the displacement map, create two video tracks and place the `stone.png` image on track 2. Create a title using a bold font on track 1.

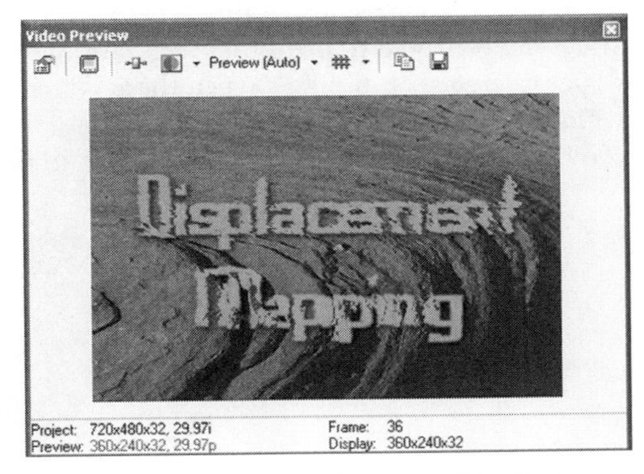

8.61 Three tracks were used to create this effect.

Duplicate track 2 and slide the new track so that it is track 1, the text is on track 2, and the original image is on track 3.

Now make track 1 parent to track 2, which will cause the text to be masked.

Check the Composite Mode button and choose the CUSTOM | SONIC FOUNDRY DISPLACEMENT MAP from the menu that opens.

In the dialog box, change both channels to Red. In the Horizontal axis, set the value to 0.018 and the Vertical axis to 0.026. The appearance of the text in the Preview window will vary depending on the font used. Experiment with the slider values to find the appropriate setting. Notice how the lettering conforms to the rocks in the photo. Using a text color slightly lighter in hue over the rocks completes the effect.

Height Map

This mode uses the parent image as a controller to cause pixels in the composited child tracks to appear closer to or farther away from the viewer. Adding a height map is a great tool for creating the appearance of moving water or viewing an image through glass or water, creating shimmers, fire, or smoke over another image. This mode can be added to gradients, masks, or other images above lower images.

To learn more about the height map, create a new project. Insert two video tracks.

Place the image `rings.png` from the CD in this book onto track 1 and place a video or generated media event on track 2. Now make track 1 parent to track 2 by selecting the Parent/Child button on the left side of the Track Control pane.

On track 1, click the Compositing Mode button on track 1. From the menu, select CUSTOM | SONIC FOUNDRY HEIGHT MAP.

Enter the following settings in the dialog box:

- Amplitude—0.023
- Elevation—0.319
- Height Scale—0.320
- Smoothness—11
- Source Channel—Intensity (menu selection)
- Edge Pixel Handling—Wrap pixels around (menu selection)

Add a pan/crop to the event on track 1, zooming in by 40 percent. Experiment with the Amplitude and Elevation settings in the Height Displacement dialog box in order to better understand how this plug-in functions.

Any track-level keyframes can be viewed, moved, and edited directly from the timeline. By expanding the keyframes in the track view, keyframes can be slid forward or backward in time and double-clicked to open the track motion or compositing keyframes. This feature provides a visual reference at all times when there is track motion or compositing information on a track, which allows for cursor placement on musical beats, specific musical or dialog points, or simply provides a reference to what is being seen in the Preview screen.

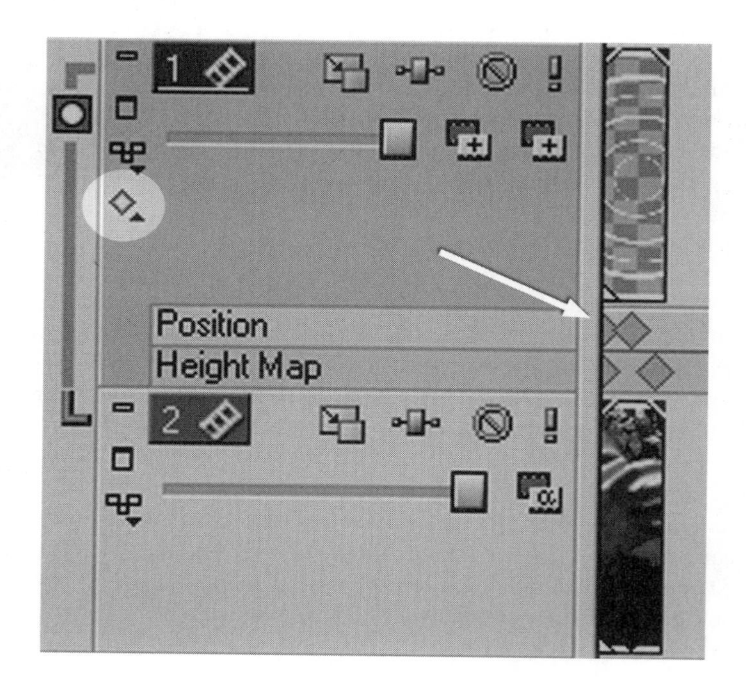

8.62 The Expand Track Keyframes button opens a keyframe pane directly on the timeline.

In nearly all of the previous projects, a gradient mask is used. Gradients and other generated media forms are very useful as masks. As an example, try this method of creating a mask:

1. Insert a gradient mask.
2. For mask shape, select rectangular.

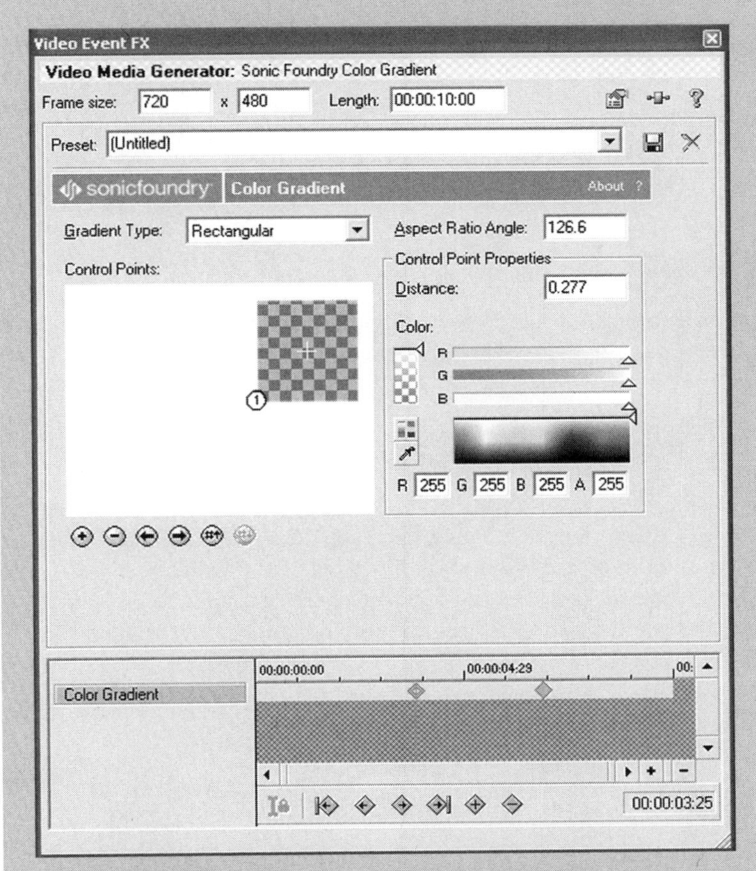

8.63 Masks can be any shape at all.

3. In the Gradient Preview window, slide the color placement control points so that they are lying over top of each other with very little overlap.

This process takes some attention, because if the control points are not placed correctly, the mask will reverse itself and change from an outer mask to an inner mask. This element can also be used as a valuable tool. A color with reduced opacity can also be added to the mask's outer area, acting as an overlay, if desired. This type of mask is similar to the way the Cookie Cutter filter functions; however this method has greater flexibility in some instances.

Practice using the black-to-white gradient as a mask in projects, as it is one of the most powerful masking tools Vegas offers.

Open the file called `jellyfishswim.veg` from the CD in this book. It is a fairly complex project and contains many attributes found in Vegas 4.0.

Let's look at each track individually.

- **Track 1**—contains generated media and height map and is at 20 percent. The generated media acts as a guide to shift pixels generated on track 2.
- **Track 2**—contains generated media from the new Noise Texture generator. The Noise Generator is shifting in bias, causing it to create ripples on the "water."
- **Track 3**—is a gradient mask over track 4. It's composited in an Add mode, which brightens the image and adds contrast to allow the jellyfish to stand out over the other elements.
- **Track 4**—is another Noise Generated event, with color shift and pan/crop to create motion in the water.
- **Track 5**—is a simple blue generated media overlay at 50 percent opacity.
- **Track 6**—is another gradient, in Subtract compositing mode, acting as a mask for track 7. This feature helps bring out the colors in the jellyfish by masking brighter color values. With some work, a similar effect can be had with a secondary color corrector.
- **Track 7**—is a Noise Generated event, shifting across the Frequency, Noise, Offset, and Amplitude settings. Masked by the gradient on track 6, this color sets the shimmer over the jellyfish.
- **Track 8**—is the still image of the jellyfish, using pan/crop to move the fish, while two instances of the Spherize plug-in create the motion in the jellyfish. There is also a minor adjustment to the color of the still, reducing the exposure of the flash in the original shot. It also contains a height map to give the jellyfish texture and shine and is mapped to the event on track 9.
- **Track 9**—is a Noise Generated event, creating a texture of movement under the water, beneath the jellyfish.

Underneath all these layers is a motion blur added only to the jellyfish and Noise Generated layer found on track 9. Motion blur is applied at a value of 15 percent.

With this project open, change the compositing modes, opacities, and generated media gradients or colors. See how these changes affect the jellyfish appearance. This demonstration provides a visual overview of the function of each aspect of the tools.

Bump Map

This mode uses the texture of the parent image to create a map of light and dark information. This tool is very handy for creating 3-D titles, creating textures over an event, and mapping a texture to an underlying event.

Lighting angles are definable with the Lighting Type menu. Use the Intensity slider to increase or decrease the presence of a light. The X, Y, and Z parameters can be controlled by inputting values to the input field or by moving the target dot in the Bump Placement window. Reset the X-Y parameter by double-clicking the yellow dot or by right-clicking in the Bump Placement window and selecting Reset.

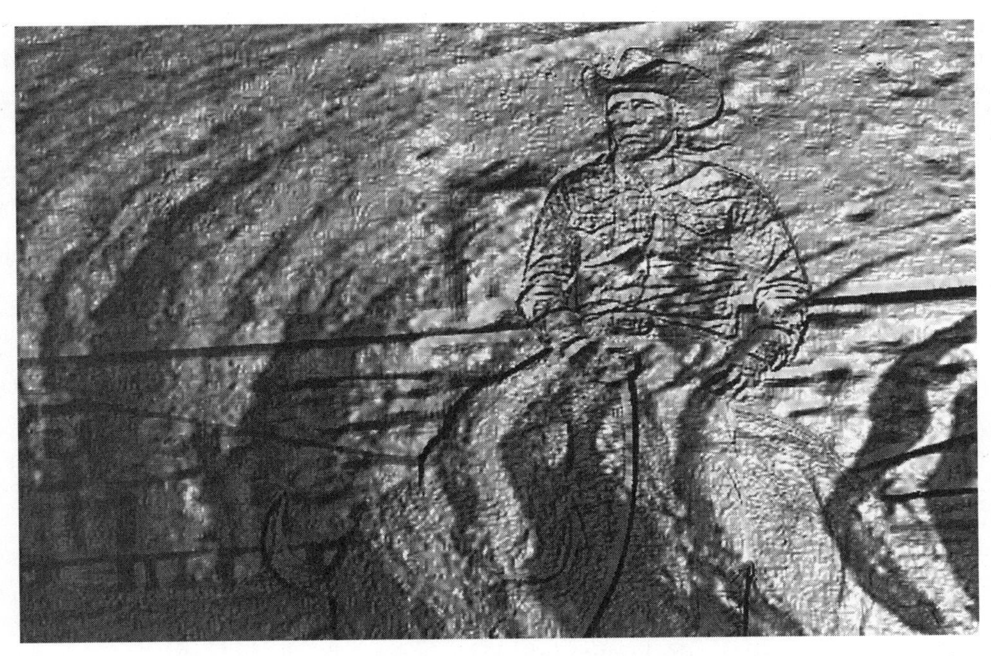

8.64 In this image, the stone creates a texture, mapped to the outlines of the cowboy, creating a hand-drawn effect on the image.

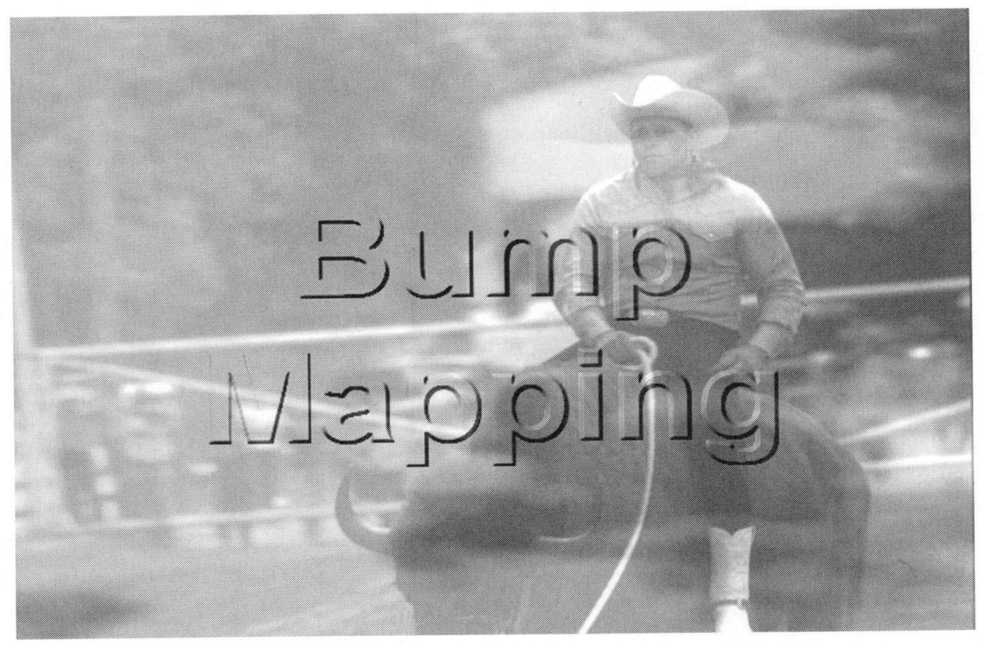

8.65 In this image, a simple text event takes on 3-D attributes with definable lighting, depth, and bump height.

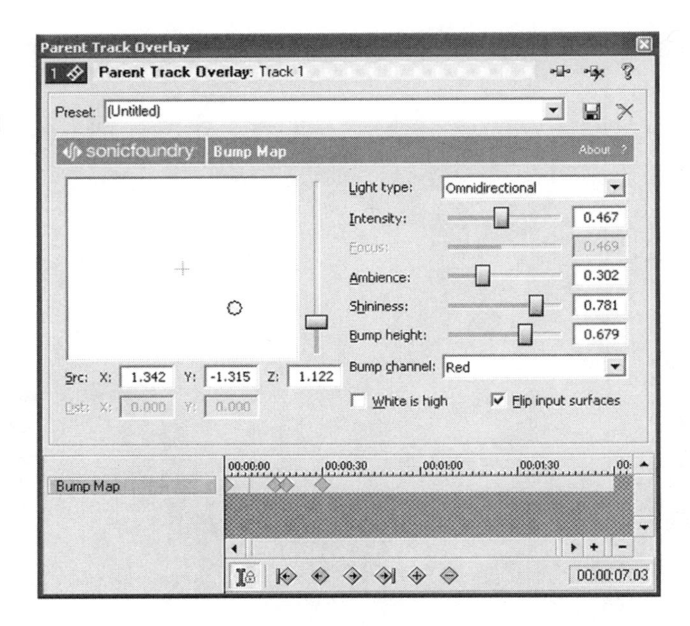

8.66 Controls found in the Bump Map dialog box control light, focus, depth of mapping, and light placement.

The Z slider (vertical) controls the height of the light on the mapped image. When Spotlight is selected from the menu, the distances of X and Y are controllable, and the focus slider is enabled as well. Focus controls the tightness or breadth of the light image. Omni-directional and Directional menu options automatically disable the focus and X-Y distance controls. The Destination of X-Y can be adjusted by moving the X in the Bump Placement window.

The Shininess slider controls the reflective surface of the map, and the Ambience slider controls the amount of light seen in the mapped image.

The White is high check box causes the bump map to view white areas as the highest areas of the bump texture. Leaving this box unchecked causes the black areas to define the high areas of the texture.

Finally, flipping the input surface reverses the two images, causing the upper and lower tracks to flip images.

Open the `bumpmap.veg` file found on the CD in this book for examples of the bump mapping shown in the images earlier.

Learning to work with these tools helps build titles, scene atmospheres, and magical shots quickly and powerfully. Compositing itself is an art form and requires a great deal of time and experimenting. Vegas 4.0 has some of the most powerful compositing tools in its class of editing systems and, over all, has more compositing tools than any NLE system that doubles as a finishing editor. On the CD in this book, several compositing projects are available, although not described in this chapter.

Using Chromakey Tools in Vegas 4.0

Chromakey, also referred to as *greenscreening* or *bluescreening*, is an art form that must first be practiced at the camera and shooting stage of a production. Even fairly weak footage, however, can be usable in Vegas 4.0 when filtered with the color correction tools and chromakey tools.

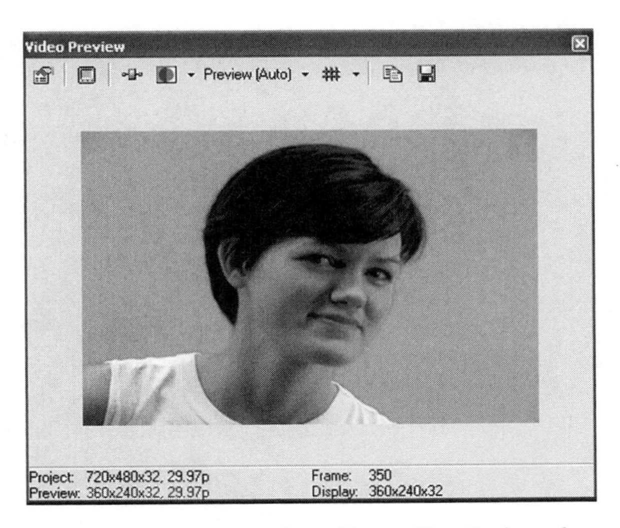

8.67 A clean greenscreen shot, with no spill on the face of subject.

Chroma (color) and key (mask) come together to create a mask from selected color(s). A quality mask or key can also be made from nearly any color or reasonable gradient of color. Selectable colors/chroma can be keyed or masked out, so that other events can show through those sections. Although industry tradition uses a screen of blue or green, any color can be keyed. Due to the luminance and chroma sensitivity of the bright green used in screens, however, in most situations, green is the best choice for DV.

(For more information on the relationship between chromakey and DV, visit `http://www.sundancemediagroup.com/help/kb/` and read the tutorial on 4:1:1 DV.)

Shooting a clean screen is critical to having a relatively smooth experience with keying. Many tutorials are available on how to shoot a clean screen, and the Sundance Media Group V.A.S.S.T. training course teaches this as well (`www.sundancemediagroup.com/aboutvasst.htm`). Some basics to know are presented here to assist the inexperienced cameraperson/editor in creating a clean shot to be keyed in the editing stage.

Use Rosco green paint or a commercially produced greenscreen. We recommend the Photoflex dual-color screens that have flex hoops in them for easy storage. When shooting on DV, green is the only safe choice. Blue can be used in analog shooting.

Good lighting is critical. To shoot a clean screen requires at least three lights: a key light for the talent and two lights for the screen itself. A reflector for the key light to light the opposite side of the subject is valuable too, as are flags (masks) to keep the key light from spilling onto the screen.

Shoot the subject at a minimum of six feet from the screen. This practice is necessary to prevent green reflection from spilling onto the subject. This kind of reflective spill is very difficult to remove in editing, regardless of the application used for creating a key or matte.

Vegas 4.0 has excellent chromakey tools that allow for rapid and clean keying of colors. A successful key, however, is also dependent on practice and experience on the part of the user.

From the CD, open the `greenscreen.avi` file on the timeline in Vegas 4.0. Select the Event FX on the file and open the Sonic Foundry Chromakeyer tool.

In the Chromakey dialog box, uncheck the check box in the upper-left corner. The Chromakey dialog defaults to blue, and unchecking this box allows Vegas to sample an accurate replacement color. Select the Eyedropper tool, found in the left center of the Chromakey dialog. With the Eyedropper enabled, click the color to be removed from the event, in the Preview window.

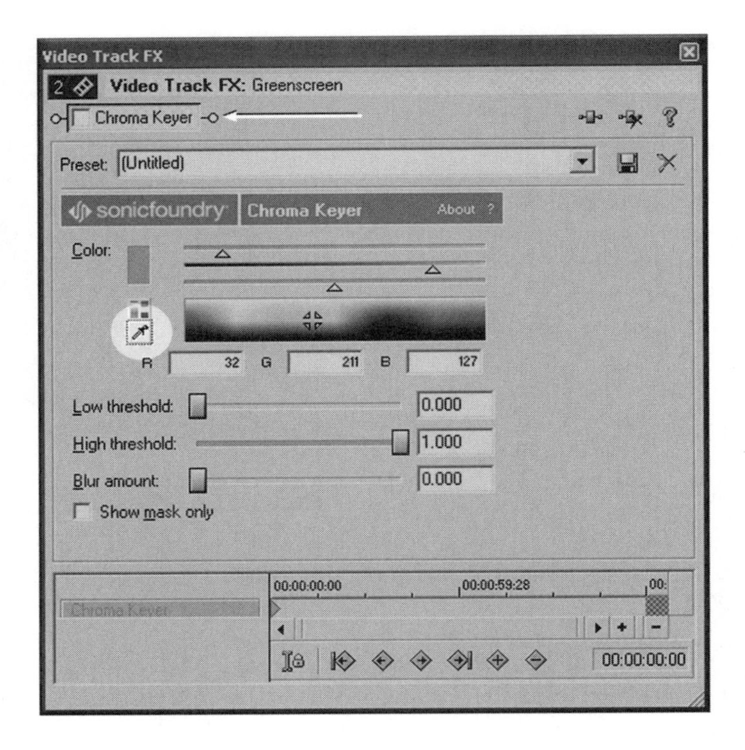

8.68 Chromakey tool at default setting. Uncheck the Chromakeyer box in the upper-left corner.

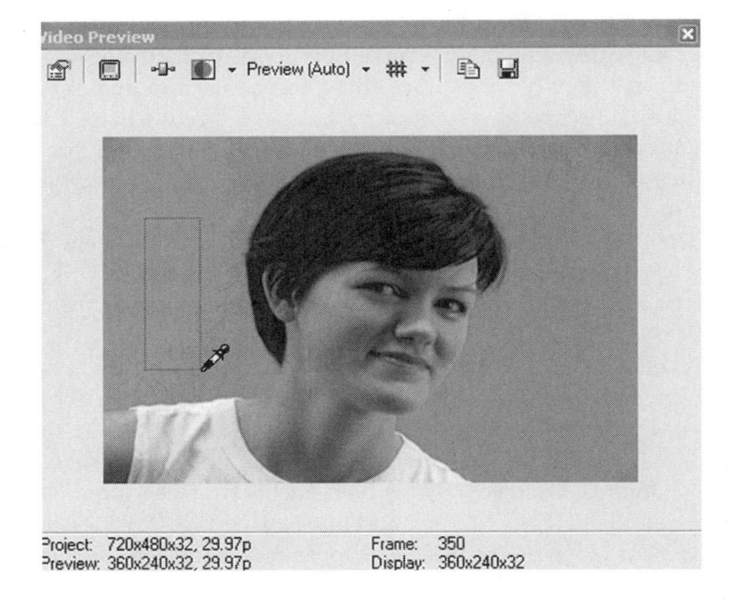

8.69 Draw or click in the Preview window to select color for removal.

Enable the Chromakey tool by placing a checkmark in the check box in the upper-left of the Chromakey dialog. A portion of the selected color will immediately change in the Preview window. Now the Chromakey tool must be adjusted to create a clean key.

Start by placing the event to be seen beneath the keyed event. This process helps adjust the key for best matting. Generated media is great for this, especially generated media that is a similar color to the replacement event. Of course, the replacement event can also be used. If it's a fairly busy event with lots of detail, however, it is sometimes difficult to create clean lines, as the detail in the background event hides lines during the key process.

Tip

Viewing the key process on an external monitor is helpful to ensure a clean key process. During the color selection, however, the preview must be on the computer screen. After key color is selected, video can be sent to an external NTSC or PAL monitor.

After an event is placed beneath the event being keyed, slide the Low Threshold slider to the right and slide the High Threshold slider to the left. On the `greenscreen.avi` file, a clean key will be achieved with the Low Threshold slider at 0.400 and the High Threshold at 0.800. The Blur setting is dependent on the background media used. In the instance of the greenscreen.veg file found on the CD, the Blur setting is at 0.015 to soften the harsh transition between the background color and the edges of the key. Some keyed events require a fair amount of blur while others can require no blur whatsoever.

Creating a Garbage Matte in Vegas 4.0

A *garbage matte* or mask is a simple mask, created from an event or image in Vegas, which removes one element from other elements on the screen. This option assists in the keying process, particularly where more than one color must be keyed out. The earlier project does not require a garbage matte; however, many instances exist in which a garbage matte is beneficial.

To create a garbage matte, click the Show Mask Only box in the Chromakey tool. This step displays the areas being masked and quickly shows a clean or unclean key.

A popular effect is to create an image that is entirely black and white, except one or two colors. This is part of a *color pass*, which allows a color to pass through an image that is otherwise black and white.

To create this effect quickly, place the events on the timeline to be color passed. Duplicate the track so that two identical tracks exist. On the upper track, insert a Chromakey and Black and White filter in order.

Using the eyedropper in the Chromakey tool, select the color to be passed, or allowed to be shown. Use the View Mask Only check box to see what areas will be knocked out in black and white. Adjust the Black and White filter to the desired level of black and white blend. For maximum effect, the lower track can have a slight blend of black and white with the color. This option reduces the amount of contrast between the color and black and white areas if this is the desired effect.

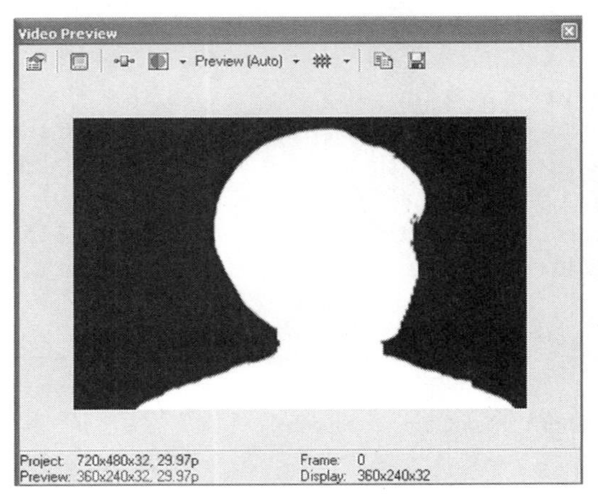

8.70 A clean and simple garbage matte/mask.

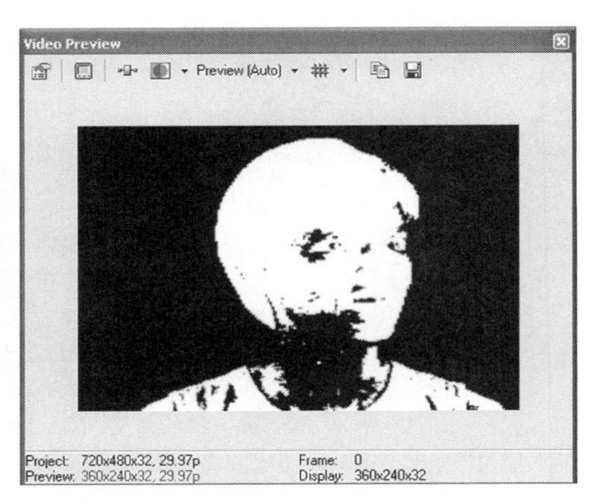

8.71 An improper mask/matte.

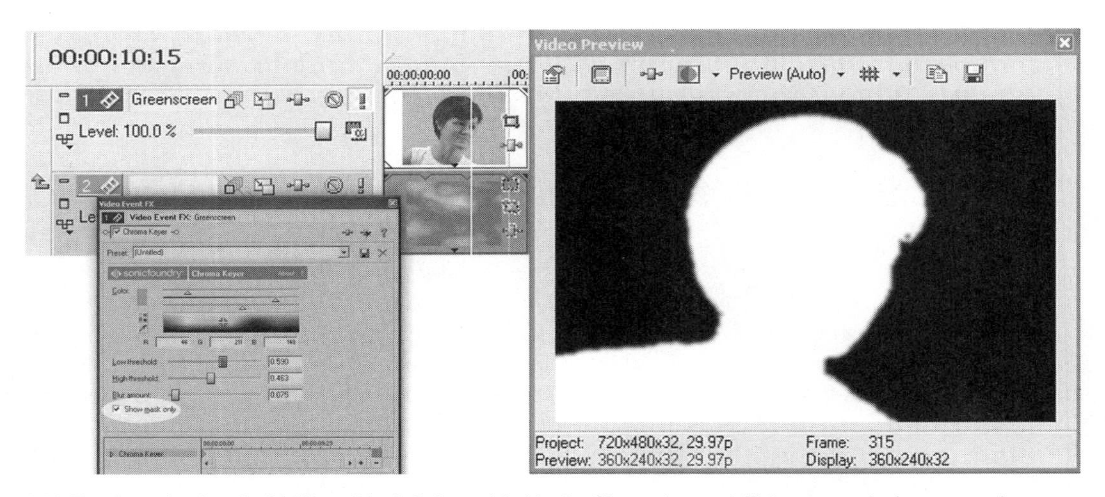

8.72 Track 1 soloed and with Show Mask Only enabled in the Chromakey tool. This process is done to render a matte/mask to a new track.

This matte/mask can be rendered to a new track, which is then parented to the original event, saving processor time and allowing multiple masks to take place on the timeline/project. To render the matte to a new track, solo the track containing the mask. Double-click the event to be rendered to a new track, creating a selection. Select TOOLS | RENDER TO NEW TRACK. In the dialog box, select Custom and select the Audio tab.

After the new track is rendered, it should be parented to the original track. Click the Event FX button on the original track and disable the Chromakey tool by un-checking the check box in the upper left of the Chromakey tool. Parent the mask event to the original event by clicking the Parent/Child button to the left of the original track.

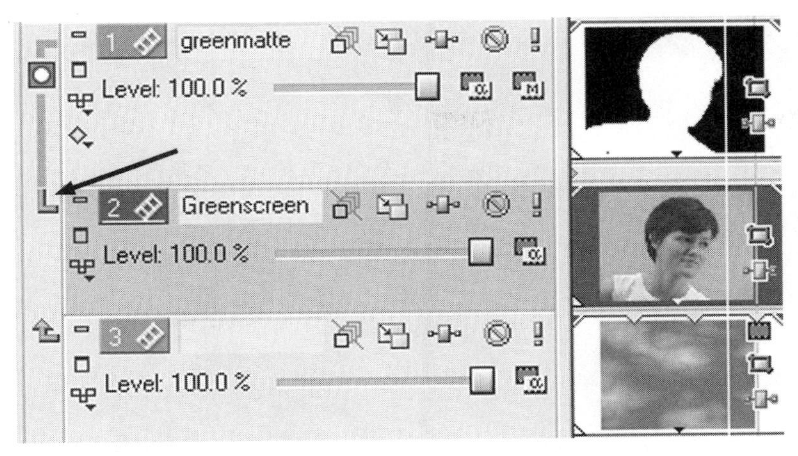

8.73 Parent track 1 to track 2 by using the Parent/Child button on track 2.

8.74 A clean composite.

The two events over the third track, which contains the background information, create a clean composite over the background image if all parameters are adjusted correctly, as illustrated in Figure 8.74.

8.75 Identifying problem areas in a screen isn't always easy or obvious.

The original image can be color corrected, blurred, or otherwise filtered for creative presentation if desired.

Figure 8.75 shows three problem areas in the screen.

1. The corners of the screen are visible. These areas will not mask cleanly.
2. Lens flare/reflection shows across screen. This will not mask cleanly.
3. There is spill getting past a flag, allowing backlight spill on the screen and creating two colors of green. This will not mask cleanly.

These problems can be overcome, but the time involved, coupled with the loss of resolution, quickly demonstrate the necessity and time savings of having media shot correctly in the production stage.

To see how to correct the poorly shot footage, open the `badscreen.veg` file on the CD in this book.

A clean mask has been rendered to track 1. This garbage matte is required to create a clean composite. Track 2 is parented to the mask/garbage matte, which is similar to the way the first greenscreen project was done. To be able to create a clean mask, however, a secondary color corrector and black restore were used to cause the mask to render cleanly. The event is also panned/cropped to remove the corners from the scene and to create a clean masking area. Still the original event requires color correction to blend into the background with a believable presentation. The original settings for the color correction and chromakey have been left in place on the original event in region one. Click the Event FX to enable these two FX, allowing you to see how the mask was created. Color correction has been added to the event as well. Disable this by unchecking the check box in the upper-left corner of the FX dialog. A very small amount of blur has been added to the mask to smooth over the transition from the masked area to the background area.

These are two examples of how a chromakey can be used to repair a background image or to insert new media. There are other instances where a screen is valuable however.

Open the `tvscreen.veg` file on the CD in this book.

The `tvscreen.veg` file contains a still image: `tvshot.png`. This image is an old photograph of a child watching television. The television screen has been replaced with green, allowing for a motion picture to be placed in the television screen. This method can also be used to replace screens in television studio-type shots or military establishment-type shots to create a sense of realism. There are more examples of this sort of masking on the `tvscreen.veg` file. Play with them and notice how the Track Motion tool is used to fill in the masked sections.

Never be afraid to experiment using various FX, track Motion, Pan/Crop, and other tools on composited media. Turn lots of knobs, slide the sliders, and check the check boxes. Various combinations turn up lots of artistic possibilities, and, as mentioned at the front of this chapter, compositing is an art form. Be creative!

NEW! Motion Blur

Vegas 4.0 has a feature called Motion Blur that allows an envelope to be applied controlling the duration and shape of the blur applied to the motion of video. What makes motion blur unique is that it applies blur at the frame level, creating a blur over time (temporal) as the video shows movement. This option makes zooms, pans, and motion in the image blur smoothly. A *standard blur* applies a blur in 2-D space, or a spatial blur. A *motion blur* acts more like a wide-open shutter, allowing for a greater exposure time.

Applying the envelope is performed by selecting **VIEW | VIDEO BUS TRACKS** (**SHIFT+CTRL+B**) and then right-clicking the video control pane and inserting a Motion Blur envelope. Double-clicking the envelope and adding handles/nodes controls the amount of blur.

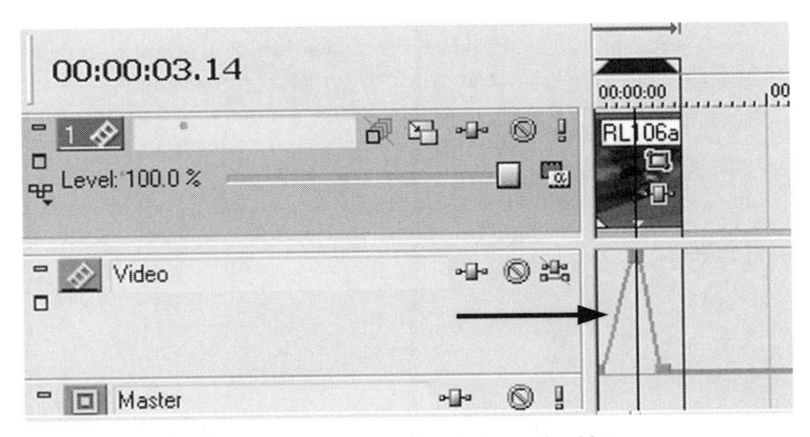

8.76 Using a video bus track to create and control a motion blur.

8.77 Motion blurred at full value with the Motion Blur tool.

Applying this filter to sequences of still images, fast or slow motion video, and animations can add a sense of realism to the media. It can also be inserted with great intensity to generate a dream-like quality.

One really great effect to try with motion blur is to render an event with lots of motion to a new track and to add a motion blur to suit the eye. In that new track/event, right-click and select Undersample. Undersample the event by at least .750, which will create a unique and stuttered look with blur shifting. The most unusual effect comes from using extremely low undersampling/ frame rate in which the frames skip enough to cut the blurred frames to the point that they are almost unrelated in motion.

Another terrific use of this tool is to blur a camera pan when transitioning from one scene to another. A motion blur, coupled with a velocity filter and transition, is a good combination for the popular effect of stuttered motion or stopped-motion blurring during a pan into another scene or shot sequence. Motion blur does add considerable time to a rendered section, so make sure the envelope is completely off when not being used.

Every track will be blurred when this envelope is up, unless the Bypass Motion Blur button is chosen. This option allows you to selectively blur tracks. The button is next to the Scribble Strip on each video track in the Track Control pane. Leaving it enabled causes the video track to be blurred.

Super Sampling

Super sampling resamples each pixel of video information in Vegas 4.0 and significantly improves the smoothness of the pixel edges and matches to adjoining pixels. This feature is primarily valuable when used in conjunction with the Motion Blur envelope but is also beneficial when used in upsampling smaller resolutions to DV or higher levels. Super sampling will not be as noticeable with video containing high motion or with video that is not being changed in output size or resolution. Use super sampling in connection with Motion Blur, upsampling small resolution video, or when working with extremely low resolution stills.

The way this works is that Vegas creates interpretive frames based on the difference between the project frame rate and the frame rate of the media, or the computer-generated imagery. This process also creates smoother flow and edges for generated behaviors, such as Pan/Crop, Track Motion, Transitions, and other new media created in the timeline during the editing stages.

Insert a Super Sampling envelope in the same manner as inserting a Motion Blur envelope by selecting VIEW | VIDEO BUS TRACKS (SHIFT+CTRL+B) and inserting a Super Sampling envelope. Use the envelope to control how many interpolative frames are created. For instance, by using an envelope setting of four, four times as many frames are rendered as existed in the original event. This means that render times are significantly slowed, so don't plan on doing a super sample while you are answering a phone call or grabbing a quick bite to eat. Super sampling, however, is a fantastic tool when shifting resolution, frame rate, or both.

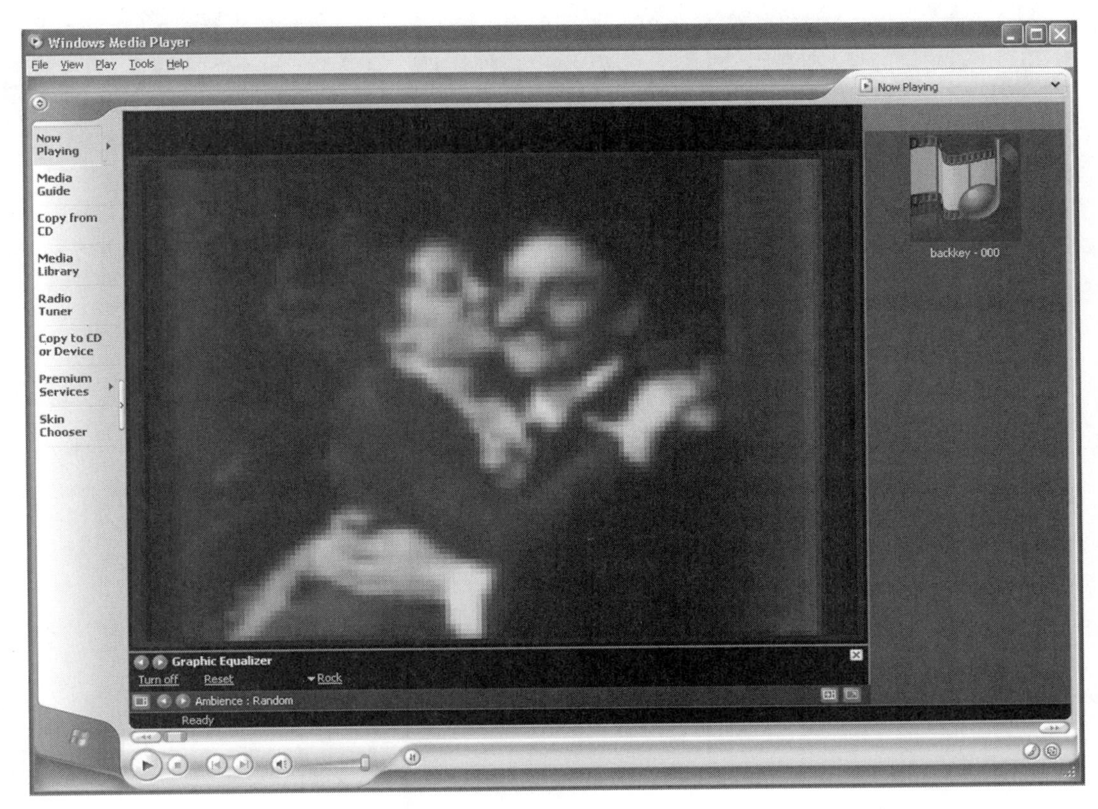

8.78 The original image from Thomas Edison's 1900 film, *The Kiss.*

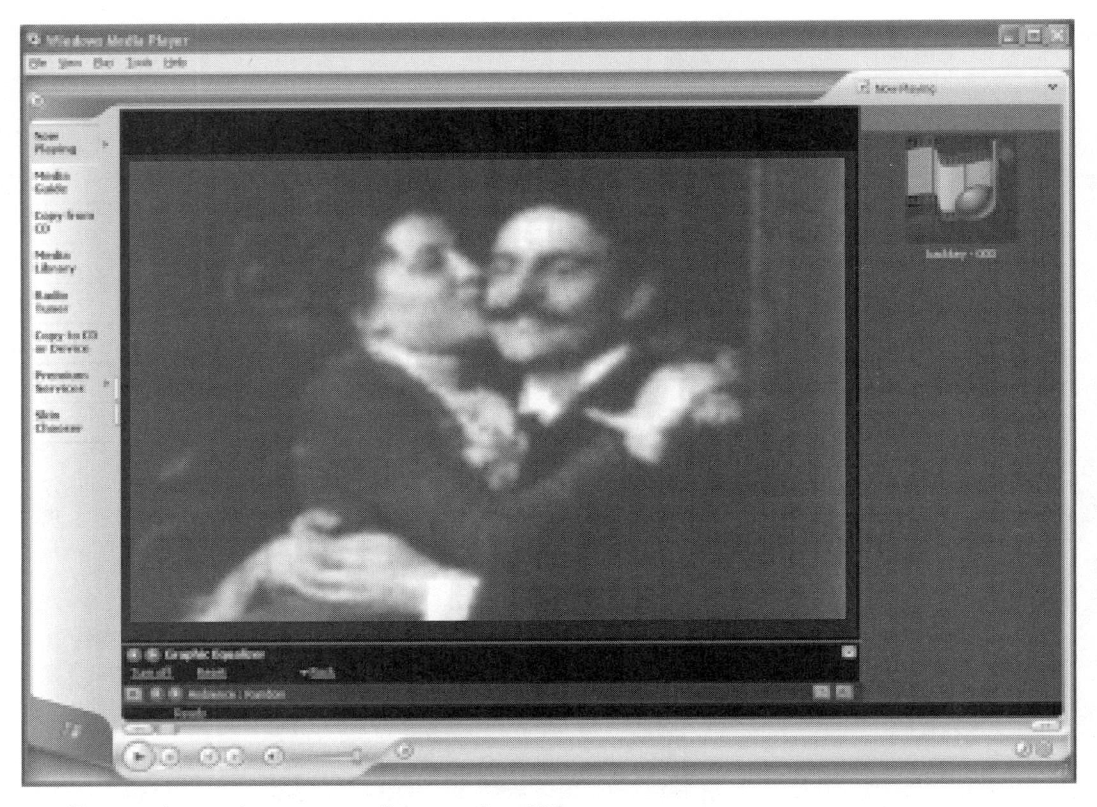

8.79 The same image after super sampling at a value of 2 frames.

Color Correction and Manipulation

Understanding the Color Correction Tools in Vegas 4.0

Vegas 4.0 has many color correction tools that previous versions of Vegas did not have. A new color correction toolset, a secondary color correction toolset, waveform monitor (WFM), vectorscope, parade scope, histogram view, and optimized hue saturation luminosity (HSL), color curves, broadcast filter, and more make Vegas one of the more powerful color correction–capable NLEs available today. (See Figures 9.1, 9.2, 9.3, 9.4, and Color Plate 9.1.)

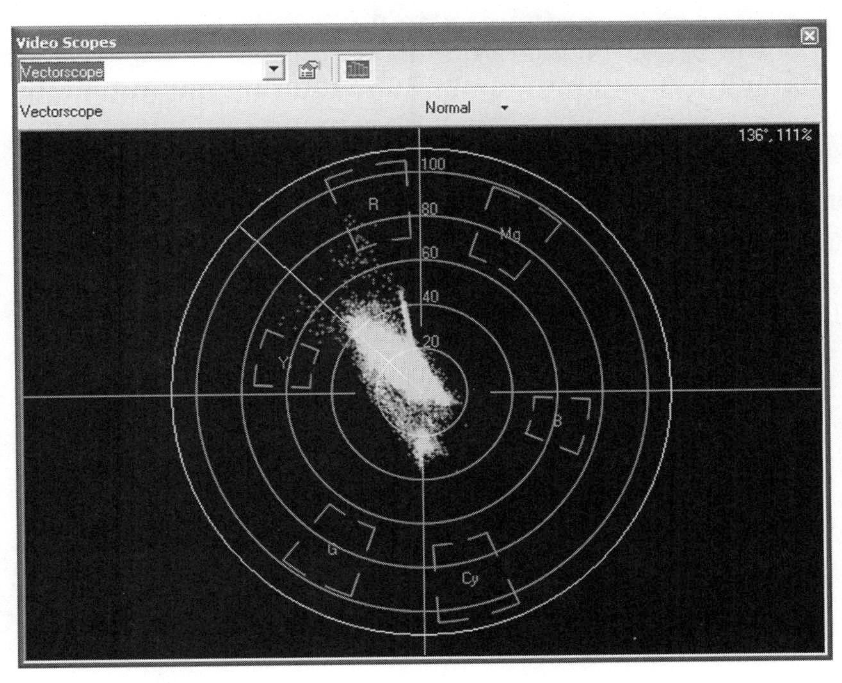

9.1 Vectorscope.

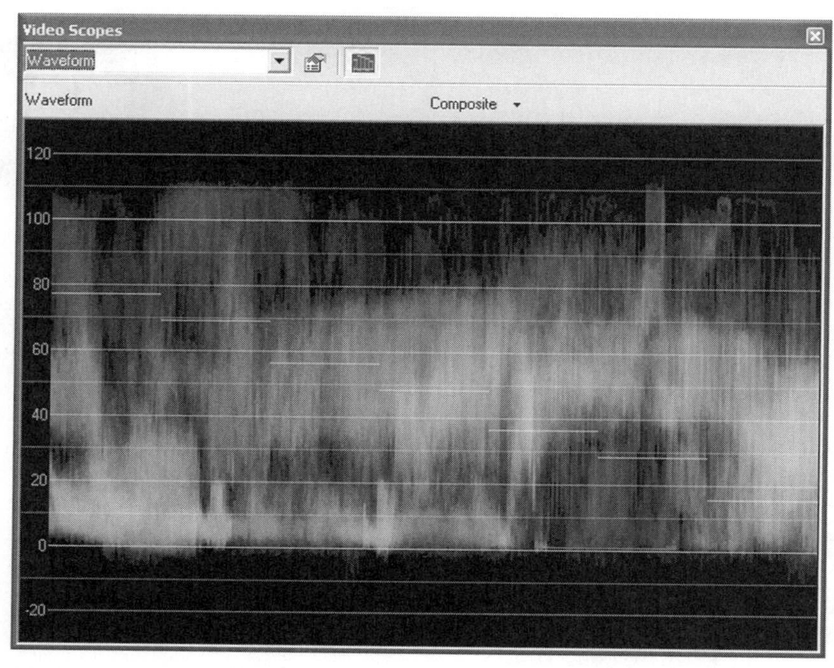

9.2 Waveform monitor.

Tools that can be used for color correction include:

- Black Restore
- Brightness and Contrast
- Broadcast Colors
- HSL plug-in
- Color Curves plug-in
- Color Balance
- Color Corrector
- Color Corrector—Secondary
- Gradient Mapping
- Pixelan Chromawarp (third party)

The new color correction tools found in Vegas 4.0 provide the ability to set apart colors or a range of colors and to command isolated colors in order to suit emotional, artistic, accurate, or whimsical expressions.

Also, tools are available to maintain color shift from event to event in Vegas, which assists in matching shots from two different cameras or from DV to other video sources. Finally, Vegas provides tools that can limit or force colors to remain legal or become broadcast legal in Vegas.

These scopes and monitors are all tools to help identify color values and luminance values and to assist in color matching. Just having them doesn't correct color; they are for reference. Although we'll briefly discuss how to use them, this chapter is not intended to teach you how to read them and gain an in-depth understanding of how they work.[1]

This chapter, however, provides a basic overview of how these monitoring and display tools function within Vegas and the editing world.

Broken down to the most simplistic form, a waveform monitor is for reading *luminance*, or brightness, and most other tools are for reading *chrominance*, or color. Keep this information in mind while working with the tools. In the majority of color correction use, both the WFM and vectorscope are open at the same time, which is why menu choices for both are in the Video Scopes menu.

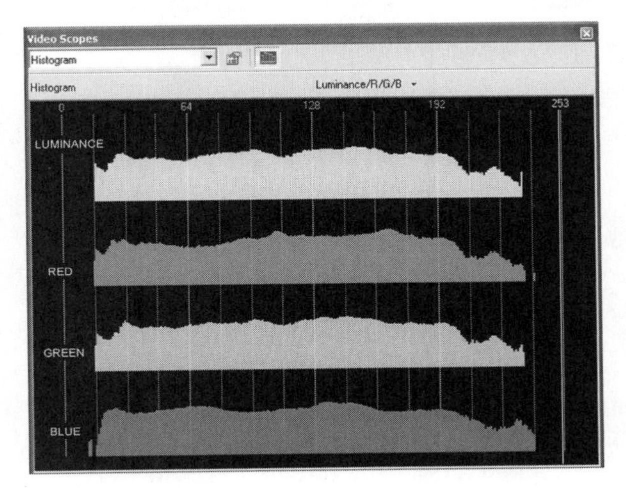

9.3 Histogram.

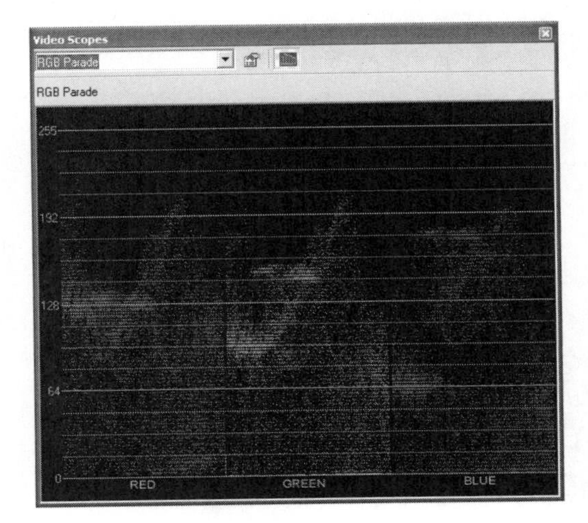

9.4 RGB Parade display.

Displays and Scopes

The Histogram tool is one of the most valuable tools in Vegas. A histogram gives a good indication of color balances overall, as shown in Figure 9.5.

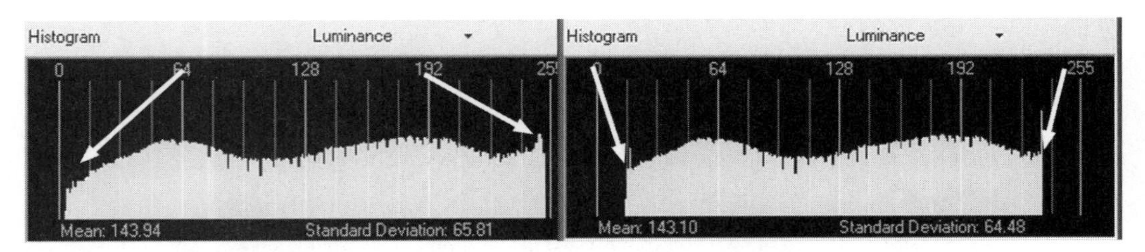

9.5 The histogram on the left contains illegal colors; however, the histogram on the right is legal, using the Broadcast Colors filter.

A histogram also provides a very fast way to view illegal colors. Illegal colors are usually created by inserting text, adding color-altering filters, or using improper color correction. Use the Broadcast Colors filter to bring colors into legal status.

The parade display in Vegas 4.0 shows which colors are within individual limits and also demonstrates the level of the total video output, as shown in Color Plate 9.2. This important tool can help you discover color distortion at output. A standard parade display shows level information on RGB, instead of showing the total level of the modulated video output.

Illegal colors are colors that are not compliant with NTSC standards. Illegal colors over-modulate the broadcast signal, causing distortion of video and audio. No one will put you in jail for using illegal colors, but the strain on the eyes and ears in most instances, will drive viewers crazy. If you've ever seen a low-budget car-lot advertisement on television with buzzing audio and distortion on the announcer's sibilants, chances are great that the commercial contains illegal colors, most likely the whites used in titles. (Distortion is actually across the entire audio signal but seems more apparent with an overmodulated color signal.)

Color value in the NTSC world is measured in IRE. Color value in the RGB world is measured in RGB component values. For example, in the IRE world, extreme black is 7.5 IRE. Extreme black in the IRE world, however, is really just a very dark gray, measuring at 16.5 in the RGB world. Extreme white in the IRE world is 100 IRE. Again, this is really just very bright gray, measuring at 234 in the RGB world. RGB extreme black is 0, while extreme white is 255. Any color value in the RGB world that goes below 16 or above 234 is illegal by NTSC broadcast standards. Any color value in the IRE scale that goes below 7.5 IRE or above 100 IRE is also illegal.

Legal colors are related to NTSC broadcasting. Because most video equipment in the NTSC realm is calibrated or manufactured to the same specification, however, practicing legal color on DVDs, video tapes, and hard drives is a good practice. The Internet currently has no color limits.

9.6 Shows the video event with illegal colors.

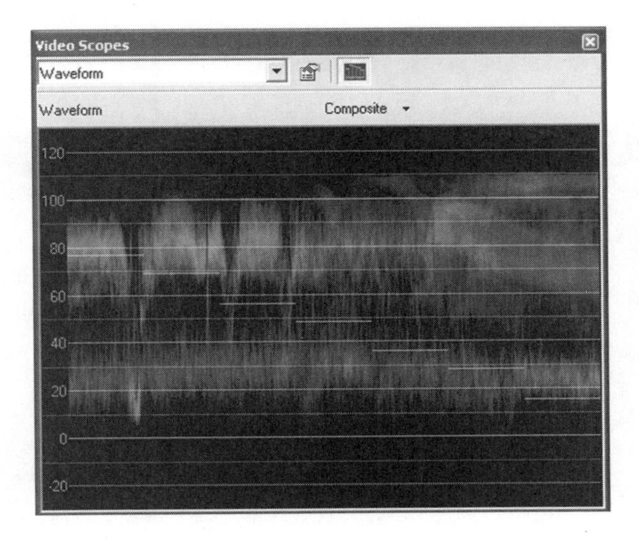

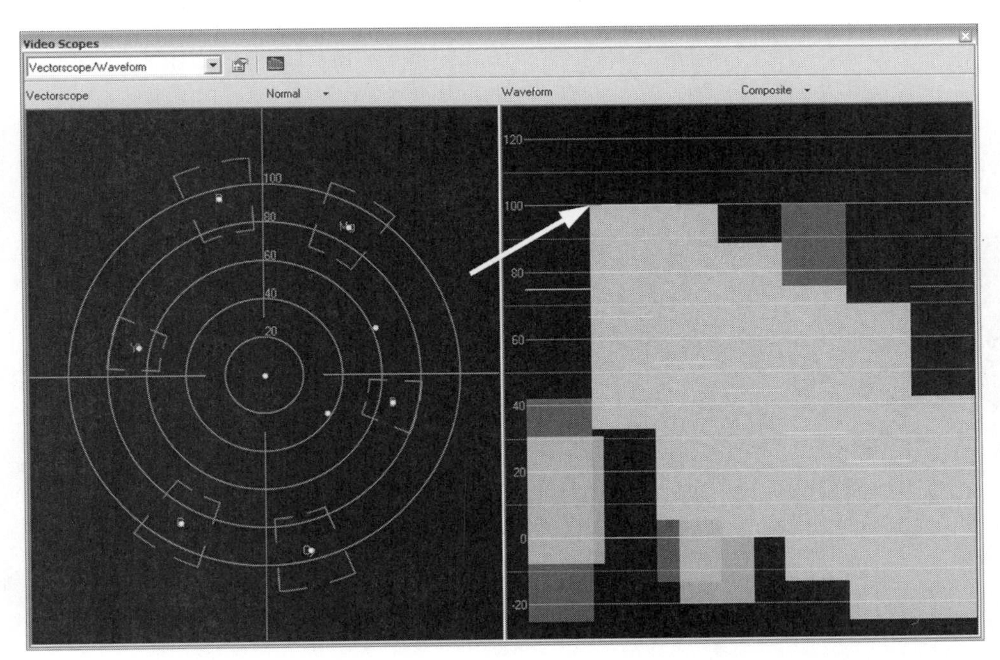

9.7 Standard SMPTE colorbars, inserted as generated media in Vegas 4.0. The white dots represent positions of the individual colors found in SMPTE colorbars (WFM in Composite mode, with indicators enlarged for illustration).

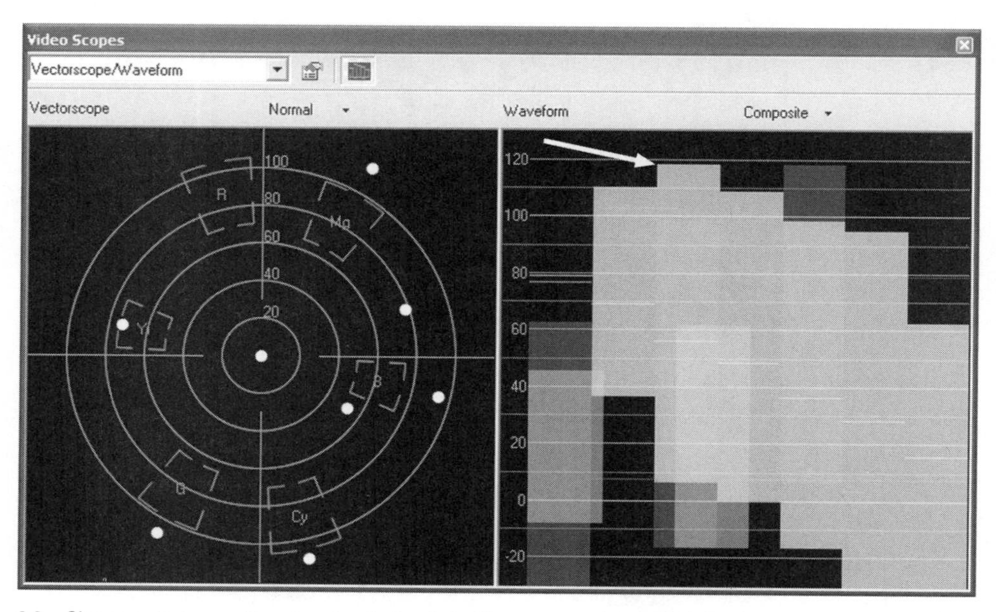

9.8 Shows colorbars with saturation in the Color Corrector tool at 1.500 (all other settings reset to null).

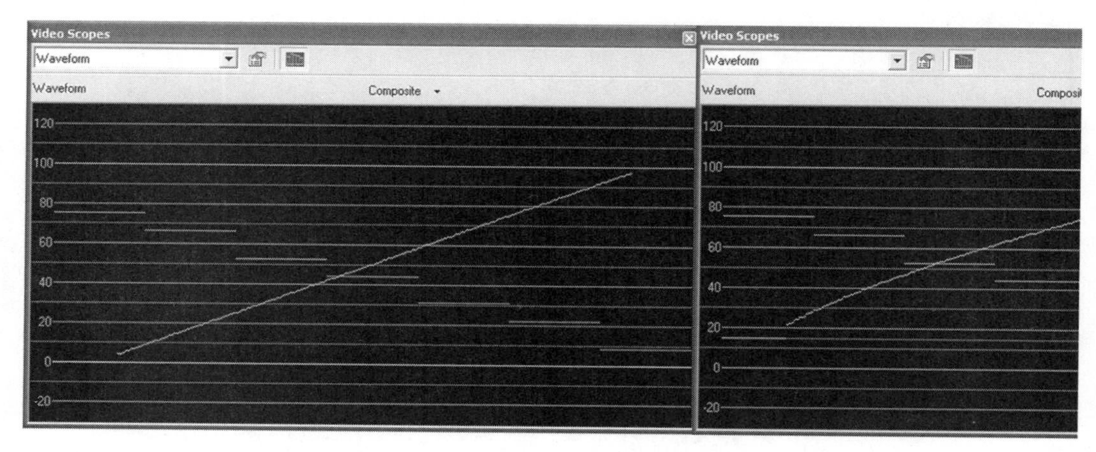

9.9 Shows ramp with gamma at 1.500. Notice that the curve of the waveform display has changed significantly in the right image (all other settings reset to null). The ramp is used to demonstrate gamma, as colorbars aren't a good choice to express gamma.

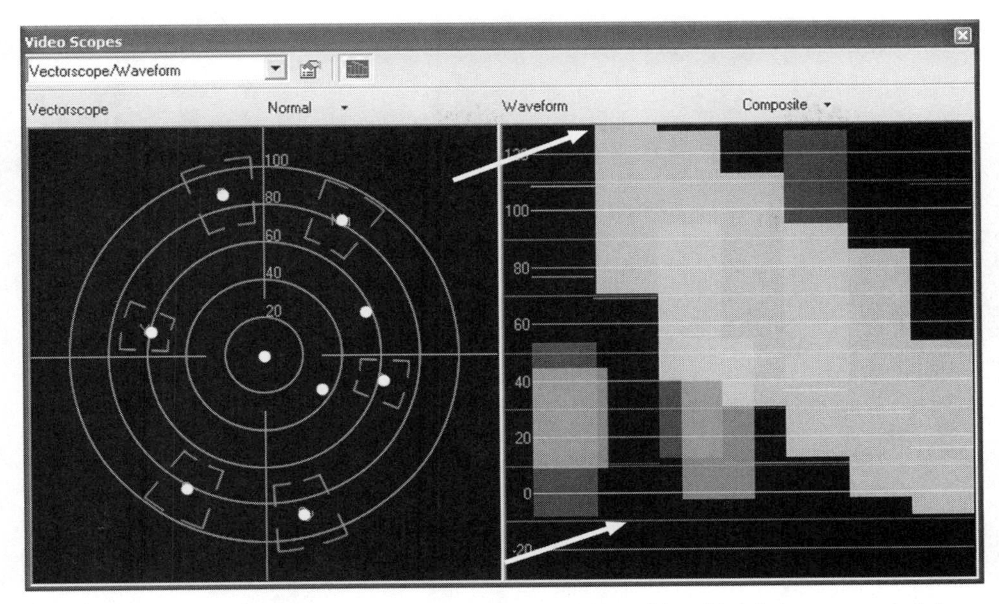

9.10 Shows colorbars with gain at 1.500. Notice that the WFM breadth and height have expanded significantly (all other settings reset to null).

The parade display can be updated by enabling the "Update scopes while playing" button. Place the cursor over the highest part of a color display in the parade scope to display the RGB value of that color point.

Figure 9.6 demonstrates the waveform view of the same video event, which is showing illegal colors. This instance shows how the waveform monitor can demonstrate illegal or clipping colors.

Figure 9.7 shows the same event with the Broadcast Colors clamp set to Lenient. This tool brings the colors to a legal status, but the blue channel is still clipped.

Working from an SMPTE colorbar pattern, examine how various changes affect the display of the WFM/vectorscope. (The waveform monitor displays IRE in percentages, not IRE levels. Use the Broadcast Colors clamp/plug-in if a 7.5 IRE pedestal is required.)

The images shown in Figures 9.7 through 9.11 are in `colorbars-scopes.veg` on the CD in this book.

These figures demonstrate how the sliders in the Primary Color Corrector tool affect the display of images and how they appear in the WFM/vectorscope. Knowing how each of these controls works is key to using the tools correctly and to achieving the desired image quality.

Pedestal, also called *setup* or *lift*, is a base line for indicating black. Without knowing exactly where black exists in relationship to other colors, it's difficult to process colors accurately. Each video display (histogram, WFM, vectorscope, parade) has a Settings button in the upper center. Select options from Setting to determine a Studio RGB setup (RGB-16 for black, 7.5 IRE for black, or none).

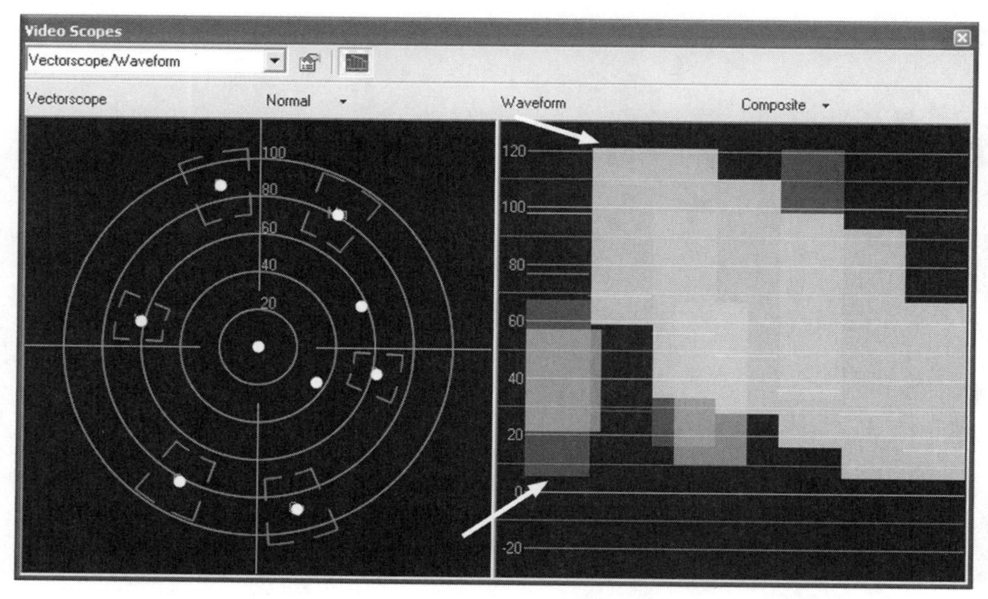

9.11 Shows colorbars with offset at 50.0. Notice the positioning of the colorbars (all other settings reset to null).

9.12 The new Color Corrector tool found in Vegas 4.0.

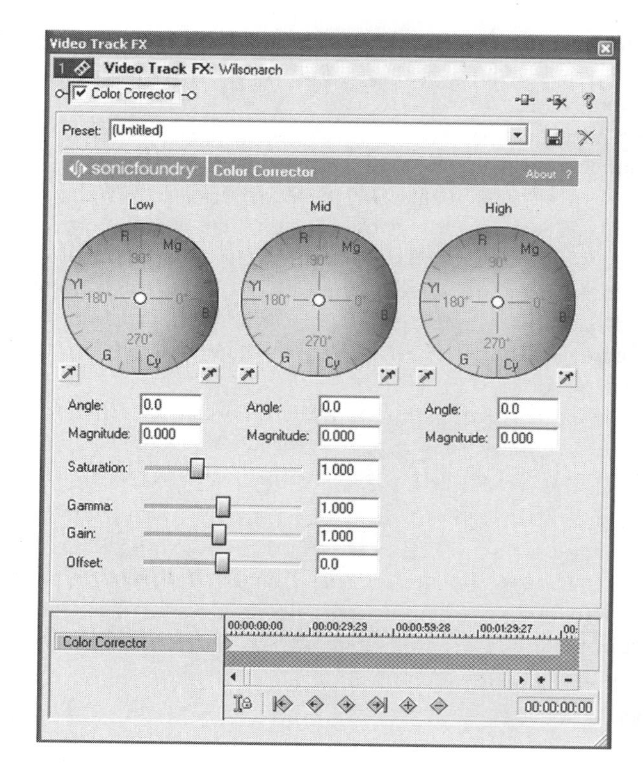

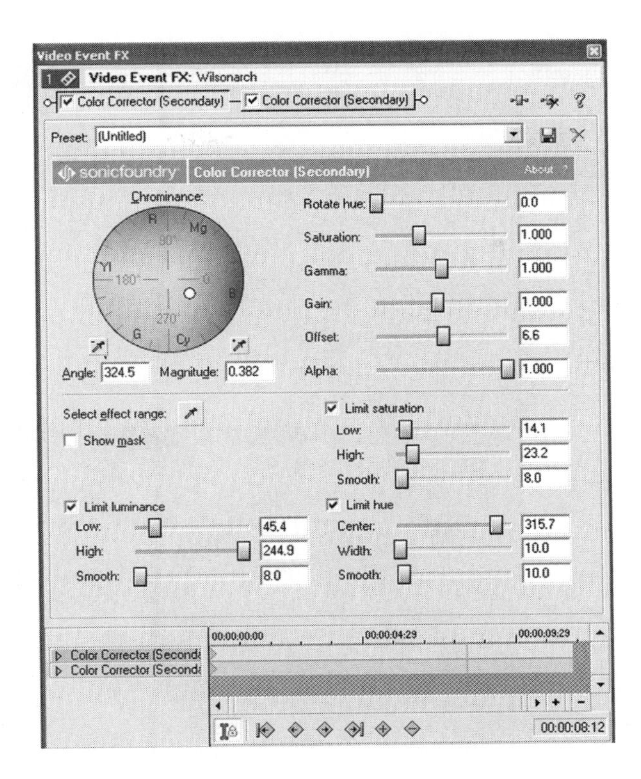

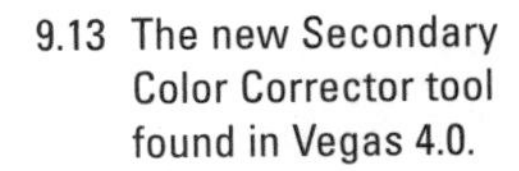
9.13 The new Secondary Color Corrector tool found in Vegas 4.0.

These new tools allow concise and accurate color selection in Vegas to be isolated and managed. Each of the color wheels is controllable with a mouse or a joystick, making it easier to fine-tune each color. The color wheels match the position of colors on the WFM.

The Primary Color Corrector tool works much like an equalizer in an audio system. The three wheels control lows (blacks), mids (largest range of color information), and highs (whites). Thinking in this mode may help some editors work with the colors more comprehensively.

Shadows and contrasts fall into the category of black. Facial colors, clothing, paints, and predominantly everything focused on in a frame, all fall into the midrange color category. Whites are the sparkle and the high end of various hues of color, providing detail in most instances. Dulling the high end, as in audio, generally results in less clarity of a picture image. Sometimes illegal colors, however, function much like distortion and need to be reduced to bring colors to a more reasonable and appreciable level. Depending on how the image was captured/acquired and with what, resolution levels might not allow colors to be intensified without seriously degrading the image.

All plug-ins in Vegas are now dockable with scroll bars. This feature means that the workspace is less cluttered, yet all controls are accessible from within the docking space (see Color Plate 9.3). This process makes color correction more efficient when working on a single screen, such as a laptop, as all correction parameters are accessible without having the screen covered with the correction tools.

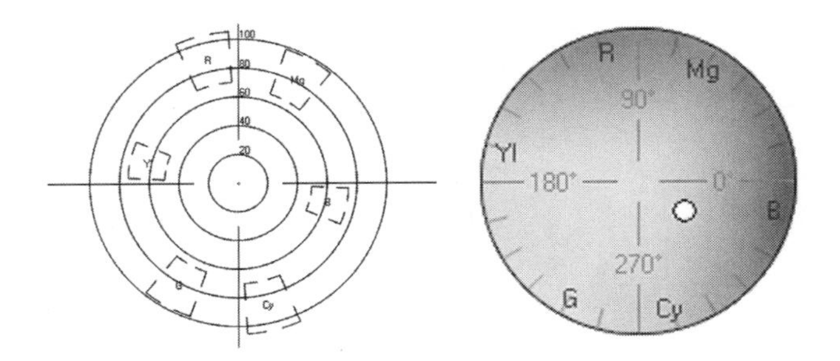

9.14 Color wheel and vectorscope color distribution is the same.

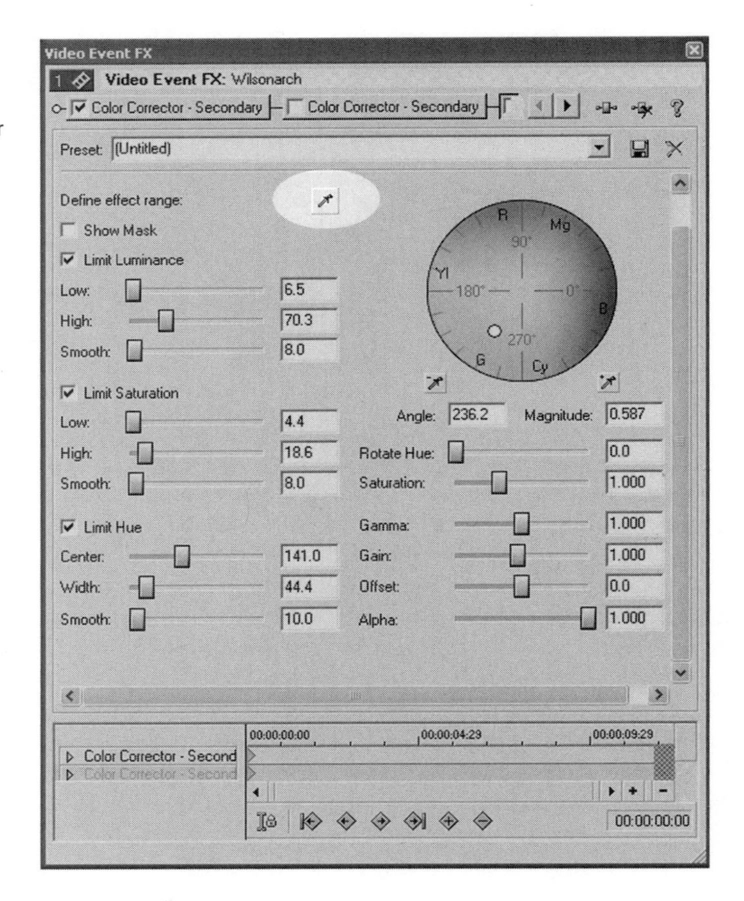

9.15 The Define Effect Range is enabled by using the Eyedropper tool to select the color range to be affected.

It should also be understood that when working with most of the color correction tools, manipulating images in the midrange will usually have some impact on the blacks/lows, the whites/highs, or both.

We'll use the color image in Color Plate 9.4 as a color correction starting point. This image is flat and dull. In reality, however, we know that the rocks of Moab should be red and vibrant, and the sky should be deep blue, with white clouds. A brightness/contrast filter could be applied to the image to bring up the brightness; however, that process would then brighten the shadow and landscape areas as well, causing detail to disappear. The overall image would appear washed out and soft in appearance. Using the Secondary Color Corrector, colors can be selectively adjusted.

Open the `rocks.veg` file found on the CD in this book.

Four views of the same event are provided in the `rocks.veg` file found on the CD that accompanies this book. Version one has no color correction whatsoever. Version two has red added to create a richer and deeper color. Version three has the red added in version two, and the sky has been brought to a deeper color for greater depth. Finally, version four has green added to the trees to set them more in the foreground of the shot. In Color Plate 9.5, the left half of the screen is the original shot, and the right half is the color corrected shot.

Let's look at how this was done.

Apply a secondary color corrector by selecting the Event FX button on the event, which opens up the Secondary Color Corrector dialog. First, select the color intended to be isolated and to be enhanced or edited. Use the Eyedropper tool found in the Secondary Color Corrector tool to select the rocks and define the effect range.

After the color range is selected, move the color wheel target around in the color wheel. This step causes colors in the Preview window to change with the rotation, demonstrating how the targeted colorspace is affected. Open the scopes view (Ctrl+Alt+2) and look at how the image shows up in the vectorscope.

With the orange of the rocks selected as the target color, move the target dot to an angle of 117.5 and the magnitude of .480. You can either key the numbers in the dialog box or drag the target dot to that quadrant/location. Adjust the Saturation slider to 1.00. The rocks take on a reddish/orange hue. Adjust the saturation to the point that it is satisfying to your eye. These numbers are relative to a computer monitor and not previewed on an external NTSC monitor.

After the color of the rocks is brought to a point where you are satisfied, click the Event FX button on the same event and insert another Secondary Color Corrector tool by clicking the tool and the Add button.

Use the Eyedropper tool again to select the blue in the sky. Now set the angle to 324.5 and the magnitude to 0.382. Adjust saturation to 1.00. Adjust the Limit Saturation/Low setting to 20.0 and the Limit Saturation/Smooth setting to 50.0. Finish the correction on the blue/sky area by adjusting the Offset setting to 6.5. Now the rocks and the sky should be a pleasing color and depth. Again, keep in mind that various monitors display the images differently, so check this image on an external monitor or adjust to your eye and satisfaction.

Notice that after the red and blue have been corrected, the trees in the foreground are darker and inconsistent in color with the rest of the image. Another Secondary Color Corrector tool is needed, so click the Event FX again, select Secondary Color Corrector tool, and click Add to insert it. Using the Eyedropper tool, select the green trees and bushes in the foreground. Set the angle to 211.0 and the magnitude to 0.485. Set the Saturation level to 1.070. In the Limit Luminance, set the High level to 76.0. Notice how the color of the green pops out to match the color levels of the rocks and sky.

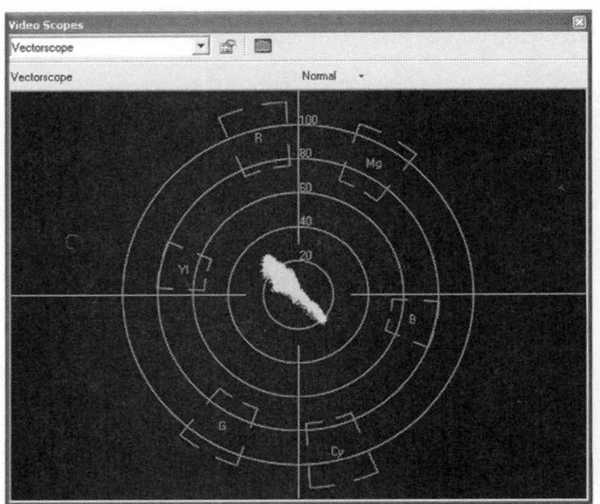

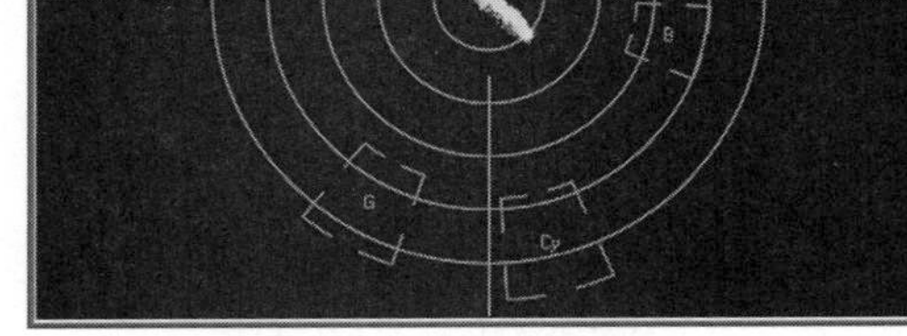

9.16 Original color in vectorscope.

9.17 Color display for corrected version. Notice the extended blue, red, and yellow/green display.

On the timeline is a still frame event that must be color-matched as closely as possible to the video event preceding it. This process is difficult to do because of differences in resolution, location, and time of day. A reasonable compromise, however, can be made between the two images, making a transition between the two less objectionable. Secondary Color Corrector tools and Color Curves tools are used to bring the two events closer together in color value. The vectorscope is used to match the colors more closely together in terms of relevance. The rocks and sky are the two common subjects in the two events, so we'll use those to make our corrections and changes (see Color Plate 9.6).

Finding a middle ground for the two images, particularly when the original video image is washed out (see Color Plate 9.7) and has been corrected to bring out more color, means that the still image event must be washed out and colors reduced to match the video event. Using the vectorscope, we can see a footprint of the two images. Because little can be done to realistically bring the video image up any further in resolution quality and color depth, the still image event must be corrected and brought to resemble the footprint displayed by the video event.

The Primary and Secondary Color Corrector tools are exceptionally powerful. However, don't forget the other correction tools. Color Balance, Color Curves, HSL, and Saturation Adjust are all tools that are equally powerful. Color Plate 9.8 has a Secondary Color Corrector tool, Gaussian Blur, Color Curves, and Saturation Adjust on the image to bring it as close as possible to the video event transitioning into it. Select the Event FX button on the still image event to see how the image is brought to corrected color values.

Open `wedding.veg` from the CD found in this book.

In this wedding scene, we can see how the opening shot is lacking contrast, depth, and personality (see Color Plate 9.9). It's also fairly dark, which would make most of us reach for the brightness/contrast tool.

Using the Primary Color Corrector tool, highlights and color can be brought out while maintaining warmth and personality. In this example, the face tones, the green in the trees, and the red in the bunting can be brought out, which will make for a warm and interesting image (see Color Plate 9.10).

When matching two events on a timeline, one technique is to do a screen capture of one of the two events, preferably the correctly colored event. Move the cursor to the second event so you can see it in the Preview window. Select SPLIT SCREEN | CLIPBOARD and draw a rectangle around the area of the second event on the timeline. The screenshot will be shown next to the second event, making it easier to color match/correct the second event to the first event.

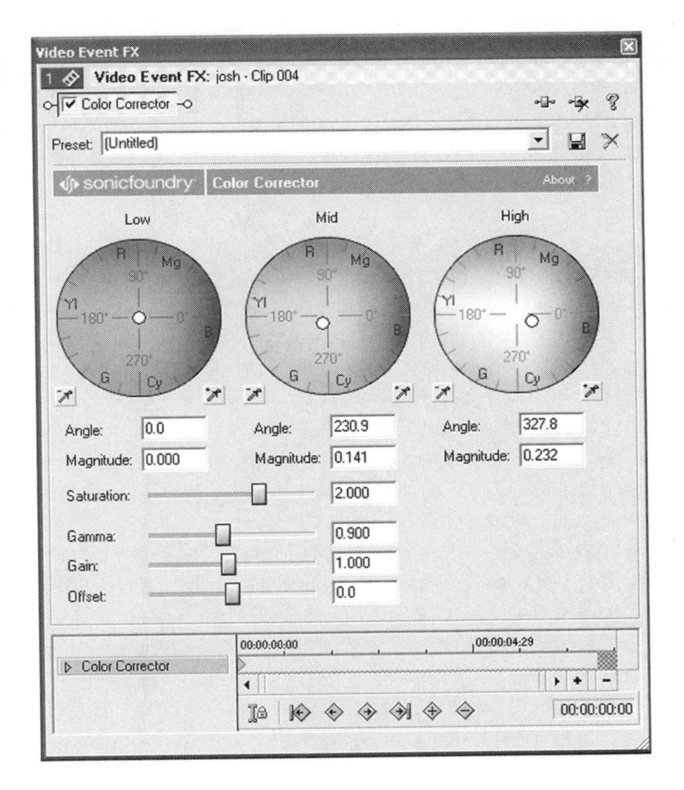

9.18 Use the Select Adjustment Color eyedropper to select colors to be restored/repaired.

On this same event, several individual Secondary Color Corrector tools can be used to bring out the same features while maintaining the mountain shadow in the background. This particular image has many approaches, depending on whether you are matching other imagery to the shot. In this case, many images needed the same level of color correction, so the Primary Color Corrector was applied to a single video track, affecting all events.

Another way the Primary Color Corrector tool is used is to sample low, mid, and highlight colors. Sometimes, this process is all that is needed to correct a color.

Select the Event FX in the first event on the `wedding.veg` timeline and insert a Color Corrector tool. Using the Eyedropper tool, choose Complementary Color from the Highlight tool and sample the color of the shirt collar from the man in the Preview window. Notice how this sets the neutral white for the entire shot. Using the Midrange Color Selector, use the Select Complementary Color eyedropper to sample the green in the evergreen bush. Sample a brighter portion of the green for best results. The color will immediately pop and balance the overall image. Because the colors are still not as saturated as they should be, increase saturation to a value of 2.00 and decrease gamma to a value of .900. These settings create a warm and pleasant color space without washing out the resolution of the shot (see Color Plate 9.11).

As you can see, the Primary Color Corrector tool is useful and powerful with the ability to repair or improve errors or difficult shot situations. For example, white balancing a camera is often overlooked, or in many cases, not possible with lower-quality DV cameras. Consequently, images might be too blue or green from fluorescent or mercury vapor lighting, yellow from tungsten lighting, or washed out from too much light.

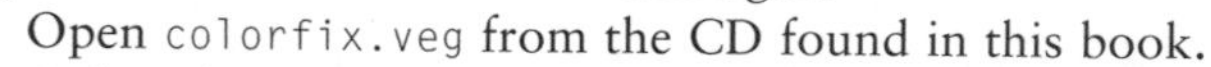

Open `colorfix.veg` from the CD found in this book.

Color Plate 9.12 shows how tools in Vegas can compensate for a forgotten white balance in a classroom containing fluorescent lighting. Skin tones, neutral whites, and color depth are all regained by using this tool.

To understand how this process works, insert a Color Corrector tool on the first event on the `colorfix.veg` file.

Using the Select Complementary Color eyedropper in the high tones, draw a square on one of the ceiling tiles on the upper part of the image in the Preview screen. The angle/magnitude settings should automatically set to approximately 130/0.584. Using the Select Complementary Color eyedropper on the High color wheel, select the ceiling where it's slightly darker. A sample can also be taken from the lightest portion of skin. The angle/magnitude settings should automatically be approximately 130/0.450. The skin tones immediately become warm, the surrounding colors are no longer blue from the lack of white balance, and the image is much deeper and interesting to look at. This same method works equally well with footage shot under tungsten lights that is overly yellow in color. The Primary Color Corrector tool is powerful and may be turned to for quick color repair on almost any image.

Some images simply cannot be brought to a correctable state, however. Sometimes, poor quality shots must be used. In this case, adding noise, saturation, and generally unacceptable color management helps rather than trying to hide poor color balances in the image (see Color Plates 9.13 and 9.14).

Another use of the Color Corrector tool is to apply a color gradient over an image. In buildings with large walls or obstacles, it is sometimes difficult to get a good balance with lighting if it's a live shot or a run-and-gun. Using a gradient over the image creates a balance and offers unlimited possibilities (see Color Plate 9.15). Used like a Neutral Density filter on a camera, gradients can be used to restore or enhance images.

When working with the Color Corrector filter/tool, remember this order of use:

1. Select the brightest point of an image with the High Complementary tool.
2. Select skin tones or mid-range colors with the Mid Complementary tool.
3. Select the darkest point of an image with the Low Complementary tool.

By doing so, color balancing is almost automatic, white balancing a shot that was incorrect is virtually hassle-free, and a great starting point is established for correcting colors.

This element can be built with any kind of image—simply create a gradient, such as the one in the `gradient.veg` file.

In Color Plate 9.16, notice how the warmer look of the original image seems to be flat on the left side, while the bluer image created by the gradient on the right side appears to have more depth.

In addition to using a gradient to correct a color space, gradients can be used to enhance an image overall. Warm images generally seem flatter, shallower than cooler, more contrasted images. Our eyes tend to believe that warmer colors have less detail.

Sonic Foundry has a Gradient Filter plug-in that provides exceptionally fast and simple changes to the emotion of an image (see Color Plate 9.17).

Using Color Correction as FX

Color correction tools can also be used for artistic expression, creating certain moods and emotions that cannot be obtained with lighting or camera lens filters in a low- to mid-budget project, or in some cases, cannot be obtained in the real world at all.

Inserting or replacing colors that aren't part of the original shot is fast and easy with Vegas. Using the Secondary Color Corrector tool, colors can be selectively replaced, graded, or enhanced.

In Color Plate 9.18, the original black-and-white photo is given a sense of warmth from the Secondary Color Corrector tool. A similar effect is possible with a Sepia filter or by layering a sepia-generated color on a track above and reducing the opacity. When you use the Secondary Color Corrector tool, however, you can insert keyframes, changing the transparency and chroma value of the overlaid color.

Color Plate 9.19 shows how colors may be desaturated using the Secondary Color Corrector tool. This particular image has the "Desaturate all but red" preset selected. We'll see how this same effect can be managed with another tool later in this chapter.

Color Plate 9.20 demonstrates a shot taken with a low-end camera, which did not allow compensation for a bright sky. Using the Secondary Color Corrector tool, the sky was brought to a richer and more vibrant color. Without using a glass filter on the camera lens, this shot would have been difficult for even the most high-end camera.

As an example, the movie *Pleasantville*, along with various commercials on television, have made the all-black-and-white-except-one-color look very popular. This type of work is nearly impossible to create with only a camera without doing lots of preproduction work. In Vegas, it's a few steps to create this look cleanly and quickly.

You can use a Saturation Adjust filter in Vegas to single out a color quickly, while desaturating all remaining colors. Color Plate 9.23 demonstrates the Saturation Adjust filter, applied in order to remove the color of the rocks and surrounding color. Only the climber's clothing is in color. The Saturation Adjust filter works much like an audio equalizer, allowing colors to be selectively reduced or enhanced in order to create an artistically pleasing image. Of course, the Saturation Adjust filter may be applied to any image to bring down oversaturated colors, pop up undersaturated colors, or bring an overall balance to images on the screen (see Color Plate 9.24).

When using the Secondary Color Corrector tool in Vegas, use caution, as overextended colors can quickly become illegal, washed out, or destructive to detail in an image. Color correction is meant to be a corrective tool, not a fix-all for poor camera skills or poorly shot footage.

The wedding examples are just salvageable but can't really be brought to full life, as the video camera was not white balanced, the backlight was brighter than the foreground, and the entire scene was shot in the shadow of a building. Sometimes footage can only be saved, not enhanced. Plan shooting schedules accordingly and remember that footage shot well saves a tremendous amount of time in the postproduction process.

Making Digital Video Look Film-Like

Notice that this section heading says "film-like" and not "like film." Video cannot look just like film. It's a different medium. It can be made softer, with adjusted gamma, grain, and color saturation, but it still will not look the same as 16mm or 35mm film. I'll preface this section by saying that if you are interested only in shooting media that looks like it was shot on film, shoot on film. If you are interested in exploring how to make DV more palatable to the eye, read on.

Making DV look more film-like requires starting at the lens and shooting aspects of the production. Using nothing but prime lenses, using dolly shots rather than lens zooms, and shooting with lighting intended for film are all part of the process. Shooting through filters, such as the Tiffen Black Mist series filters, helps warm the image as well. Practice shooting in a progressive scan or frame-based mode rather than shooting interlaced images, and, if the camera is capable of multiple frame rates, shoot at 24p, shoot with a PAL 25 frames per second (fps) camera, or learn how to operate the camera properly in progressive scan mode. In progressive scan mode, the camera must be handled differently from when shooting in interlaced modes. Pans can easily become mush and blur in the hands of the inexperienced. Hand-held shots become a wash of colors in those same hands. Managing the camera correctly is half the battle in getting a good film-like appearance from the digital information.

Interlaced or Not?

One of the first exercises in the process of making DV look film-like is to deinterlace. We discussed interlacing in Chapter 2, but to revisit, interlacing is the process in which lines, known as fields, are drawn for every frame of video. NTSC-DV has a frame rate of 30 fps or 29.97 fps. PAL-DV has a frame rate of 25 fps. This information means that NTSC DV has 30 half frames of lower fields/lines and 30 half frames of upper fields/lines. PAL has 25 half frames of upper fields/lines and 25 half frames of lower fields/lines. These lines generally should be removed or blended to gain the smooth look of film. Video shot in progressive scan mode does not have these temporally offset fields. Be certain that when editing progressive scan footage in Vegas 4.0 the project properties are set for Progressive Scan. If the setting is not correct, Vegas might insert the fields in transitions or other generated media.

Removing interlacing can be done within the project itself, by setting Vegas to the project settings of progressive scan versus interlaced. Several ways to accomplish a properly deinterlaced image are available. The first and fastest way is to set the project properties to Progressive Scan. To do so, open the FILE | PROPERTIES | ADVANCED dialog, and in the Deinterlace menu, select Blend Fields. This method is fast and easy and assures continuity.

Another way, or "look," is to blend fields manually. This process creates a slightly softer image and may be preferable to your eye.

When manually deinterlacing, create a new project. Set the Project Properties to Progressive Scan, and insert a new video track (CTRL+SHIFT+Q). Place events on the timeline. To deinterlace

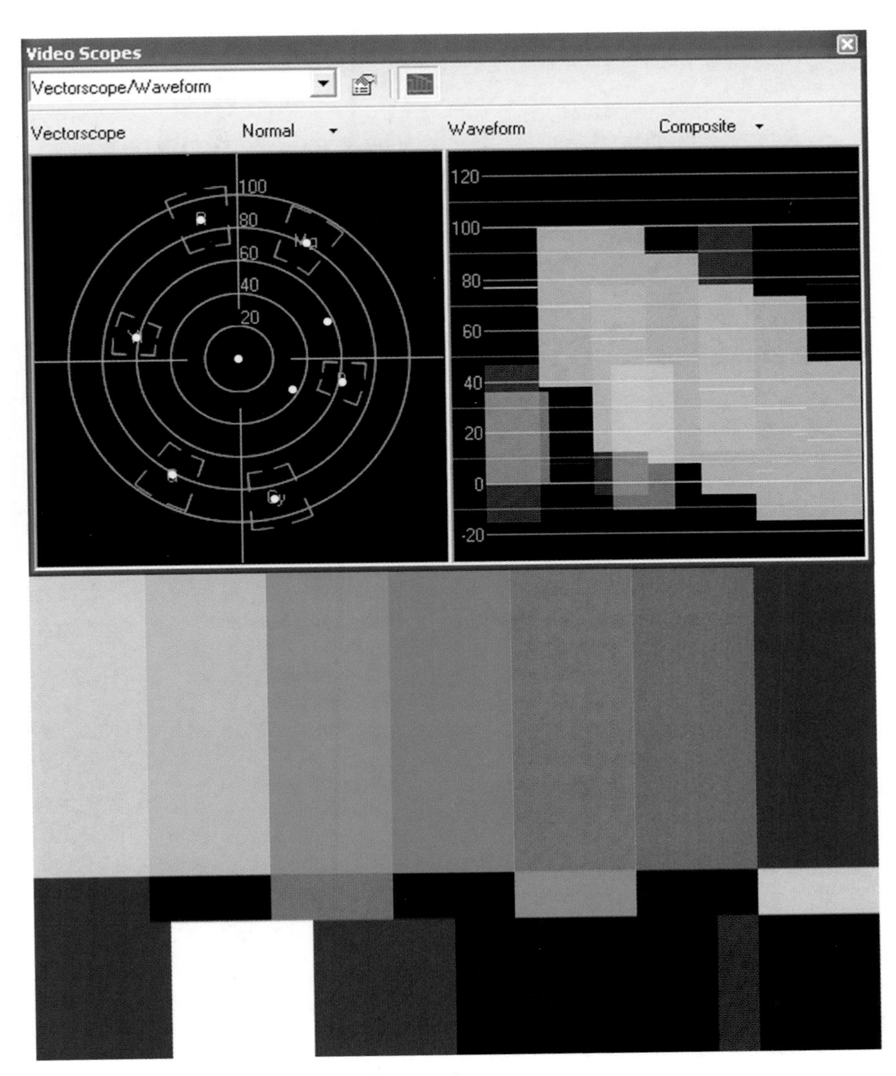

Color Plate 9.1 Vectorscope/waveform monitor.

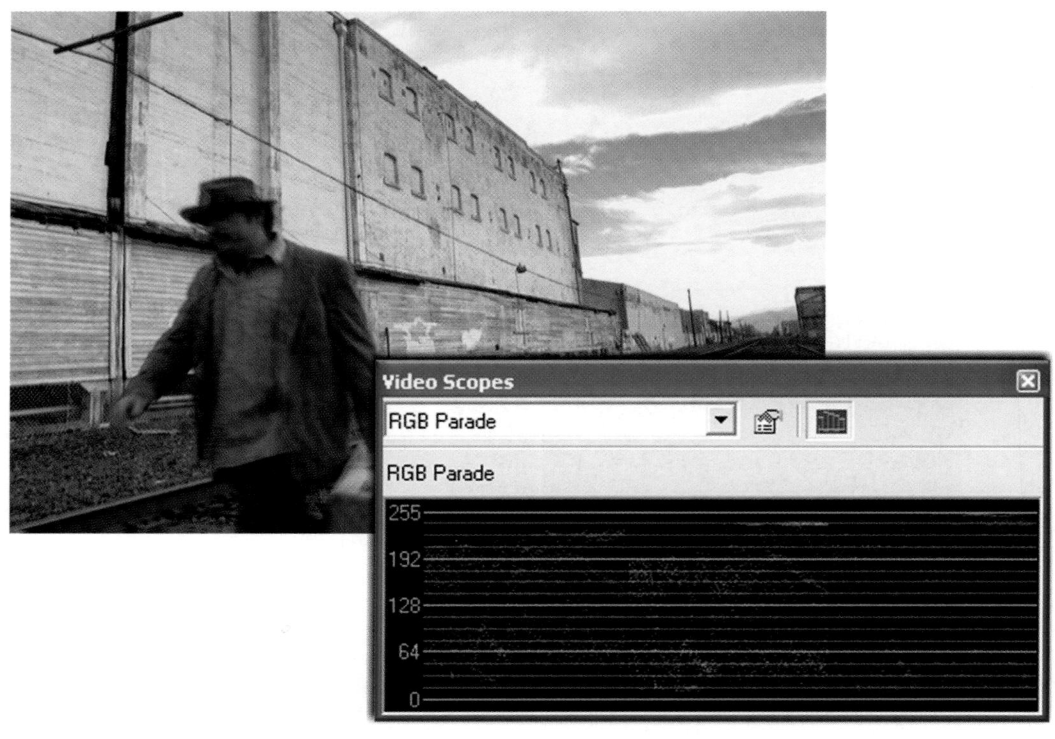

Color Plate 9.2 In this screenshot, the blues are clipping in the image, even though the eye doesn't see extreme blues. This result occurs because a blue gradient is being overlaid. The gradient is only five percent opaque, yet that is enough to clip the blue channel.

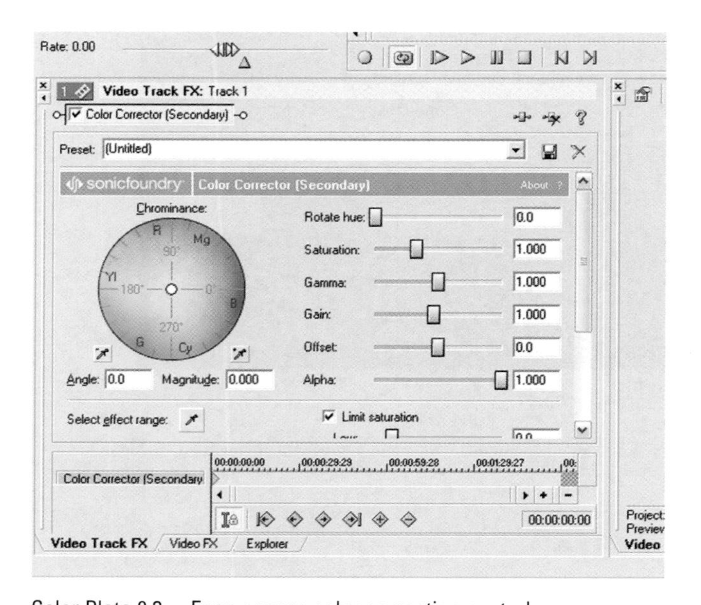

Color Plate 9.3 Easy-access color correction controls.

Color Plate 9.4 The original shot.

Color Plate 9.5 The image on the left is dull and uninteresting. The image on the right has depth and more interesting detail.

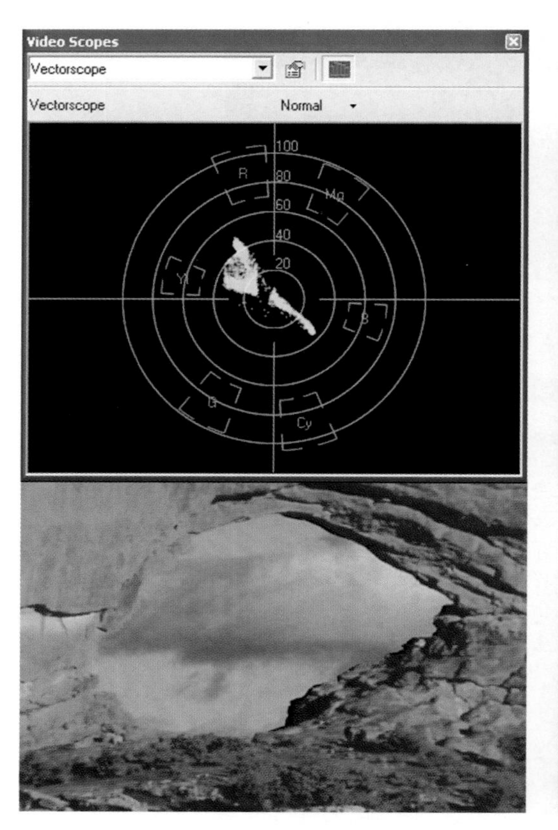

Color Plate 9.6 Video event on scope.

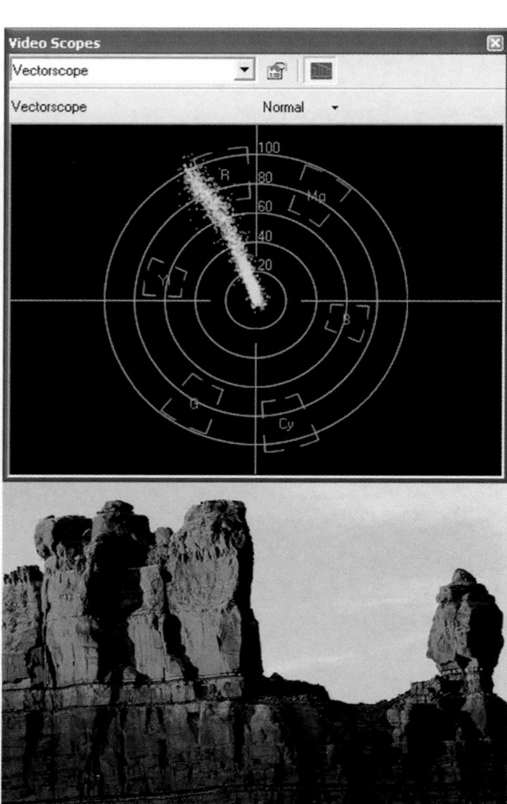

Color Plate 9.7 Still image event on scope before color correction.

Color Plate 9.8 Still image event on scope after color correction.

Color Plate 9.9 This image is dull and lifeless and lacks contrast.

Color Plate 9.10 Using a brightness/contrast filter, the image now has contrast and color but appears cold, harsh, and unfriendly.

Color Plate 9.11 The left half of image is improved, or corrected. The right side is the original image. The Split-Screen Preview tool is invaluable for comparing events while editing.

Color Plate 9.12 The original image coloring is on the left, and the corrected image is on the right.

Color Plate 9.13 The original image. White balance was not taken before the shoot, hence the white spotlight on performer's face on a dark stage. Overblown whites have destroyed any hope of detail.

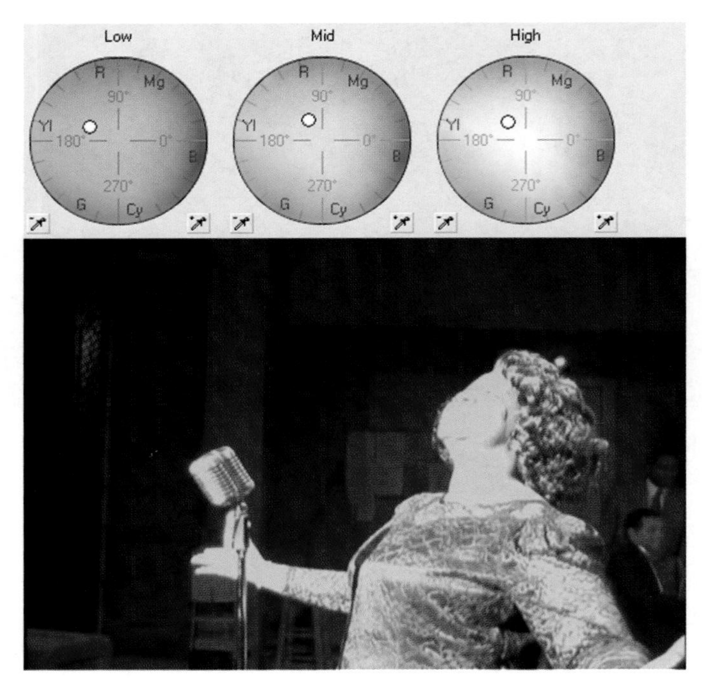

Color Plate 9.14 The image after correction. Rather than trying to restore detail that doesn't exist, bring up the surrounding colors to make the image less objectionable, more realistic, and potentially usable in the absence of more acceptable footage.

Color Plate 9.15 The left side is the original image, and the right side is the gradient-covered image. Notice the male dancer on the right image is clearer, and the colors from the right to left side of the screen are more balanced.

Color Plate 9.16 Gradient applied to the image using the Gradient Map plug-in.

Color Plate 9.17 Which contains more detail—the warmer images on the right or the cooler images on the left? The images are exactly the same. The image on the left has the Sonic Foundry Gradient plug-in with the Warm preset, and the image on the right has the Sonic Foundry Gradient plug-in with the Cooler preset.

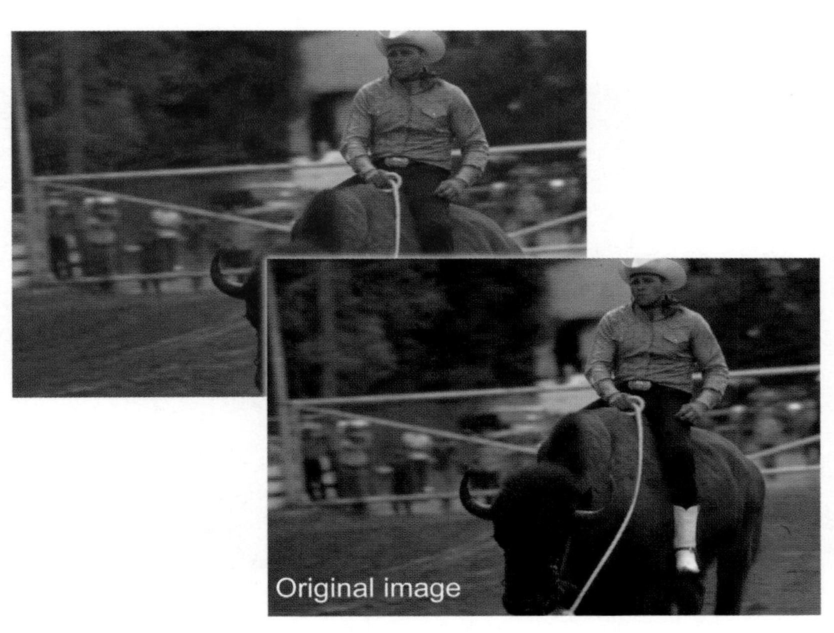

Color Plate 9.18 The original and color-enhanced image using the Secondary Color Corrector tool.

Color Plate 9.19 The left half is the processed image, and the right half is the original image. The Secondary Color Corrector tool removed all color except for the red in the jersey.

Color Plate 9.20 The processed image is on the left, and the original image is on the right.

Color Plate 9.21 The left side is deinterlaced using duplicate tracks.

Color Plate 9.22 Gamma reduced, saturation increased, dual tracks of interlaced video, top track at 50 percent opacity.

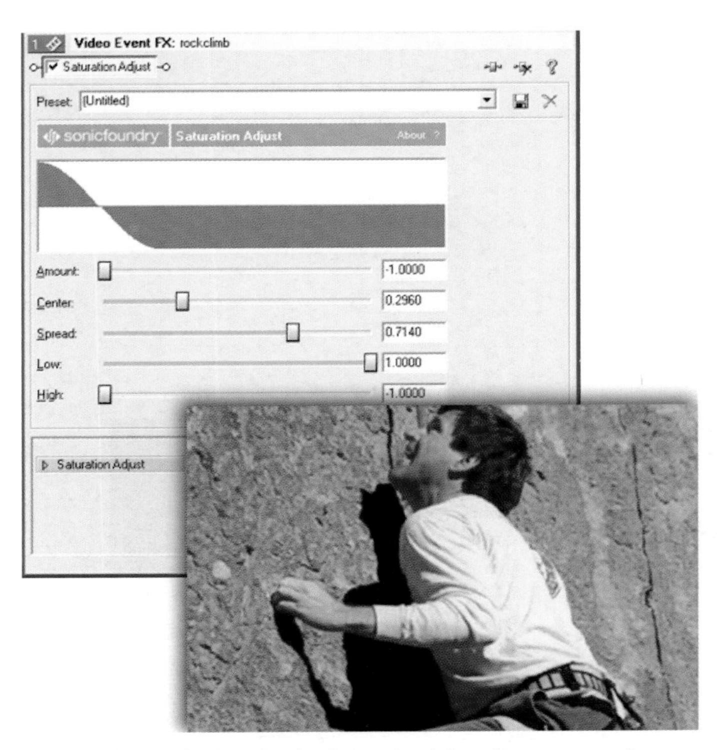

Color Plate 9.23 Settings for the Saturation Adjust filter remove all saturation except for lower mid-levels and leave the clothing in color while everything else is desaturated.

Color Plate 9.24 By blowing out or overextending the gain and luminance of an image, a look of infrared or night goggles can be created.

for a film-like appearance, duplicate the video track. Select an event on the top track, right-click, and then select PROPERTIES | FIELD ORDER | UPPER FIELD FIRST. Using the Track Opacity/Level slider on track 1, set opacity to 50 percent. On track 2, be certain that events are lower-field first. This process deinterlaces the footage by drawing all parts of the frame, rather than drawing only half the frame (see Color Plate 9.21).

There is no wrong way to achieve deinterlaced footage with Vegas to obtain a film-like appearance.

Another big difference between DV and film is the gamma curve. In Vegas, the gamma can be reduced using a color correction tool (see Color Plate 9.22). Increasing saturation is also good practice for obtaining a more film-like look.

You need to know a few things about shooting for a film-like appearance. First, if the video will ever be transferred to film, do not shoot with filters such as the Black Mist 2 from Tiffen. Shoot clean and with clarity. The same goes for using film-look plug-ins, such as grain. All of the digitally created film look is created in the transfer from DV to film. You need to understand going in that DV-to-film transfers are exceptionally expensive. An average 90-minute project will cost a minimum of $30,000 to transfer from DV to film. `http://www.dvfilm.com` has lots of information on how to make this happen if this is your end-goal. If video will be shown on a large screen, digitally projected, test the footage before sending out the entire project as an uncompressed file. Projection can create problems, such as pixilation or overblown colors, so the entire project should be checked over before public viewing. Rendering uncompressed footage, which is generally advisable for large-screen projections, can take a long time, particularly when motion blurs and super-sampling are applied. Be sure to have everything checked out before starting this process to avoid long periods wasted because of haste in the postproduction process.

Legal Colors

Video that will be broadcast or displayed on NTSC-display equipment must be brought within legal color guidelines to avoid clipped colors and clipped/distorted audio and to avoid embarrassment as a professional. The best means of handling illegal colors is to have color correction, titles, still images, and filtering monitored so that the video isn't allowed to become illegal in the first place. Fulfilling this requirement, however, sometimes can be a challenge. Vegas provides a Broadcast Colors clamp that ensures that colors are legal. Although Vegas 2.0/3.0 had the Broadcast Colors clamp as well, Vegas 4.0 brings a number of parameters to this tool that were previously unavailable. The Broadcast Colors plug-in can be considered in a similar light as a limiter in the audio world. The new tools can be considered more of a video compressor as the new tools allow some elasticity as to how Vegas treats illegal colors.

These options give editors many choices in working with color values. With technology changing rapidly in the broadcast industry, these limitations are less stringent, although the standards will most likely persevere for many years to come.

Rather than dropping the Broadcast Colors filter on an entire project, which would absolutely assure that the project is legal, consider inserting the plug-in on events that are demonstrably beyond legal colors. Use the histogram or waveform monitor to display and indicate luminance issues.

Inserting the Broadcast Colors filter on troublesome events rather than on an entire project will speed render times. More importantly, adding the filter to individual events will not clip or affect color values that may border on the edge of legal and then have a potentially negative impact on the overall picture if much of the project is at near-illegal levels.

9:19 The new Broadcast Colors tool has a broad palette of options.

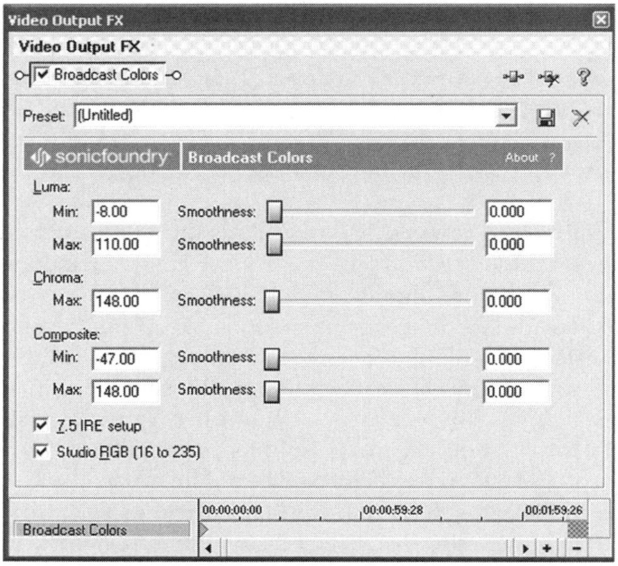

The Broadcast Colors plug-in presets bring colors to within 7.5–100 IRE levels or 16–234 RGB levels. Luma and chroma (light/color), however, are adjustable separately, as is the composite blending smoothness of luma/chroma.

> Even though the Internet does not have color limitations, inserting a Broadcast Color, Black Restore, or Levels filters on events can indeed benefit video streams over the web. Encoding is an art form, have no doubt. The fewer color variations in a stream, the better the stream. Redundant frames are the key to good streams. The Levels plug-in can limit colors. This process would not be useful if the HSL and Restore Black filters were used to crush blacks and reduce color saturation. Be aware, however, that the Broadcast Colors plug-in will not ever add anything to colors but only assure that colors are within legal ranges.

Endnote

[1] For more in-depth reading on WFM/vectorscopes, pick up *Color Correction for Digital Video* by Steve Hullfish and Jaime Fowler. Available from CMP Books.

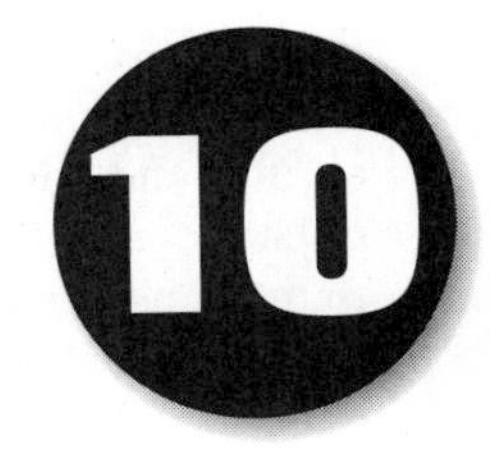

Streaming Media Tools

Creating High-Quality Streaming Media

Just as television was to radio, webstreaming is to television today. Television, in its infancy, delivered exceptionally poor information with fairly poor quality. Streaming media is much like that today; while the information might be high quality, the quality of the view is compromised because of the delivery medium. Until greater bandwidth is available for the public, a need will always exist for editors that have the ability to edit just for the web.

Vegas 4.0 is one of the most powerful web-delivery authoring and editing tools available today. Preproduction, however, is equally important as the editing and encoding ability when it comes to creating great streaming media. While many resources are available for great streaming media authoring techniques, I've included some tips in this chapter for shooting and producing great media.

Shooting, Editing, and Delivering

Perhaps the most important part of streaming production is knowing more about video than simply knowing how to turn on a camera. Understanding the basics of production is equally important to the flow of the production process and to the quality of the media itself.

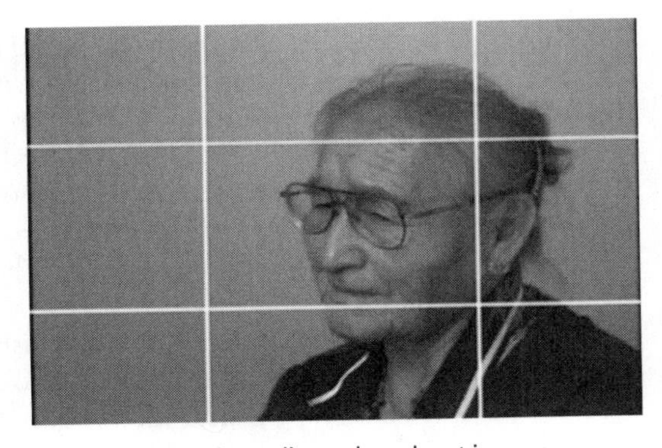

10.1 Rule of thirds applies to broadcast images.

People trained for working directly in broadcast are often the worst at generating quality streaming media shoots. Much of what is known about broadcast shooting is diametrically opposed to shooting good web video. Keep in mind that television is a *lean-back* experience, in which viewers are laid back, perhaps with a bowl of popcorn in hand, relaxing in some form. Streaming media, or media on a computer, is generally known as a *lean-in* experience, in which the viewer is close to the computer monitor, leaning in to view a fairly small image. This issue alone forces a completely different perspective for production of media. Of course, with the convergence of television and streaming media close at hand, this issue might well change in the near future.

Both REAL and Windows Media support high-definition streams with surround sound ability, and the quality of these streams is outstanding. For the moment however, more than 80 percent of the world is still on dial-up, so bandwidth will continue to be an issue in the near future.

Three processes are in the chain of web video production:

- Production
- Editing
- Encoding and delivering

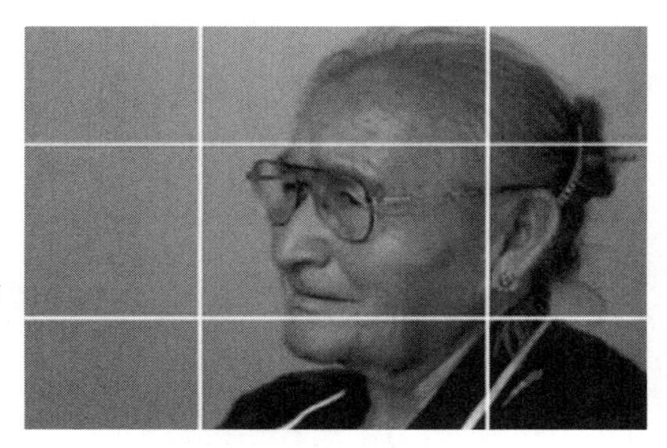

10.2 Streaming media works best using more of the screen space or the Rule of Two Thirds.

Production

Rule number one for shooting good web video is to always shoot with a tripod. Handheld camera work, regardless of the ability of the cameraperson, simply has too much movement to be encoded well. Redundant information in the frame encodes well, whereas moving media, even if it's only slight movement, encodes poorly.

Shoot close in. Normally in broadcast-destined footage, shots are based around the rule of thirds. The rule of thirds is simple—divide the viewfinder into a grid that looks like a tic-tac-toe game. Any two points where lines intersect become the focal point of your subject matter, preventing the subject from being in the exact center. With streaming media, however, the message is more important than is the aesthetic of the picture. Clarity and a less informational background is the goal. Shooting for two-thirds imagery means bisecting three lines rather than only two in the framed image.

Be as tight as practical. Limit any background movement. Never shoot in front of moving trees, roads, or other backgrounds where there is movement. In fact, shooting against a backdrop, green-screen, or bland fixed set is generally best. Keep background information/imagery to its simplest form. This process makes it easier for the encoder to process information.

Don't forget audio. Audio is even more important in the streaming world than in the broadcast world, and it's pretty important in the broadcast world. Viewers of streaming media will forgive the poorer video quality of a stream. They generally will not forgive bad audio. Always use lavaliere or boom microphones. Never use the microphone built onto the camera. The proximity of the microphone to the subject is too distant, and the sound is muffled and full of reflections, noise, and ambience, which is difficult to remove. Use Sonic Foundry Noise Reduction if necessary, and it usually is. Unless the production is being delivered over the Internet or if it's absolutely live-or-die necessary, audio should be in mono rather than stereo. Aside from phasing issues, stereo audio is more difficult to encode than is mono audio, and the audience can rarely appreciate stereo audio on most computer systems.

Limit or curtail reverb decay times, and cut off bottom- and top-end frequency information with an audio equalizer. This process leaves less information for the encoder to deal with. Normalize or maximize audio to assure that it's as loud as possible. The encode process works best with highly compressed, loud media. Large dynamic ranges (quiet to very loud) do not encode well. Try to limit dynamic range to 12dB or less.

10.3 Well-lit face for streaming. High contrast and a tight shot make for a great stream at any speed. If the background is static, of one or two colors, and contrasted highly against the subject, Windows Media 9 is capable of delivering even a full-screen stream at 56 Kbps. This image is a 56 Kbps stream projected in a training session. (Photo by R. Grandia.)

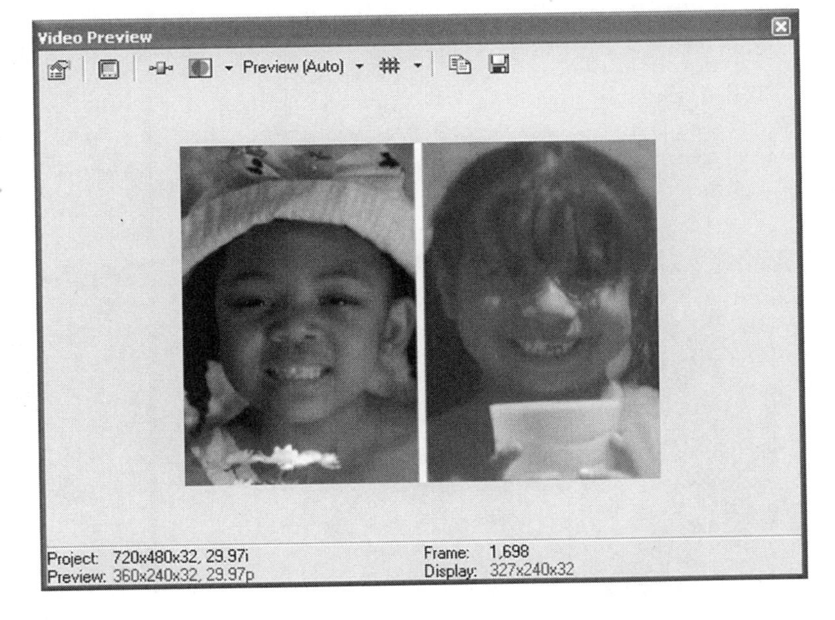

10.4 Using a split screen allows camera angles to be less important and still keeps a two-part conversation visible and clean.

Light well. Having great lighting is critical to any web-video production. This issue is true more so than in the broadcast world because encoders have a difficult time distinguishing as many colors as the broadcast equipment does. High contrast and low saturation always work best for web media.

Editing

Edit for clarity, content, and cropping in mind. Beautiful panoramic shots are wonderful for demonstrating high-definition TV (HDTV) or for displaying the great outdoors. Such shots are horrible for the web, however. Detail becomes so small as to be invisible. Is that detail really critical to the message that's being delivered? The message is the most critical consideration, and anything not directly related to the message should not be found in the stream.

In the edit process, if two talking heads are needed, rather than showing the two heads in one frame, consider the CNN-style appearance where each head has its own space, divided by a clean line down the middle. This setup accomplishes two tasks. First, it keeps the background frames redundant. Second, it requires less distance between the heads, so both heads may be cropped in more tightly.

Removing unnecessary information limits what the encoder has to look at. Encoding tools are not capable of separating a few pixels of a reflection on a window from the blowing hair of the subject. To the encoder, it's all just moving pixels, and nothing is prioritized over anything else. Cropping helps alleviate this concern because cropping allows for less information to be on the screen.

Less saturation is best in streaming media. While the RGB color space allows for millions of colors, keeping colors to a minimum is best because it offers fewer color choices to the encoder. Reduce colors whenever possible. The Hue, Saturation, Luminance (HSL) filter in Vegas is excellent for this concern. Insert the Black Restore filter as well to ensure that black is really black. Gradients of black are difficult for the encoder to compress and difficult for the eye to see in the production stage. This filter, used correctly, flattens out black in the stream.

When editing for streaming media, forget about long, beautiful dissolves and fades. These sorts of transitions do not encode well. Use linear wipes, hard cuts, or other straight-line transitions. Don't use titles with serifs, but do use clean-cut title fonts and keep them as large as possible. Sometimes no title is necessary if metadata is used to convey text information.

10.5 The broadcast shot allows for more background to set the stage and location of the shot.

10.6 Cropped for the web, this shot is more interesting when viewed from very close up, such as with a computer monitor.

10.7 Streaming media should not have as much background. This example shows a poorly cropped screen. Too much information is in the background, and the subject takes up too little of the viewing space.

Also, in the edit stage, be aware that extreme contrasts are harder to encode, so keep extreme variants in saturation limited. Remember that placing media on the web is about messaging, not brilliant colors and flashy editing.

To summarize:

- Use clear fonts for titles.
- Limit dissolves; use cuts or hard wipes instead.
- Avoid moving paths when possible. Use picture-in-picture or split screens.
- On headshots, crop to half, rather than thirds.

- When you want to deliver to low-band systems, keep audio in a mono setting. For both low- and high-bandwidth delivery, consider using both stereo and mono audio. Does stereo really add value?
- Don't limit yourself to the standard 4.3 aspect. If your video is about something involving heights for instance, consider scaling the video to be tall and narrow. This process might also use less bandwidth and might allow for better audio or compression. It also might be that the stream quality will suffer, depending on the aspect selected. Either way, do not be afraid to experiment.
- Try a variety of codecs to see what allows for the best look. Perhaps put low bit rate media in one codec and high bit rate media in another codec.
- Try frame rates of 12 fps rather than 15 fps when bandwidth or file size is a concern. For slide and still-type presentations, use even slower rates, such as 5 fps.
- For headshots, try using color key or chroma key tools to create background masks. Then fill in with high-speed, blurred imagery. This process keeps the headshot clean, while causing the eye to focus on the headshot and not the black behind it. It also lends an MTV-type quality to the interview or presentation.
- Impress viewers by embedding metadata into the end or middle of the video files that direct their browser to open at the end, so that they are given other images to view following the video/audio presentation. If you are selling a product, the metadata could potentially send the viewer to a "purchase me" type of page.

Audio for Internet Delivery

Preproduction of audio for web-streaming or Internet distribution is often overlooked in the streaming preparation processes. Oftentimes, videographers intent on placing media on the Internet or an intranet use tools built into the applications that are being used for editing media for the web. If these tools are the only ones available, little can be done to prepare the audio for the web without going through multiple renders and loss of video quality. The processing described in the next paragraphs apply equally to audio only, as well as audio for video.

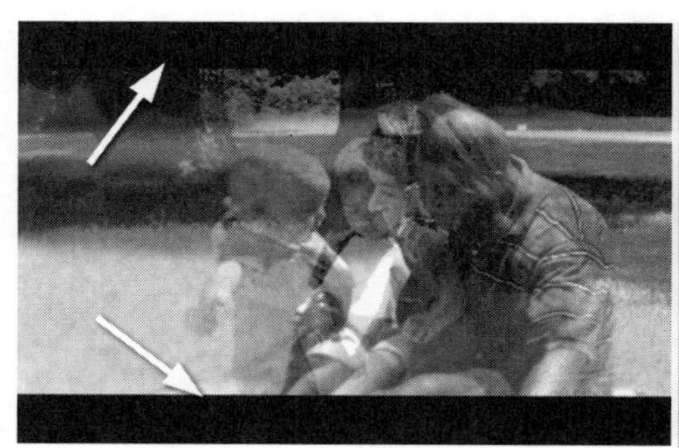

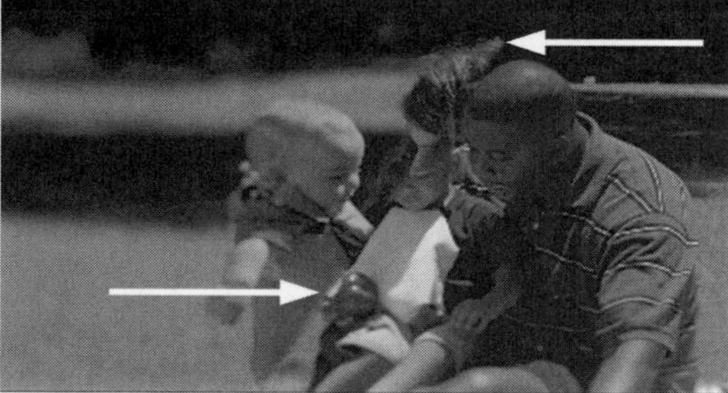

10.8 The image on the left shows a slow fade and letterboxing. The image on the right incorporates a hard wipe and no letterboxing. This image will encode/compress much better.

If third-party editing tools are available, use them to the best advantage possible. Most nonlinear editing packages come bundled with Sonic Foundry's XP audio editing software. For this book, I've used Sound Forge 6.0, a full version. The tools used in this book, however, are all found in the XP version.

Audio takes a tremendous amount of unnecessary bandwidth in the streaming world. Most computer speakers can't reproduce sounds of any quality above 10,000 Hz (10 KHz) or below 125 Hz. If unnecessary audio is taken away, it leaves room for more efficient use of both video and audio bandwidth.

All video and audio compression tools use both high- and low-pass filters of some sort. The question is what is the quality of the filters, and where are they cutting? Video tools simply aren't designed well for most audio processing chores.

How Is Audio Preprocessed for the Web?

First, start by fully rendering your file. If you are working with most audio editors, you'll want to be sure your file is smaller than 2GB, as current limitations prevent larger WAV files from being opened in audio editing tools on FAT32 drives. Moreover, you probably don't want to create too large a file, as large files are a bane in streaming media. If you want to save a step in the production process, render your file in an AVI format that is the view size for streaming media, such as 320 × 240 at 15 fps (using square pixels).

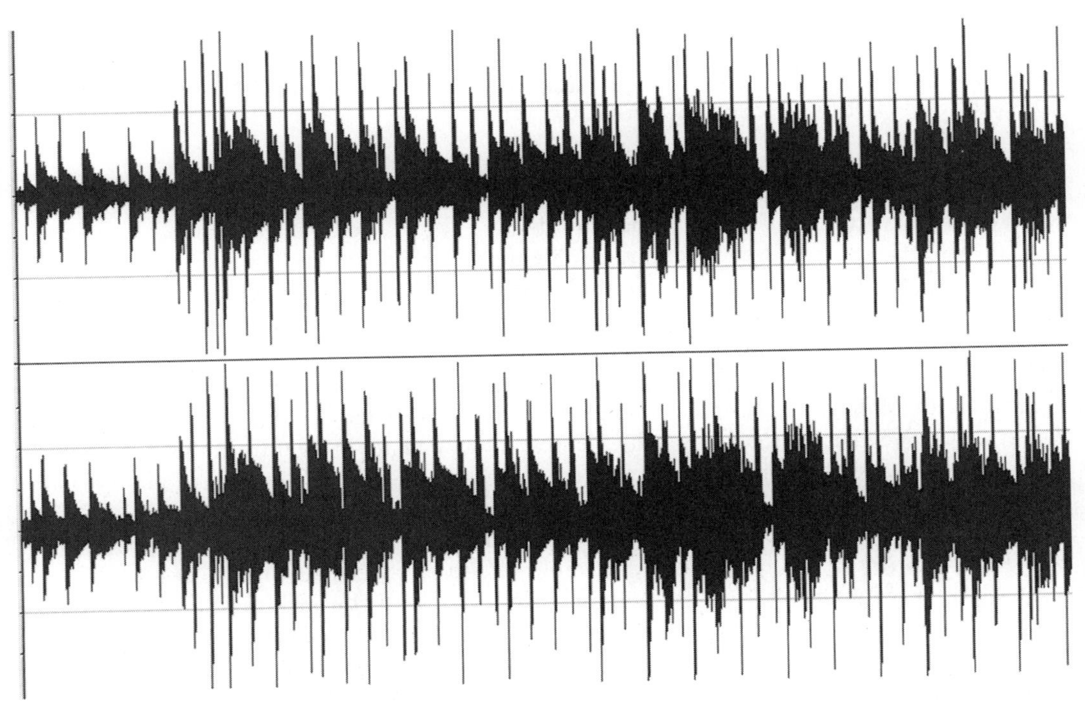

10.9 Notice the dynamic range of audio in the waveform.

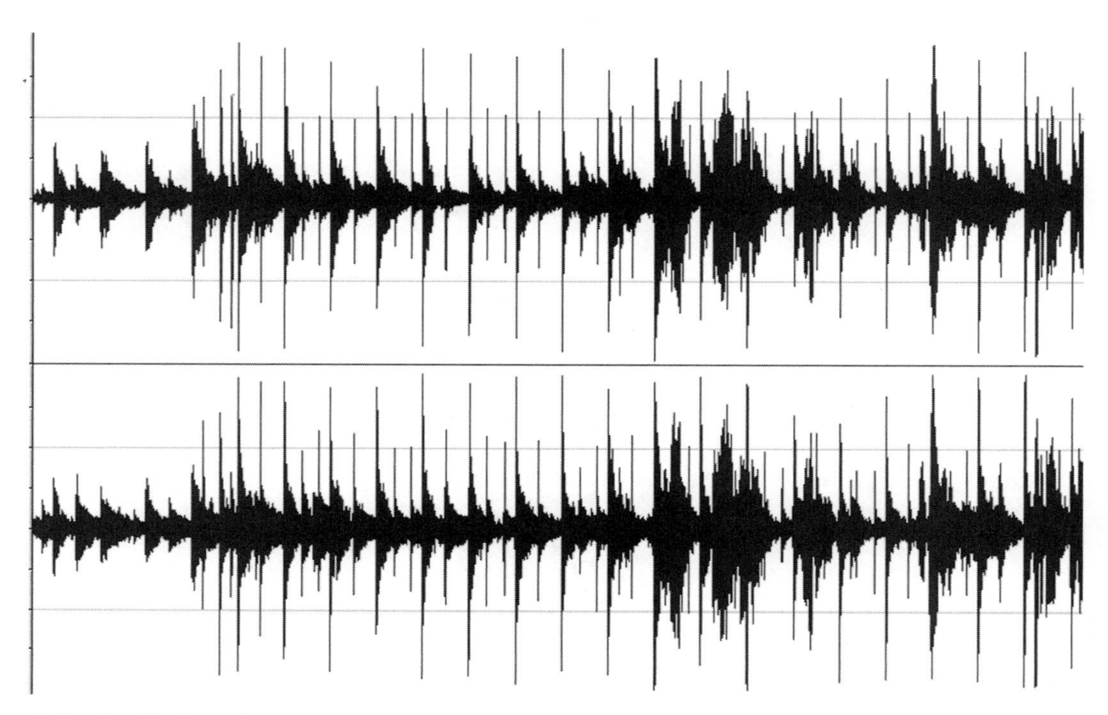

10.10 After EQ, the audio waveform contains less transient information because extreme low/bass frequencies and extreme high frequencies have been cut.

Open the rendered file in your audio editor. Notice how the audio dynamic is fairly consistent across the spectrum. This audio has full dynamic range, containing both frequencies that multimedia speakers can't reproduce and also frequencies that can't be heard. If it's a live recording, sounds, such as microphone stand rumble, use up to 100 times as much power to reproduce as the breath in a singer's voice. Those frequencies use bandwidth/streaming space that can better be used for video or audio enhancement/quality.

Using the graphic, paragraphic, or parametric equalizer in the audio editing tool, locate the high and low roll-off controls. Set the low-pass filter at 10 KHz and the high-pass filter at 100 Hz. These settings will cut off the extreme high-end and low-end audio right from the start. Just this preproduction task alone makes a tremendous difference in how the compression application accepts files for compression.

At this point, it might be advisable to do any final tweaks and enhancements with the audio. If you have any stereo widening, reverb, or ambience adding effects, now is the time to add them. Time delays and long reverbs do not compress well. Care and consideration, therefore, should be used at this point, so that streaming media will sound the best it possibly can on the receiving end. Multimedia speakers have a tendency to have a bump in the 5 KHz region, so if your media is music intensive, it might be advisable to reduce this frequency area by as much as -3dB.

Notice the difference in the audio file's appearance in Figure 10.11 following the compression change. The waveform has less mass, and the transients become less apparent.

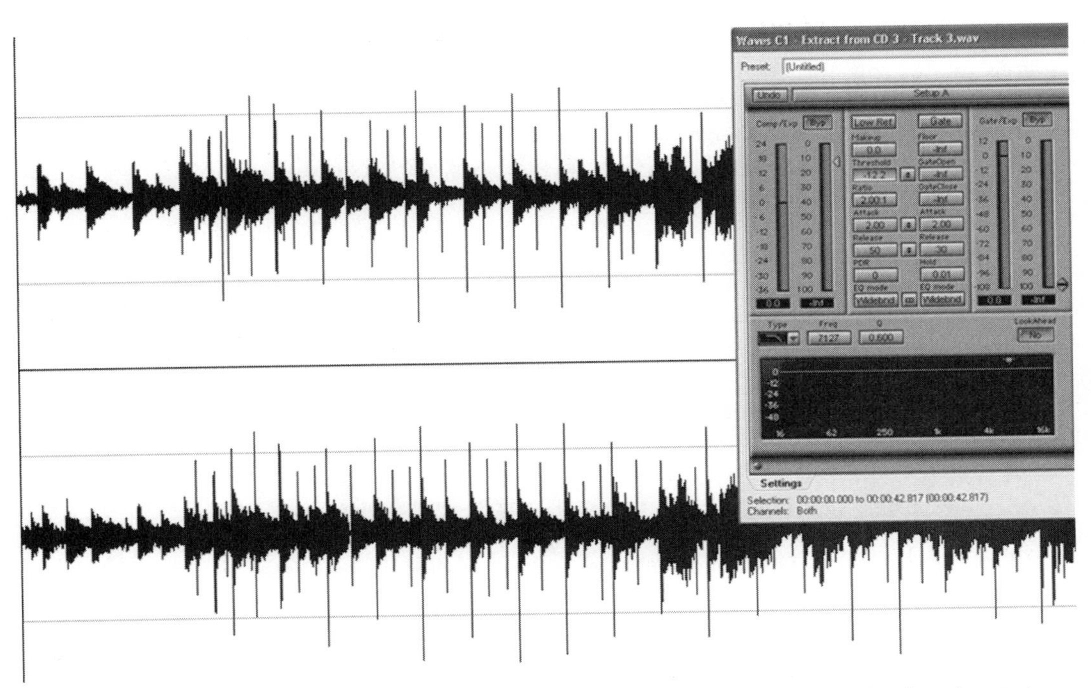

10.11 Use light compression to smooth out waveforms. In this illustration, a 2.1 compressor has been inserted, with compression starting at -12dB.

Be sure to have a preview listen to the file before moving on, making certain that the audio is the way you would like it to sound when it is received on the listening end of the stream.

Often in multimedia creation, particularly dealing with files generated directly in the computer environment, audio levels are not consistent across the project nor are the audio levels at optimum settings. These audio files may still be used and brought to maximum levels by normalizing them to an average level across the median plane. Two approaches to normalizing are used—normalize to peak values of XdB or normalize to an average RMS value. Use a normalization of an average RMS value if you wish to match the apparent overall loudness of a sound file. The Normalize to Value selects the volume to which the sound file will be normalized. As an example, when using peaks, if the peak level is -10dB and the Normalize to setting is set to -3dB, a constant boost of 7dB will be applied to the entire file.

When using RMS levels, normalizing to 0dB means boosting the signal so that it has the same apparent loudness as a 0dB square wave. This process usually creates a file that is very loud and often clipped or distorted. Normalizing to values between -12dB and -6dB is recommended.

Normalizing on a peak value allows for more dynamic interplay in the audio file. The file is scanned for its maximum peak value, which is then raised to -.03dB. This process applies equally weighted amplitude across the audio file. Most normalizing functions allow for the introduction of dynamic compression at any stage in the normalization process. Using this tool is advisable as it prevents clipping at all stages, while allowing the normalize tool to bring audio up to maximum levels. This is especially desirable when the sampling frequency or bit rate of audio will be changing. Audio files should always be maximized before converting the format.

After the file has had the extreme low and high frequencies removed, been equalized to the desired sound, and normalized, the file is ready for encoding/compression for streaming formats. If the file is an audio-only file, it is a simple matter to encode to REAL, Microsoft media, or QuickTime formats using the tools built directly into your audio application or using third-party tools. Be aware of the ability to insert command markers or instructional markers into the audio or video file. These command markers can display copyright information, authoring information, HTML redirections, and much more. If the processed audio file is part of a video file, it is an equally simple process to encode the video file for streaming. Reducing the audio content and maximizing the audio volume both make for more encoding bandwidth for the video portion of the file and provide a more efficient use of available bandwidth. These benefits are especially useful when dealing with low-speed connections, such as dial-up service.

Until bandwidth and connections bring broadband to every home and business in the world, and even after that point for purposes of bandwidth and storage, content creators/providers need to learn every trick in the book and invent some new ones in order to keep media files appearing and sounding the best that they can. Audio is generally an afterthought in the video creation/editing world, although that is rapidly changing. Experiment a bit, listen closely, and you'll be surprised how great audio can really help out a less-than-great visual presentation. Don't forget metadata can be a great tool for audio-only delivery, as well as video delivery over the web.

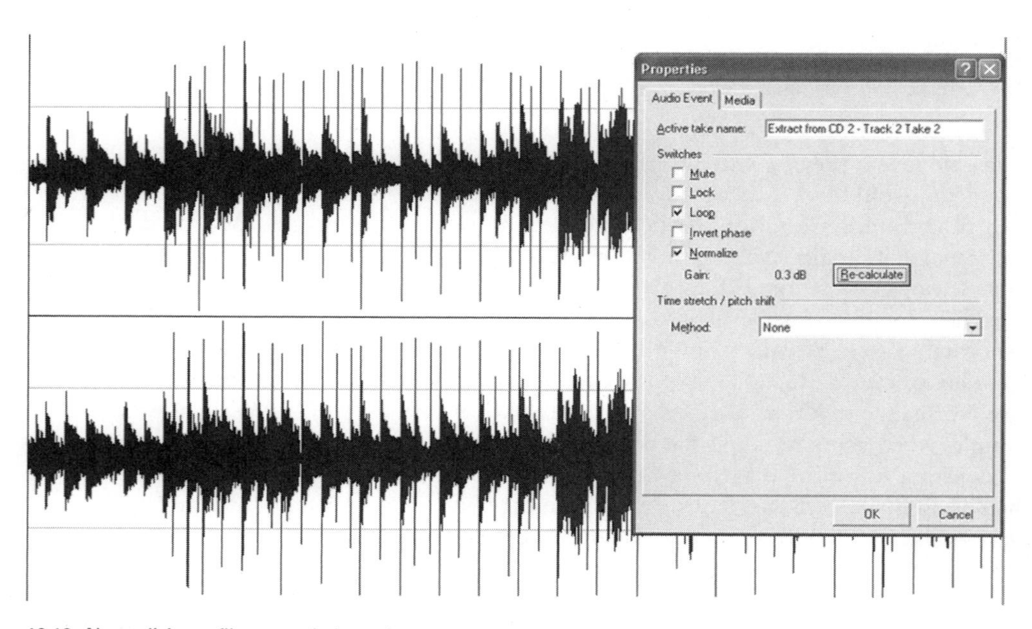

10.12 Normalizing a file to maximize volume.

Delivery

Encoding is not much of a challenge if the techniques mentioned earlier are employed in production and editing stages. Vegas 4.0, however, offers a number of choices for the encoding/output side of media.

An exceptionally valuable tool found in Vegas is the ability to encode portions of the timeline for preview, rather than rendering out entire files, encoding, and if dissatisfied, repeating the entire process. Vegas is one of the only tools available today that can encode directly from the timeline, complete with metadata insertions. This Preview from Timeline tool is valuable in that producers are able to see exactly what their media will look like when viewed from the web, directly in the media player of their choice.

Open the Streaming Project on the accompanying CD.

The project file already has three sections of metadata embedded: two text instruction lines and one URL locater line. When viewed in the media player, the text data shows up at the bottom of the screen (in the Windows Media Player, select VIEW | NOW PLAYING TOOLS | CAPTIONS). Place the cursor on the timeline at the first marker, labeled Insert Command Here. Press C, and a command dialog box will pop up. In the Command dialog box, use the menu and select Text. In the parameter box, type the words, *Then I go and spoil it all by saying something stupid*.

Select a small section of media on the timeline over the top of the section containing at least one metadata flag. This section will be highlighted in gray (or blue if looping [Q] is enabled). Open the TOOLS | PREVIEW IN PLAYER dialog (CTRL+SHIFT+M). In the dialog, choose Windows Media V9 (**.wmv*) for the codec and select 256 Kbps for the streaming bandwidth. Click the Custom button, and in the audio attributes, choose the 48 Kbps, 44 KHz Mono (A/V) CBR setting. Click the Video tab, and in the Format window, choose Windows Media Video 9.

Now click the Summary tab and note the text boxes that may be filled out. This data is embedded with the file and shows up in the player on the client (viewer) side when the file is streaming. This dialog also provides a great means of data management for later use, as information about the file can be embedded in the file for database searches. Be aware that the last information entered in this dialog carries to the next time the tool is opened, so be certain to check this dialog with each stream.

WHAT IS METADATA?

Metadata is what makes streaming reach out and touch someone. Metadata is embedded information in a streaming file that opens secondary browser windows, directs viewers to web pages, provides lyrics or text information to a song or informational media piece, provides advertising opportunities, all from a streaming file. The author of the streaming file places the metadata markers on the timeline, and viewers have a more in-depth experience as a result. Vegas has many metadata authoring tools found in the Command key (C).

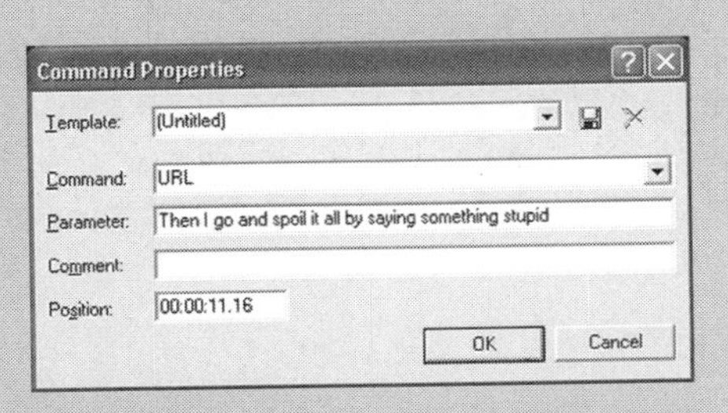

10.13 Metadata may include song lyrics as displayed text.

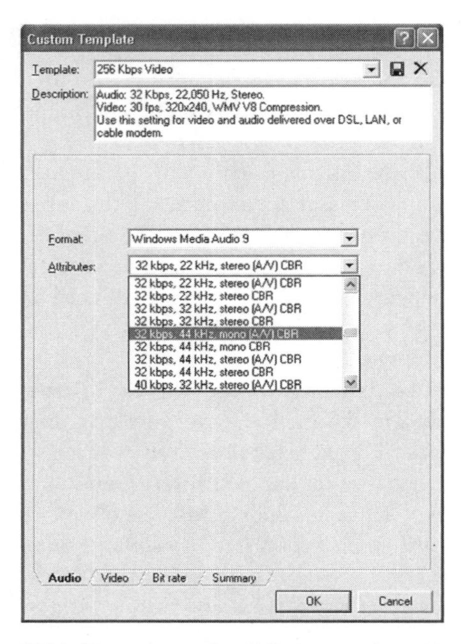

10.14 Streaming audio attributes may be set to user preferences.

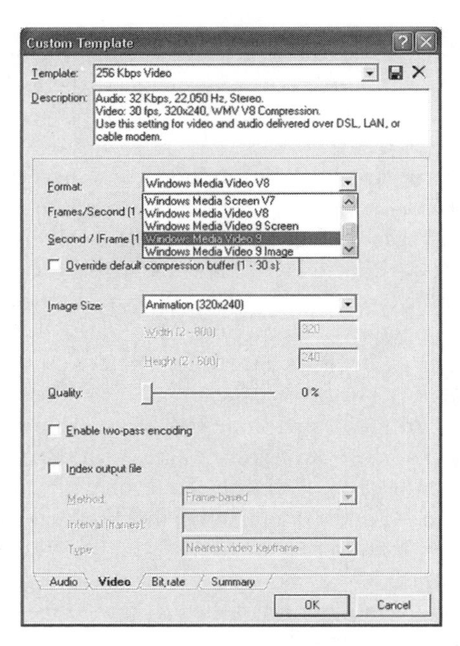

10.15 Select the video codec and other attributes in this dialog.

Click OK in the Custom Template dialog and click OK in the Preview in Player dialog. Vegas will begin to render the file as a WMV file. After the file has rendered, Vegas will automatically call on the Windows Media Player to open and play the file. If the Captions option is enabled in Windows Media Player, you'll see the captions on the bottom of the Windows Media Player screen. This will show only the selected portions of the timeline. In addition, only a temp file has been written for the streaming file. A full render will still need to be completed for a final output file.

 ***Tip***

Preview in Player is a powerful tool. Use it to identify trouble spots in the streaming media by previewing high motion/intense color areas. This process helps you determine how the video should be handled. Prerendering trouble spots at various frame rates before rendering a streaming file often helps make high-motion media look more polished. Tight cropping while following motion often helps as well, as it reduces background imagery.

Importing a Spreadsheet

Vegas has the ability to import a scripted spreadsheet. This asset is very important for the content creator who has concerns about complying with Rule 508 Assistive Technology laws. For any government-funded streaming endeavor, these laws almost always must be complied with. Colleges, training facilities, hospitals, corporate environments, and federally subsidized businesses, among others, must comply with these laws.

Vegas' streaming toolset enables closed caption-text to be embedded in the video or audio stream for hearing-impaired viewers. To do this, start by creating a four-column, tab-delimited spreadsheet in your spreadsheet application of choice and continue the process as follows:

1. In the first column cells, type 00:00:00:00.
2. In each of the second column of cells, type the word *text* to specify the command format.
3. In the third column, type the text messages that you wish to be seen, for example, *When I was 17.*
4. In the fourth column, type marker information that you can use to identify the caption information. For example, they might be called *cap 1* or *line 1* in the cell.
5. In the last row of cells, leave the third column empty, which will clear the last caption text from the screen of the media player.
6. Select all information on the spreadsheet, or press CTRL+A in Microsoft Excel. Copy all information.
7. In Vegas, select VIEW | EDIT DETAILS | SHOW | COMMANDS. Right-click in the gray box in the upper-left corner and choose Paste. Vegas will paste the spreadsheet information into the Edit Details window.
8. In the Edit Details window, click the first column, which will sort the captions by line number as named in step 4. Position the cursor where you wish the first caption to occur. Select the first row in the Edit Details window and play the video/audio file by pressing the SPACEBAR. Press CTRL+K to drop a marker for the first caption. Repeat for all subsequent captions. Caption markers can be dropped on the fly and moved on the timeline by clicking and holding on the marker and then dragging to the desired position.

Be sure Captioning is enabled in Windows Media Player by selecting VIEW | NOW PLAYING | CAPTIONS. Otherwise captions will not be previewable nor can they be seen in the playback of a rendered file.

After all edits have been made, all metadata inserted, screen markers completed, and various points in the timeline checked for quality, the file is ready to be rendered. Open the FILE | RENDER AS dialog, select the format of media you wish to stream, the bit rate at which to stream, and any summary information you wish to appear. If custom selections, such as variable bit rate (VBR) streaming or custom audio settings, are needed, click the Custom button. This action opens the same dialog shown during the Preview in Player exercise.

The Audio tab opens by default. As mentioned earlier, audio should be kept in mono to provide greater bandwidth for video. The Video tab is next on the menu tabs; settings can be adjusted according to need. Experiment with settings that best suit your requirements.

When you select various bit rates in the Bit rate tab, the file rendered by Vegas will be a multiple bit rate (MBR) file. If only one bit rate is selected (the default), the file is rendered as a constant bit rate or variable bit rate (CBR/VBR) file. Viewers also generally receive a more stable stream with CBR instead of MBR files.

The Summary tab allows for information, such as copyright information and author information, to be embedded with the file. This information shows up in the information bar of a REAL, Windows, or QuickTime media player, and in most third-party media players as well. Other than re-rendering the file, this information is difficult to remove and is a starting point in the concern for digital rights management (DRM).

After the custom selections/setups have been accomplished, you may wish to save the template settings for future use. A menu is available at the top of the Custom Templates dialog. Clicking the cursor in this menu allows you to rename the template, and clicking in the Description field allows you to create a new description of the template. After defining the name and description of the new template, click the icon that resembles a floppy disk, which saves the template for future use. After the template has been saved, the final render can take place.

Click OK in the Custom Templates dialog, which closes. Define where you want the final rendered file to be saved/rendered to and click the Save button. The render begins. After the render is complete, Vegas provides an option to open the folder in which the rendered file was stored. If you do not wish to see the Open Folder dialog in the future, you can use the check box that Vegas provides to disable this pop-up screen in the future.

The finished file is now ready to be uploaded to a website and streamed. Bear in mind that with REAL and QuickTime media, unique streaming server software is required for streaming, as opposed to download-and-play files. REAL's new Helix server is capable of serving all three stream formats, in addition to many other formats in the streaming world. A non-commercial free version with a very limited number of streams is available at `www.real.com`.

Don't be afraid to use streaming media for more than streaming over the Internet. It's a great tool for delivering high-quality media on a CD to a potential client. With the ability to burn a high-definition file to a CD and to include a player-installer on the CD for the client, this tool is very powerful in creating workflows between client and editor, while not taking inordinate amounts of time to show the client what the video project looks like.

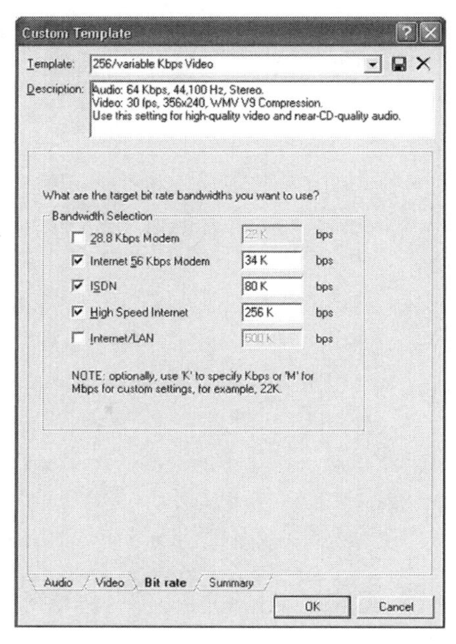

10.16 Checking multiple bit rates creates a multiple bit rate (MBR) file.

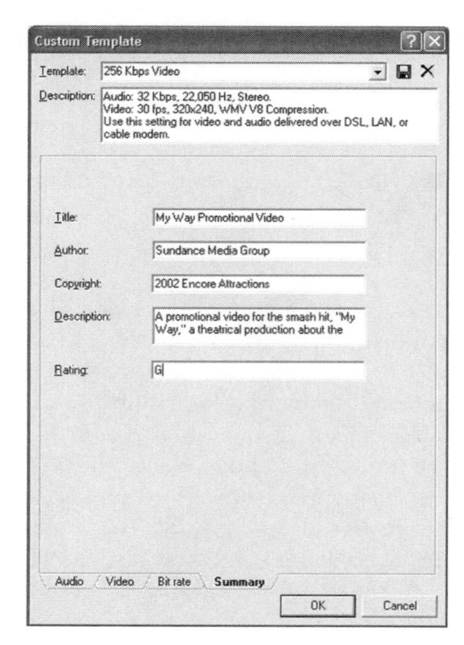

10.17 Use the Summary tab to embed copyright, author, and other information that can be seen in the stream/media player.

DVD Architect

Setting Up a DVD Project

DVD Architect (DVDA) is an extension of the Vegas 4.0 NLE system. Until now, authoring a DVD has been a fairly complex and expensive undertaking; however, with DVDA, Sonic Foundry has taken DVD authoring to a new level.

In keeping with the format-agnostic/resolution-independent attributes in Vegas, DVDA also allows nearly any format of media to be placed on the workspace. DVDA encodes files directly from the workspace, saving render time and helping prevent bad DVD burns because of errors in the editing process. Additionally, DVDA is capable of creating a DVD slideshow as a drag-and-drop process, having motion backgrounds, motion buttons, and motion submenus and menus. All aspects of the DVD can be previewed before burning the DVD.

Menu Flow in DVD Architect

Menu flow is a critical component to having a clear message in the DVD content. This aspect should be planned out before inserting media into the DVD project. With a weak flow, viewers of the DVD can easily be confused. It's usually helpful to create a workflow chart on paper before starting the layout of a DVD project, much like the flow chart shown in Figure 11.1.

Giving advance thought to layouts also helps build comprehensive menus as to how they should be laid out in DVDA, saving unnecessary steps in the authoring process. Compiling all media to one folder or having a project folder is helpful so that all media is accounted for in the authoring process.

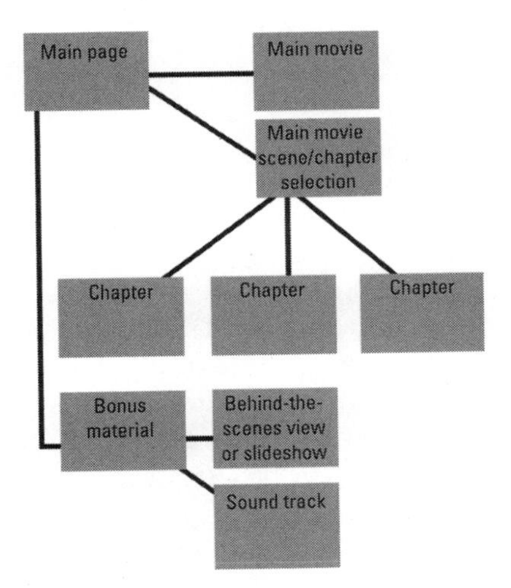

11.1 Layout of a simple DVD project, created before inserting media into the DVDA project.

Creating the Project

To get started, open the DVDA application.

Before creating the DVD, the properties for the project should be set. The default settings are for an NTSC DVD; if PAL is the desired format, it should be set in the Project Properties dialog.

Open the Project Properties dialog by selecting FILE | PROJECT PROPERTIES or pressing ALT+ENTER.

11.2 This is the Main menu/main workspace in DVD Architect.

In the PROJECT PROPERTIES | GENERAL dialog, a menu allows choices for NTSC-Standard, NTSC-Widescreen, PAL-Standard, and PAL-Widescreen. You can also select an Introductory Video, which is the same as a First Play video in most other DVD authoring applications. This option is useful for providing a copyright notice, a station or production house ID, or just a welcome screen to the DVD project.

Audio and video for the introductory video must be specified separately, even if the audio for the introductory video is part of the selected video stream. Select the introductory video in the video browse/path dialog. If audio is desired, browse to the same location and select the same file for the audio. DVDA syncs the two together.

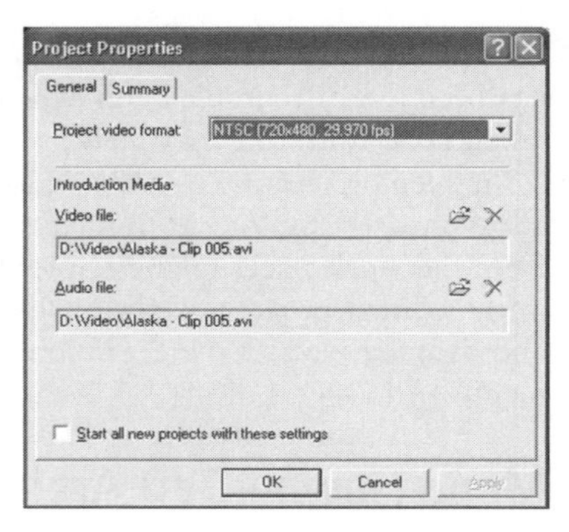

11.3 Setting up the introductory video with audio and video specified separately.

The Project Properties dialog box also has a Summary tab. This tab allows information to be recorded about the project, such as copyright, author, and project information. This information does not become part of the DVD but is used for file management.

To start a project using templates, select FILE | NEW PROJECT or press ALT+ENTER. This step opens up a template dialog box in DVDA. Menu-based, Music Compilation, Slideshow, and Single Movie choices all appear in the dialog box. DVD formats are also selectable as PAL or NTSC in this dialog box.

Creating a single movie disc, or a DVD with no menu structure, just a playback disc, is as simple as selecting the media file that you want to play. This disc will behave like a VHS tape, and once inserted into a DVD player, will simply play with no other options. The media file can either be a standard DVD or a video file with surround sound (AC-3 Surround) embedded.

Slideshow and music compilation discs are discussed later in this chapter.

Selecting the Menu-Based project settings opens a standard workspace in DVDA. At the top of the workspace, there is a drop-down menu similar to the Windows Explorer. This menu allows fast navigation between layers/menu pages in DVDA. After menus and scenes are created, this bar is the fastest and most efficient method of moving between the various menu pages.

Clicking the drop-down menu button on the right side of the navigation bar allows for rapid movement between layers in the DVD project.

The toolbar is on the left side of DVDA. This toolbar is where tools to edit placement of buttons, selection tool, and sizing are located. A wide selection of alignment tools is also available.

The right side of DVDA's main workspace contains the rest of the tools needed for creating a DVD, labeled as the Page Properties window. This area is where background media, thumbnails, button shapes, and more are created in DVDA. This area is also where the Object Properties tab is found and where thumbnails, links, frames, and scenes are edited.

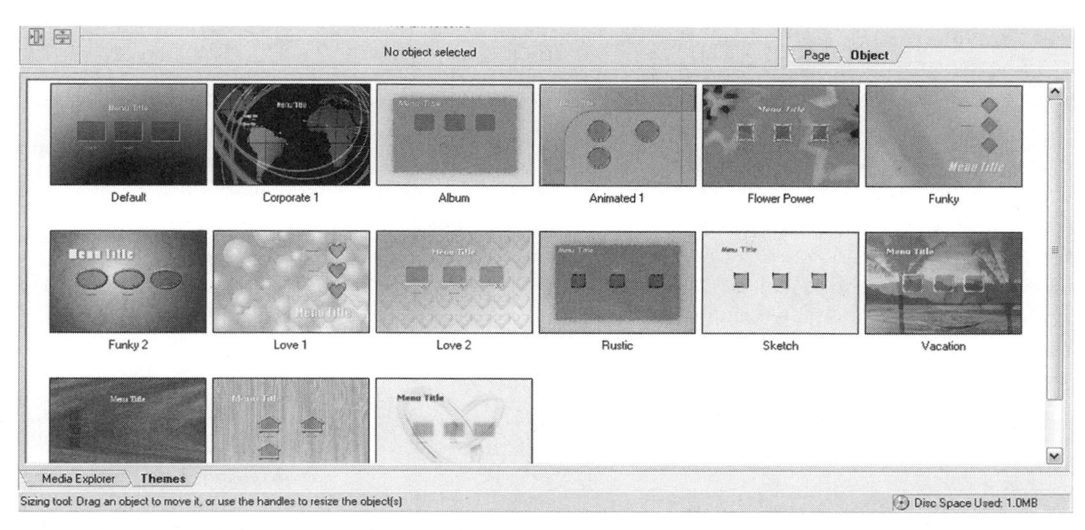

11.4 Prebuilt templates for creating different DVD types are available by pressing **ALT+ENTER** in DVD Architect.

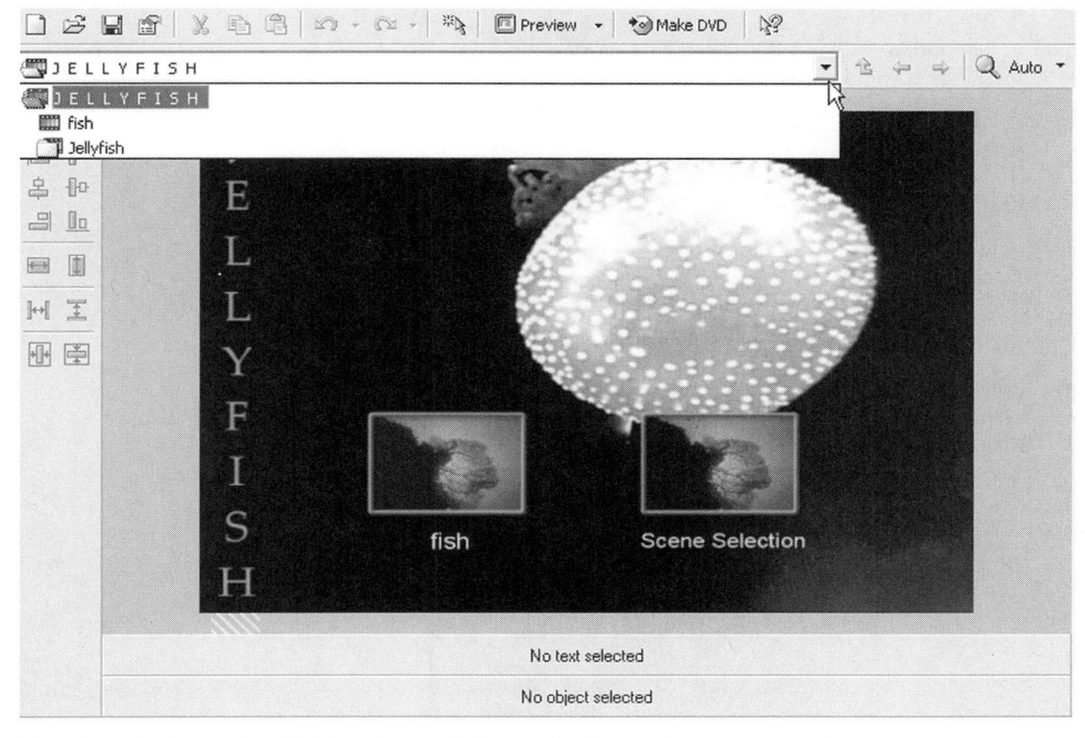

11.5 The navigation bar in DVDA functions much like the Windows Explorer navigation bar.

In the bottom of DVDA, an Explorer identical to the Explorer in Vegas is found and is where nearly all media is added. Next to the Explorer window is a Preview window, where video can be auto-previewed in the same manner it is previewed in Vegas.

To create a basic DVD in DVDA, this screen is where you start. Begin the authoring process by inserting a background for the DVD menu. A background can be inserted by clicking the Background File icon in the Page Properties view or can be dragged on to the work surface from the Explorer. If the image is not 655 × 480, the image will need to be sized with one of three methods.

- **Letterbox**—allows the image to be its actual size and creates borders on the sides or top and bottom of the screen.
- **Zoom-to-fit**—zooms the screen on the picture, so that the picture retains the correct aspect but is cropped to the correct view. This action can cut top, bottom, or sides from the image.
- **Stretch-to-fit**—picture is not cropped at all but is stretched, causing the aspect ratio of the original image to be disregarded.

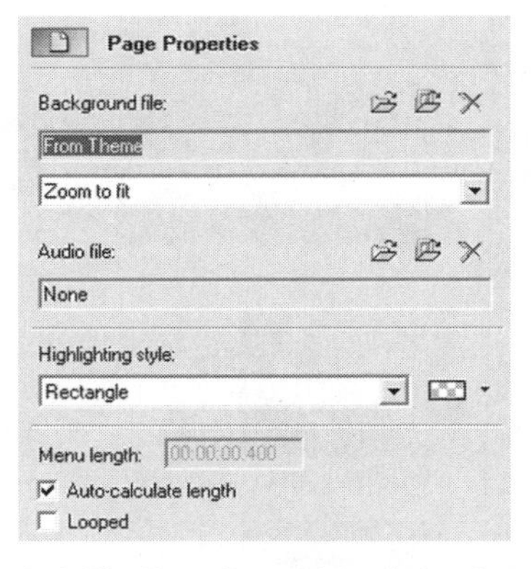

11.6 The Page Properties window is where thumbnail highlighting, background audio, and background images are selected.

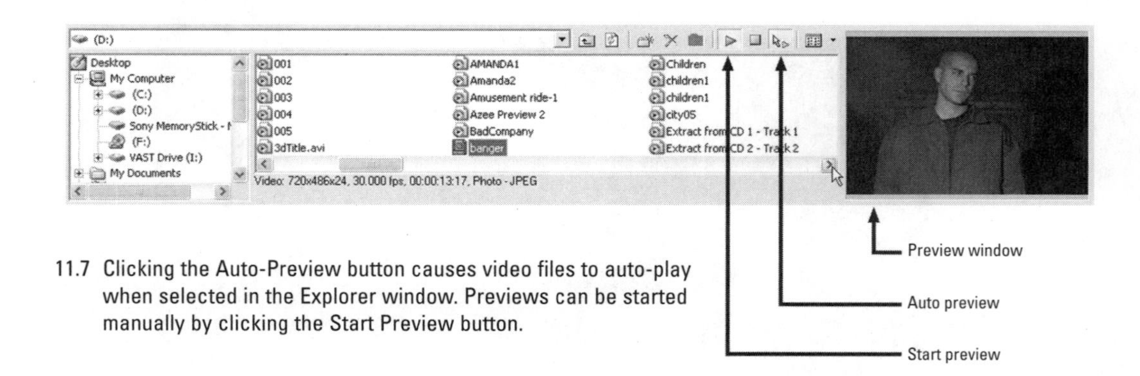

11.7 Clicking the Auto-Preview button causes video files to auto-play when selected in the Explorer window. Previews can be started manually by clicking the Start Preview button.

11.8 Letterboxed image.

11.9 Zoom-to-fit image.

11.10 Stretch-to-fit image.

Using and Creating Themes in DVD Architect

Themes allow starting users to create a professional appearance in their DVDs without any experience in graphic editing or creation. DVDA ships with some themes for basic use to get users going in DVD authoring. These themes are basic, but effective. To use a theme in a project, look at the bottom of the workspace, beneath the Explorer. Two tabs are available: one for Explorer and one for Themes. Click the Themes tab, which displays several themes to choose from. Double-clicking a theme assigns that theme's attributes to the workspace. If an existing project is in place, the theme replaces all background media with the theme media. Buttons and/or links from the existing project remain intact. New projects using the theme will use the theme attributes when creating buttons. By default, double-clicking a theme in DVDA's Theme menu applies the theme to all pages in the project. This option can be turned off and made page-specific by selecting OPTIONS | PREFERENCES | GENERAL and clearing the check box.

If an image is zoomed to fit and is not the correct aspect, the background image might not be adjusted in DVDA. A photo editor is required to place the aspect and crop, before being inserted as a background image. An alternative is to open the still image in Vegas and save as a still image after pan/cropping the image there. On the Video Preview window in Vegas, click the "Save Timeline Snapshot to File" button.

If an image is to be saved from Vegas, be sure that the Preview window settings in Vegas are at Best/Full to capture the highest resolution possible.

If editing in an external photo editor, background stills should be 655 × 480 for NTSC and 704 × 576 for PAL, at a dpi of not greater than 150. If the image is too high in resolution, the risk of collecting artifacts is fairly high. Remember, the image will be compressed to MPEG.

11.11 A snapshot can be saved from Vegas and used as a background image.

Themes can also be created/edited by advanced users, with some knowledge of XML. To edit a theme, it's easiest to work from an existing theme. Find existing themes at `C:\Program Files\Sonic Foundry\DVD Architect 1.0\_Themes\` or search for **.thm* in the Windows Search/Find function. Open the `\_themes\` folder, locate a theme to be edited, replace images in the theme with desired images, and open the XML document, changing all references to the old images to references to the new images. When finished, save the new file as a ZIP file and rename the ZIP file to a THM file format, saving it in the `\_themes\` folder. More information on theme creation and editing can be found in the help documents. Sonic Foundry has more themes available as downloads from the Sonic Foundry website.

Placing Text

After placing the background, you can place the text. On the Main Menu page, the default Main Menu text appears when DVDA is open. Right-click the text, and, at the bottom of the menu that appears, select Edit Text or press F2. Now the text in the main menu can be edited to create your own title. When text editing is enabled, the text is surrounded with a scalable window, as shown in Figure 11.12.

When Edit Text is enabled, the Text toolbar is at the bottom of the screen. You can use this toolbar to set fonts, typestyles, size, color, shadow, and position. Text can also be justified for vertical or horizontal positions.

Right-clicking anywhere on the Main Menu page provides a menu where text can be inserted. Text can be used as a link to video, audio, or a slideshow, in addition to being a scene-selection link. The quickest means of creating a text-only link to media is to drag media to the workspace on any menu page and then in the Object Properties dialog, choose Text-Only from the drop-down menu.

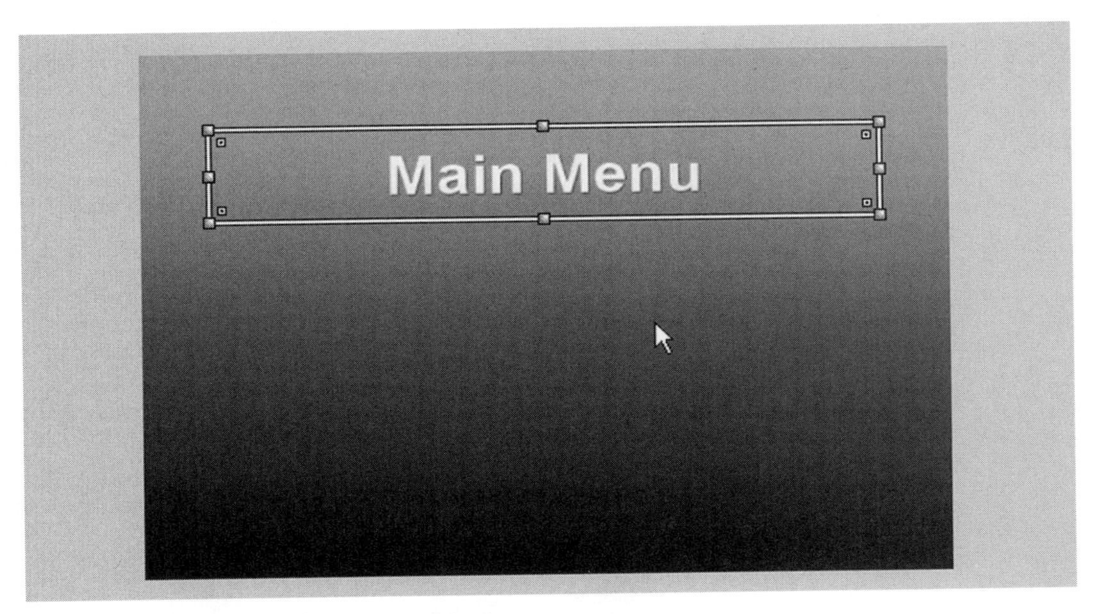

11.12 Selected text is indicated by a surrounding box.

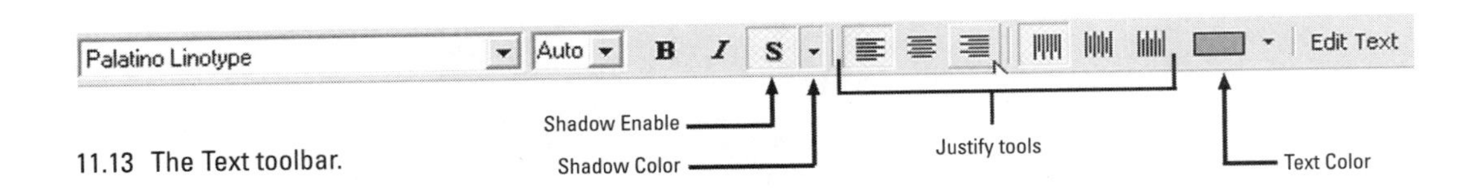

11.13 The Text toolbar.

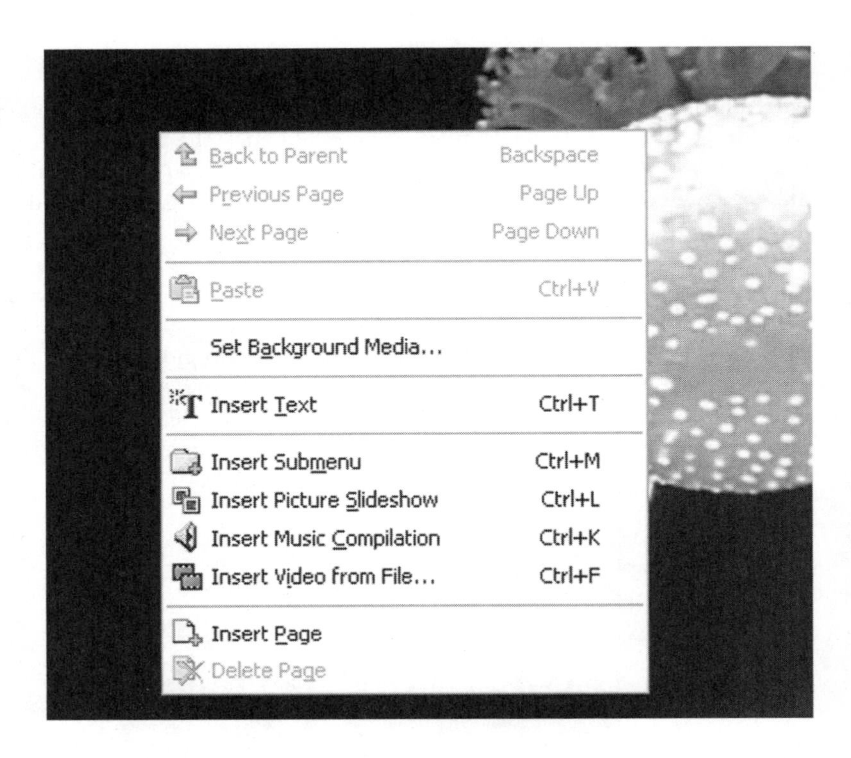

11.14 Right-clicking the workspace calls a menu for placement of text, video, or image.

When working with text or images, it's a good idea to check for safe viewing areas by turning on the Safe Viewing Area grid, which is found in the Options menu. Notice that in Figure 11.15, the word *Jellyfish* is outside the safe title area. Therefore, when the word is displayed on most NTSC or PAL television sets, the text is not completely visible. On a computer screen, it is always visible.

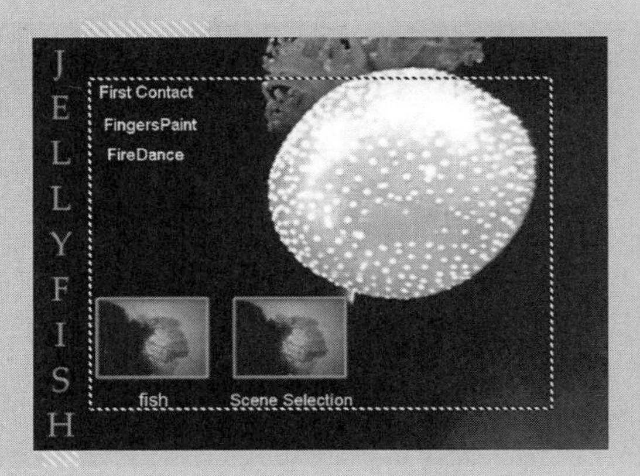

11.15 Checking the safe viewing area.

Insert text to create the name of the DVD title. Select the font, color, and shadow for the text. You don't need to be concerned about text size at this point, as text can be dragged to any size at any point. Be aware that pressing ENTER after creating a text box creates a new line in the text box. Click outside of the box instead.

To create a drop-shadow on the text, click the Shadow Enable button on the Text toolbar. Clicking the down arrow next to the Shadow Enable button provides a drop-down menu from which you can select the offset, blur, and color of the shadow.

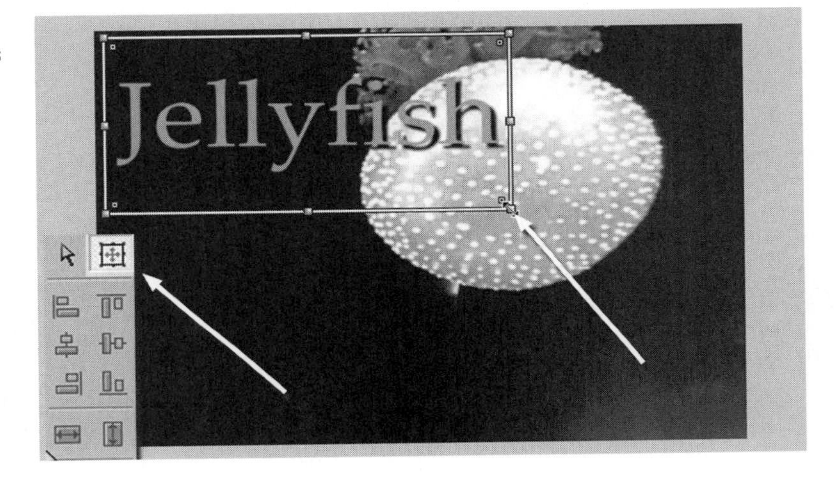

11.16 Using the Sizing tool allows text or images to be resized.

To create a text event that is vertical on the screen, type the desired letter, press ENTER, and type the next letter on the next line. Letters can be justified for center, right, and left and can be justified to the top, bottom, or center of the screen.

11.17 Creating vertical text events.

Setting Properties/Preferences in DVD Architect

Selecting PREFERENCES | OPTIONS opens a dialog box with four tabs: General, Editing, Preview, and Burning. In the General tab, the default settings are highly recommended. While you can experiment with these, the only box that may be safely unchecked in all instances is the "Show splash screen on startup" box. This option prevents the DVD Architect splash screen from being displayed on startup. Disabling this preference improves the load time of the application by a small factor.

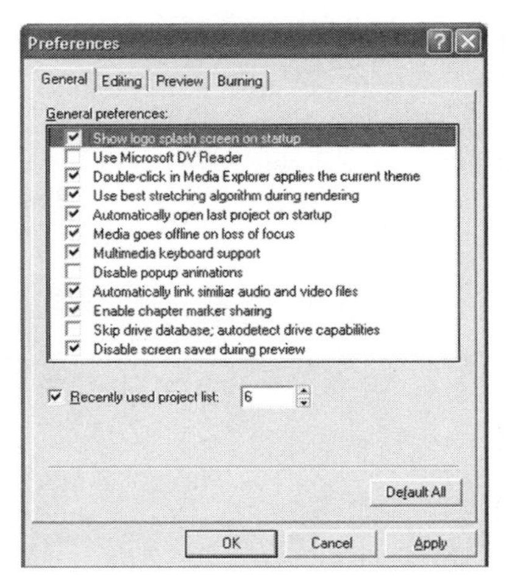

11.18 User preferences can be selected in DVDA, offering performance tweaks for users.

In the Editing tab is an Undo Options option. By default, this setting is set to 1,000 undo points. For faster system performance, reduce the number of time undo can be performed to a more reasonable level, such as 10–25.

In the Preview tab, there is little to change, and nothing that affects the quality or behavior of a DVD burn. This tab controls the speed and amount of rewind/fast forward times in the Preview.

In the Burning tab, Media Size settings can be applied, up to 4.7GB of storage. Near the bottom of the dialog is a Temporary Files folder setting. For faster performance, set this temporary setting to a drive other than the default `C:\ drive`.

Creating Buttons and Linked Events

Creating chapters, buttons, and link points are some of the primary functions of authoring a DVD. DVDA brings these to a new level of function and efficient DVD authoring by continuing Sonic Foundry's format-agnostic/resolution-independent application protocol. DVD authoring applications are typically capable of only recognizing a limited group of formats, and successfully writing a DVD has been based on meeting these format requirements. However, DVDA recognizes AVI, MPEG, JPEG, WMV, WMA, MP3, MOV, and other file formats. All of these formats encode to an MPEG-2 file when the authoring stage is complete.

In the Explorer window, locate media to be used in the DVD. Drag it to the main workspace. A button is automatically created. The button can be edited later on; for the moment, we'll work with the button as it appears.

The button and text field are separable; we'll come back to this later in the chapter. The button will show the first frame of the media. This option is adjustable in the Object Properties window. Click the Object tab found in the Page Properties window, which opens the Object Editing window.

In the top of the Object Properties window, the title and location of the file linked to the new button are displayed. A Browse button is also available that allows a link to be edited quickly. In addition, a slider and Thumbnail Start selection tool are located just below the Browse/File Location window. Moving this slider moves the image in the button on the main workspace. This action allows a specific frame of video to be selected as the thumbnail seen in the button on the desktop. Selecting the Animate Thumbnail check box below the slider automatically shows the video in the thumbnail, creating a Motion Menu button.

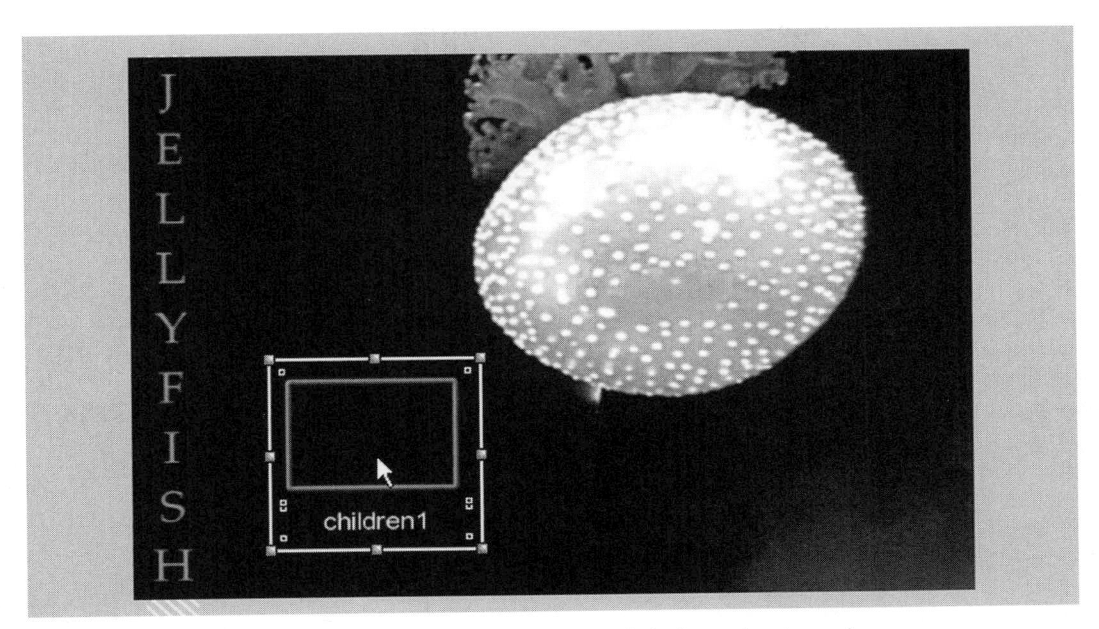

11.19 A button and text field are automatically created when media is dragged to the workspace.

A menu button can be resized at any time by using the Resize tool found in the toolbar on the left of the screen. Select a menu button or text and click the Resize tool at the left of the workspace. Click the edge of a menu button or text selection and drag the menu button or text to the desired size.

The Object Frame menu allows different button shape selections to be made. Animated buttons and static buttons are available in this selection of buttons through the drop-down menus.

In the event of an animated button, the opening frame of the button can be selected with the slider in the same manner as the animated button. If a button is animated, the animation will begin at the selected start frame; however as the animation loops, it will loop over the beginning frame if the selection is longer than the menu length.

Beneath the Object Frame selection is a Modify Link button. This button opens the selected object properties in an editing window. A timeline is available for inserting chapter points and in/out points of the media. Double-clicking the media accomplishes the same task.

Notice that a Start Chapter marker is already inserted in the timeline. This marker may be moved to a different location with different in/out points on the timeline. Placing the cursor at the desired start point and pressing S will create a new start chapter point. To create an in/out point, place the cursor at the desired in point and either press I or click the In Point button on the toolbar at the bottom of the Preview screen. Place the cursor at the desired out point and press O or click the Out Point button on the toolbar.

In this window, markers can be inserted to create chapter points. To insert a marker, press M. In the toolbar at the bottom of the screen is a Marker Insert button as well. Markers can also be inserted by double-clicking at the desired point, automatically inserting a marker/chapter point. Markers can also be auto-named if this option is selected in the OPTIONS | PREFERENCES | GENERAL dialog.

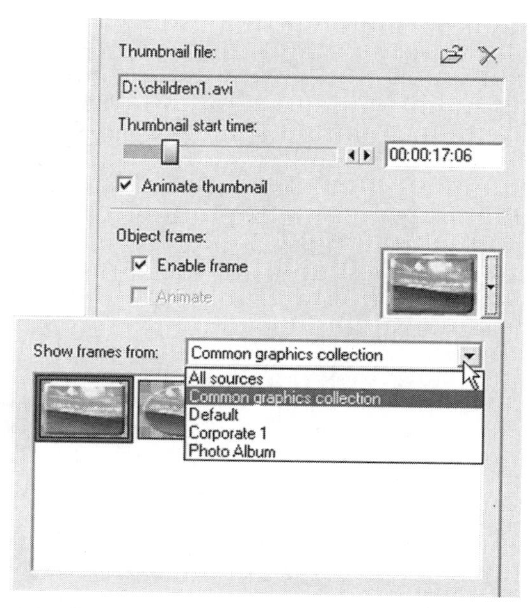

11.20 Choose a button shape and style from the drop-down menu.

Markers inserted in Vegas are also recognized by DVDA, as they are embedded in the file. Clicking the Load Markers button on the toolbar loads all markers inserted in Vegas and makes them recognizable as chapter points. Markers inserted in DVDA in a video file and saved by clicking the Save Markers button on the toolbar can be seen on the file when opened in Vegas' Trimmer tool, where they can be edited for position in the file.

Place at least two markers in the timeline in order to proceed to the next section. (This step is not necessary for a project but is necessary to understand how to create scene menus.)

Clicking and dragging in the timeline view of this window allows video to be scrubbed for in/out points, marker points, or thumbnail points. Place the cursor on a frame that contains an image that should be the thumbnail view and press T. A Thumbnail View button is also available on the toolbar at the bottom of the screen. When a thumbnail is selected, a blue marker is inserted in the timeline, indicating the location of the marker. This location can be edited by placing the cursor on another frame and pressing either T or clicking the Thumbnail button.

In the lower-right corner of the timeline is a time indicator, which shows total length of the timeline media, selected time, and cursor position.

Creating Scenes

Return to the main workspace by pressing BACKSPACE or clicking the up arrow at the top of the Preview window, shown on the right.

Scenes are chapters or marked spaces in a main video. Typically, the DVD hierarchy works in the following order:

- The Introductory video leads to
- The Main Menu page containing links to
- The Main Video containing links to
- The individual chapters, also called scenes, and/or slideshows, music compilations, or other videos, such as bonus videos, behind the scenes, interviews, director's cuts, deleted scenes, and other bonus material

Upon return to the main workspace or parent window, right-click the button inserted previously. Select Insert Scene Menu from the menu. DVDA asks for a Scene Page title and the number of scenes desired. DVDA by default selects six scenes as the default number of scenes. If fewer scenes are needed, however, DVDA does not insert them. Additional scenes can easily be inserted as well. Click OK when the scene page title is created and number of scenes has been selected.

A new button will appear on the workspace. This button can be animated and selected as a thumbnail by moving the slider in the Object Properties window.

Double-click the Scene button, which opens a new workspace, showing all the chapters that were selected in the timeline view.

As with both menu buttons and scene buttons, each chapter button can be selected for thumbnail view, animation, and button type in the Object Properties window. Click a button, move to the Object Properties window, and apply properties to the button as desired.

You'll want to save your project at this point if you haven't been saving all along. DVDA saves the project as a DAR file. If you need to locate a missing DVDA file, search in the Windows Search/Find feature by using **.dar* and Windows will locate all files created by DVDA.

When saving a file, DVDA only saves references to media in the project and does not save the media itself. The only time DVDA saves media is when a project is ready to be rendered and/or burned to a disc.

If creating a basic DVD is the goal, DVDA now has enough information to burn a disc. However, read on, as DVDA is much more powerful.

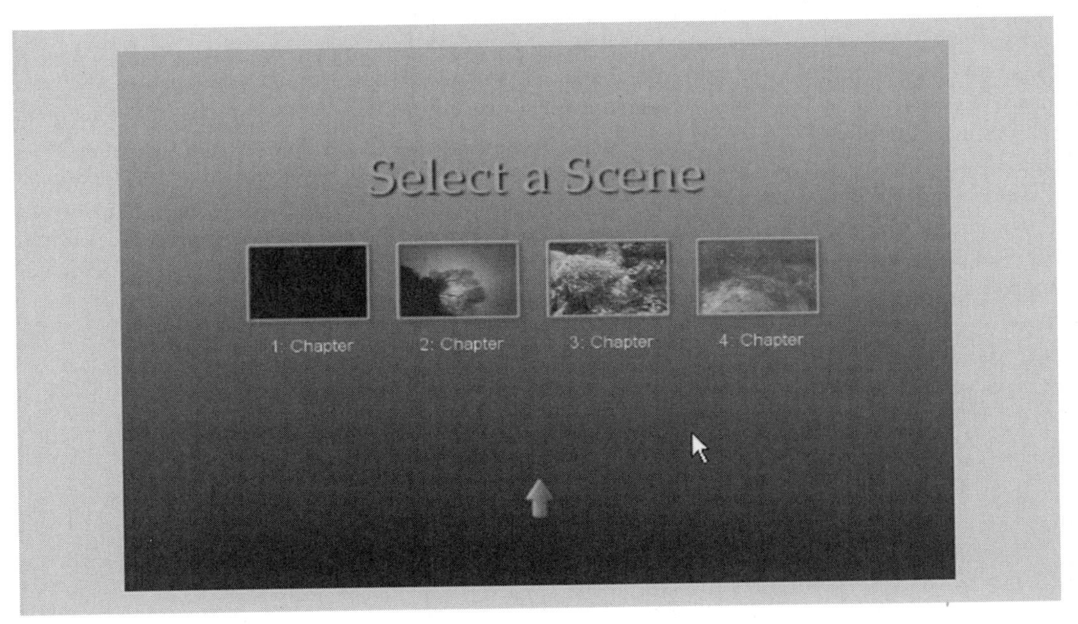

11.21 Scene buttons are automatically created with chapter markers.

Advanced Tools in DVD Architect

DVDA has tools to allow for very complex menus with motion and background music. As shown earlier, creating a motion menu is quite simple and can be done by checking a box in the Object Properties window. Your own photographs or other video objects, however, can be inserted on the timeline and linked to project media.

Inserting a background image on scene pages is as simple as dragging an image to the workspace in the scene menu. Right-clicking the workspace in any scene or workspace page allows a background image to be selected and placed. Backgrounds can also be selected in the Page Properties dialog box at the right side of the main workspace.

If background music is desired, locate the audio file to be used as a background file in the Explorer and drag it to the workspace, or select a file in the Page Properties/browse window. Audio in the project will either play at full length or can be time-selected as a loop. In the Page Properties window, uncheck the Auto Calculate Length box to allow the menu length to be specified. Using short audio clips, such as those found in the Sonic Foundry ACID loop libraries, is great for this sort of menu selection, as the clips are pre-loop edited for seamless menu looping. Music is encoded to the DVD as an AC-3 file, saving space on the DVD disc and allowing for longer video files.

As media files are dragged to the workspace surface and a button/text indicator is auto-created, either the button or text link may be eliminated, leaving only a button or text event as a link. Initially, the button and text are linked together. You can remove either link, however, by clicking one or the other and removing it using Delete or pressing CTRL+X.

In the Object Properties tab, image or text linking can be specified. A Modify button that opens the link in its related page is contained in this same window. If a button linked to media is selected and the Modify button clicked, the media opens in the timeline view for chapter placement, thumbnail selection editing. If the button is linked to a menu or scene selection, the Scene Selection editor opens. This same task can be accomplished by double-clicking either button style.

Scenes may be linked to submenus as well. This process is useful in the event that multiple videos comprise the DVD contents and each video is broken down into individual chapters. Right-click the scene button and insert a submenu from there. Double-click the new submenu, and the media opens in the timeline editor to insert chapter markers.

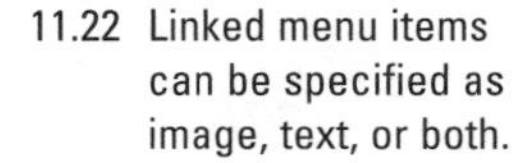

11.22 Linked menu items can be specified as image, text, or both.

Create a menu button in DVDA and double-click to open the editing window. This menu button may be either a submenu or main page menu. Insert markers (M) for scenes, giving each chapter marker a name. Now press BACKSPACE to return to the workspace where the menu button lies. Right-click the menu button and select Insert Scene Selection Menu. A new page opens, containing all the scenes for which markers were created, including the names entered for the chapter markers. If it is decided later that these chapter markers should appear on the Main menu page, select the chapter markers that should appear on the main page and copy/paste them to the main page. The submenu button may be deleted or may still link to the submenu containing identical chapter buttons. This is one method of bringing several individual chapters to the main menu space in DVD Architect.

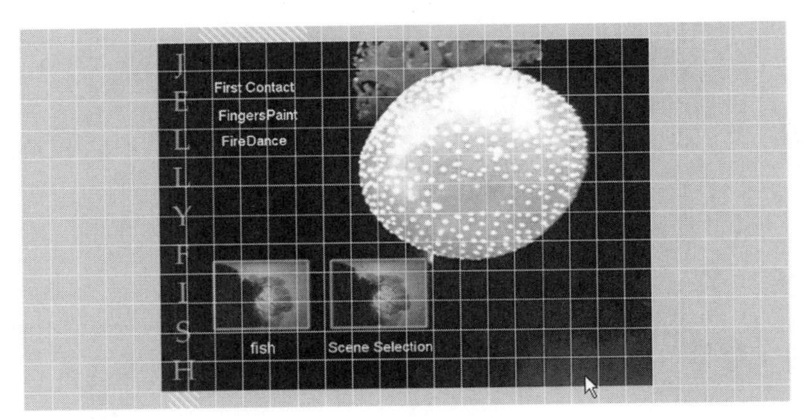

11.23 Grid alignment assists in having menu items maintain straight lines.

All buttons and text can be snapped to a grid in DVDA. This process helps ensure that all images, buttons, and text boxes will be aligned. Turn the grid on by pressing CTRL+G or by selecting OPTIONS | SHOW GRID. This step creates a grid on the screen that images and text can be snapped to, when snapping is enabled (press F8). Grid parameters are adjustable (press CTRL+SHIFT+G) from the OPTIONS | GRID SETTINGS dialog.

You can quickly align menu buttons, text, or images by clicking and drawing a square around multiple items and selecting the alignment tools on the left of the screen. You can align items to the left, right, top, or bottom.

Returning to a main menu from any submenu is a typical function of a DVD. This function is often referred to as a *back button* in various applications. In Vegas, this function is known as a *go up* action. This action can be set up by selecting OPTIONS | PREFERENCES | GENERAL and checking "Enable Go Up button in videos." This option is not set by default because some DVD players do not support this function, and timecode displays and chapter lists might not be viewed with this function enabled. This button is not active during video, slideshows, or music compilations. The button is a navigational aid.

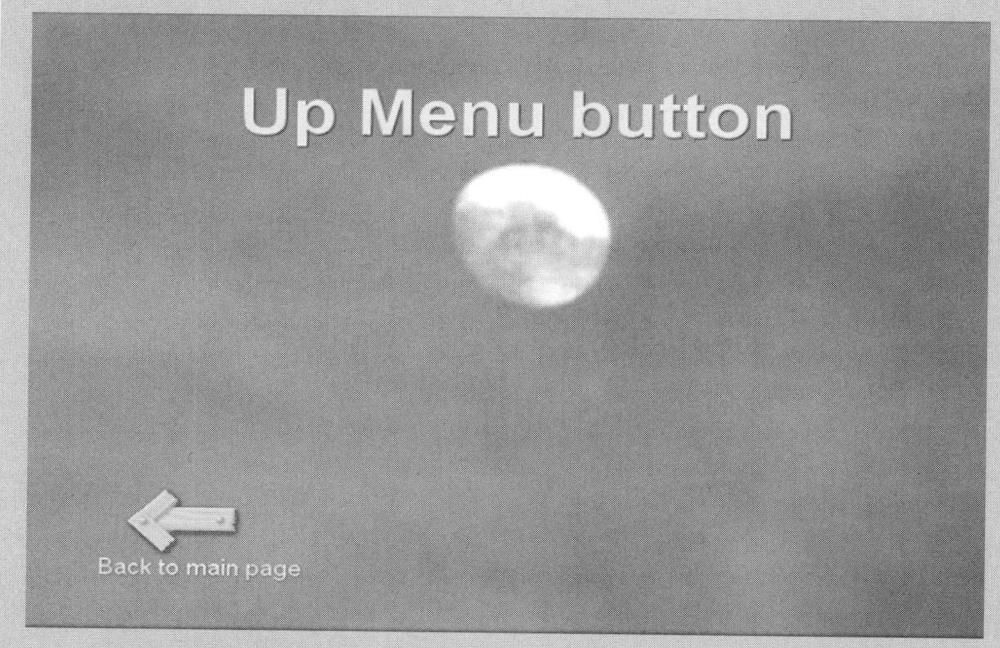

11.24 Setting a Go Up button in place on a menu. This button returns viewers to the last menu linked.

Working with 24p Media in DVD Architect

If you're using the MainConcept MPEG-2 encoder in Vegas 4.0, use the DVD Architect NTSC video stream or DVD Architect 24p NTSC video stream template to render your video stream. (You'll need to render your audio stream separately according to the parameters listed in the AC-3 audio or PCM audio headings.) The MPEG-2 files are not recompressed, and AVI files are compressed at the correct rate when selecting the 24p template. Previews of 24p media are correct in DVDA and are previewable with either AC-3 or PCM audio.

24p files allow for up to 20 percent greater disc efficiency, as the files are smaller. 1080i and 720p support coupled with 24p makes for an awesome picture quality when encoded to MPEG-2.

Check the Help section of DVDA for more information relating to 24p in DVDA, as updates have become available since the publication of this book.

Inserting Slideshows and Soundtracks

DVDA can display a slideshow on a DVD or provide the ability to play sound track or audio information. Either can be accessed using a separate button or as a single button that plays a slideshow or playlist.

To insert a slideshow in DVDA, right-click the main workspace or submenu and select "Insert picture slideshow" or press CTRL+L. This step inserts a new menu button on the workspace. Double-click the new button and a dialog is displayed in which photo images can be placed. DVDA recognizes most common photo formats such as JPEG, GIF, BMP, TGA, PSD (no layer detection), TIFF, and PNG. PNG format files are generally recommended, as they are an uncompressed format and do not require an external reader such as TIFF files do. Images should be 655 × 480 (NTSC) or 704 × 576 (PAL) for the slideshow, unless photos are to be zoomed or letterboxed or the aspect ratio is to be disregarded.

Drag images from the Explorer to the Slideshow dialog box. When photos are dragged, the name of the file is automatically displayed.

The filename can be deleted from the slideshow manually or automatically. To manually remove text from the slide, select the text and Delete. To prevent filenames from appearing on any slide, select OPTIONS | AUTO INSERT SLIDE TEXT. This step un-checks the preference.

Each slide may be shown for a preselected length of time or for the default length of time of five seconds. To select a time for each slide, double-click the time selection box and input a new time. An alternative is to press CTRL+A, double-click the time selection box, and insert a new time. All slides will now appear for that length of time.

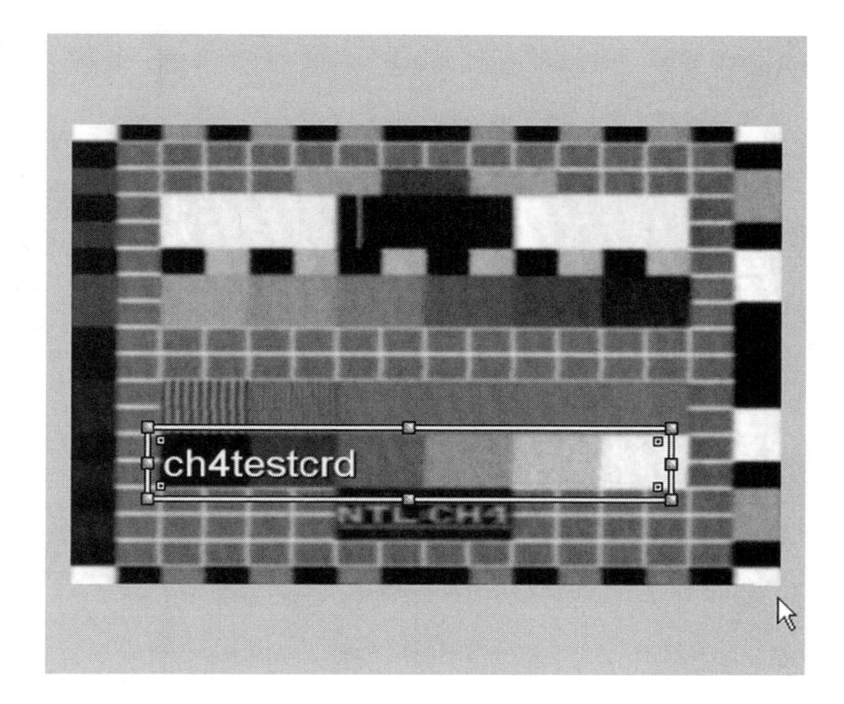

11.25 Filenames appear in the slideshow automatically.

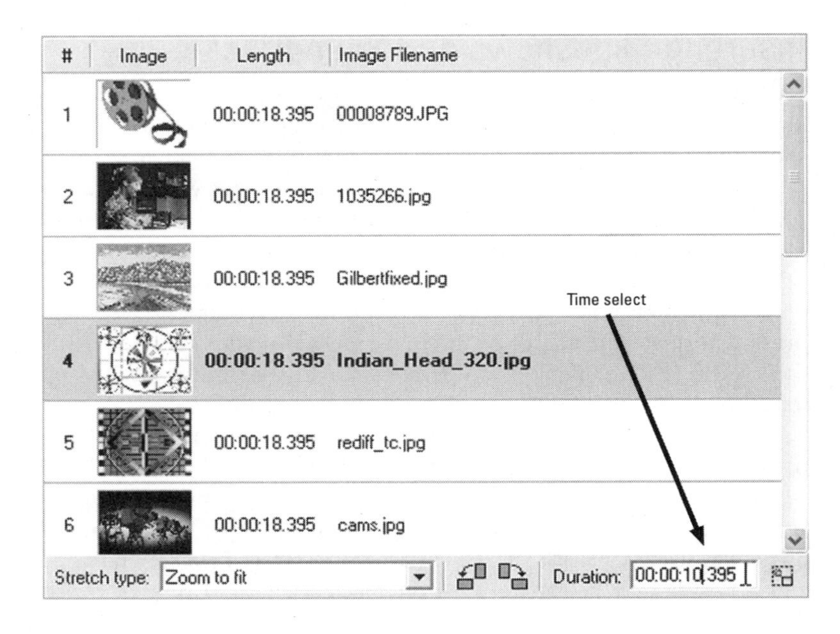

11.26 Time selection is managed in the Slideshow Properties page.

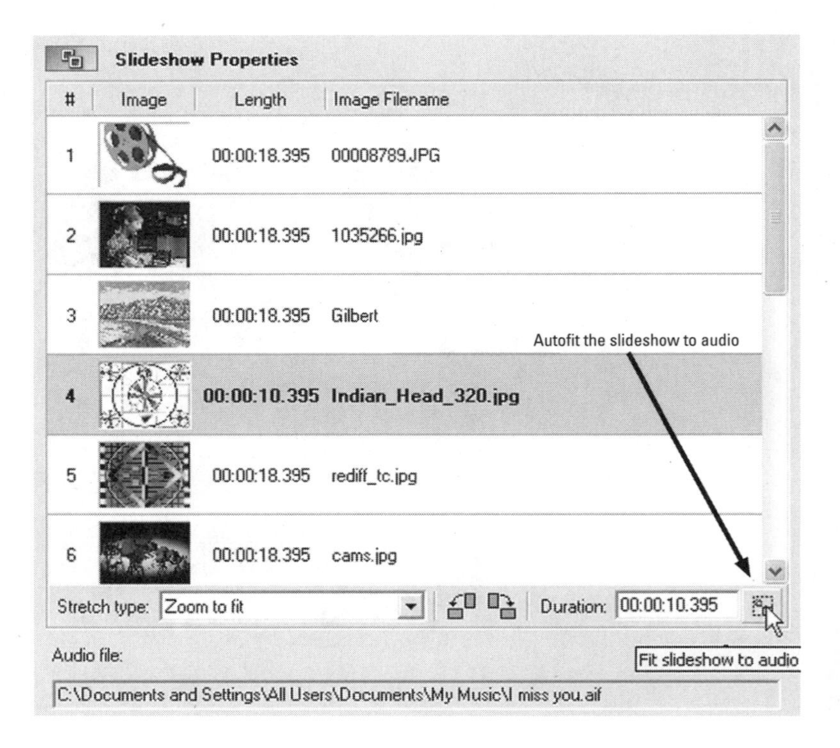

11.27 The length of time that an image appears can be controlled by the length of an audio file.

A group of images, regardless of format, can be selected in the Explorer and dragged to the Slideshow Properties box. Press CTRL+A to select all photos in the Properties box and select the cropping format, Letterbox, Zoom-to-fit, or Stretch-to-fit. This step adjusts all images in all formats to the desired size.

Images can also be rotated to fit the screen or to be viewed at the correct aspect. To rotate an image, click the Rotate Clockwise or Rotate Counterclockwise buttons in the Slideshow Properties window next to the Letterbox/Zoom/Stretch drop-down menu. Pressing CTRL+A rotates all images simultaneously. This workflow is very fast when placing several images that might have an aspect ratio more conducive to viewing vertically rather than horizontally.

Audio can be selected to accompany the slideshow. Audio can be inserted into the slideshow using two different methods. Select an audio file in the Explorer window and drag it to the slideshow workspace. This step inserts the audio into the slideshow. An alternative is to select an audio file in the Slideshow Properties page, in the lower-right of the page. DVDA accepts WAVE, AIFF, MP3, PCA, or WMA as audio file types.

The slideshow can be auto-timed to the audio file by clicking the "Fit slideshow to audio" button. The length of the file is divided by the number of slides, and each slide/image show is divided by the sum of the divided audio length. For example, a 4-minute slideshow with 100 images displays each image for a length of 2.4 seconds (240 seconds [4 minutes] ÷ 100 images = 2.4 seconds per image). A common length of a slide view is four to five seconds in the average slideshow. Using the Slideshow tool coupled with narration, however, is great for a corporate presentation, an electronic travel brochure, or other description of the image on the screen.

Inserting an Audio Playlist

DVD Architect also has the ability to create an audio playlist, known as a *music compilation*. Obviously, this playlist does not necessarily have to be music; it can be spoken word, sound effects, or any other audio format. Files can be individually linked, or managed as a compilation of files playing one after the other, depending on user preference.

To create a button linked to audio files, right-click the main workspace or page workspace of a submenu or scene. Select Insert Music Compilation. This step places a frame/button on the workspace. A graphic can be dragged to the newly created button to provide an image button. Double-click this button to open the Music Compilation workspace.

Multiple tracks can be dragged to the Compilation Properties window and reordered to suit personal preference. The title of each file is displayed automatically in the Preview window while each audio track plays. The text can be edited for name, position, font, and color, or deleted from the Preview. To prevent track names from appearing in a Music Compilation, while in the Music Compilation workspace, select OPTIONS | AUTO-INSERT TRACK TITLE. Uncheck the box, and the filenames will no longer appear on audio tracks.

Images can be inserted to be viewed during playback of the audio track. The image remains static during playback of the audio. To insert an image, select a file in the Music Compilation Properties window. In Explorer, locate the desired image and drag to the workspace window.

The image can be letterboxed, zoomed-to-fit, or stretched-to-fit, which is selectable by user preference.

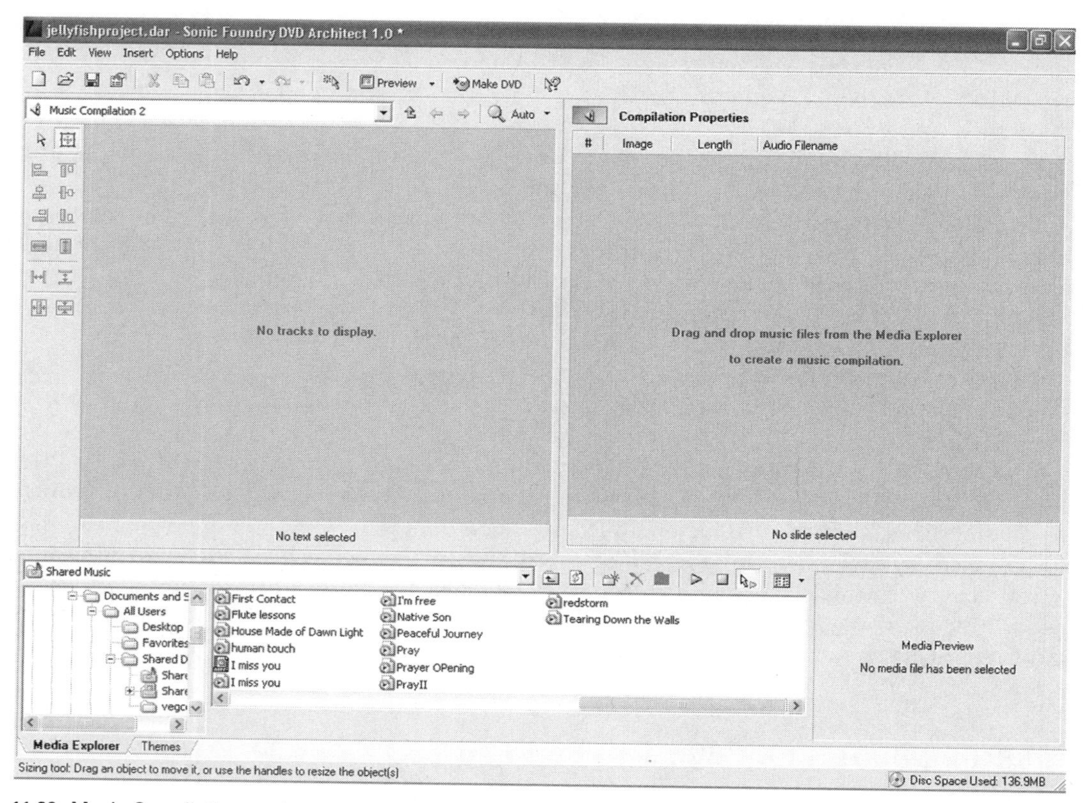

11.28 Music Compilation workspace.

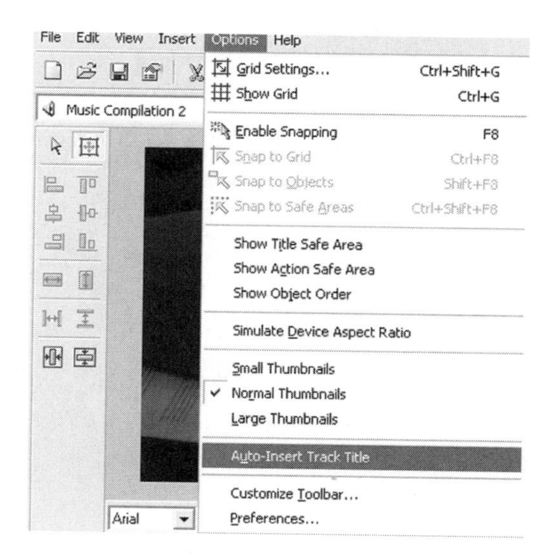

11.29 The preference to view audio track filenames on screen during playback can be selected.

11.30 Images can be displayed during audio file playback.

11.31 Multiple graphics can be inserted over background images during music compilation playback.

Tip

The Music Compilation tool can also be used as a sequential video player, where at the end of a video event, another video starts from the sequence. This feature is useful for training videos requiring chapter menus or sequential playback in which the main video is a separate project from the chaptered videos.

Multiple graphics can be displayed on top of the background image during audio track playback. This feature is great for displaying an album cover over an artist's image or multiple artists' images above an album cover. To insert an image over the background, right-click the workspace/background image and select Insert Graphic. This graphic may not be linked to media; it acts only as a displayed graphic image. As many graphic images as desired can be placed on the background image.

Clickable links to audio can also be created as text-only, image-only, or text-and-image links. This feature is useful for creating user-selectable playlists or single-song playlists. A musical artist, spoken word book author, or public relations project would find this useful, allowing clients or users of a DVD to select what songs they'd like to hear.

You can insert clickable audio links by right-clicking the workspace and selecting Insert Media from the menu or by pressing CTRL+F. Multiple audio files can be dragged from the Explorer to the workspace, creating buttons and text links automatically. (If a single audio file is dragged to the workspace, the file becomes a background audio file. To insert a single file as button-linked, right-click and drag the audio from the Explorer.) If text-only links are desired, select all audio files by pressing and holding down SHIFT and clicking each file or by clicking and drawing a square around each audio file link. In the Object Properties view, select Text-only from the drop-down menu. Select Image-only if an image is desired for the link display.

Inserting an AC-3 Audio File

DVDA supports 5.1 surround and Stereo AC-3 audio file formats. Files can be encoded to AC-3 in Vegas as a separate audio file that is a companion to video files. Audio is imported as a separate file format in DVDA. The video file is rendered as an AVI or MPEG in Vegas, and the audio is rendered separately in Vegas as an AC-3 file format. If video is rendered as an MPEG video/audio file in an NLE application, such as Vegas, the audio is rendered as an MPEG Audio format. DVDA requires that this audio be re-encoded as AC-3 or as a PCM audio file. It is recommended that files encoded outside of DVDA be encoded without audio, as an elementary stream. The extension associated with the elementary stream is .M2V, or MPEG-2/video. Audio should be rendered as an AC-3 in the editing application when rendering elementary streams.

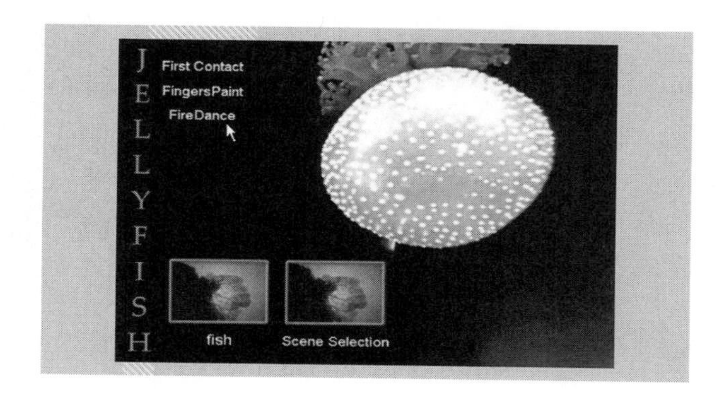

11.32 Text-only links to audio files can be created quickly by dragging several audio files to the workspace and selecting Text-only from the Object Properties drop-down menu.

Tip

Remember to check button locations to ensure that buttons and linked text are inside the safe viewing zone.

When you insert video into a DVDA project, the video is imported to DVDA by dragging the video to the workspace, and the accompanying audio is replaced by the AC-3 file in the Media Properties menu. Double-click the media file to navigate to the Media Properties menu.

Audio files encoded at bit rates higher than 448 Kbps are not recognized by DVDA, as the DVD standard does not allow for bit rates higher than 448 Kbps.

Tip

When audio and video-only files are of the same name and in the same root directory, DVDA automatically opens the AC-3 audio file when the related video-only file is opened. This attribute can be selected by visiting the OPTIONS | PREFERENCES | GENERAL dialog and checking the "Automatically link similar audio and video files" check box.

Tip

PAL DVD authors should be aware that AC-3 5.1 surround files might not play in all PAL-format DVD players This issue predominantly affects older players. If a warning appears in your project when it is time to burn the DVD, and authors are certain that DVDs will only be played in compliant PAL DVD players or NTSC DVD players, this warning may be ignored. Otherwise, PAL DVD authors should use LPCM (WAV) files at 48KB/16bit settings.

Replacing Button Images on Menus

While buttons linked to video and slideshows display the first frame or image in the video or slideshow, the button image can be easily changed to any image desired. To replace the image properties of a button, click a button containing an undesired image. Locate the replacement image in the Explorer. With a button/link selected on the workspace, drag the replacement image to the selected button. The image in the frame will be replaced with the new image.

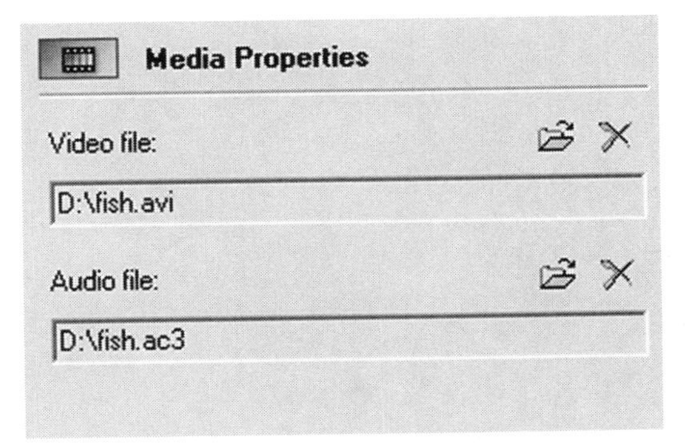

11.33 AC-3 files are imported as audio files in the Media Properties menu.

Fading menus into video after a button is selected is a creative way to make a transition between the menu and the selected video. In DVDA's Preview window, copy the menu as a still shot and place it on the clipboard. Open it in your favorite photo editor, save, and place on the timeline in Vegas. Fade the menu image into the opening of the video. When a button is clicked from the menu, the menu appears to fade seamlessly into the video. Chapter points can be placed on the video after the menu fade, so that when the chapter locator is selected, the viewer does not see the faded video.

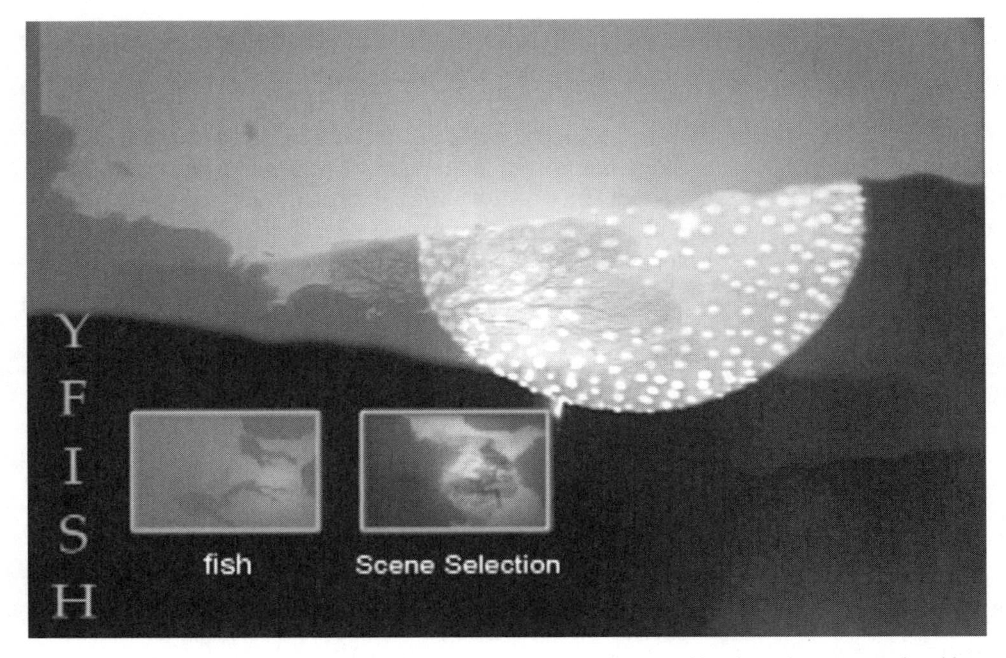

11.34 Menus can be made to fade into video to create a seamless transition from the menu to the video.

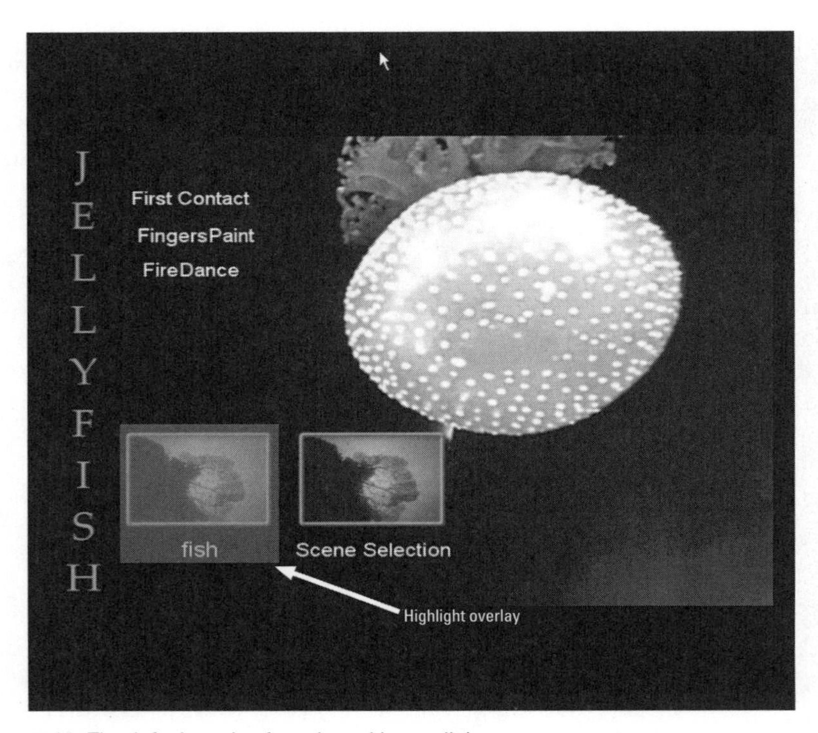

11.35 The default setting for selected button links.

Buttons are by default highlighted with a semi-transparent overlay. This setting indicates to viewers which button link is selected. This setting can be edited, however, to provide a mask overlay, an underline, or other indicator choices. Changing the highlight to a simple image mask is common on many professional DVDs. The user can also define highlight color and opacity.

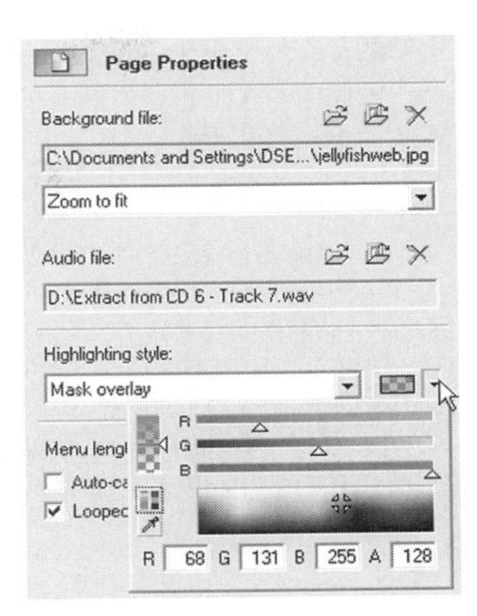

11.36 Highlight style, color, and opacity can be selected in the Object Properties menu.

Previewing the DVD Project

DVDA allows the project to be previewed in active form at anytime by clicking the Preview button at the top of the toolbar above the workspace area, or by pressing F9. This step opens the active workspace in the Preview Project screen, where the page is shown in the same way that it will appear on a television or computer screen. Animated buttons, menus, background images, and music can all be previewed in this view.

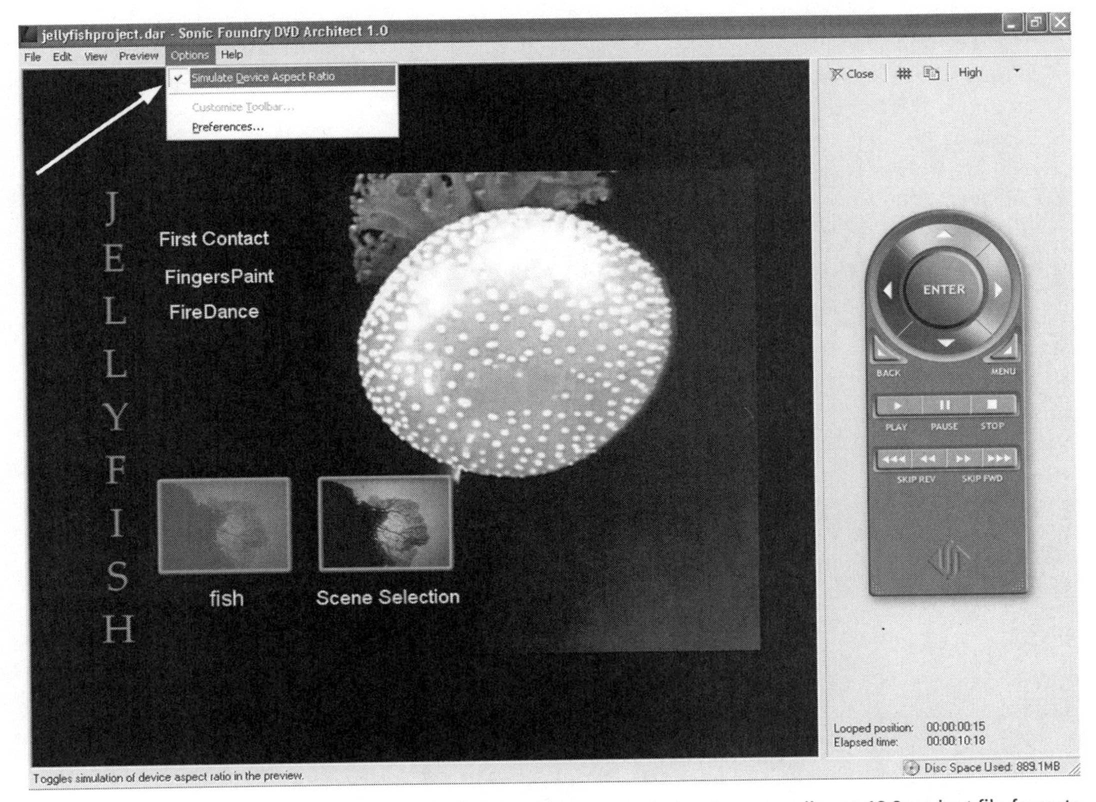

11.37 Be sure that Simulate Device Aspect Ratio is enabled, particularly when encoding to 16:9 project file formats to ensure correct previewing.

The remote control acts the same as most DVD remote control devices, allowing you to skip, fast forward, reverse, return, and navigate menus. Use the remote control to navigate through various layers in the DVDA project, so that all aspects of the project can be checked for consistency, placement, and color.

If you encounter difficulty when previewing AVI files, select OPTIONS | PREFERENCES | GENERAL and click the Use Microsoft DV Reader check box. This setting usually alleviates preview issues by forcing the use of the Microsoft DV codec to read files.

The DVDA Preview Project window allows for a wide variety of functions and testing. Under the Options menu is a menu selection for viewing the project at correct aspect ratio. On the right side of the Preview Project, a toolbar is displayed. Clicking the View Safe Areas button displays a safe grid. In addition, a Copy to Clipboard button is next to this grid, allowing a screen shot to be exported to an image editing application or to Vegas 4.0.

A valuable use of the Copy to Clipboard feature is the ability to preview a DVD menu on an external monitor before rendering and burning a disc.

With a menu on the screen, set the preview quality to Best/Production Quality in the Quality menu in the upper-right corner of the Project Preview screen. Click the Copy to Clipboard button.

Open Vegas 4.0, create a new track, and insert the video or an image on that track. It is necessary to have an image in the Preview window of Vegas to accomplish this trick.

In the Vegas Preview window, enable the Split Screen tool and click the menu, selecting Clipboard. The DVD menu screen now shows in open areas of the Vegas Preview screen and can be viewed on an external monitor. Drawing a full square in the Vegas Preview window that extends from top-to-bottom and side-to-side of the Preview area fills the screen with only the DVD menu.

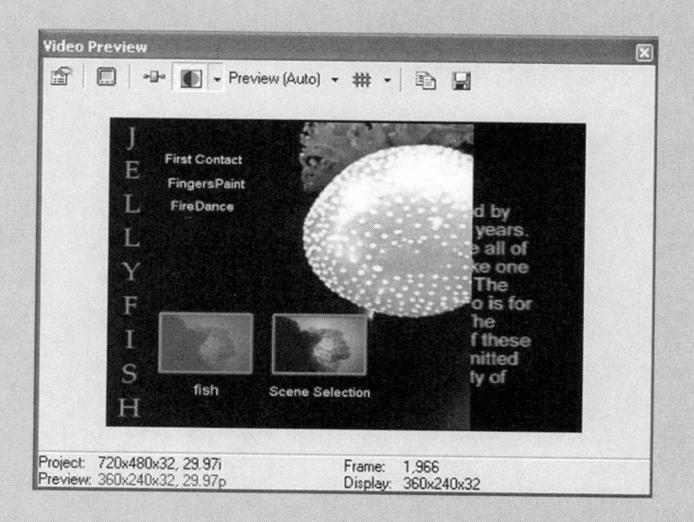

11.38 DVD menus can be previewed on an external monitor from the Vegas Preview window.

This feature is valuable as it allows a menu screen to be previewed on a monitor before encoding and displays any artifact or anomaly in the menu.

Next to the Copy to Clipboard button is a Preview Quality button. This button selects the preview output quality at preview but has no bearing on the final render. Viewing a project at Best/Production Quality may be too high a resolution for slower processors; use the default high quality if you can view the video at normal motion and frame rate.

Determining Bit Rates

With DVDA, determining bit rates is not necessary as DVDA figures the bit rate for you based on the project. If this figure is to be overridden, however, the formula is 600 divided by project minutes. (This is a good rule of thumb for the total bit rate in Mbps of a 4.7GB disc. Note that the rule of thumb is only relative to 4.7GB discs.) The reason it works is the following:

$$4.7\text{GB} \times (1 \text{ min} \div 60 \text{ sec}) \times (8 \text{ bits} \div 1 \text{ B}) = 624 \text{ Mbps} \times \text{min}$$

As an example, for a 90-minute project, 600 ÷ 90 = a bit rate of 6.6 Mbps.

Also, the bit rate of AC-3 stereo can range from 0.064 Mbps to 0.448 Mbps depending on how it was encoded. Dolby's recommended bit rate for stereo is 0.192 Mbps and for surround is 0.384 Mbps. That's the bit rate at which DVDA encodes AC-3 audio.

Burning the DVD

After all aspects of the DVDA project have been previewed and are complete, the project is ready to burn to disc. Before burning, the project should be checked for optimization. This step is done by selecting **FILE | OPTIMIZE DVD.**

Files show either exclamation points or check marks beside them. Files with a check mark will be left as is. Files with exclamation points will be recompressed for the burn to DVD disc. As illustrated in Figure 11.39, all files will require recompression with the exception of the fish. AC-3 file.

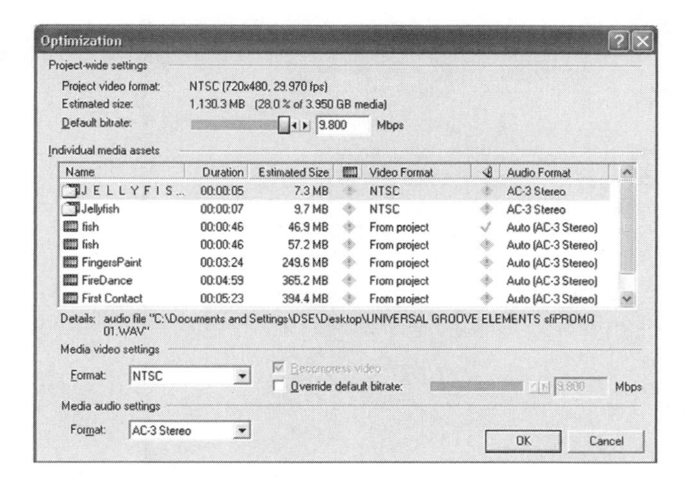

11.39 Nearly all files in this instance shall require recompression as indicated by the exclamation points.

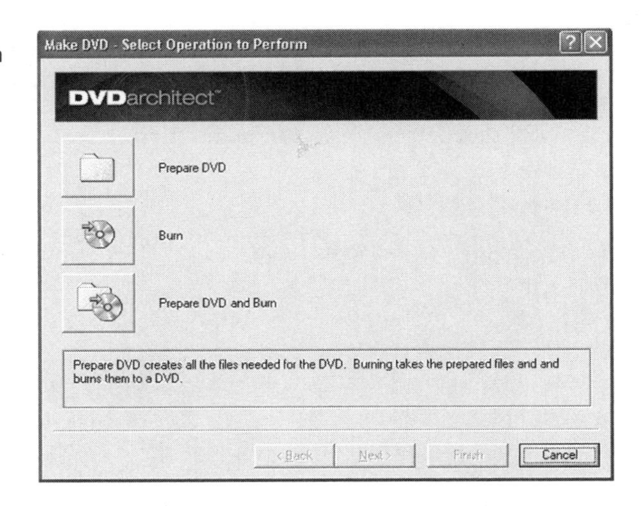

11.40 The first screen seen in the burn process.

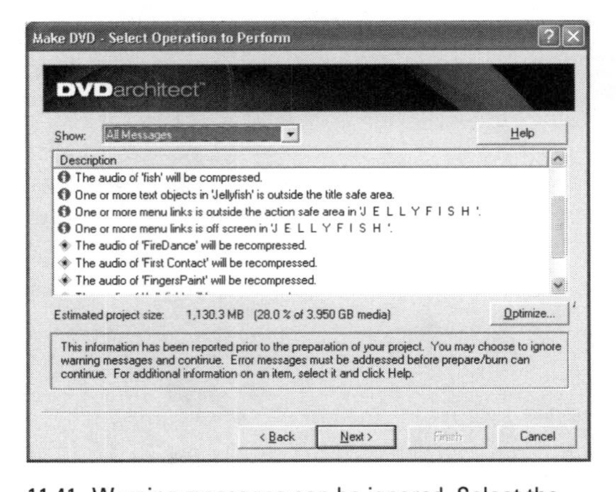

11.41 Warning messages can be ignored. Select the Show Only Errors from the drop-down menu. If errors occur, they must be corrected before a successful burn can happen.

11.42 The final screen before burning the DVD.

To prepare for the burning process, click the Make DVD button on the top of the toolbar. This will open a dialog asking how to proceed with the burn. The choices are:

- Prepare
- Burn
- Prepare DVD and Burn

Preparing the DVD opens a dialog that provides information about any inconsistencies or potential errors in the burn. Warning messages may be ignored; errors must be dealt with before burning.

Selecting Burn prepares the project for burning by building a temp file. Temp files may be pointed to any drive containing enough space to handle the entire file. For best results, point the temp files to a drive other than the boot/application drive.

Prepare and Burn combines the two actions, opening a dialog that displays errors and warnings and offering an option to optimize the disc for size.

After optimizing the media, click Next on the dialog. This step takes you to the final screen. Here, a device and speed can be selected. The project disc can also be named here. This name will be displayed in some set-top units on an LCD display and will be displayed in the drive properties of any computer DVD player. While most media formats are safe to burn at high speeds, fewer errors are found with 1x speed burns. If the DVD is destined for replication, using single-speed settings is best to ensure the fewest number of errors. (All disc burns contain errors, and there is an expectation of a limited number of errors on replication discs. This issue is not a concern unless the replication house error rates are exceeded. Errors are not related to software but to inconsistencies in media and DVD burners.)

The burn process can be tested before burning to check for inconsistencies or errors. Clicking the Finish button starts the preparation/rendering process and subsequently the burning process. Take the finished disc out and test it in your set-top or DVD player.

Tip

DVDA provides an option that is enabled by default that uses bi-cubic stretching during the render process. This feature increases render time while ensuring the finest possible quality of video. Turning this feature off in the OPTIONS | PREFERENCES | GENERAL dialog speeds the render process but does not ensure the highest quality render.

Output and Export

Last Steps with Vegas 4.0

Vegas is unique in that it is currently the only end-to-end NLE or digital audio workstation (DAW). Vegas contains tools for output to every web format, CD/VCD format, serial digital interface (SDI), high-definition broadcast quality, and multiple audio format. Vegas is formidable. No wonder it's the choice of many large encoding and finishing houses!

As mentioned previously, Vegas started life as an audio-only tool. As a result, the audio features in Vegas are what first brought attention to the application years ago. Development for audio has continued, and Vegas is currently one of the most open-ended output DAW systems in existence on any platform.

Output Formats

After a mix is completed, it needs to be output in a format that others can hear or distribute or that a replicator can work with.

Open a multitrack mix in Vegas. Assuming all parts of the mix are satisfactory, the multichannel mix needs to be mixed to a two-track mix that can be put on a CD and prepped for distribution. This process is called the *mixdown*. Mixing down multiple tracks is the next-to-last step in the finishing process.

To mix the multichannel mix to a two-track mix, select FILE | RENDER AS and render the mix to a Stereo WAV mix.

When rendering the multiple tracks to a final two-track mix, be certain that if any buses were used they all point to the master output. This issue only occurs if multichannel hardware has been used, with outputs assigned to other devices.

If the project has been recorded at any sample rate other than 16 bit, 44.1 KHz, the mixdown process is not where the sample rate should be dithered to 44.1. Keep files in their higher resolution state where applicable. They'll be dithered to a 16/44.1 state later in the process.

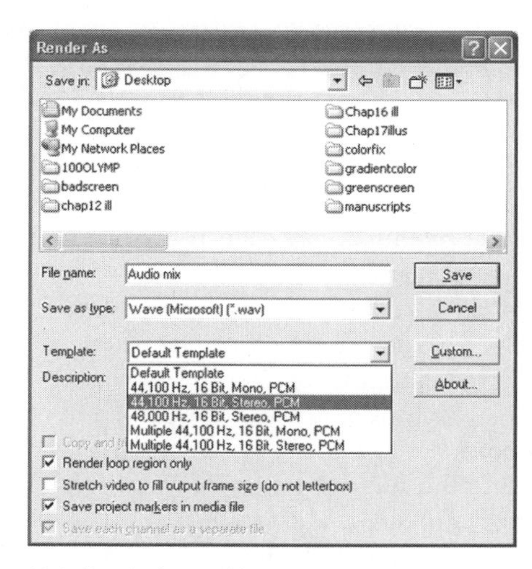

12.1 Rendering multitrack to a two-track mix, ready to be mastered.

After this process has been performed with each song, or multichannel audio project, the multiple two-track mixes are ready to be mastered. Vegas has excellent mastering tools—tools that have been used to master/produce my Grammy-winning and Grammy-nominated recordings.

Mastering

The mastering process is not to be taken lightly. This stage is the final one before sending the master disc to a replicator. It is generally advisable to do the mastering in a familiar room, using familiar monitors. It's rarely a good idea to master on the same monitors that were used in the mixing stage, as it is entirely possible for monitors to mask or enhance problems in a mix. The entire point of mastering is to discover anomalies or errors in a mix and to bring all cuts of a project to a state of consistency so that regardless of sequencing, tracks compliment each other and flow comfortably from one cut to another. Mix on a different set of speakers if possible. Move to a different room, if possible. If nothing else, check the mastered mix multiple times on different stereo systems to assure a consistent and balanced master. In one studio years ago, we installed an FM transmitter that was fed by our master tape deck. We then went out to the car in the parking lot and listened to the mix over an FM receiver. To this day, all mixes at Sundance Media Group are checked on an old Craig boom box, a standard home stereo system, and a high-end car audio system before being sent off for replication.

Mastering is not mixing. *Mixing* is blending several channels of audio together, placing elements in different portions of the sound field while using EQ, reverb, delay, and other tools to create a sense of depth and space. An analogy might be that mixing is like baking a cake, whereas mastering is the decorating part that pretties things up.

Mixing is an entire book in itself, so this section covers mastering only. See Chapter 7 for more information on mixing.

Mastering is taking the 2-channel, 4-channel, 5.1, 7.1, and 16.1 final mix and finishing it for the audio space/delivery mechanism. The primary preparation is making sure that all audio is of maximized level, in phase, properly faded in/out, with all hum, pops, and noises removed, proper compression applied where/if needed, relative level adjustment between cuts of audio matched, and other tweaks applied to give the audio that final sparkle. Mastering is rarely done in the same room as the audio mix was done in, as a different set of speakers in a different environment tells the tale when poor audio practices, sneaky noise, or other anomalies are present but were not heard in the original recording environment. Mastering is also generally done by a mastering engineer, rather than by the engineer who mixed the audio. A fresh set of ears is invaluable at this point, even if it's just a friend with good ears who can make comments on how something sounds.

If you'll be mastering for surround 5.1 or 7.1, be certain that you have a high-quality audio tool, such as the Echo Audio Layla, Mona, Midiman Delta Series, or a similar tool. A low-quality audio card will create problems in any instance at this point. Higher-quality cards allow for synchronous playback of multiple channels of audio, whereas a low-end sound card with surround ability will have slur and phase issues in most instances. Worse still, many low-end audio cards induce noise into the audio path. (With Microsoft's Windows Media 9, you'll need a surround-capable card in order to prep audio for the new codec/delivery tools.) Remember, this stage is the last stage in preparing your audio for the world to hear. If you've worked hard to have a mix, video, or other project that sounds great, you'll want to put your best foot forward, especially when presenting the product for sale or broadcast.

12.2 Delta 1010.

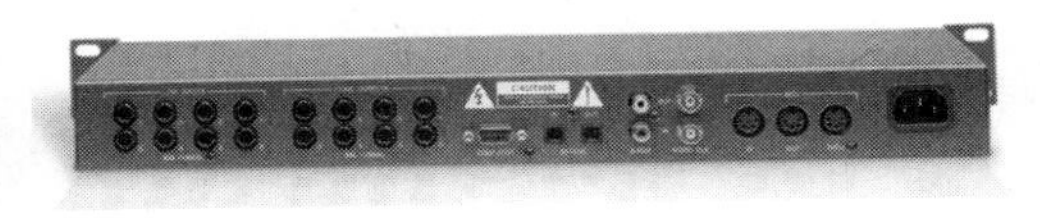

12.3 Echo Audio Layla (back view).

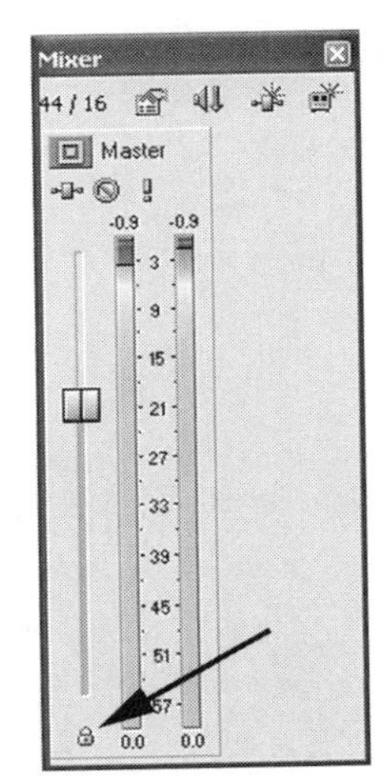

12.4 Disable locks on master faders if separate control of right/left channels is necessary.

Meters are critical to this endeavor. Vegas has two meter display spaces: one is the master and the other is for buses/auxes. For this section, we're only concerned with the master volume meters. Notice the small lock icon on the master meters. This locks right and left together, so that both channels are identical. They may be unlocked by clicking the icon.

Ideally, during the mix process, the audio was mixed to a level not exceeding 0dB and not less than -6dB at *peaks*, which are the loudest points in the audio mix. Audio that is too quiet won't be heard well over the television or stereo system on which the video is played or on which the CD is played. In addition, in the event of poor playback equipment, audio that is too quiet can cause any noise associated with the poor playback equipment to shine through, making your finished product sound unfinished or of poor quality.

Many ways exist to bring up audio levels in the digital world. One is to apply a simple volume adjustment. In Vegas, volume can be increased by 12dB at the track slider and decreased infinitely until to the point of no volume whatsoever. Vegas' track pane setting for volume is 0dB, at which point there is no increase or decrease in the volume of the audio.

Another means of raising volume is to normalize the audio levels. *Normalizing* in Vegas instructs the application to search the various peak levels of the audio information and set the loudest point to a prespecified level. To normalize in Vegas, right-click the audio file and select PROPERTIES | NORMALIZE. This step can also be done by selecting EDIT | SWTCHES | NORMALIZE from the Vegas menu bar.

Change the Normalize Default Level in the OPTIONS | PREFERENCES | AUDIO dialog. Normalize defaults to -.1dB in Vegas 4.0.

Yet another method to bring audio to the maximum level is to use a maximizing tool, such as Sonic Foundry's Wave Hammer or the industry favorite, WAVES Ultramaximizer. Maximizers are fantastic for bringing audio to an optimum level without distortion. These are usually applied last in the chain, however.

12.5 WAVES Ultramaximizer plug-in is great for mastering output levels.

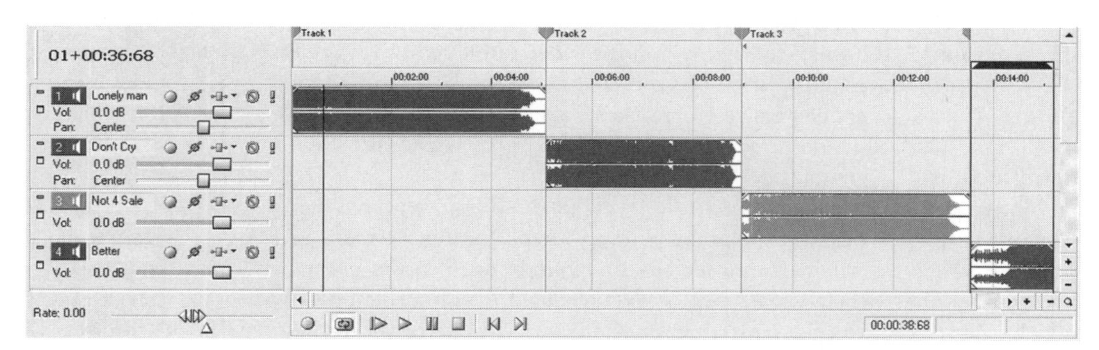

12.6 All audio cuts on separate tracks provide the greatest control.

During mastering, effects, such as delay and reverb, are rarely applied. These are more mix tools than master tools. However, EQ, compression, and audio wideners are all commonly applied. Filters/processes can be applied either at a track-level or at an event-level in Vegas. Generally, it is best to apply filters at the track-level, only for purposes of maintaining consistency and clear processing paths.

To begin the mastering process, place all audio to be included in the final output on the timeline in the desired order. This process can be done either with each event on its own track or each event on the same track. It makes no difference other than catering to individual workflow preferences. (Leave exactly two seconds of blank space at the head of the first track, if the audio will be burned to a Redbook/duplicatable CD.) Vegas can auto-insert these two seconds as well, which is defined in the **OPTIONS | PREFERENCES | EDITING** dialog. Audio CDs have specific areas for specific info, such as in the first 46mm of the CD, where the lead-in occurs. *Lead-in* is where indexing, table of contents, data information, logical block address (LBA), and length of track information is stored. If anything more or less than two seconds is defined at the head of the CD, Vegas will give an error message indicating that the error must be repaired.

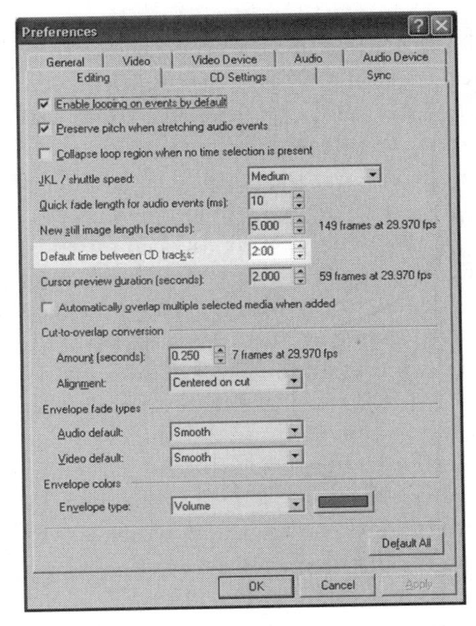

12.7 Set default time between tracks at 2.00 seconds in the **OPTIONS | PREFERENCES | EDITING** dialog.

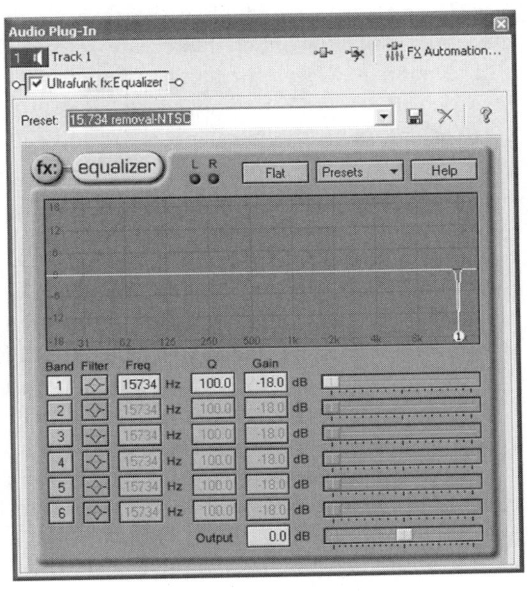

12.8 Ultrafunk contains tools specifically for video editors, as well as being very musical and warm.

Listen to the timbre/tonal differences between pieces of audio. Perhaps one has more or less bass than another, or more high-end or less mid-range information. This might be desirable or it might not be. The first part of the process is to determine how to make each event complement the next, without appearing to have jumps in bass, midrange, treble, volume, or pan/right-left information. The transition from one event to the next should be fairly seamless, regardless of whether it's musical or not.

To add EQ to an event at the event level, select **TOOLS | AUDIO | APPLY NON-REALTIME EFFECTS**, choose an EQ, and click Add. Graphic or parametric choices are available and depend predominantly on the engineer's requirements. Graphic EQs are broader in terms of frequency adjustability, and parametric EQs allow for specific frequency controls.

Here is where plug-ins really shine. While Vegas does come with a selection of audio plug-ins, many third-party plug-ins are available. For instance, the WAVES Q series or Renaissance plug-in tools are very natural, yet powerful tools. They are very musical and have great control.

Ultrafunk also has some great plug-ins available as well, and their EQs are very musical and warm-sounding. Apply EQ sparingly and lean more toward smoothing out harshness, rather than attempting to fix a problem with the overall EQ from the mix. If the track requires a tremendous amount of EQ at this point, it's probably a good idea to revisit the mix. If the master file is audio from a video file, pay attention to the weaknesses in EQ and compensate in future times with better mic techniques and attention to room acoustics.

Applying a touch of bass at the 150 Hz setting will add punch to the mix, while adding any frequencies below 90 Hz will most likely enhance any low-end rumble or noise. It's generally a good idea to roll these lower frequencies off with an EQ, using the high-pass filter setting. Adding 1 to 3dB of gain in the higher ranges, 8 KHz to 12 KHz, will give a little sparkle to the overall mix, but again, too much can create resonant frequencies that are painful to the ear and difficult to control. Watch for harshness in the 1 KHz to 2 KHz regions. When working with audio recorded in a large reflective room, such as a warehouse, bathroom, or conference room, pay close attention to the frequencies around the 500 Hz point. This combination often creates a boomy sound in the audio and should be compensated for or removed if overt and noticeable to the average listener.

Listen to the applied filter and determine if the sound is to satisfaction. If so, then move on to the next event. Notice that when the non–real-time plug-in is applied, a new file is written for the event. Track-level mastering/processing becomes valuable at this point, as real-time effects can be applied without writing a new file. To apply filters/processing at the track-level, open the track filter/processing dialog by clicking the FX Filter button found on the Track Control pane. The dialog box will show a noise gate, compressor, and EQ already in place, with check marks showing that the processors are selected. There is also a Plug-In Chain button in the upper-right corner, which allows you to insert other real-time effects and processes. Click this button to add a real-time effect, from the Cakewalk, Sonic Foundry, WAVES, Ultrafunk, iZotope, or other DX plug-in. Keep in mind that this effect will affect every piece of audio on the track, which is why placing each piece of audio on its own track is a good idea for the mastering process.

When all audio is finished, it is ready to be output. To output to an MP3 format file, select FILE | RENDER AS, and choose MP3 from the Save As Type drop-down menu. The file will be rendered as an MP3 file. Using this same method, Vegas can output AIFF, Ogg Vorbis, or Windows Media Audio files.

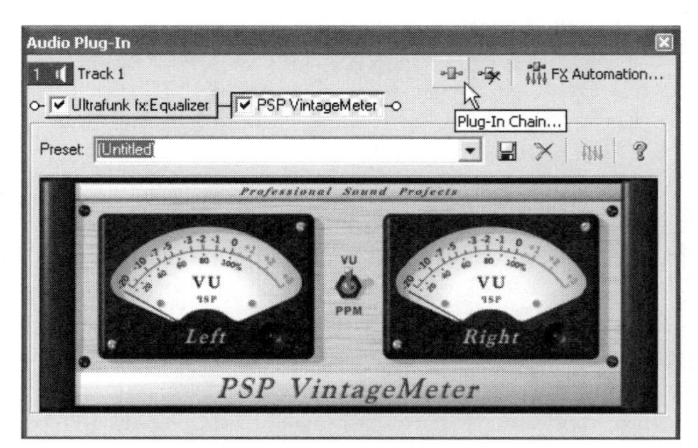

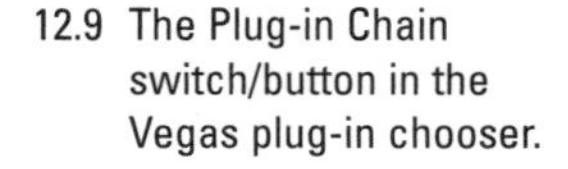
12.9 The Plug-in Chain switch/button in the Vegas plug-in chooser.

Burning to a Master CD

With all tracks processed to satisfaction, all levels checked, compression at correct settings per cut, and a master compression or EQ setting for the overall project (if it needs a master compressor or EQ), the project is nearly ready to be burned to a CD. Track indexes must be inserted so that listeners have track numbers to skip around to. This process is quite easy in Vegas; just double-click each audio event/track event and press N for a new index point. If subindexes/chapter markers are required, use SHIFT+N to insert subindex markers. Index points can also be laid out automatically in Vegas 4.0. Select TOOL | LAY OUT AUDIO CD FROM EVENTS. Vegas will inspect events on the timeline and auto-create index markers. However, if events overlap, Vegas will not see the individual beginning/end of events and therefore not assign a track index marker. Check each event beginning to ensure it has a track index marker if this is desired. Auto-layout can also be achieved by right-clicking the Marker toolbar, visible only when at least one marker is on the timeline.

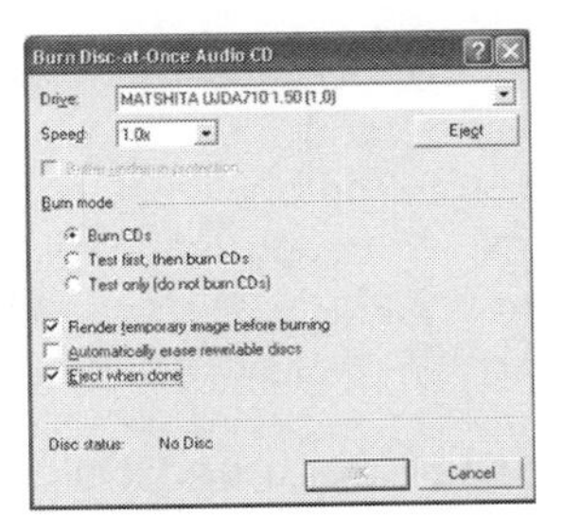

12.11 The Burn Disc-at-Once Audio CD dialog offers speed, testing, and performance options.

12.10 Vegas is capable of automatically creating track indexes for audio CDs.

With markers inserted to indicate track indexes, Vegas is ready to burn a CD. Vegas is capable of Disc-at-Once (DAO) CD burns or Track-at-Once (TAO) CD burns. If a Redbook Master is required for duplicating, Disc-at-Once is the only option. Redbook standards do not support TAO-burned discs. (Redbook is a mastering/duplicating standard.)

 ***Tip***

All CDs have errors, which is a fact of digital media creation regardless of the authoring application used. However, errors can be minimized. CDs destined for a duplicator will contain fewer errors when burned at lower burn speeds. Although errors cannot usually be heard, they are seen by replicating equipment in most professional duplication houses. Minimize burn errors by burning CDs at the 1x speeds where possible.

Audio CDs have specific areas for specific information, such as the lead-in in the first 46mm of the CD. For burning a Redbook-compliant CD, there must be exactly two seconds of dead space/silence before the beginning of the first track. Some burners will allow more than two seconds but all require not less than two seconds of blank space. Up to 99 tracks are allowed on a Redbook-compliant CD.

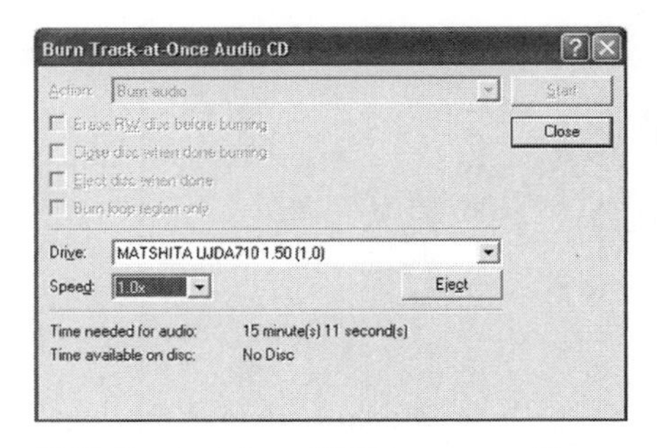

12.12 Track-at-Once (TAO) burns are efficient for getting a session burned to disc for referencing the session's work.

Track-at-Once CDs

TAO CDs can be burned from Vegas. TAO burns are handy for capturing tracks as they are mixed for reference purposes or when merely getting the audio to a CD is necessary.

To burn a TAO, have the track/cut to be burned to a disc on the timeline and select TOOLS | BURN DISC | TRACK AT ONCE. Select the burn speed and burn device, and finish.

In TAO mode, Vegas can also burn a region to a disc from the timeline. Select the Burn Loop Region Only check box. This option is useful for authoring Sonic Foundry's ACID loops, for sending an advanced digital recording (ADR) loop section out, or for burning a section of a track that needs to be reviewed without cutting and pasting sections of the track.

Mastering Audio for Video

Mastering for video broadcast is different from mastering for audio CD or DVD. Broadcast standards require deep bass and ultra highs to be rolled off or they'll be chopped off at the limiter/compression stage at the broadcast facility. It's better to have control over the amount of cut off and roll off at the editing/mastering stage than to be unhappy about processing during the broadcast stage. While no set standard exists for audio roll off in the NTSC or PAL broadcast world, the average is an -18dB per octave roll off at 80 Hz and a -9dB per octave roll off starting at 12 KHz in most houses. Going beyond these ranges almost ensures that you'll have frequencies cut off and compressed at broadcast. DVD and CD playback allow for almost any levels of audio, so long as levels don't exceed 0dBfs. (See Chapter 7 to refresh yourself on the differences between analog and digital zero.)

Using Compression in Vegas for Video

Audio for video should not have a dynamic range of greater than 20dB. Generally, a dynamic range of less than 15dB is preferable, depending on the content or attitude of the video and the desired effects. A scary visual might call for a 20dB dynamic range, going from sublime to extreme to elaborate a particular shot, but in the general norm, dynamic ranges for video should be fairly limited.

Vegas comes with a good compressor. Here again, however, is where a third-party tool can really shine. The Ozone compressor from iZotope has a very warm and broad sound, as they use analog emulation in their application. The maximizer, stereo widener, and compressor alone are worth the price of admission with this tool.

Ozone has to be the sexiest-looking plug-in around, worthy of a Hollywood appearance. SonicTimeworks Compressor is also fantastic, with a warm sound, soft or hard compression, and a great look and ease of use, too, as are the WAVES, Ultrafunk, and Cakewalk plug-ins.

Compression must be used sparingly at the mastering point; generally settings of less than 3:1 are used at the mastering stage, as audio should have been compressed in the mix stage if the mix process was done to standards. Compression is applied to smooth out dynamic range. If desired, a compressor can bring quiet passages up to a nominal level, while reducing loud passages to a manageable point. Compression is used to get maximum levels while maintaining control over distortion and output levels.

A compression tool can be used at this stage as well when only one or two frequencies are causing the audio to go beyond rational norms. A dynamics compressor, which is a compressor that works only on specific frequencies, is invaluable at this point. Troublesome frequencies can be specified, and the amount of compression of those specific frequencies can be controlled. This option helps smooth out sibilant or plosive sounds from a poor interview mic or helps take down the overdriving pulse of a train that might be in the background of a wide shot mic'd with a handheld or shotgun.

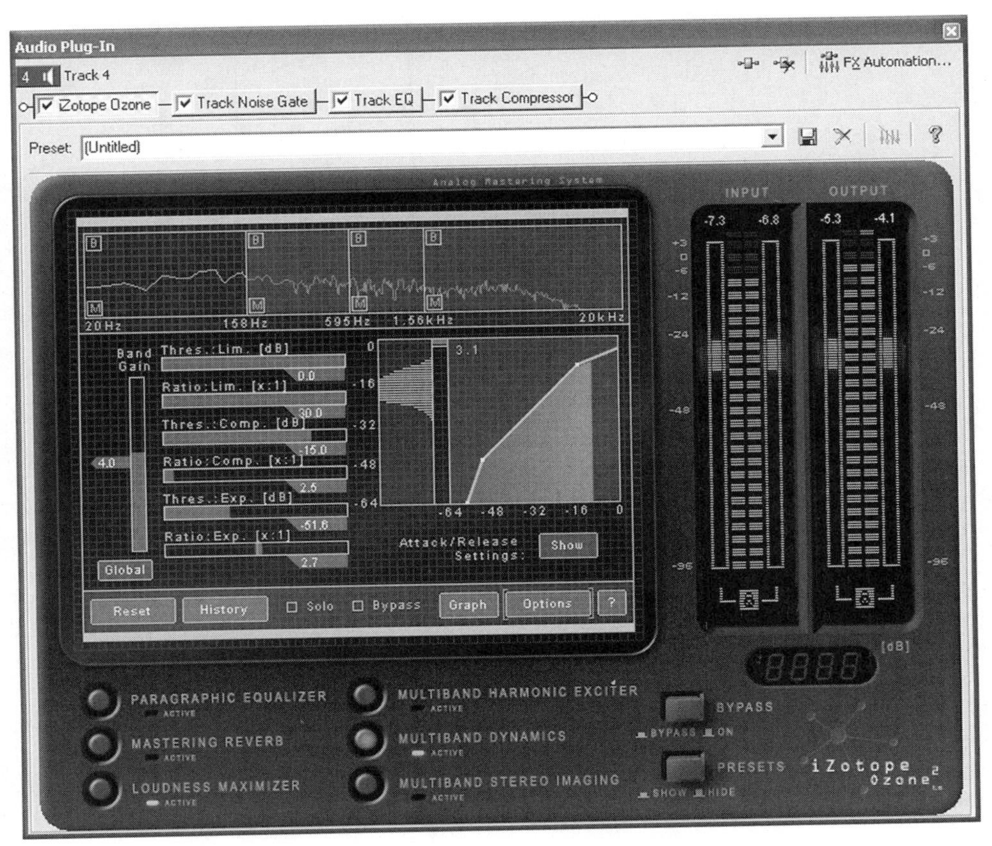

12.13 The Ozone from iZotope is a terrific compression and mastering tool.

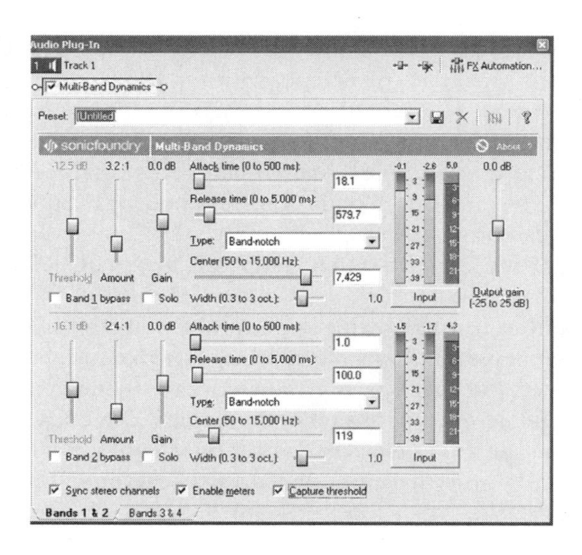

12.14 A dynamics compressor is excellent in the finishing stages to suppress irritating frequencies, particularly from background sounds interfering with primary dialog or primary audio information.

To tune a compressor, listen to the audio closely and notice the peaks and valleys in the spoken word or music. These can be watched at the same time as being listened to. Notice how the peaks seem to jump out and how the valleys need special attention to be heard.

The space between these peaks and valleys is *dynamic range*. Too much dynamic range creates fatigue on the ear and creates a sense of overdriven sound, even though the audio might be clean. It's a quick way to tire out viewers.

Using Equalizers

A touch of EQ may be added at the final stage as well to bring the overall product to a consistent EQ attitude. The addition of EQ at this last stage, however, should be very minimal and only used to balance out inconsistencies that might arise or to create a final personality in the mix. Most professionals, however, do not add a final EQ as the proper EQ will have been applied during the initial stages of mastering. Keep in mind, there have been two previous opportunities to add EQ to each event/track in the mixing and mastering processes, and it's usually a mistake to apply much EQ to a final master stage. Often, EQs are mistakenly used here to compensate for error in the initial master, mix, or recording stage. If EQ is used in the finishing/mastering stages for video, it should be sparing and applied as an overall mix setting.

Dithering Audio in Vegas

Dither is valuable when audio is going to be printed to a CD, and the original recording was at a sampling rate higher than 44.1 KHz or at a bit rate of other than 16 bit. Of course, this is valuable when working with audio for video as well, rather than simply truncating bits in the sample rate change. This process is not the same as resampling but is part of the resampling process. Sonic Foundry has a dithering plug-in, and a demo of the iZotope Ozone plug-in is on the CD in this book.

The dithering plug-in should be applied to the master bus in order to ensure that the last audio process is dither.

Sonic Foundry's dithering plug-in is Type triangular and the noise shape high pass contour should be similar to the Ozone Type 2, simple shape. The Ozone Type 1 is rectangular; Type 2 is triangular. Triangular is recommended regardless of which dither you're using, and high-pass shaping is good in both plug-ins. For more advanced noise shaping, the Clear shape for acoustic stuff is highly recommended. It's not as aggressively shaped as the psychoacoustic shapes. The Clear shape sounds natural, but it still has less perceived noise than a simple high-pass shape.

Fade in/out will be the most noticeable applications of dithering. Highlight the last part of a fade, turn it up fairly loud, and compare right when the audio fades into the noise floor. With the in/out check boxes on plug-ins, it's pretty easy to A/B. No dither (just truncation) will make the end of the note fade sound like a buzz, harmonic distortion, or metallic. It will most likely have a zipper sound before it fades out completely. Add dither and get a smoother tail but more noise. Add dither and some noise shaping, such as the Sonic Foundry or Ozone Type 2, Clear shape, and you'll get a smooth tail but less noise on the fade. (iZotope has excellent information on dithering on their website at www.izotope.com. There's also a dithering guide from iZotope on the CD found in this book.)

Rendering Options in Vegas

Vegas is amazingly flexible when it comes to outputting video whether to tape, CD, DVD, the Internet, or a hard drive. AVIs, QuickTime, Windows Media, REAL media, MPEG-1, MPEG-2, MPEG-4, and any other codec installed on the computer can be used for output. (See Chapter 10 for information on outputting streaming media files.)

Vegas has the option of either rendering an entire project or merely doing a print to tape. The difference is that with a print to tape, Vegas temp-renders sections of the project that require recompression. These sections are written as multiple AVI files. A full render writes an entirely new file as one long file. Either method works as well as the other, and the choices are up to the user's workflow and time constraints. A video that needs to be moved off the hard drive and back to tape quickly might be best handled as a print to tape. For final archiving or for multiple file-type outputs, a full render is recommended. If the same media will be output to Windows Media, QuickTime, REAL, and DVD, all of these processes will go much faster if working from an AVI file, rather than from a project file.

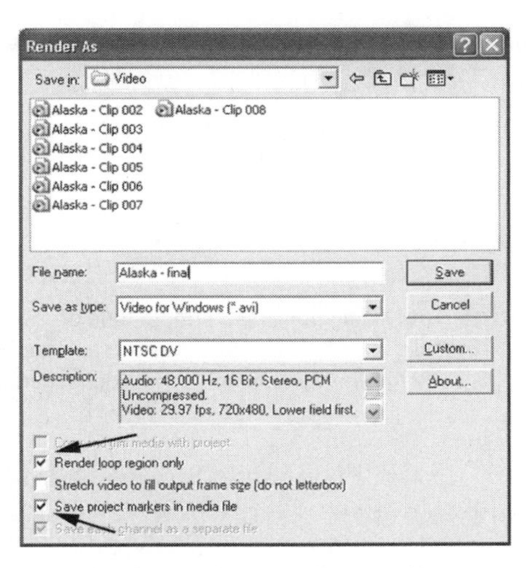

12.15 The Render As dialog box provides a choice for the destination of the rendered file.

Save the file before rendering, just as a precaution. Writing long video files is taxing on a hard drive and processor, and some older systems cannot handle multiple drives getting very hot and will shut down.

Rendering to an AVI

Select **FILE | RENDER AS** to render to an AVI format file. In the Render As dialog box, click the Custom button.

In the Project tab of the Custom Template dialog, select Best. This option will slow down the rendering time slightly, yet ensures that the best rendering options will take place. Under the Video tab, the "Create an OpenDML compatible file" check box may be unchecked. Unchecking this option will prevent users from rendering a file larger than 2GB. You might need to uncheck this box to be compatible with some third-party applications. Leave this box checked if the file contains alpha channels that will need to be accessed later in another editing session.

For a standard AVI file to be printed to tape using the Vegas Capture tool, leave the settings under the Audio tab, including the codec and aspect settings, at their default positions for a successful print to tape. The video must be 720 × 480 for an NTSC-DV print to tape and 720 × 576 for a PAL-DV print to tape, and audio must be 48 KHz/16 bit. Otherwise, the camera, converter, or whatever is being used to get the video to a DV tape will not be capable of accepting the media.

To print to tape without rendering the entire project, select **TOOLS | PRINT VIDEO TO DV TAPE.** This option opens a series of dialogs that make sure the project is prepped correctly to print.

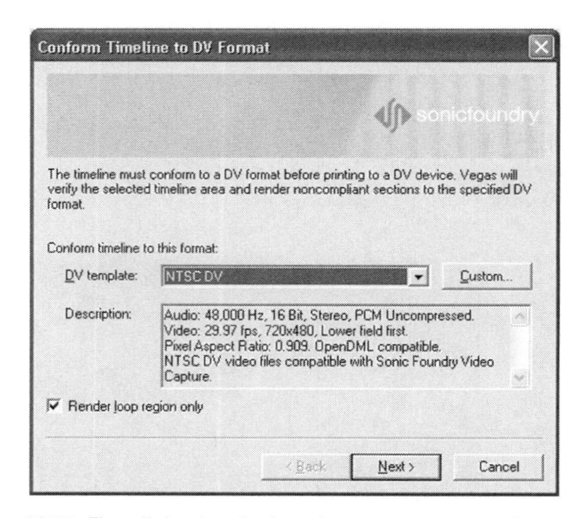

12.16 First dialog box in the print to tape process. Be sure you have the entire region selected if leaving the Render Loop Region box checked.

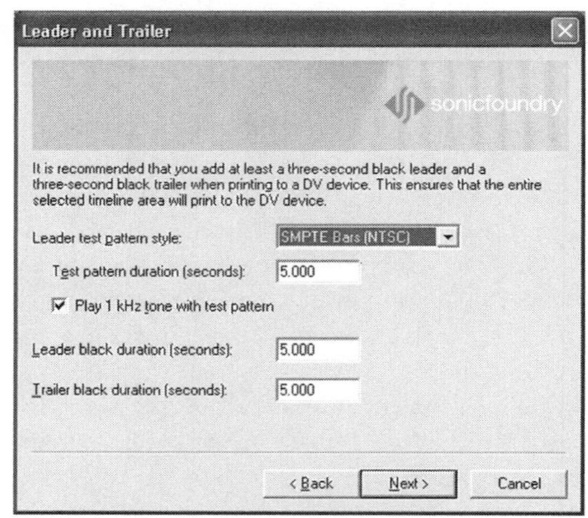

12.17 Second dialog box in the print to tape process. Here is where color bars and test tones are chosen.

Clicking Next in the first dialog leads to the second dialog. In the second dialog, test tones and color bars can be specified. Color bars are standard SMPTE color bars, and the test tone is a 1 KHz test tone, output at -20dB. Color bars can be substituted for any test pattern by using the choices in the menu in this second dialog. These bars and tones are necessary, even required by most replication houses and broadcast houses. If this is a print to tape for personal use or to a VHS tape deck, the tones/bars are not required.

The tones in previous versions of Vegas were at -12dB.

Vegas 4.0 has output tones in alignment with the new Advanced Television Standards Committee (ATSC) output standards (www.atsc.org). If any products are calibrated to the old output standard, recalibration will be necessary.

Click Next in the second dialog; a third box is displayed. This box asks whether you want automatic control of a tape machine or manual control. Automatic control will start printing to tape after a series of prerenders are finished wherein all recompressed media on the timeline is written to a temp file. This option should be selected when long prerendering times are anticipated. This option is useful for the *overnight renderer*, or for those users who edit during the day or evening and then render and print to tape late at night while all the world is sleeping.

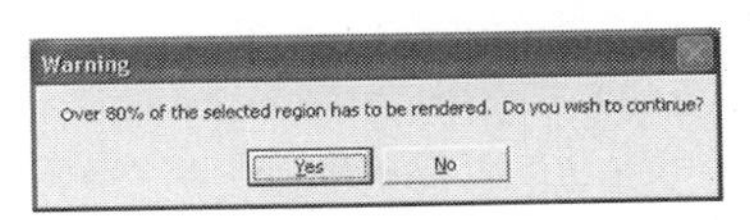

12.18 Vegas warns users if the render process appears to be a lengthy task.

Manual tape control is useful for DV decks that cannot be controlled by the computer or for printing to an analog tape deck, such as a VHS or Beta deck. Selecting the manual option still allows users to walk away from the computer during renders, as Vegas gives a countdown option to printing manually.

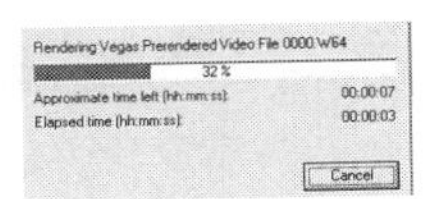

12.19 The rendering progress window indicates how long a render will take.

After this selection is made, click the Finish button. Vegas then does one of two things:

- Vegas will start to print to tape immediately if no prerenders are left to complete, or if there are very few prerenders, it will complete those and begin the printing or countdown process.
- Vegas will give a warning that more than 80 percent of the project must be rendered and will ask if you wish to proceed. In either event, instruct Vegas to complete the process.

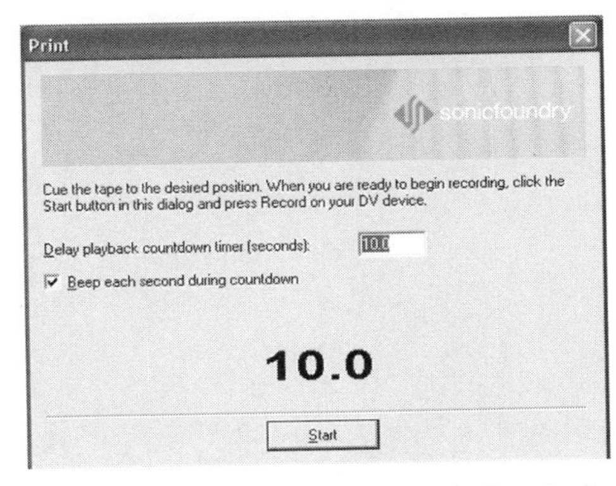

12.20 Vegas allows users to define the length of a countdown.

Click OK from the warning, unless a standard render is determined to be more time-efficient.

After Vegas has completed the render process, if automatic control was selected in the Print to Tape dialogs, Vegas will start the DV machine automatically and print the project to tape with no further input/involvement from the user. In a situation in which manual control has been selected in the Print to Tape dialog, Vegas will prerender the recompressed sections and then wait for a click in a countdown box. Vegas will then count down from ten, providing a beep at each second, until it reaches three seconds. It will then start to print to tape. To effectively use this manual feature, allow Vegas to count down and, when the counter reaches three and no more beeps are heard, click Record on the tape machine. These steps usually provide enough leader time for various tape machines to get up to speed and begin the recording process.

Exporting Events to Third-party Applications

Sometimes it is necessary to export video events from Vegas to use in a third-party application, such as Adobe After Effects or Pinnacle Commotion. When exporting, files should be exported at the best possible resolution, and choices are therefore limited. If files are exported from Vegas using the standard NTSC-DV settings, most third-party applications can read the files, but until Sonic Foundry opens their codec up to other applications to use, those third-party applications can't write back to the Sonic Foundry codec. Using the Microsoft codec really isn't a solution either, as it limits quality output too much. Therefore, an uncompressed codec is the answer.

In the FILE | RENDER AS | CUSTOM dialog, an uncompressed setting can be chosen. Understand that using this output method will exponentially increase render times.

Be certain to uncheck the "Create an OpenDML compatible file" option. Most third-party applications cannot read these file types yet.

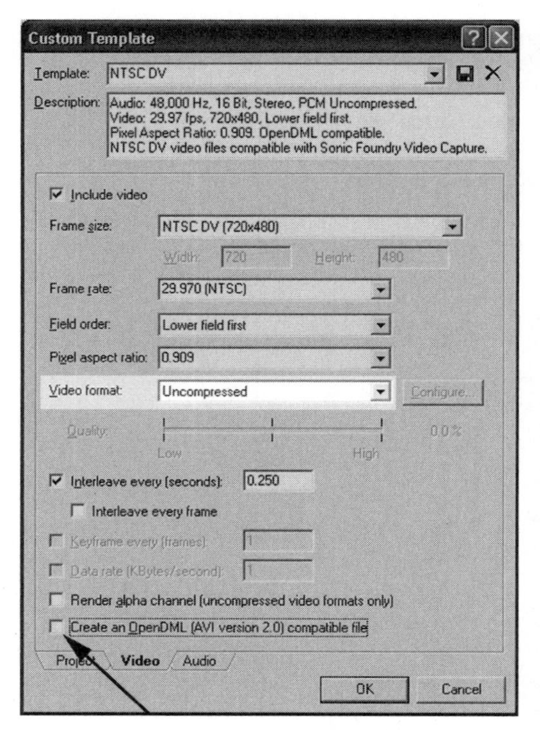

12.21 Select Uncompressed in the Video tab in the rendering options. This option also allows the uncompressed file to retain an alpha channel, should one exist or have been created in the editing process.

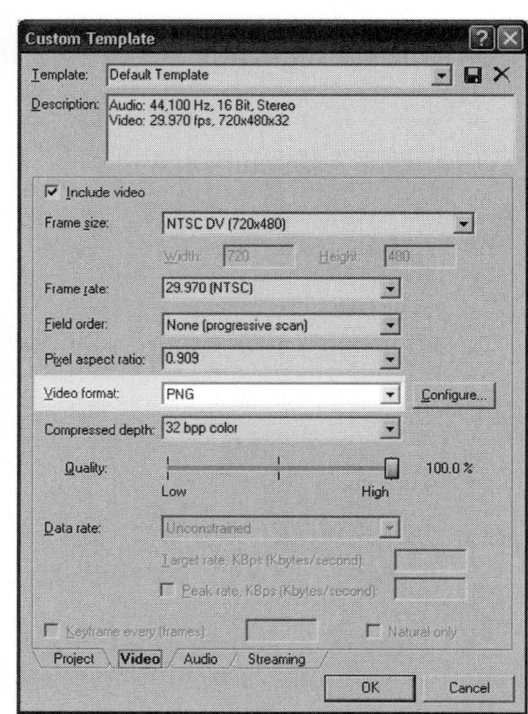

12.22 Rendering to a PNG/QuickTime format is a high-quality output method for exporting to a third-party application.

Another option is to use QuickTime for output, with its numerous options. A few users of Vegas have found great success using the PNG codec as an output option that both Vegas and most third-party applications can read. The PNG format is a lossless compression format that keeps file sizes more manageable and render times slightly faster than using a standard uncompressed output.

To render to the QuickTime format and its various codecs, select the part of the file or event that requires export. Select FILE | RENDER AS, choose QuickTime/.mov from the Save As menu. Click the Custom button. In the General tab, select Best. In the Video tab under the Video Format menu, select PNG. Adjust the quality slider to 100 percent.

Animators and rotoscope artists often require Targa files (TGA) for their editing applications. Render to QuickTime format, select the Video tab, and select FORMAT | TGA. Set the frame rate, pixel aspect, and frame size in their menus according to the requirements of the importing application.

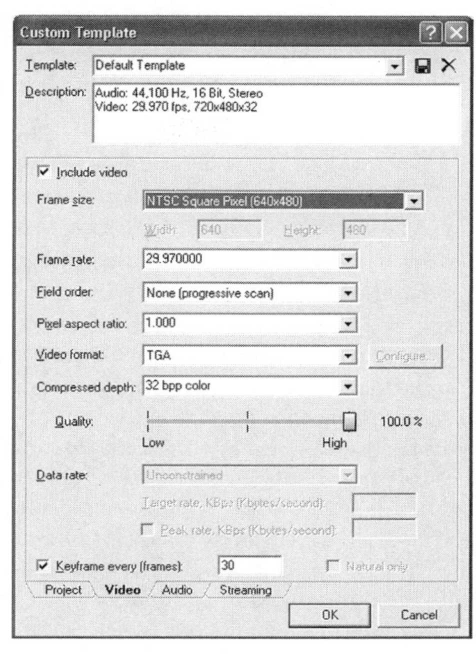

12.23 Render a Targa file by using the QuickTime option in the Render As dialog. The QuickTime authoring tools must be installed on the system to access this menu.

If audio is part of the export and sync is important, move to the Audio tab and change the default output sample rate to 48 KHz/16 bit, rather than 44.1/16 that QuickTime asks for.

Files exported and imported in these formats will be clean and without quality loss. Using any codec or compressed format between applications runs a risk of loss of quality, so experiment with various outputs to discover what is best for each particular situation. Upsampling/downsampling is a foolish thing to do, as video always loses some information in the up/downsample. Some applications upsample imported DV as 4:2:2 or 4:4:4. Use caution and check files carefully for color accuracy when upsampling or downsampling between images.

Outputting Other than AVI Files

Vegas has output capability for nearly every output format, including MPEG-1 and MPEG-2.

While Vegas does a reasonable job of outputting MPEG-1, the primary MPEG format output focus is on MPEG-2, which is used for DVD encoding

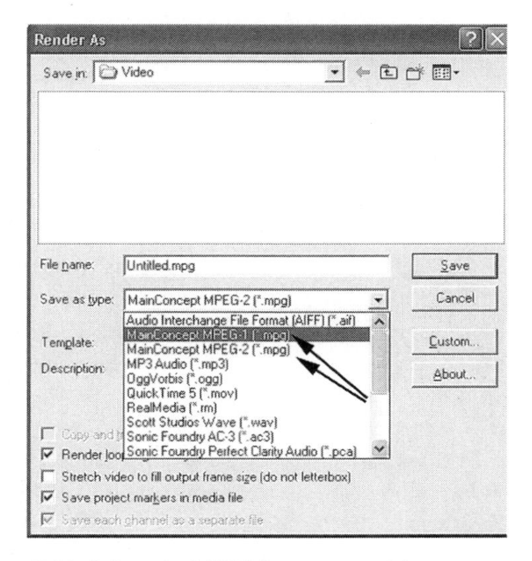

12.24 Select the MPEG format you wish to work with from the Save As Type drop-down menu.

and authoring. The primary benefit of an MPEG-1 file is that although it's an old codec and not of top quality, MPEG-1 files do indeed play on just about any player in the world.

Tip

If working with DVD Architect to author DVDs, rendering to an MPEG file is not necessary at this time. DVD Architect works with multiple file formats, and users of Vegas+DVD might find it in their best workflow interest to work with AVI files or native file formats until the project is ready to burn. Vegas 4 provides a render template for DVD Architect in the render settings, while DVD Architect has similar encoding qualities to Vegas 4.0. See Chapter 11 for more information.

MPEG-2 file types are fairly ubiquitous, and it is the file format of choice for DVD. Vegas uses the MainConcept codec to encode files to the MPEG format, although the Sonic Foundry version of the MainConcept codec should not be confused with other applications using this codec. Each group of software engineers from the various NLE manufacturers have written individual interfaces to the codec.

To render to MPEG codecs, select **FILE | RENDER AS** and choose MPEG-1 or 2 from the Save As Type drop-down menu. Selecting MPEG-1 renders an MPEG-1 format video that can be placed on a CD or on the web for download. Selecting the MPEG-2 option renders the MPEG-2 file for use on CD or for sending to a DVD authoring application. (When using Vegas with DVD Architect, files are not required to render to MPEG until DVD Architect prepares the media for the DVD burn.) Audio files in the MPEG-2 format are rendered to an MPEG-2 audio format.

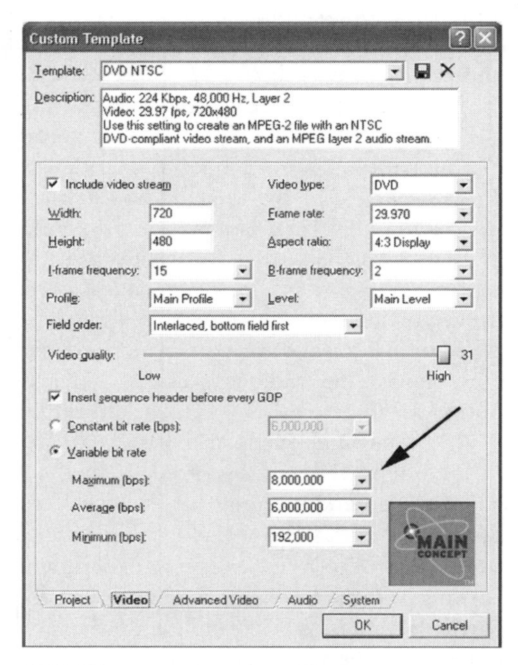

12.25 Bit rate can be selected in the **CUSTOM | VIDEO** tab.

MPEG files can be rendered as progressive scan or as interlaced files. MPEG files should be rendered to a progressive scan output for best appearance if the targeted display is a computer screen or progressive scan television. If the file will predominantly be played on television sets, leave the file interlaced. If the primary target is for computer viewing, set the file format as progressive scan. In the event that a progressive scan file is displayed on a television that does not have progressive scan settings, the television or output device will interlace the video. The settings for progressive scan/interlaced are found under the Custom button in the Render As dialog box.

The bit rate (encoding depth) can be selected in the Custom dialog as well. This option is found in the Video tab. Bit rate for standard DVDs should be 8 Mbps variable bit rate (VBR). Using VBR allows the encoder to determine the quality of frames based on motion in the various frames.

Some DVD authoring applications require separate video and audio streams. Vegas 4.0 can output separate streams to meet these requirements. To output separate video/audio streams, select the FILE | RENDER AS | MPEG-2 | CUSTOM | SYSTEM tab. Place a check mark in the "Save as separate elementary streams" box in this dialog. Vegas will create two separate files: one labeled *m2v* for the video stream and one labeled *mpa* for the audio stream. Import these two streams into the DVD authoring application.

Unless you are exceptionally well informed about MPEG encoding functions, you shouldn't change anything else in the MPEG output settings. For more MPEG setting information, see the Sonic Foundry MPEG white paper on the CD included with this book.

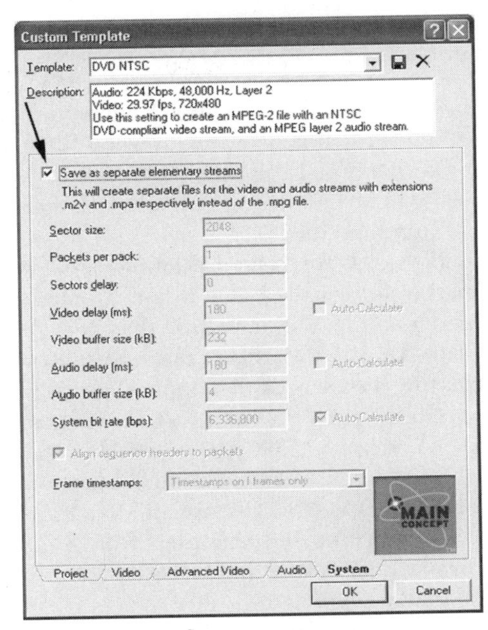

12.26 Create separate audio/video streams for DVD authoring applications that require separate streams.

Outputting an AC-3 Format File

One of the new features of Vegas 4.0 is the ability to output AC-3 audio files. AC-3 is a compression scheme devised and overseen by Dolby Laboratories (www.dolby.com). AC-3 can be output in stereo or 5.1 channel formats. The primary benefit of AC-3 is that it allows for audio to be compressed, leaving more space for video on a DVD. AC-3 audio delivers high-quality audio in an encoded format that most consumer and computer DVD players can decode. This option gives viewers the best possible audio experience.

To output AC-3 files, select FILE | RENDER AS and select AC-3 from the menu. Vegas will ignore the video file and render the audio as a stereo or 5.1 audio file, depending on project settings. If the project is a stereo project, audio will be rendered as a stereo/two-channel AC-3 file. If the project is a 5.1 surround project, the AC-3 file will render as a 5.1 file.

12.27 Stereo or 5.1 surround, Vegas offers a number of choices for output formatting.

For most purposes, the AC-3 presets in Vegas are designed to function as they are built. Experienced AC-3 encoders, however, might find some of the menu options quite valuable for creating a better listening experience. Changing AC-3 settings could possibly create an invalid render in the output file, so experiment on projects that are not time-critical.

Creating a VCD/SVCD

Burning a super video compact disc (SVCD) or video compact disc (VCD) is a one-step process in Vegas, with or without a rendered project. VCD is only capable of reading an MPEG-1 file. SVCD is capable of reading several file formats.

With a project on the timeline, rendered or not, select TOOLS | BURN CD | VIDEO CD. This selection will open a dialog that allows either a non-rendered project or a previously rendered project to be specified.

If the project has not been rendered to an MPEG-1 format file, Vegas will render to this format automatically when it is instructed to burn a VCD. The new file will need to be named, which will be the name/title that appears in the header file of the burned CD when it is complete. Vegas will indicate the required amount of disc space that the finished render will use. If the file size exceeds the amount of disc space, Vegas will indicate this before the burn.

The menu choices in the VCD dialog are few, as the VCD spec is very narrow. The only options are to burn an NTSC CD or a PAL CD.

Burning an SVCD is a much more open option. QuickTime, REAL, Windows Media, AVI, MPEG-1, and MPEG-2 are all SVCD-capable formats. To burn the SVCD, the project does not need to be rendered separately from SVCD authoring. The file will require rendering in its final format; this can be done directly from the SVCD burn dialog. Select TOOLS | BURN CD | MULTIMEDIA CD.

In the dialog that opens, select the destination for the final render. Then select the file format to which the SVCD file will be rendered. Select a file template based on the file type chosen. NTSC and PAL variants are available for AVI and MPEG file formats.

As with all other render formats, a loop region can be defined as the only area to be rendered.

Defining a loop region allows you to select a smaller area of a larger file to be rendered without splitting, copying and pasting, or rendering new sections.

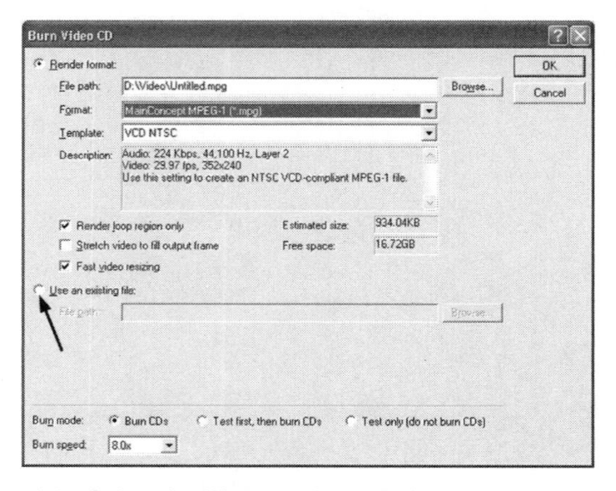

12.28 Select the "Use an existing file" check box and browse to locate and burn a previously rendered MPEG-1 file to create a VCD if the project is finished.

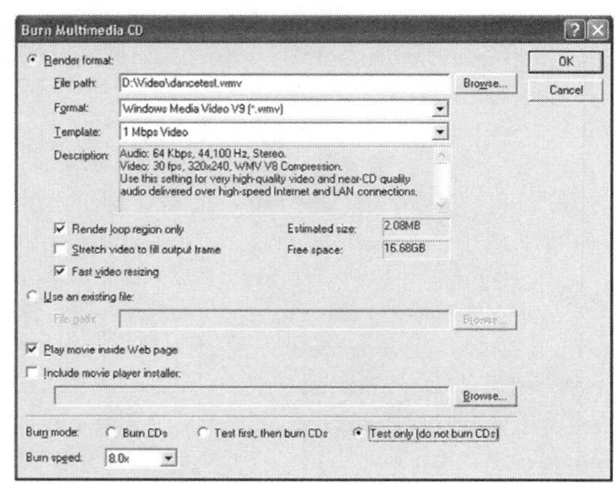

12.29 Menu options include playing video inside a web page and including a player/installer with the SVCD.

12.30 The video can be played inside a company web page or inside any form of HTML document.

If including a player/installer on the disc to be distributed, be sure that the executable file (EXE) is the extension seen on the file. Players can be found at `www.microsoft.com`, `www.quicktime.com`, `www.real.com`, or at several third-party player sites. For any distribution other than personal/internal distribution, a license is required from the appropriate owner of the player software.

12.31 Viewers of the file who do not have an appropriate player will receive this message complete with links to downloadable installers.

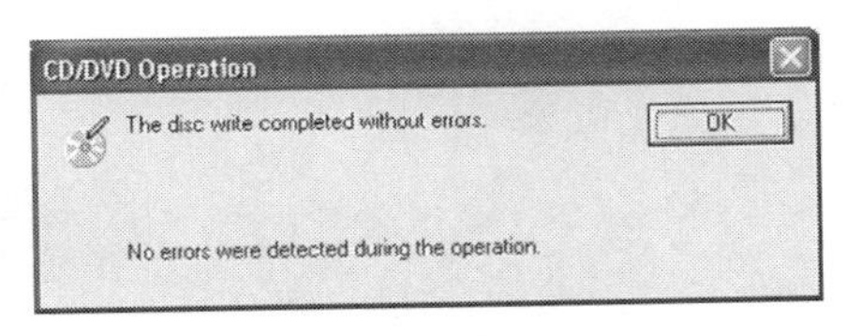

12.32 When the disc has been burned, a successful burn message is displayed.

Additional Rendering Options

Vegas has additional rendering options available, giving users a wide variety of choices. As an example, a loop region can be defined on the timeline by creating a selection and pressing Q, which will turn on the Looping tool. In the Render As dialog, rendering only a loop region is an option that is valuable when only selections need to be rendered out.

Another option is to stretch the video to fill the screen. This option prevents black bars from being added to the top, bottom, or sides of a video render that is of a different aspect ratio than the project. Check this box in the Render As options if you want to stretch to fill the aspect ratio.

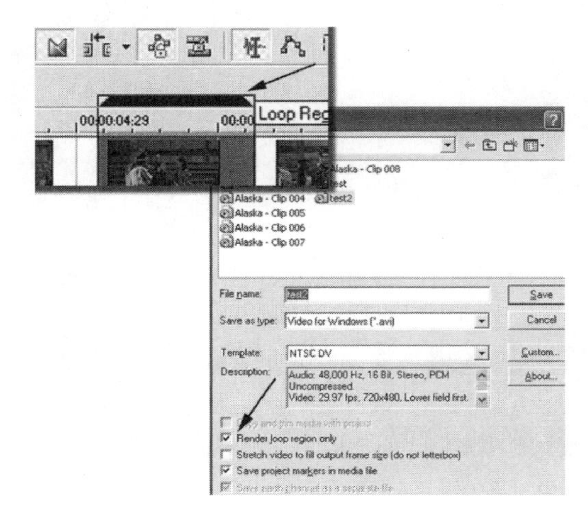

12.33 Defining a loop region for rendering allows select sections to be rendered for output or export.

12.34 Letterboxed image, rendered for RGB output/Windows Media.

12.35 Stretched image, using the Stretch to Fill Screen option.

NEW! Saving markers in a file takes on a new meaning with Vegas 4.0. Markers can be inserted into a rendered file to be used as chapter markers that DVD Architect recognizes. For long-form video projects, this option can be invaluable, saving time. Markers can be edited in the Trimmer after the render has taken place.

High-Definition Video

NEW! Vegas 4.0 also brings the ability to edit and output high-definition media in 720p, 1080i, and 1080p formats. Currently, Vegas 4.0 does not have the ability to capture high-definition media, so any hardware card used for the capture of high definition will require the use of any capture utility associated with the hardware card.

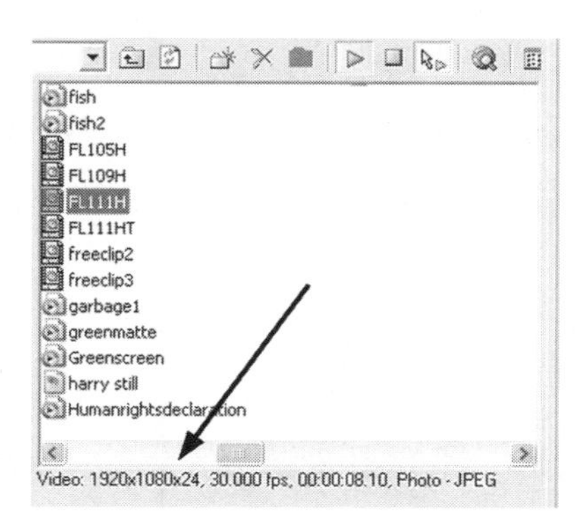

12.36 Project Properties in Vegas 4.0 now include high-definition settings.

24p has some very significant benefits and support in Vegas. Transitions, titles, and filters are optimized in Vegas for 24p. Vegas also has specific support for the new Panasonic DVX100 camera and cameras with similar features. Vegas auto-detects 2-3 and 2-3-3-2 pulldown at capture and previews and prints to tape in the 24p format when using the Panasonic DVX100 camera. The 2-3 and 2-3-3-2 pulldowns are performed on the fly in Vegas with no additional hardware required. Vegas also allows for preview on external monitors at 24p.

On the output side, Vegas can output a still image sequence up to 2048 × 2048, is capable of outputting 1080i or 720p at the 24p frame rate, and can print to tape at 24p as well. Windows Media 9 has support for 24p/HD, and streaming file sizes are smaller at 24p than at 29.97. 24p also offers up to 25 percent additional disk-space savings on a DVD.

Given that 24p renders twice as fast as 60i (standard NTSC frame rate) and that 24p offers a look closer to film, this format will continue to grow.

Vegas edits and prints media at 720p with frame rates of up to 60 fps, at resolutions of 1080p with frame rates of 24 fps and 29.97 fps, and at 1080i with frame rates of 60 fps (29.97 interlaced). With the output capability of 24 fps/progressive scan, video shot with the newer 24p will stay in their native format with no pulldown.

When working with high-definition media, full frame rate playback is very difficult to achieve because of the load on the processor. Be aware that playback at full frame rate is unlikely on all but the fastest systems. On a Pentium 3 with 1.8 GHz processor with 1GB of RAM, the best playback achieved was less than 15 fps. Rendering to RAM allows for full motion playback; however, for most circumstances, a different workflow may appeal to most users.

When working with high-definition media, current IDE drive speeds will prove to be unsatisfactory for most users. SCSI drives/RAIDS with a 320-controller card are considered good starting points for most high-definition editing practices.

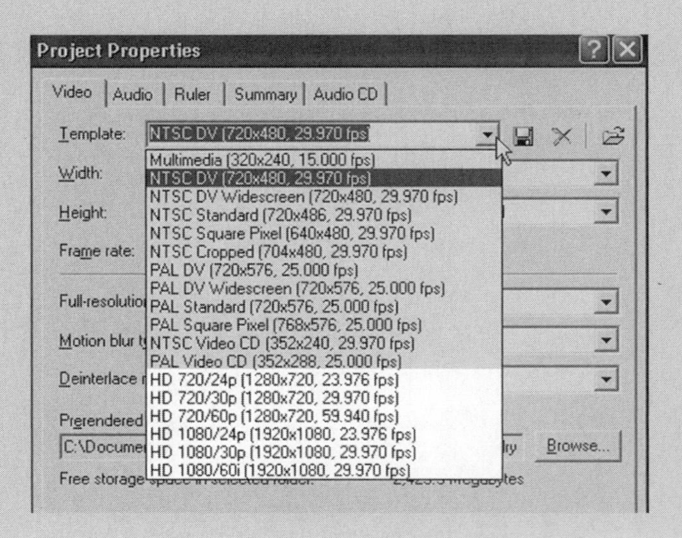

12.37 Rendering/output settings include high-definition capability in Vegas 4.0.

With Vegas' ability to recapture media via the Media Pool or to work with originally captured media, try to capture using a dual-stream card where possible, so that a low-resolution/compressed capture takes place at the same time as the high-definition media is being captured. Most high-definition cards allow for this to take place. Edit on the timeline using the lower-resolution media and replace media with the high-definition video in the Media Pool when you finish editing. This process will allow editing to take place with full motion and full resolution without taxing the processor as heavily as native high-definition media. Rendering to a RAID, the media can then either be printed to tape using a high-definition card, capture tool, and high-definition tape machine, or using the RAID delivered to a service bureau for final print and replication for distribution. This is a form of offline/online editing but can be done natively with Vegas as both a rough cutting and finishing tool. By using the Super Sampling Envelope tool, even DV can be effectively upsampled to HD output. Bear in mind that render times will be fairly slow on longer projects. Upsampling from resolutions of 720 × 480 to 1920 × 1080 will more than double the resolution of DV.

12.38 This high-definition footage from Artbeats fills most of a standard monitor even at three-quarter size (Artbeats Family Lifestyles HD library).

Scripting

NEW!

Vegas 4.0 has a feature that is terrific for users needing customized output and processing options. This feature uses JavaScript to define these behaviors. Vegas can literally be instructed to render several file types, to batch process, to match aspect ratios, even to build a video at random with random transitions from media in the Media Pool! Scripts basically provide an open-ended function for Vegas users to implement code in order to instruct Vegas to provide unique behaviors.

To use scripting, you'll need to have the Microsoft .NET framework installed. The .NET framework can be downloaded at no cost from Microsoft. Visit `http://www.microsoft.com/net/` to order a CD or download the framework.

Many scripts are available on the Sonic Foundry website, as well as on the Sundance Media Group website (`http://www.sundancemediagroup.com/help/kb/`). Writing scripts is a challenge for those not familiar with Visual Basic or JavaScript authoring. You can write scripts with any plain text editor, even Notepad. The Microsoft developer's website (`msdn.microsoft.com`) probably has the most authoritative reference for the .NET scripting languages, but other tutorials are available on the web. In addition, several books on JScript .NET and Visual Basic .NET are available from Amazon.com and other bookstores.

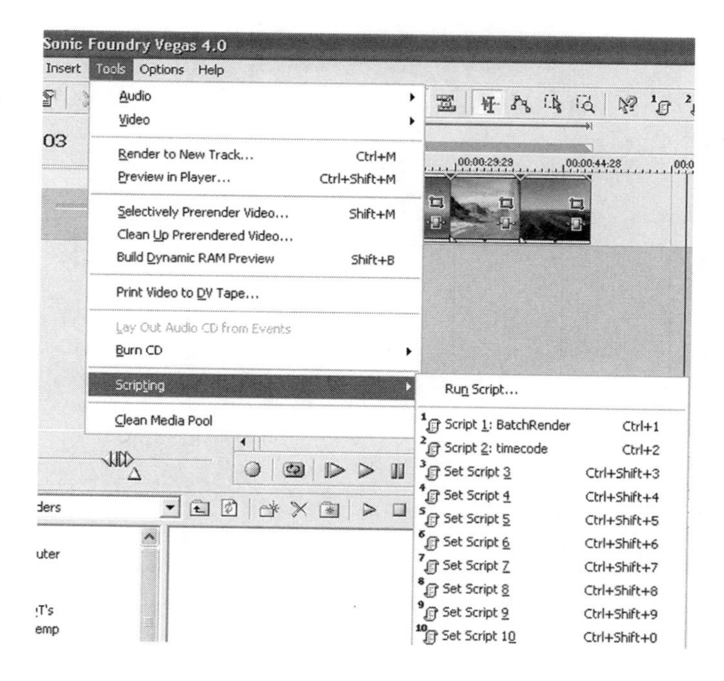

12.39 Access the scripts by selecting **TOOLS | SCRIPTING | RUN SCRIPT.**

Here is a sample .js script that opens ten tracks with blank events on them:

```
* Sample script that script creates 10 tracks each containing 10
 * events. No media is assigned to the events.
 *
 * Revision Date: Feb. 04, 2003.
 **/

import SonicFoundry.Vegas;
import System.Windows.Forms;

var trackCount = 10;
var eventCount = 10;
var eventLength = new Timecode(10000); // 10 seconds

var mediaType = MediaType.Audio;

var track, trackEvent;
var i, j;
var startTime;

for (i = 0; i < trackCount; i++) {

    // create a track
    if (mediaType == MediaType.Audio)
        track = new AudioTrack(i, "Audio " + (i+1));
    else
        track = new VideoTrack(i, "Video " + (i+1));

    // add the track
    Vegas.Project.Tracks.Add(track);

    startTime = new Timecode();
    for (j = 0; j < eventCount; j++) {

        // create an event
        if (mediaType == MediaType.Audio)
            trackEvent = new AudioEvent(startTime, eventLength, "Audio Event " + (j+1));
        else
            trackEvent = new VideoEvent(startTime, eventLength, "Video Event " + (j+1));

        // add the event to the track
        track.Events.Add(trackEvent);

        // increment the start time
        startTime += eventLength;
    }

    // toggle the media type
    if (mediaType == MediaType.Audio)
        mediaType = MediaType.Video;
    else
        mediaType = MediaType.Audio;
}
```

Running scripts can also appear to be intimidating, when in truth, running scripts is quite straightforward. After locating/downloading scripts, place them in a directory. You might find it efficient to create a folder named Scripts and access them all from one location. To run a script, select TOOLS | SCRIPTING | RUN SCRIPT, which will open a browse dialog box. Browse to where the .vb or .js scripts are stored.

Select the script you wish to run, which will start the script. Scripts are like trading cards that are available from many Vegas users, and a small cottage industry has sprung from this concept of having an open engine.

Scripts will continue to add to the power of Vegas and offer users a tremendous number of options.

One aspect of scripting that creates exceptional ease of use is to create a button on the toolbar for the script. This option allows rapid access to commonly run scripts. Up to ten scripts may be associated with ten buttons on the toolbar.

To place a button/script on the toolbar, double-click the toolbar, which will open the Customize Toolbar dialog. Select a script button to add to the toolbar. In Figure 12.42, Script 3 is selected to be placed in the toolbar. This step places a button on the toolbar that can be associated with a script by pressing CTRL+SHIFT+3 or by clicking the Script button. Vegas will open a browse window that allows you to select a specific script to be assigned to that button.

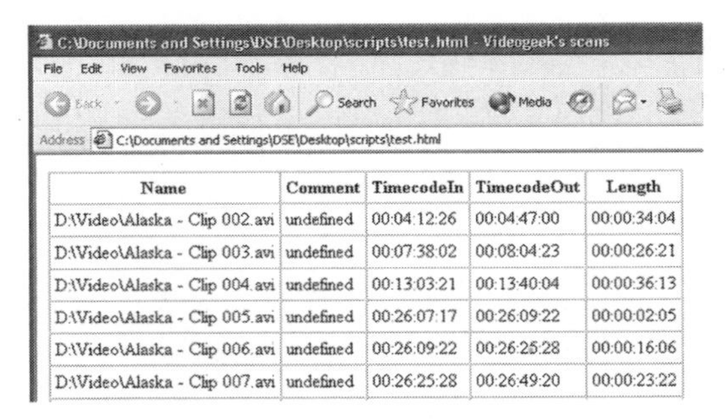

Name	Comment	TimecodeIn	TimecodeOut	Length
D:\Video\Alaska - Clip 002.avi	undefined	00:04:12:26	00:04:47:00	00:00:34:04
D:\Video\Alaska - Clip 003.avi	undefined	00:07:38:02	00:08:04:23	00:00:26:21
D:\Video\Alaska - Clip 004.avi	undefined	00:13:03:21	00:13:40:04	00:00:36:13
D:\Video\Alaska - Clip 005.avi	undefined	00:26:07:17	00:26:09:22	00:00:02:05
D:\Video\Alaska - Clip 006.avi	undefined	00:26:09:22	00:26:25:28	00:00:16:06
D:\Video\Alaska - Clip 007.avi	undefined	00:26:25:28	00:26:49:20	00:00:23:22

12.40 The script used in this instance locates in and out timecodes for all files on the timeline, lists their running time/length, and saves them as an HTML file.

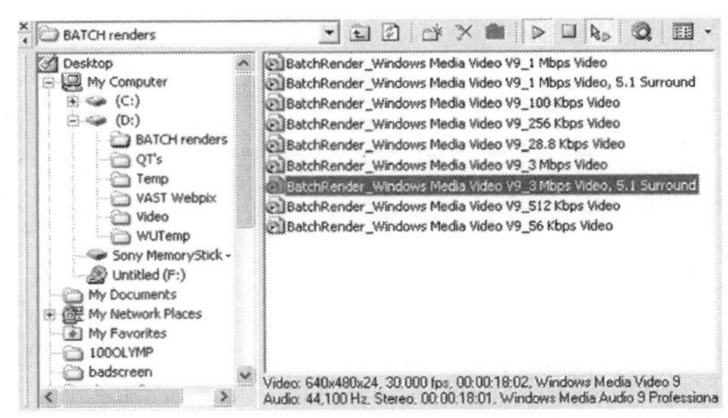

12.41 The Batching script renders multiple forms of Windows media from one timeline and one action.

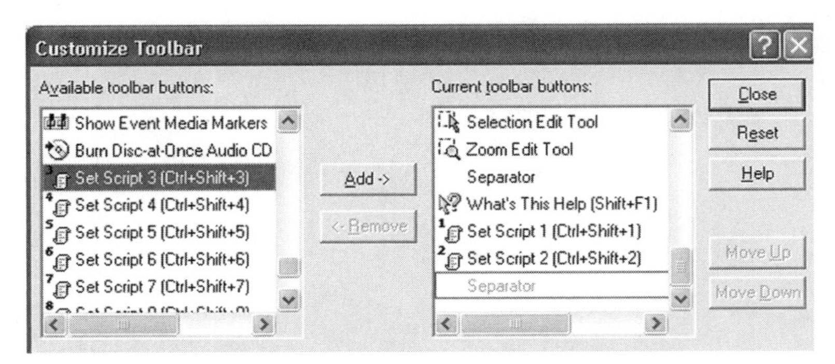

12.42 Assign a script to a button on the toolbar in Customize Toolbars.

Another method of assigning a script to a toolbar button is to select TOOLS | SCRIPTING | SET SCRIPT and choose the script number to which you wish to assign a script. This action will open the Browse Scripts window as well, from which a script can be assigned to a script number. This same script number can be associated with a button by following the steps outlined earlier.

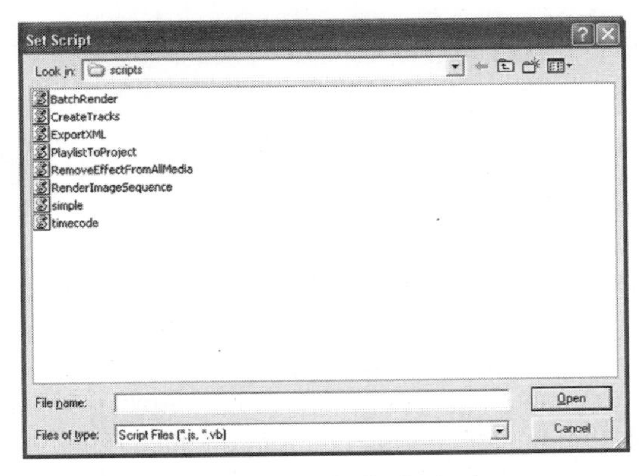

12.43 Select scripts for use in the Script browser.

Using Vegas with DVD Architect

To use Vegas with DVD Architect, render the project as either an MPEG-2 or as an AVI file. If the project is to be 5.1 at final output, render the file as an MPEG-2 file. Insert markers in the file for chapter points if desired. DVDA will recognize these markers as chapter points and will insert them as such if instructed to do so in DVDA.

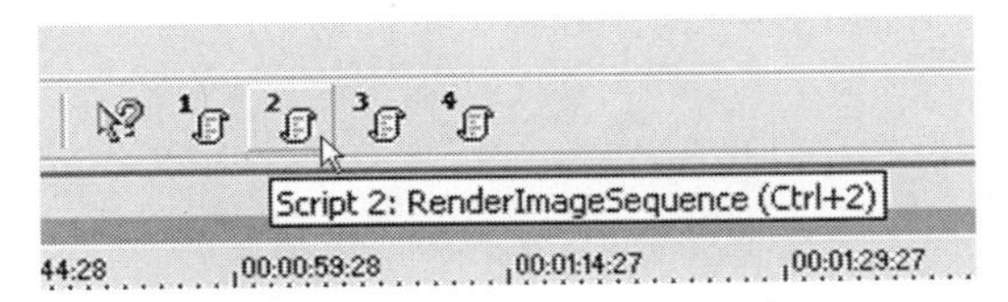

12.44 Buttons can be created on the toolbar for scripts.

Audio that is to be associated with the MPEG file, regardless of whether it is stereo or mono, should be rendered as an AC-3 file. DVDA will automatically convert stereo audio to an AC-3 file during the DVDA render/burn process. (See Chapter 11 for more information.) Audio in a 5.1 channel project file will not be embedded in the MPEG file in any form. (See the previous sections in this chapter on creating AC-3 audio for DVD Architect.)

Files that are rendered as 24p files will not be transcoded/recompressed in DVDA; they will stay in their native format. This feature allows 20 percent more video media to be placed on the DVD, which can be monitored/previewed in DVDA as well. Export of 24p NTSC MPEG-2 for DVD Architect is also supported as an output format in Vegas 4.0b.

Keyboard Shortcuts

Project File

Command	Keyboard Shortcut
Create new project	CTRL+N
Create new project and bypass the Project Properties dialog	CTRL+SHIFT+N
Exit Vegas	ALT+F4
Close the current project	CTRL+F4
Open existing project	CTRL+O
Project properties	ALT+ENTER
Save project	CTRL+S

Magnify and View Commands

Command	Keyboard Shortcut
Set focus to track view	ALT+0
Show Explorer window	ALT+1
Show Trimmer window	ALT+2
Show Mixer window	ALT+3
Show Edit Details window	ALT+4
Show Media Pool window	ALT+5
Show Video Preview window	ALT+6
Show Transitions window	ALT+7
Show Video FX window	ALT+8
Show Media Generators window	ALT+9
Show Plug-Ins window	CTRL+ALT+1
Show Video Scopes window	CTRL+ALT+2
Show Surround Panner window	CTRL+ALT+3

Command	Keyboard Shortcut
Show/hide audio bus tracks	B
Show/hide video bus track	CTRL+SHIFT+B
Show next window	F6 or CTRL+TAB
Show previous window	SHIFT+F6 or CTRL+SHIFT+TAB
Toggle focus between track list and timeline (and bus track list and timeline if bus tracks are visible)	TAB
Switch focus to previous/next track or bus track	ALT+SHIFT+UP/DOWN ARROW
Decrease height of all tracks or bus tracks (depending on which has focus)	CTRL+SHIFT+DOWN ARROW
Increase height of all tracks or bus tracks (depending on which has focus)	CTRL+SHIFT+UP ARROW
Minimize all tracks	`
Display all tracks at a uniform height	~ (SHIFT+`)
Set track heights to default height	CTRL+`
Minimize/restore the window docking area	F11 or ALT+`
Maximize/restore the timeline vertically and horizontally (window docking area and track list will be hidden)	CTRL+F11 or CTRL+ALT+`
Minimize/restore the track list	SHIFT+F11 or SHIFT+ALT+`
Zoom in/out horizontally in small increments (if timeline has focus)	UP/DOWN ARROW
Zoom in/out horizontally in large increments or zoom to selection (if one exists)	CTRL+UP/DOWN ARROW
Zoom in time until each video thumbnail represents one frame	ALT+UP ARROW
Zoom waveforms in/out vertically (audio only)	SHIFT+UP/DOWN ARROW
Center view around cursor	\
Go to	CTRL+G
Set end of time selection	CTRL+SHIFT+G
Go to beginning of time selection or viewable area (if no time selection)	HOME
Go to end of time selection or viewable area (if no time selection)	END
Select loop region	SHIFT+Q
Restore previous five time selections	BACKSPACE
Go to beginning of project	CTRL+HOME W

Command	Keyboard Shortcut
Go to end of project	CTRL+END
Move left by grid marks	PAGE UP
Move right by grid marks	PAGE DOWN
Move left/right one pixel (when Quantize to Frames is turned off)	LEFT/RIGHT ARROW
Move to previous/next marker	CTRL+LEFT/RIGHT ARROW
Move one frame left/right	ALT+LEFT/RIGHT ARROW
Move left/right to event edit points (including fade edges)	CTRL+ALT+LEFT/RIGHT ARROW
Jog left/right (when not in edge-trimming mode or during playback)	F3/F9
Move cursor to corresponding marker or select corresponding region	0 to 9 keys (not numeric keypad)
Move to previous CD track	CTRL+,
Move to next CD track	CTRL+.
Move to previous CD index or region	,
Move to next CD index or region	.
Set in and out points	I (in) and O (out)
Create time selection while dragging on an event	CTRL+SHIFT+drag with mouse

General Editing Commands

Command	Keyboard Shortcut
Cut selection	CTRL+X or SHIFT+DELETE
Copy selection	CTRL+C or CTRL+INSERT
Paste from clipboard	CTRL+V or SHIFT+INSERT
Paste insert	CTRL+SHIFT+V
Paste repeat	CTRL+B
Delete selection	DELETE
Mix to new track	CTRL+M
Undo	CTRL+Z or ALT+BACKSPACE
Redo	CTRL+SHIFT+Z or CTRL+Y
Rebuild audio peaks	F5
Switch to normal editing tool	CTRL+D
Switch to next editing tool	D

Command	Keyboard Shortcut
Switch to previous editing tool	SHIFT+D
Run script 1 to 10	CTRL+1 to 0
Set script number 1 to 10	CTRL+SHIFT+1 to 0

Event Selection and Editing Commands

Command	Keyboard Shortcut
Select a range of events	SHIFT+click events
Select multiple events	CTRL+click individual events
Select all	CTRL+A
Unselect all	CTRL+SHIFT+A
Cut selection	CTRL+X or SHIFT+DELETE
Copy selection	CTRL+C or CTRL+INSERT
Paste from clipboard	CTRL+V or SHIFT+INSERT
Paste insert	CTRL+SHIFT+V
Paste repeat	CTRL+B
Delete selection	DELETE
Split events at cursor	S
Trim/crop selected events	CTRL+T
Enter edge-trimming mode and select event start; move to previous event edge In this mode, 1, 3, 4, and 6 on the numeric keypad trim the selected event edge	Numeric keypad 7+[
Enter edge-trimming mode and select event end; move to next event edge In this mode, 1, 3, 4, and 6 on the numeric keypad trim the selected event edge	Numeric keypad 9+]
Trim left/right (when in edge-trimming mode)	F3/F9
Exit edge-trimming mode	Numeric keypad 5
Move or trim selected events one frame left/right	Numeric keypad 1/3
Move or trim selected events one pixel left/right	Numeric keypad 4/6
Move selected events up/down one track	Numeric keypad 8/2
Slip: move media within event without moving the event	ALT+drag inside the event

Command	Keyboard Shortcut
Slip trim: move the media with the edge as it is trimmed	ALT+drag edge of event
Trim adjacent: trims selected event and adjacent event simultaneously	CTRL+ALT+drag edge of event
Slide: trims both ends of event simultaneously	CTRL+ALT+drag middle of event
Slide cross-fade: moves cross-fade	CTRL+ALT+drag over a cross-fade
Stretch (compress) the media in the event while trimming	CTRL+drag edge of event
Select next take	T
Select previous take	SHIFT+T
Convert cut to transition	Numeric keypad /, numeric keypad *, or numeric keypad –
Convert transition to cut	CTRL+numeric keypad /
Open in audio editor	CTRL+E

Playback, Recording, and Preview Commands

Command	Keyboard Shortcut
Start/stop playback	SPACEBAR
Stop playback	ESC
Play from any window	CTRL+SPACEBAR or F12
Dim (attenuate) mixer output	CTRL+SHIFT+F12
Play/pause	ENTER
Loop playback	Q
Preview cursor position (You can specify the length of the time Vegas will preview using the Cursor Preview Duration box on the Editing tab of the Preferences dialog)	Numeric keypad 0
Record	CTRL+R
Arm track for record	CTRL+ALT+R
Arm for record and set recording path	CTRL+ALT+SHIFT+R
Scrub playback	J, K, or L
Selectively prerender video	SHIFT+M
Preview in player	CTRL+SHIFT+M
Build dynamic RAM preview	SHIFT+B
Generate MIDI timecode	F7

Command	Keyboard Shortcut
Generate MIDI clock	SHIFT+F7
Trigger from MIDI timecode	CTRL+F7

Timeline and Track List Commands

Command	Keyboard Shortcut
Insert new audio track	CTRL+Q
Insert new video track	CTRL+SHIFT+Q
Mute selected tracks	Z
Mute selected track and remove other tracks from mute group	SHIFT+Z
Solo selected tracks	X
Solo selected track and remove other tracks from solo group	SHIFT+X
Override snapping	SHIFT+drag
Enable/disable snapping	F8
Snap to grid	CTRL+F8
Snap to markers	SHIFT+F8
Quantize to frames	ALT+F8
Post-edit ripple affected tracks	F
Post-edit ripple affected tracks, bus tracks, markers, and regions	CTRL+F
Post-edit ripple all tracks, markers, and regions	CTRL+SHIFT+F
Auto-ripple mode	CTRL+L
Automatic cross-fades	CTRL+SHIFT+X
Render to new track	CTRL+M
Group selected events	G
Ungroup selected events	U
Clear group without deleting events	CTRL+U
Select all events in group	SHIFT+G
Insert/show/hide track volume envelope	V
Remove track volume envelope	SHIFT+V
Insert/show/hide track panning envelope	P
Remove track panning envelope	SHIFT+P
Cycle through effect automation envelopes	E or SHIFT+E
Insert command marker	C

Command	Keyboard Shortcut
Insert marker	M
Insert region	R
Insert CD track region	N
Insert CD track index	SHIFT+N

Trimmer Commands

Command	Keyboard Shortcut
Add media from cursor	A
Add media to cursor	SHIFT+A
Transfer time selection from timeline to Trimmer after cursor	T
Transfer time selection from timeline to Trimmer before cursor	SHIFT+T
Toggle selected stream: audio/video/both	TAB
Next media file in Trimmer	CTRL+TAB
Previous media file in Trimmer	CTRL+SHIFT+TAB
Toggle audio/video stream height	CTRL+SHIFT+UP/DOWN ARROW

Surround Sound Commands

Command	Keyboard Shortcut
Constrain motion to 45-degree increments	SHIFT+drag the pan point (only when Move Freely is selected)
Constrain motion to a constant radius from the center	ALT+drag the pan point (only when Move Freely is selected) or ALT+mouse wheel
Constrain motion to the maximum circle that will fit in the Surround Panner	ALT+SHIFT+drag the pan point (only when Move Freely is selected) or ALT+SHIFT+mouse wheel
Move the pan point forward/back (when the pan point is selected)	UP/DOWN ARROW, PAGE UP/PAGE DOWN, or mouse wheel
Move the pan point left/right (when the pan point is selected)	LEFT/RIGHT ARROW, SHIFT+PAGE UP/PAGE DOWN, or SHIFT+mouse wheel forward/back
Move the pan point to a corner, edge, or center of the Surround Panner (when the pan point is selected)	Numeric keypad
Move the pan point to a corner on the largest circle that will fit in the Surround Panner (when the pan point is selected)	CTRL+Numeric keypad 1,3,7,9

Miscellaneous Commands

Command	Keyboard Shortcut
Open Online Help	F1
Open What's This help	SHIFT+F1
Shortcut menu	SHIFT+F10
Rebuild audio peaks	F5
Make fine fader/slider adjustments	CTRL+drag
Change relative keyframe spacing	ALT+drag keyframes
Move region without changing length	ALT+drag region tag

Mouse Wheel Commands

Command	Mouse Shortcut
Zoom in/out	Wheel
Scroll vertically	CTRL+wheel
Scroll horizontally	SHIFT+wheel
Move the cursor in small increments, trim the selected event edge one pixel (if you're in edge-trimming mode), or adjust scrub rate during playback	CTRL+SHIFT+wheel
Move the cursor by frames, trim the selected event edge one frame (if you're in edge-trimming mode), or adjust scrub rate during playback	CTRL+ALT+SHIFT+wheel
Auto-scroll	Click wheel button and move mouse
Move fader/slider or, in plug-in windows, click the control first to give it focus	Hover over handle and use wheel
Move fader/slider in fine increments	CTRL+hover over fader and use wheel

Bibliography

Ascher, Steven and Edward Pincus. *The Filmmaker's Handbook: A Comprehensive Guide for the Digital Age*. New York, NY: Plume, 1999.

Billups, Scott. *Digital Movie Making*. 2nd ed. Studio City, CA: Michael Wiese Productions, 2003.

Button, Bryce. *Nonlinear Editing: Storytelling, Aesthetics, and Craft*. Lawrence, KS: CMP Books, 2002.

Creative Cow. http://www.creativecow.net (March 2003).

Hullfish, Steve and Jaime Fowler. *Color Correction for Digital Video: Using Desktop Tools to Perfect Your Image*. San Francisco, CA: CMP Books, 2003.

LaBarge, Ralph. *DVD Authoring & Production*. Lawrence, KS: CMP Books, 2001.

Mack, Steve. *Streaming Media Bible*. New York, NY: Hungry Minds, 2002.

"mir DMG: What Is Y'CbCr?" http://www.mir.com/DMG/ycbcr.html (March 2003).

Panasonic. "Panasonic Consumer Electronics—Progressive Scanning Explained." http://www.panasonic.com/consumer_electronics/dvd_players/progscan.asp (March 2003).

Poynton, Charles. *Digital Video and HDTV: Pixels, Pictures, and Perception*. 1st ed. San Diego, CA: Academic Press, 2001.

Pure Motion. http://www.puremotion.com/editstudio/manual/videoediting/dvfiletypes/ (March 2003).

Rose, Jay. *Audio Postproduction for Digital Video*. San Francisco, CA: CMP Books, 2002.

Sonic Foundry. http://www.sonicfoundry.com (March 2003).

Video University. http://www.videouniversity.com (March 2003).

Virtual Studio Systems, Inc. "Recording Tips & Tricks." http://www.virtualstudiosystems.com/tipstrix.htm (March 2003).

Wilt, Adam. http://www.adamwilt.com (March 2003).

Zettl, Harold. *Television Production Handbook*. 8th ed. Belmont, CA: Wadsworth/Thomson Learning, 2002.

Index

Numbers

A

B

E

F

G

N

O

P

Q

R

S

U

V

W

X

Z

Want to download the source code or get periodic updates? Send a blank email to `vegas4@news.cmpbooks.com`**. If you have a suggestion or correction, please send your comments to** `info@sundancemediagroup.com`.

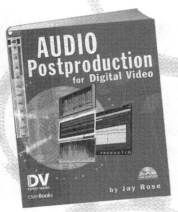

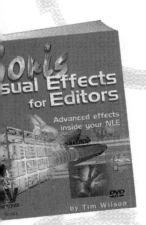